WITHDRAWN
UTSA LIBRARIES

Tree-Rings, Kings, and Old World Archaeology and Environment:

*Papers Presented in Honor of
Peter Ian Kuniholm*

Edited by

Sturt W. Manning & Mary Jaye Bruce

Oxbow Books
Oxford and Oakville

Published by

Oxbow Books, Oxford, UK

© the individual authors 2009

Library of Congress Cataloging-in-Publication Data

Tree-rings, kings, and Old World archaeology and environment: papers presented in honor of Peter Ian Kuniholm / edited by Sturt W. Manning & Mary Jaye Bruce.
    p. cm.
Papers from a conference held Nov. 3–5, 2006, at Cornell University.
Includes bibliographical references.
ISBN 978-1-84217-386-2
1. Mediterranean Region—Antiquities—Congresses. 2. Mediterranean Region—Civilization—Chronology—Congresses. 3. Mediterranean Region—Environmental conditions—Congresses. 4. Dendrochronology—Mediterranean Region—Congresses. 5. Archaeology and history—Mediterranean Region—Congresses. 6. Climatic changes—Mediterranean Region—History—Congresses. 7. Human ecology—Mediterranean Region—History—Congresses. 8. Social archaeology—Mediterranean Region—History—Congresses. 9. Kuniholm, Peter Ian—Congresses. I. Manning, Sturt W. II. Bruce, Mary Jaye. III. Title: Tree-rings, kings, and Old World archaeology and environment.
    DE60.T74 2009
    930—dc22   2009038639

ISBN 978-1-84217-386-2

This book is available direct from:

Oxbow Books, Oxford, UK
Tel: +44 (0)1865 241249, Fax: +44 (0)1865 794449
Trade Sales: +44 (0) 1865 256780
Email: oxbow@oxbowbooks.com

and

The David Brown Book Co, PO Box 511 (28 Main Street), Oakville CT 06779
Toll-free: 800 791 9354
Tel: 860 945 9329
Fax: 860 945 9468
Email: queries@dbbconline.com

Or from our website:

www.oxbowbooks.com

Printed in Malta at
Gutenberg Press

Contents

FOREWORD
    A. Colin Renfrew vii

PREFACE AND ACKNOWLEDGMENTS ix

CONTRIBUTORS TO THE VOLUME xi

BIBLIOGRAPHY OF PETER IAN KUNIHOLM xv

PETER KUNIHOLM'S DENDRO TIME
    Fritz H. Schweingruber 1

PERSPECTIVE: ARCHAEOLOGY, HISTORY, AND CHRONOLOGY FROM PENN TO THE PRESENT
    AND BEYOND
    James Muhly 3

EXCURSIONS INTO ABSOLUTE CHRONOLOGY
    M. G. L. Baillie 13

ONE HUNDRED YEARS OF DENDROARCHAEOLOGY: DATING, HUMAN BEHAVIOR, AND PAST
    CLIMATE
    Jeffrey S. Dean 25

THE ABSOLUTE DATING OF WASSERBURG BUCHAU: A LONG STORY OF TREE-RING RESEARCH
    A. Billamboz 33

IS THERE A SEPARATE TREE-RING PATTERN FOR MEDITERRANEAN OAK?
    Tomasz Wazny 41

DENDROCHRONOLOGICAL RESEARCH AT ROSSLAUF (BRESSANONE, ITALY)
    Maria Ivana Pezzo 51

THE DEVELOPMENT OF THE REGIONAL OAK TREE-RING CHRONOLOGY FROM THE ROMAN
    SITES IN CELJE (SLOVENIA) AND SISAK (CROATIA)
    Aleksandar Durman, Andrej Gaspari, Tom Levanič, Matjaz Novšak 57

DENDROCLIMATOLOGY IN THE NEAR EAST AND EASTERN MEDITERRANEAN REGION
    Ramzi Touchan and Malcolm K. Hughes 65

A 924-YEAR REGIONAL OAK TREE-RING CHRONOLOGY FOR NORTH CENTRAL TURKEY
    *Carol B. Griggs, Peter I. Kuniholm, Maryanne W. Newton, Jennifer D. Watkins, and
    Sturt W. Manning* 71

DENDROCHRONOLOGY ON PINUS NIGRA IN THE TAYGETOS MOUNTAINS, SOUTHERN
PELOPONNISOS
Robert Brandes *81*

COULD ABSOLUTELY DATED TREE-RING CHEMISTRY PROVIDE A MEANS TO DATING THE
MAJOR VOLCANIC ERUPTIONS OF THE HOLOCENE?
Charlotte L. Pearson and Sturt W. Manning *97*

DENDROCHEMISTRY OF PINUS SYLVESTRIS TREES FROM A TURKISH FOREST
D. K. Hauck and K. Ünlü *111*

NEUTRON ACTIVATION ANALYSIS OF DENDROCHRONOLOGICALLY DATED TREES
K. Ünlü, P. I. Kuniholm, D. K. Hauck, N. Ö. Cetiner, and J. J. Chiment *119*

THIRD MILLENNIUM BC AEGEAN CHRONOLOGY: OLD AND NEW DATA FROM THE
PERSPECTIVE OF THE THIRD MILLENNIUM AD
Ourania Kouka *133*

MIDDLE HELLADIC LERNA: RELATIVE AND ABSOLUTE CHRONOLOGIES
Sofia Voutsaki, Albert J. Nijboer, and Carol Zerner *151*

ABSOLUTE AGE OF THE ULUBURUN SHIPWRECK: A KEY LATE BRONZE AGE TIME-CAPSULE
FOR THE EAST MEDITERRANEAN
*Sturt W. Manning, Cemal Pulak, Bernd Kromer, Sahra Talamo, Christopher Bronk Ramsey,
and Michael Dee* *163*

HOW ABOUT THE PACE OF CHANGE FOR A CHANGE OF PACE?
Jeremy B. Rutter *189*

ARCHAEOLOGISTS AND SCIENTISTS: BRIDGING THE CREDIBILITY GAP
Elizabeth French and Kim Shelton *195*

CENTRAL LYDIA ARCHAEOLOGICAL SURVEY: DOCUMENTING THE PREHISTORIC THROUGH
IRON AGE PERIODS
Christina Luke and Christopher H. Roosevelt *199*

THE CHRONOLOGY OF PHRYGIAN GORDION
Mary M. Voigt *219*

THE END OF CHRONOLOGY: NEW DIRECTIONS IN THE ARCHAEOLOGY OF THE CENTRAL
ANATOLIAN IRON AGE
Geoffrey D. Summers *239*

THE RISE AND FALL OF THE HITTITE EMPIRE IN THE LIGHT OF DENDROARCHAEOLOGICAL
RESEARCH
Andreas Müller-Karpe *253*

AEGEAN ABSOLUTE CHRONOLOGY: WHERE DID IT GO WRONG?
Christos Doumas *263*

THE THERA DEBATE *275*

COLD FUSION: THE UNEASY ALLIANCE OF HISTORY AND SCIENCE
Malcolm H. Wiener *277*

SANTORINI ERUPTION RADIOCARBON DATED TO 1627–1600 BC: FURTHER DISCUSSION
Walter L. Friedrich, Bernd Kromer, Michael Friedrich, Jan Heinemeier, Tom Pfeiffer, and Sahra Talamo 293

DATING THE SANTORINI/THERA ERUPTION BY RADIOCARBON: FURTHER DISCUSSION (AD 2006–2007)
Sturt W. Manning, Christopher Bronk Ramsey, Walter Kutschera, Thomas Higham, Bernd Kromer, Peter Steier, and Eva M. Wild 299

THERA DISCUSSION
Malcolm H. Wiener, Walter L. Friedrich, and Sturt W. Manning 317

Peter Ian Kuniholm

Foreword

Dendrochronology and Peter Ian Kuniholm have become synonymous in the eyes of Aegean archaeologists. This volume is a celebration of what has been achieved in the Aegean and in Turkey, and in many areas beyond, through his pioneering efforts in the Dendro Lab at Cornell, now the Malcolm and Carolyn Wiener Laboratory for Aegean and Near Eastern Dendrochronology. As befits what is now a new scientific field within the broader realm of the archaeological sciences, it is a celebration also of what we hope can be achieved in the future, under the leadership of the laboratory's new director, Sturt Manning.

For many of us working in the Mediterranean and south-east Europe, tree-ring dating first became really interesting at the time of the 'second radiocarbon revolution,' associated with the calibration of the radiocarbon time-scale through the dendrochronology of the Californian bristlecone pine (*Pinus aristata* or *Pinus longaeva*). The first revolution, initiated in 1949 by Willard Libby had already had a major impact in Europe, where the radiocarbon time scale for the European neolithic and copper age pushed the chronology back (i.e. earlier) by at least a millennium. Then came the tree-ring calibration based on the dendrochronological work of Charles Wesley Ferguson, of the Arizona Tree Ring Lab at Tucson, and the radiocarbon analyses of Hans Suess, then based at La Jolla. By setting the calibrated dates up to nine centuries earlier than the uncalibrated dates this caused something of a sensation. For the uncalibrated dates had already suggested a chronology which was much earlier in some areas (including neolithic Europe) than traditional archaeological reasoning had concluded. There was quite a fuss, with Vladimir Milojčić and other senior scholars calling the whole radiocarbon method into question. Professor Stuart Piggott had called some of the uncalibrated dates for neolithic Britain "archaeologically unacceptable." But unlike Milojčić he was impressed by the effects of the calibration and came to accept the efficacy of radiocarbon dating. It became possible to recognise an "archaeological fault line" across Europe. In parts of the east Mediterranean (including Egypt), to the south-east of the fault line, the dates actually conformed better, after calibration, to the existing historical chronology. In much of Europe, on the other hand—to the north and west of the fault line—the traditional dates, based upon diffusionist thinking, had to be set earlier by up to two millennia to come into agreement with the new radiocarbon chronology based upon the calibration. Some basic assumptions had to be changed: the chronology of prehistoric Europe could no longer be based on application of the principle of *Ex Oriente lux*. Hence the fault line.

The tree-ring dates over the past two thousand years, based on a Sequoia trunk preserved in Cambridge, had already been used to test the short-term fluctuations in the concentration of atmospheric radiocarbon over that time range. So began the long partnership between radiocarbon dating and dendrochronology in chronological studies. It was realised, moreover, by R. M. Clark as well as Hans Suess, that "wiggle matching," lining up the kinks in the master calibraton curve with those derived from radiocarbon determinations from a floating tree-ring sequence, might, in favourable cases, allow such a floating sequence to be assigned an accurate calendar date. This was achieved by the systematic comparison of the wiggles with those on the master calibration curve. These possibilities gave Aegean prehistorians, as well as classicists and Byzantinists, good reason to interest themselves in tree-ring dating, even though there seemed at that time no prospect of setting up a continuous tree-ring chronology for the Aegean, stretching back millennia before the present, such as existed for the bristlecone pine, and later for Irish oak and for south German oak also.

The absence of much serious tree-ring work in the Aegean before the inception of Kuniholm's work explains his splendidly omnivorous hands-on approach. No waterlogged timbers in some classical site, no promising roof beams in some obscure Byzantine church were too modest to escape his attentions. And so it came about that over the years an astonishingly good coverage was built up of samples of wood dating back over the past nine thousand years—although not, unfortunately, in unbroken sequence. The great site of Gordion in western Turkey was one where Kuniholm, building on the work of Bryant Bannister, was able to establish a bronze age/iron age chronology which now covers more than 2000 years. This has been followed by work at a number of important sites, many of them in Turkey, so that real chronological issues are now being addressed.

The long-term prize, of course, is to find timbers constituting the "missing links" in the continuous tree-ring chronology that one would like to see for the Aegean, stretching back without a break from the present to the early neolithic and perhaps beyond. The gaps have narrowed, but some of them are still there. And of course, when that great day comes, one might in principle be able to date wooden structures, and so the events with which they are associated, reliably to within a single year. That is the dream, and how nice it would be to establish chronologies in the classical period, let alone the early bronze age, down to a single calendar year! But Peter Kuniholm would be the first to point out that, even with the continuous tree-ring master chronology, there would be many practical problems in using it to date specific buildings and sites in that way. How old was the timber when it was used? Do the outer rings relate to the felling of the tree preserved? And so on. But it is not too optimistic to hope that such a time, and such a time-scale will come. And when it does it will be based in large measure on the work of Peter Kuniholm and his associates in the Cornell Dendro Lab.

And perhaps even before that time it may be possible to give a secure and accurate calendar date to that most controversial of "global events," the great Thera eruption. Various scholars have demonstrated that it was a massive eruption whose effects would have been noticed on a worldwide scale. So it is with bated breath, year after year, that we have awaited its secure recognition, by some incontrovertible marker, in some reliable and long-lasting time scale based upon the annual climatic fluctuations of the earth. Since the Aegean dendrochronology for the late bronze age is still a floating one we have to look to other regions. But tree-ring indications of climatic events observed in the Californian bristlecone pine have proved to be ambiguous. Reports of Theran tephra in the polar ice cores have been refuted. Possible correlations with floating tree-ring sequences in Turkey have not yet been substantiated. The matter is made more tantalising by the archaeological correlations for Thera, particularly in respect of the wall paintings, with the important excavations at Tell el-Daba'a in the Nile delta, and indeed with the Egyptian historical chronology in general. It seems astonishing that, after so many years of study, there are still discrepancies between the various attempts to use the resources of archaeological science—long compendia of radiocarbon dates, wiggle matching with olive wood, dubious identifications of tephra in the ice cores—to match those of the best scholars working on the basis of Egyptian historical data. There is even the tantalising obstacle of a discrepancy between the calibrated radiocarbon dates from Tell el-Daba'a itself and the established Egyptian (historical) chronology.

Peter, thou should'st have worked hereafter! These are issues which cry out for resolution. Indeed we hope that you will, and for many years, contribute to the work of the Malcolm and Carolyn Wiener Laboratory, under the leadership of your successor and participate in that much-needed resolution. There is work to be done! And it is a tribute to the enormous contribution which you have made that the issues in Aegean and Mediterranean chronology have become so sharply defined, and their resolution so tantalisingly close, even if still beyond our grasp.

Many other themes in Aegean and Near Eastern archaeology are, of course, being illuminated by the ongoing work of the Dendro Lab. New techniques are being developed and we can cheerfully anticipate that problems now too difficult to resolve will find resolution, especially as the continuous master sequence is extended.

It was a pleasure to participate in the meeting to celebrate the achievements which gave rise to this volume. They are reflected and exemplified in the quality and diversity of the papers here, and above all in the continuing vitality of the work of the Malcolm and Carolyn Wiener Laboratory for Aegean and Near Eastern Dendrochronology.

A. Colin Renfrew

Preface and Acknowledgments

On 3–5 November 2006 we held a conference entitled "Tree Rings, Kings, and Old World Archaeology & Environment: Cornell Dendrochronology-Archaeology Conference in Honor of Peter Ian Kuniholm." The conference formally marked Peter's retirement and the beginning of his new active emeritus career at Cornell. The gathering aimed to celebrate Peter's achievements and to provide stimulating assessments of a number of important topics associated with and inspired by his research over the years as the founder and director of the Aegean Dendrochronology Project and, later, the Malcolm and Carolyn Wiener Laboratory for Aegean and Near Eastern Dendrochronology. We hope that this volume will stand as a tribute to Peter as it attempts to capture in print the essence of the conference lectures and the spirit and energy of its honoree.

In addition to the lectures given by contributors to this volume, several other talks or poster presentations were offered at the conference by the following: Carin Ashjian; John Brinkman; Benjamin Cavallari; Paolo Cherubini; Otto Cichocki; Cari Sasser Furiness; Sevil Gülçur; Rachel King; Walter Kutschera; Christine E. Latini; Ezra Marcus; Stephen E. Nash; Michael Rafferty; Michael F. Rosenmeier; Catherine C. Thompson; Ken Wardle; James Weinstein; and Harvey Weiss.

The Conference was opened by Peter Lepage, the Dean of the College of Arts and Sciences, followed by a standing-room-only keynote lecture by Lord Colin Renfrew, and closed with a presentation by Cornell President emeritus Hunter R. Rawlings III. We were privileged to have with us at the conference Malcolm H. Wiener, whose nameplate honors the lab door. It was a further pleasure to have so many of the former members of the Laboratory (the "dendroids") return to Ithaca for the occasion. Indeed, there was lively competition to recall the best On the Road with PIK adventure story, but with such a cast of characters (see below), all inevitably came away winners.

Support for the conference was gratefully received from the Cornell University Lectures Committee, the College of Arts and Sciences, the Department of Classics, the Department of the History of Art, and the Office of the Vice Provost for Research.

Finally, we wish to thank Peter Brewer for his very great assistance with the technical aspects of getting this volume into production. We would still be slogging away in Windows without him.

Sturt W. Manning
Mary Jaye Bruce

"Dendroids" who attended the conference for Peter.

Contributors to the Volume

M. G. L. Baillie
School of Geography, Archaeology, and
Palaeoecology
Queens University
Belfast BT7 1NN
Ireland
m.baillie@qub.ac.uk

André Billamboz
Landesamt für Denkmalpflege
Arbeitsstelle Hemmenhofen
Dendochronologisches Labor
Fischersteig 9
D 78343 Hemmenhofen
Germany
andre.billamboz@rps.bwl.de

Robert Brandes
Geographical Institute
University Erlangen-Nuernberg
Kochstr. 4
91054 Erlangen
Germany
Dr-Robert-Brandes@t-online.de

N. Ö. Cetiner
Radiation Science and Engineering Center
Breazeale Nuclear Reactor
The Pennsylvania State University
University Park, PA 16802

Jeffrey S. Dean
Laboratory of Tree-Ring Research
The University of Arizona
Tucson, AZ 85721-0058
jdean@ltrr.arizona.edu

Michael Dee
Oxford Radiocarbon Accelerator Unit
Research Laboratory for Archaeology
Dyson Perrins Building
South Parks Road
Oxford OX1 3QY
United Kingdom
michael.dee@rlaha.ox.ac.uk

Christos G. Doumas
Emeritus Professor
University of Athens
27 Lambrou Photiadou
116 36 Athens
Greece
doumases@otenet.gr

Aleksandar Durman
Department of Archaeology
University of Zagreb
I. Lučića 3
10000 Zagreb
Croatia
adurman@ffzg.hr

Elizabeth French
British School at Athens
26 Millington Road
Cambridge CB3 9HP
United Kingdom
Lisacamb@aol.com

Michael Friedrich
Institute of Botany
Hohenheim University
D-70593 Stuttgart
Germany
michaelf@uni-hohenheim.de

Walter L. Friedrich
Department of Earth Sciences
University of Aarhus
C.F. Moellers Allé 1120
DK-8000 Århus
Denmark
geolwalt@geo.au.dk

Andrej Gaspari
University of Primorska
Scientific Research Centre
Institute for Mediterranean Heritage
Garibaldijeva 1
SI-6000 Koper
Slovenia
andrej.gaspari@siol.net

Carol B. Griggs
Malcolm and Carolyn Wiener Laboratory for
Aegean and Near Eastern Dendrochronology
Cornell Tree-Ring Laboratory
B-48 Goldwin Smith Hall
Cornell University
Ithaca, NY 14853
cbg4@cornell.edu

Danielle K. Hauck
Archaeometry Laboratory
University of Missouri Research Reactor
Columbia, MO 65211
hauckd@missouri.edu

Jan Heinemeier
AMS 14C Dating Centre
Institut for Fysik og Astronomi
Aarhus Universitet
Ny Munkegade, Bygn. 1520
University of Aarhus
DK-8000 Århus C
Denmark
jh@phys.au.dk

Thomas Higham
Oxford Radiocarbon Accelerator Unit
Research Laboratory for Archaeology
Dyson Perrins Building
South Parks Road
Oxford OX1 3QY
United Kingdom
thomas.higham@archaeology-
research.oxford.ac.uk

Malcolm Hughes
Laboratory of Tree-Ring Research
University of Arizona
Tucson, AZ 85721
mhughes@ltrr.arizona.edu

Bernd Kromer
Heidelberger Akademie der Wissenschaften
Institut für Umweltphysik
INF 229
D-69120 Heidelberg
Germany
bernd.kromer@iup.uni-heidelberg.de

Ourania Kouka
Archaeological Research Unit
Department of History and Archaeology
University of Cyprus
P.O. Box 20537
1678 Nicosia, Cyprus
ouraniak@ucy.ac.cy

Peter Ian Kuniholm
Malcolm and Carolyn Wiener Laboratory for
Aegean and Near Eastern Dendrochronology
Aegean Dendrochronology Project
B-48 Goldwin Smith Hall
Cornell University
Ithaca, NY 14853
pik3@cornell.edu

Walter Kutschera
Institut für Isotopenforschung und
Kernphysik
VERA Laboratorium
Universität Wien
A-1090 Wien
Austria
walter.kutschera@univie.ac.at

Tom Levanič
Slovenian Forestry Institute
Večna pot 2
SI-1000 Ljubljana
Slovenia
tom.levanic@gozdis.si

Christina Luke
Department of Archaeology
Boston University
675 Commonwealth Ave.
Boston, MA 02215
cluke@bu.edu

Sturt W. Manning
Department of Classics
Malcolm and Carolyn Wiener Laboratory for
Aegean and Near Eastern Dendrochronology
B-48 Goldwin Smith Hall
Cornell University
Ithaca, NY 14853
sm456@cornell.edu

James Muhly
36 Proteos St.
Palio Faliron
175 61 Athens
Greece
jimmuhly@yahoo.com

Andreas Müller-Karpe
Philipps-Universität Marburg
Biegenstrasse 11
D-35037 Marburg
Germany
vorgesch@staff.uni-marburg.de

Maryanne W. Newton
6039 Bodega Ave.
Petaluma, CA 94952

Albert J. Nijboer
Groningen Institute of Archaeology
University of Groningen
Poststraat 6
9712 ER Groningen
The Netherlands
a.j.nijboer@rug.nl

Matjaz Novšak
Arhej d.o.o.
Drožanjska 23
SI-8290 Sevnica
Slovenia
arhejdoo@siol.net

Charlotte L. Pearson
Malcolm and Carolyn Wiener Laboratory for Aegean and Near Eastern Dendrochronology
Cornell Tree-Ring Laboratory
B-48 Goldwin Smith Hall
Cornell University
Ithaca, NY 14853
c.pearson@cornell.edu

Maria Ivana Pezzo
Dendrochronology Laboratory
Museo Civico di Rovereto
Rovereto
Italy
pezzoi@iol.it

Tom Pfeiffer
Department of Earth Sciences
University of Aarhus
C.F. Moellers Allé 1120
DK-8000 Århus
Denmark
tpfeiffer@decadevolcano.net

Cemal Pulak
Institute of Nautical Archaeology
P.O. Drawer HG
Texas A & M University
College Station, TX 77841
pulak@tamu.edu

Christopher Bronk Ramsey
Oxford Radiocarbon Accelerator Unit
Research Laboratory for Archaeology
Dyson Perrins Building
South Parks Road
Oxford OX1 3QY
United Kingdom
christopher.ramsey@rlaha.ox.ac.uk

A. Colin Renfrew
McDonald Institute for Archaeological Research
University of Cambridge
Downing Street
Cambridge CB2 3ER
United Kingdom
acr10@cam.ac.uk

Christopher H. Roosevelt
Department of Archaeology
Boston University
675 Commonwealth Ave.
Boston, MA 02215
chr@bu.edu

Jeremy B. Rutter
Department of Classics [HB 6086]
Dartmouth College
Hanover, NH 03755
Jeremy.Rutter@dartmouth.edu

Fritz H. Schweingruber
Gartenstrasse 19
CH-8903 Birmensdorf-ZH
Switzerland
fritz.schweingruber@wsl.ch

Kim Shelton
University of California, Berkeley
Nemea Center for Classical Archaeology
7233 Dwinelle 2520
University of California
Berkeley, CA 94720
sheltonk@berkeley.edu

Peter Steier
Institut für Isotopenforschung und Kernphysik
VERA Laboratorium
Universität Wien
A-1090 Wien
Austria
peter.steier@univie.ac.at

Geoffrey Summers
Kerkenes Project
Faculty of Architecture
Middle East Technical University
06531 Ankara
Turkey
summers@metu.edu.tr

Sahra Talamo
Heidelberger Akademie der Wissenschaften
Institut für Umweltphysik
INF 229
D-69120 Heidelberg
Germany
sahra.talamo@eva.mpg.de

Ramzi Touchan
Laboratory of Tree-Ring Research
University of Arizona
Tucson, AZ 85721
rtouchan@ltrr.arizona.edu

Kenan Ünlü
Radiation Science and Engineering Center
Breazeale Nuclear Reactor
The Pennsylvania State University
University Park, PA 16802
kxu2@psu.edu

Mary M. Voigt
Department of Anthropology
College of William & Mary
Box 8795
Williamsburg, VA 23187
mmvoig@wm.edu

Sofia Voutsaki
Groningen Institute of Archaeology
University of Groningen
Poststraat 6
9712 ER Groningen
The Netherlands
s.voutsaki@rug.nl

Jennifer D. Watkins
Malcolm and Carolyn Wiener Laboratory for
Aegean and Near Eastern Dendrochronology
Cornell Tree-Ring Laboratory
B-48 Goldwin Smith Hall
Cornell University
Ithaca, NY 14853
jdw7@cornell.edu

Tomasz Wazny
Malcolm and Carolyn Wiener Laboratory for
Aegean and Near Eastern Dendrochronology
Cornell Tree-Ring Laboratory
B-48 Goldwin Smith Hall
Cornell University
Ithaca, NY 14853
   and
Nicolaus Copernicus University
Institute for the Study, Conservation, and
Restoration of Cultural Heritage
ul. Sienkiewicza 30/32
87-100 Torun
Poland
tjw42@cornell.edu

Malcolm H. Wiener
Villa Candia
66 Vista Drive
Greenwich, CT 06830
mhwiener@villacandia.com

Eva M. Wild
Institut für Isotopenforschung und
Kernphysik
VERA Laboratorium
Universität Wien
A-1090 Wien
Austria
eva.maria.wild@univie.ac.at

Carol Zerner
Lerna Publication Team
American School of Classical Studies
54 Souidias Street
GR-106 76 Athens
Greece
czerner@triad.rr.com

Bibliography of Peter Ian Kuniholm

Few archaeological problems stimulate as much rancor as chronology, especially that of the Eastern Mediterranean. The work of the Aegean Dendrochronology Project has been for 36 years and continues to be the building of long tree-ring chronologies for the eastern half of the Mediterranean with the aim of helping to bring some kind of rational order to Aegean and Near Eastern chronology from the Neolithic to the present.

Peter Ian Kuniholm started the Aegean Dendrochronology Project as a one-man Ph.D. dissertation research project in 1973 in the basement of the Ankara Museum. Upon coming to Cornell in 1976, he founded the Cornell Dendrochronology lab in yet another basement, this time in Goldwin Smith Hall. In the three decades since, Peter created a new field: archaeological dendrochronology of the Aegean and Near East. His work has profoundly changed approaches to archaeology and several other fields.

Peter began with a study of Iron Age conifers in Phrygia and the Anatolian highlands, and work since has expanded into a much broader research enterprise. The lab work now involves the building of tree-ring chronologies for a number of species of trees spread over the Eastern Mediterranean: from Georgia on the fringe of the Caucasus to selected parts of Italy; from the Troodos Mountains of Cyprus and the cedars of North Lebanon to trees growing in the former Yugoslavia and parts of Bulgaria. This includes all of Turkey and all of Greece including Crete.

The reason dendrochronology works in such a broad geographical area is that, in addition to micro-climatic zones such as those in central Anatolia, several of the coastal regions—the Taurus, the Pontus, and so forth—possess macro-climatic zones (or a set of overlapping micro-zones) which enable us to crossdate wood over great distances. Since modern cedars at Bcharré, North Lebanon, crossdate with cedars and junipers in Anatolia, we are not surprised to find that Bronze Age cedar and juniper imported to Egypt from Lebanon also crossdate with their Bronze Age Anatolian counterparts.

Over the years, millions of tree-ring measurements have led to the successful compilation of chronologies spanning (but not continuously) 9,000 years. As Professor Emeritus at Cornell, Peter continues building chronologies in the lab, and his wife Ellie continues her support with cookies and countless other intangibles. Indeed filling gaps while fostering hundreds of students over the years remains their mutual ambition.

Articles, Papers, and Catalogues

"Gordion Müzesinden Çalınan Antik Eserler." *USIS Press*, Ankara, Turkey, 1976.

"Antiquities Stolen from the Gordion Museum." *Journal of Field Archaeology* 3:2 (1976) 209–213.

"Dendrochronology at Gordion and on the Anatolian Plateau." Ph.D. Dissertation, University of Pennsylvania 1977.

"The Tie-Beam System in the Nave Arcade of St. Eirene: Structure and Dendrochronology," with C.L. Striker. *Istanbuler Mitteilungen Beiheft* 18 (1977) 229–240.

"Ağaçların Dili," *Ufuk ABD* 3 (1977) 26–30. ("The Language of Trees," *Horizon USA*) 3 (1977) 26–30. In Turkish.

"Tree Rings as an Archaeological Dating Tool." In G.B. Stone (ed.), *Spirit of Enterprise: The 1981 Rolex Awards*: 98–99. San Francisco: Freeman and Co. 1981.

"Dendrochronological Investigations in Greece 1977–1980," with C.L. Striker. *Archaeologika Analekta ex Athenon* (Journal of the Greek Archaeological Service) XIV:2 (1981) 230–234.

"The Fishing Gear" and "Dendrochronological Analysis of the Keel." In G.F. Bass et al. (eds.), *Yassı Ada I: A Seventh-Century Byzantine Shipwreck*: 291–310 and Appendix D. College Station: Texas A & M University Press 1982.

"Dendrochronologese: Mia technike chresime ste melete tou oikologikou parelthontos." *Dasika Chronika*. ("Dendrochronology: a useful technique for studying the ecology of the past." *Forest*

Chronicle: Journal of the Greek Forestry Service 25:3/4 (March/April 1982) 85–94. Translated by Dr. C. Cassios.

"Dendrochronological Investigations in Greece: 1977–1982," with C.L. Striker. *Journal of Field Archaeology* 10:4 (1983) 411–420.

"Dating in the Aegean Region." *Mitteilungen der Bundesforschungsanstalt für Forst und Holzwirtschaft* (University of Hamburg) 141 (1983) 179–194.

"Aegean Dendrochronology Project Expedition to Greece, 1979–1980." *National Geographic Society Research Reports*, Vol. 20 (1985) 439–447.

"Dendrochronological Investigations at St. Sophia: A Preliminary Report," with C.L. Striker. *Ayasofya Müzesi Yıllığı* 10 (1985) 41–45.

"Dendrochronological Analysis of the Elaia Tumulus Kline." *Istanbuler Mitteilungen* 35 (1985) 168–182, Taf. 37.

"Demircihüyük—Dendrochronological Analysis." In M. Korfmann (ed.), *Demircihüyük II*: 1–4. Mainz: von Zabern 1987.

"Aegean Dendrochronology Project—1985," with C.L. Striker. In *Proceedings of the VIIIth Symposium on Excavation, Research and Archaeometry, Ankara, Turkey, 26–30 May 1986*: 41–47 Ankara: Türk Tarih Kurumu Basımevi 1987.

"Dendrochronological Investigations in the Aegean and Neighboring Regions, 1983-1986," with C.L. Striker. *Journal of Field Archaeology* 14:4 (1987) 385–398.

"Aegean Dendrochronology Project—1986 Results," with C.L. Striker. In *Proceedings of the IXth Symposium on Excavation, Research and Archaeometry, Ankara, Turkey, 6–10 April 1987*: 247–251. Ankara: Türk Tarih Kurumu Basımevi 1988.

"Dendrochronology and Radiocarbon Dates for Gordion and Other Phrygian Sites." *Source* VII/3–4 (December 1988) 5–8.

"Aegean Dendrochronology Project: 1987 Results," with C.L. Striker. In *Proceedings of the Xth Symposium on Excavation, Research and Archaeometry, Ankara, Turkey, 23–27 May 1988*: 97–101. Ankara: Türk Tarih Kurumu Basımevi 1989.

"Overview and Assessment of the Evidence for the Date of the Eruption of Thera." In D.A. Hardy and A.C. Renfrew (eds.), *Thera and the Aegean World III: Proceedings of the Third International Congress, Santorini, Greece, 3–9 September 1989. Vol. III: Chronology*: 13–18. London: The Thera Foundation 1990.

"The archaeological record: evidence and non-evidence for climatic change." In S.K. Runcorn and J.-C. Pecker (eds.), *The Earth's Climate and Variability of the Sun Over Recent Millennia, Phil. Trans. R. Soc. Lond.A*, 645–655, 1990.

"Aegean Dendrochronology Project: 1988–1989 Results." In *Proceedings of the XIth Symposium on Excavation, Research and Archaeometry, Antalya, Turkey 18–23 May 1989*: 87–96. Ankara: Türk Tarih Kurumu Basımevi 1990.

"Dendrochronology and the Architectural History of the Church of the Holy Apostles in Thessaloniki," with C.L. Striker. *Architectura: Zeitschrift für Geschichte der Architektur*, 1 (1990) 1–26.

"A 677 Year Long Tree-Ring Chronology for the Middle Bronze Age," with M.W. Newton. In Kutlu Emre, Machteld Mellink, Barthel Hrouda, and Nimet Özgüç (eds.), *Anatolia and the Ancient Near East: Studies in Honor of Tahsin Özgüç*: 279–293. Ankara: Türk Tarih Kurumu Basımevi 1990.

"Sakarya Nehrinden Prehistorik Bir Köprü Ayağı" ("A Prehistoric Timber from the Sakarya River)." *Devlet Su İşleri Bülteni* (Bulletin of the General Directorate of State Waterworks) 34: 345 (May 1990) 18–20. In Turkish.

"Aegean Dendrochronology Project—1989-1990 Results." In *Proceedings of the XIIth Symposium on Excavation, Research and Archaeometry, Ankara, Turkey, 28 May–1 June 1990*: 127–138. Ankara: Türk Tarih Kurumu Basımevi 1991.

"A 1503-Year Long Chronology for the Bronze and Iron Ages." In *Proceedings of the XIIIth Symposium on Excavation, Research and Archaeometry, Çanakkale, 28 May–1 June 1991* 121–130. Ankara: Türk Tarih Kurumu Basımevi 1992.

"Dendrochronological Wood from Anatolia and Environs." *Trees and Timber in Mesopotamia*: 97–98. Cambridge: Bulletin on Sumerian Agriculture VI, 1992.

"Dendrochronological Investigations at Porsuk-Ulukışla, Turkey 1987–1989," with S.L. Tarter, M.W. Newton, and C.B. Griggs. *Syria* 69:3/4 (1992) 379–389.

"A Date-List for Bronze Age and Iron Age Monuments Based on Combined Dendrochronological and Radiocarbon Evidence." In Machteld Mellink, Edith Porada, and Tahsin Özgüç (eds.), *Aspects of Art and Iconography: Anatolia and its Neighbors: Studies in Honor of Nimet Özgüç*: 371–373. Ankara: Türk Tarih Kurumu Basımevi 1993.

"Aegean Dendrochronology Project: Extensions to the Long Chronologies." In *Proceedings of the XIVth Symposium on Excavation, Research and Archaeometry, Ankara, May 1992*: 453–464. Ankara: Türk Tarih Kurumu Basımevi 1993.

"Tille Höyük (Adiyaman) Dendrochronological Report," with S.L. Tarter and C.B. Griggs. In G. Summers (ed.), *Tille Höyük 4*: 179–189. Ankara: British Institute of Archaeology at Ankara & Oxbow Press 1993.

Report on Troy I in M. Korfmann and B. Kromer "Demircihüyük, Beşik-Tepe, Troia—Eine Zwischenbilanz zur Chronologie dreier Orte in Westanatolien." In Manfred Korfmann et al. (eds.), *Studia Troica* III (1994) 135–171.

"Aegean Dendrochronology Project: 1992/1993 Annual Progress Report." In *Proceedings of the XVth Symposium on Excavation, Research and Archaeometry, Ankara, May 1993*: 281-291. Ankara: Türk Tarih Kurumu Basımevi 1994.

"Preliminary Dendrochronological Results from Amorium—1993." *Anatolian Studies* (1994) 127–128.

"Dendrochronology." In Patrick J. McGovern et al. (eds.), "Science in Archaeology: A Review." *American Journal of Archaeology* 99:1 (1995) 100–102.

"A 513–Year *Buxus* Chronology for the Roman Ship at Comacchio (Ferrara)," with C.B. Griggs, S.L. Tarter, and H.E. Kuniholm. In *Bollettino di Archeologia Soprintendenza Archeologica dell'Emilia-Romagna* (16–17–18 Luglio–Dicembre 1992) (appeared 1995) 291–299.

"Long Tree-Ring Chronologies for the Eastern Mediterranean." In Ş. Demirci, A.M. Özer, and G.D. Summers (eds.), *Archaeometry '94: The Proceedings of the 29th International Symposium on Archaeometry*: 401–409. Ankara: TÜBİTAK 1996.

"Aegean Dendrochronology Project: 1993–1994 Report." In *Proceedings of the XVIth Symposium on Excavation, Research and Archaeometry, Ankara, May–June 1994*: 181–187. Ankara: Türk Tarih Kurumu Basımevi 1996.

"Aegean Dendrochronology Project: 1994–1995 Report." In *Proceedings of the XVIIth Symposium on Excavation, Research and Archaeometry, Ankara, May–June 1995*: 189–201. Ankara: Türk Tarih Kurumu Basımevi 1996.

"Anatolian tree rings and the absolute chronology of the eastern Mediterranean, 2220–718 BC," with B. Kromer, S.W. Manning, M. Newton, C.E. Latini, and M.J. Bruce. *Nature* Vol.381 (27 June 1996) 780–783.

"The Prehistoric Aegean: Dendrochronological Progress as of 1995." *Acta Archaeologica* 67 (1996) 327–335.

"Climatology" and "Wood." In *Encyclopedia of Near Eastern Archaeology*, Vol. I, 37–38, and Vol. III, 347–349. Oxford University Press.

"Aegean Dendrochronology Project: 1995-1996 Report." In *Proceedings of the XVIIIth Symposium on Excavation, Research and Archaeometry, Ankara, May–June 1996*: 163–175. Ankara: Türk Tarih Kurumu Basımevi 1997.

"Interim Dendrochronological Progress Report 1995/6," with M.W. Newton. In Ian Hodder (ed.), *On the Surface: Çatalhöyük 1993-95*: 345–347. McDonald Institute and B.I.A.A. Monograph 1997.

"An Early Bronze Age Settlement at Sozopol, near Burgas, Bulgaria," with B. Kromer, S.L. Tarter, and C.B. Griggs. In M. Stefanovich, H. Todorova, H. Hauptmann (eds.), *James Harvey Gaul: In Memoriam*: 299–409. Sofia: The James Harvey Gaul Foundation 1998.

"Aegean Dendrochronology Project: 1996–1997 Results." In *Proceedings of the XIXth Symposium on Excavation, Research and Archaeometry, Ankara, May–June 1997*: 49–63. Ankara: Türk Tarih Kurumu Basımevi 1998.

"Sardinia and Dendrochronology: Any Potential?" *Sardinian and Aegean Chronology: Towards the Resolution of Relative and Absolute Dating in the Mediterranean (Proceedings of a Colloquium held at Tufts University 17–19 March 1995)*: 13–15. Oxbow Books 1998.

"New Dates from Old Trees (1998)." In *Proceedings of the XXth Symposium on Excavation, Research and Archaeometry, Ankara, May–June 1998*: 39–44. Ankara: Türk Tarih Kurumu Basımevi 1999.

"Wiggles Worth Watching—Making Radiocarbon Work: The Case of Çatal Höyük," with M.W. Newton. In P. Betancourt, V. Karageorghis, R. Laffineur, and W.-D. Niemeier (eds.), *MELETEMATA: Studies in Aegean Archaeology Presented to Malcolm H. Wiener as he enters his 65th Year*, 3 vols.: 527–536. Liège: Université de Liège 1999.

"Dendrochronology." In D. Oates, J. Oates, and H. McDonald (eds.), *Excavations at Tell Brak, Vol. 1: The Mitanni and Old Babylonian periods*: 127–128. London: McDonald Institute for Archaeological Research and British School of Archaeology in Iraq 1999.

"Dendrochronologically Dated Ottoman Monuments." In U. Baram and L. Carroll (eds.), *Breaking New Ground for an Archaeology of the Ottoman Empire*: 93–136. New York: Plenum Press 2000.

"Aegean Dendrochronology Project: 1998–1999." In *Proceedings of the XXIst Symposium on Excavation, Research and Archaeometry, Ankara, May 1999*: 145–147. Ankara: Türk Tarih Kurumu Basımevi 2000.

"Ège Kai to Kinton ni okeru nenrin nendai-gaku," ("The Prehistoric Aegean: dendrochronological progress as of 1995"). In M. Sawada and T. Mitsutani (eds.), *Proceedings of the International Dendrochronological Symposium, Nara, Japan, February 18–19 2000*: 38–45. Nara Cultural Properties Research Institution, Nara, Japan 2000.

"Dendrochronology (Tree-Ring Dating) of Panel Paintings." In W.S. Taft and J.W. Mayer (eds.), *The Science of Paintings*: 206–215 and 225–226. New York: Springer Verlag 2000. (Second printing with corrections, 2001).

"Aegean Dendrochronology Project: 1999–2000 Results." In *Proceedings of the XXIInd Symposium on Excavation, Research and Archaeometry, Izmir, May 2000*: 79–84. Ankara: Türk Tarih Kurumu Basımevi 2001.

"Dendrochronological Perspectives on Greater Anatolia and the Indo-Hittite Language Family." In R. Drews (ed.), *Greater Anatolia and the Indo-Hittite Language Family: Proceedings of a Symposium Held at the University of Richmond, Spring 2001*: 28–30. *Journal of Indo-European Studies* monograph series 2001.

"Dendrochronology." In D. Brothwell and A.M. Pollard (eds.), *Handbook of Archaeological Sciences*: 35–46. London: Wiley 2001.

"Aegean Tree-Ring Signatures Explained," with M.K. Hughes, J. Eischeid, G. Garfin, C.B. Griggs, and C. Latini. *Tree-Ring Research* 57:1 (2001) 67–73.

"Dendrochronological Investigations at Ayanis: Dating the Fortress of Rusa II: Rusahinili Edurukai," with M.W. Newton. In A. Çilingiroğlu and M. Salvini (eds.), *Ayanis I: Ten Years' Excavations at Rusahinili Eiduru-kai*: 377–380. Documenta Asiana VI. Rome: CNR: Istituto per gli Studi Micenei ed Egeo-Anatolici 2001.

"Absolute age range of the Late Cypriot IIC Period on Cyprus," with S. Manning, B. Weninger, A.K. South, B. Kling, J.D. Muhly, S. Hadjisavvas, D.A. Sewell, G. Cadogan. *Antiquity* v. 75 no. 288 (June 2001) 328–340.

"Regional $^{14}CO_2$ gradients in the troposphere: magnitude, mechanisms and consequences," with B. Kromer, S.W. Manning, M.W. Newton, M. Spurk, and I. Levin. *Science* 294 (21 December 2001) 2529–2532.

"Anatolian tree-rings and a new chronology for the east Mediterranean Bronze–Iron Ages," with S.W. Manning, B. Kromer, and M.W. Newton. *Science* 294, (21 December 2001) 2532–2535. And see opinion piece by Paula Reimer, same issue, 2494–2495.

"Dendrochronological Investigations at Kaman-Kalehöyük: Dating Early Iron Age Level IId," with M.W. Newton, *Anatolian Archaeological Studies* Vol. X (2001) 125–127.

"Radiocarbon and Dendrochronology," with M.W. Newton. *The Neolithic of Central Anatolia, Internal Developments and External Relations During the 9th–6th Millennia cal. BC, Istanbul: Proceedings of the Central Anatolian Neolithic e-Workshop, 23–24 November 2001* (2002) 275–277.

"Archaeological Dendrochronology." *Tree Rings and People, Conference Proceedings, Davos, Switzerland September 2001*. Special Issue of *Dendrochronologia* 20:1 (2002) 63–68.

"Lavagnone, Italy: a four-phase tree-ring chronology from the Early Bronze Age," with C.B. Griggs, M.W. Newton. In Michèle Kaennel Dobbertin and Otto Ueli Bräker (eds.), *Tree Rings and People, International Conference On the Future of Dendrochronology, Davos, Switzerland 22–26 September 2001*: 52. Abstract Volume 2002.

"Eastern Turkey: archaeology, history and tree-ring research at Urartian Ayanis," with M.W. Newton and C.B. Griggs. In Michèle Kaennel Dobbertin and Otto Ueli Bräker (eds.), *Tree Rings and People, International Conference On the Future of Dendrochronology, Davos, Switzerland 22–26 September 2001*: 53. Abstract Volume 2002.

"Dendrochronological Investigations at Işıklar Village Under Mt. Ida," with M.W. Newton and S.W.E. Blum. In *Mauerschau: Festschrift für Manfred Korfmann, Vol. 3*: 975–989. Remshalden-Grunbach: Reiner 2002.

"Dendrochronological Investigations at Herculaneum and Pompeii." In Wilhelmina F. Jashemski and Frederick G. Meyer (eds.), *A Natural History of Pompeii*: 235–239. Cambridge University Press 2002.

"Dendrochronological Investigations at Kuşaklı/Sarissa," with M.W. Newton. In A. Müller-Karpe et al. (eds.), "Untersuchungen in Kuşaklı 2001" 339–342. *Mitteilungen der Deutschen Orient-Gesellschaft* 134 (2002).

A Guide to the Classical Collections of Cornell University, with N.H. Ramage and A. Ramage, with Jane Terrell, ed. Herbert F. Johnson Museum of Art, Cornell University 2003.

"Confirmation of near-absolute dating of east Mediterranean Bronze-Iron Dendrochronology," with S.W. Manning, B. Kromer, and M.W. Newton. *Antiquity* 77: 295 (2003) http://antiquity.ac.uk/ProjGall/Manning/manning.html.

"New dates for Iron Age Gordion," with K. DeVries, G.K. Sams, and M.M. Voigt. *Antiquity* 77 (2003) 296. http://antiquity.ac.uk/ProjGall/devries/devries.html.

"Neutron Activation Analysis of Absolutely-Dated Tree Rings," with K. Ünlü, J.J. Chiment, and D.K. Hauck (poster) 6th International Conference on Methods and Applications of Radiochemistry (MARC) Kona, Hawaii, April 7–14, 2003.

"No systematic early bias to Mediterranean 14C ages: radiocarbon measurements from tree-ring and air samples provide tight limits to age offsets," with S.W. Manning, M. Barbetti, B. Kromer, I. Levin, M.W. Newton, and P.J. Reimer. *Radiocarbon* 44:3 (2003) 739–754.

"The Neolithic in Anatolia: Tracing Early Settlement in Southeastern and Central Anatolia on Temporal Scales from Radiocarbon Analysis," with M.W. Newton. In M. Özdoğan, H. Hauptmann, and N. Başgelen (eds.), *From Village to Cities: Early Villages in the Near East. Studies Presented to Ufuk Esin, Vol. 1*: 87–90. Istanbul: Homer Yayınevi 2003.

"No evidence of systematic regional ^{14}C differences," with S. Talamo, B. Kromer, S. Manning, M. Friedrich, and M. Newton, Poster presented at the 2003 Radiocarbon Conference in Wellington, New Zealand, 2003.

"Dendrochronology in North Greece," with M.W. Newton. Poster presented at the Annual Meeting of the Archaeological Society of Macedonia and Thrace, February 2004.

"Konya'da Dendrokronoloji," with M.W. Newton, D.L. Davis, M.J. Bruce, T. Lemos, and D. Bozkurt. Poster presented at the XXVIth Symposium on Excavation, Research and Archaeometry, Konya, May 2004.

"Aegean Dendrochronology Project: 2001–2002 Results." In *Proceedings of the XXVth Symposium on Excavation, Research and Archaeometry, Ankara, May-June 2003*: 1–4. Ankara: Türk Tarih Kurumu Basımevi 2004.

"A Dendrochronological Framework for the Assyrian Colony Period in Asia Minor," with M.W. Newton. Featuring Complex Societies from Early Villages to Early Towns: Studies in Memory of Robert J. Braidwood. *TÜBA-AR* 7 (2004) 165–176.

"Dendrochronological report on Arslantepe, Level VIA (Late Uruk), Temple B," with M.W. Newton. *Origini: At the Origins of Power—Arslantepe, the Hill of Lions*. Rome: University of Rome "La Sapienza" (2004) 62.

"Regional 14C gradients between Central Europe and Anatolia—Evidence, Limits and Mechanisms," with B. Kromer, S. Talamo, S. Manning, M. Friedrich, S. Remmele, and M.W. Newton, Vienna Symposium (2004).

"Regional Reconstruction of the Palmer Drought Severity Indices and Precipitation in the North Aegean Region and Northwestern Turkey from an Oak Tree-Ring Chronology, AD 1169–1984." with C.B. Griggs, M.W. Newton, A.T. DeGaetano. Poster presented at the Annual Meeting of the American Geophysical Union, San Francisco, December 2004.

"Dendrochronological Dating in Anatolia: the Second Millennium B.C.: Significance for Early Metallurgy," with M.W. Newton, C.B. Griggs, and P.J. Sullivan. In Ü. Yalçın (ed.), *Anatolia III, Der Anschnitt*, Beiheft 18. Bochum: Deutsches Bergbau-Museum (2005) 41–47.

"Meşe Yıllık Halka Kronolojileri ile Kuzey Ege ve Kuzeybatı Türkiye'nin 1169–1984 Dönemindeki Palmer Kuraklık Şiddeti İndeksleri ve Yağışının Bölgesel Düzeyde Yeniden Oluşturulması," with C.B. Griggs, M.W. Newton, and A.T. DeGaetano. Poster presented at the XXVIIth Symposium on Excavation, Research and Archaeometry (Antalya, May 2005).

"Dendrochronology and Radiocarbon Determinations from Assiros and the Beginning of the Greek Iron Age," with M.W. Newton and K. Wardle. *Archaiologiko Ergo Makedonias & Thrakis* 17 (2005) 173–190.

"Troy VIIB2 Revisited: The Date of the Transition from Bronze to Iron Age in the Northern Aegean," with K.A. Wardle and M.W. Newton. In M. Stefanovich and H. Todorova (eds.), *The Struma/Strymon River Valley in Prehistory: Proceedings of the International Conference Strymon Praehistoricos 27.10-1.10.2004. In the Steps of James Harvey Gaul, Volume 2*: 465–480. Sofia: The James Harvey Gaul Foundation 2005.

"Dendrokronoloji Yöntemiyle Tarihlenmiş Osmanı Anıtları." In U. Baram and L. Carroll (eds.), *Osmanlı Arkeolojisi*: 100–141. İstanbul: Kitap Yayınevi 2005. (Turkish Version of "Dendrochronologically Dated Ottoman Monuments" from Baram and Carroll, eds. 2000.)

"Neutron activation analysis of absolutely dated tree rings," with K. Ünlu, J.J. Chiment, and D.K. Hauck. *Journal of Radioanalytical and Nuclear Chemistry* 264/1 (2005) 21–27.

"Dendrochronological Results from the 2002 Collection at Kuşaklı," with M. Newton and N. Riches. In A. Müller-Karpe et al. (eds.), "Untersuchungen in Kuşaklı 2003" 137–172. *Mitteilungen der Deutschen Orient-Gesellschaft* 136 (2004).

"Aegean Dendrochronology Project: 2003–2004 Results." In *Proceedings of the XXVIth Symposium on Excavation, Research and Archaeometry,*

Konya, May–June 2004: 1–4. Ankara: Türk Tarih Kurumu Basımevi 2005.

"Dendrochemistry of a dendrochronologically-dated forest from Turkey," with J.J. Chiment, D.K. Hauck, C.B. Griggs, S. Burr, and K. Ünlü. Poster presented at the Annual Meeting of the Association of American Geographers, Denver, CO, April 2005.

"Ion Uptake Determination of Dendrochronologically-dated Trees using Neutron Activation Analysis," with D.K. Hauck, K. Ünlü, and J.J. Chiment. Paper given at the 8th International Conference on Nuclear Analytical Methods in the Life Sciences, Rio de Janeiro, April 2005.

"Die Datierung des Schiffswracks von Uluburun," with M.W. Newton, S. Talamo, C. Pulak, and B. Kromer. In Ü. Yalçın, C. Pulak, and R. Slotta (eds.), *Das Schiff von Uluburun: Welthandel vor 3000 Jahren, Katalog der Ausstellung des Deutschen Bergbau Museums Bochum vom 15. Juli bis 16. Juli 2006*: 115–116. Bochum: Deutsches Bergbau Museum 2005.

"Phase changes of the winter North Atlantic Oscillation recorded in spatial changes in tree-ring patterns of North Aegean Oaks, AD 1190–1967," with C.B. Griggs, A.T. DeGaetano, and M.W. Newton. Presentation at the American Geophysical Union's 2005 Fall Meeting, San Francisco. EOS Transaction 86(52), AGU Fall Meeting Supplement, Abstract PP41C-02.

"A Dendrochronological 14C Wiggle-Match for the Early Iron Age of north Greece: A contribution to the debate about this period in the Southern Levant," with M.W. Newton and K.A. Wardle. In T.E. Levy and T. Higham (eds.), *The Bible and Radiocarbon Dating: Archaeology, Text and Science*, proceedings of a conference at the Oxford Center for Hebrew and Jewish Studies, September 2004, 104–113. London: Equinox Press 2005.

"Radiocarbon Calibration in the East Mediterranean Region: The East Mediterranean Radiocarbon Comparison Project (EMRCP) and the current state of play," with S.W. Manning, B. Kromer, M. Friedrich, and M.W. Newton. In T.E. Levy and T. Higham (eds.), *The Bible and Radiocarbon Dating: Archaeology, Text and Science*, proceedings of a conference at the Oxford Centre for Hebrew and Jewish Studies, September 2004, 95–103. London: Equinox Press 2005.

"Absolute Dating at Çatalhöyük," with C. Cessford, P. Blumbach, K. Göze Akoğlu, T. Higham, S.W. Manning, M.W. Newton, M. Özbakan, and A. Özer. In Ian Hodder (ed.), *Changing Materialities at Çatalhöyük: Reports from the 1995–99 Seasons*: 78–81. McDonald Institute for Archaeological Research and the British Institute at Ankara 2006.

"Dendrochronology of Submerged Bulgarian Sites," with M.W. Newton and B. Kromer. In V. Yanko-Hombach, A.S. Gilbert, N. Panin, and P.M. Dolukhanov (eds.), *The Black Sea Flood Question: Changes in Coastline, Climate, and Human Settlement*: 483–488. Dordrecht: Springer 2007.

"Regional Reconstruction of Precipitation in the North Aegean and Northwestern Turkey from an Oak Tree-Ring Chronology, AD 1089-1989," with C.B. Griggs, M.W. Newton, and A.T. DeGaetano. *TÜBA-AR* 9 (2006) 139–144.

"Uluburun Batığı'nın Tarihlendirilmesi," with M.W. Newton, S. Talamo, C. Pulak, and B. Kromer. In Ü. Yalçın, C. Pulak, and R. Slotta (eds.), *Uluburun Gemisi: 3000 Yıl Önce Dünya Ticareti. Deutsches Bergbau Museum, Bochum, 2006*, 117–118. (Turkish version of Ünsal, Pulak, and Rainer, eds., *Das Schiff von Uluburun* 2005).

Kubadabad report in R. Arık (ed.) *Kubad Abad: Selçuklu Saray ve Çinileri*: 200, 209–211. İstanbul, Türkiye İş Bankası 2006.

"Dendrochronology." In R.G. Ousterhout (ed.), *A Byzantine Settlement in Cappadocia, Dumbarton Oaks Studies XLII*: Appendix 2: 190–191. Washington DC: Dumbarton Oaks Research Library and Collection 2006.

"Lavagnone di Brescia in the Early Bronze Age: Dendrochronological Report," with C.B. Griggs and M.W. Newton. *Notizie Archeologiche Bergomensi* 10 (2002) (appeared 2007) 19–34. Includes Italian translation by Raffaele de Marinis.

"A May–June precipitation reconstruction from Aegean oak tree-rings, AD 1089–1989," with C.B. Griggs, A.T. DeGaetano, and M.W. Newton. *International Journal of Climatology* 27:8, 30 June 2007, 1075–1089. [http://www3.interscience.wiley.com/cgi-bin/fulltext/114110986/PDFSTART].

"Regional Reconstruction of the Palmer Drought Severity Indices and Precipitation in the North Aegean and Northwestern Turkey from an Oak Tree-Ring Chronology, AD 1169–1984," with C.B. Griggs, A.T. DeGaetano, and M.W. Newton. Poster presented at "Tree-Rings, Kings, and Old World Archaeology & Environment." Cornell Dendrochronology–Archaeology Conference in Honor of Peter Ian Kuniholm, 3–5 November 2006. In English, Greek, and Turkish.

"Absolute Dating at Apliki Karamallos," with S.W. Manning. In Barbara Kling and James Muhly (eds.), *Joan du Plat Taylor's Excavations at the Late Bronze Age Mining Settlement at Apliki Karamallos, Cyprus, Part 1, Studies in Mediter-*

ranean Archaeology Vol. CXXXIV:1: 325–336. Sävedalen: Paul Åströms Forlag 2007.

"Dendrochronological Analysis of the Tatarlı Tomb Chamber," with M.W. Newton and C.B. Griggs. Appendix in Lâtife Summerer, "From Tatarlı to Munich: The Recovery of a Painted Wooden Tomb Chamber in Phrygia," in A. Dinçol, I. Delemen (eds.), *Proceedings of the International Workshop in Istanbul. The Achaemenid Impact on Local Population and Cultures in Anatolia (6th–4th B.C.), May 19–22 2005*: 153–156. Istanbul 2007.

"Evidence for Early Timber Trade in the Mediterranean," with C.B. Griggs and M.W. Newton. In K. Belke, E. Kisslinger, A. Külzer, and M.A. Stassinopoulou (eds.), *Byzantina Mediterranea: Festschrift für Johannes Koder zum 65. Geburtstag*: 365–385. Wien, Köln, Weimar: Böhlau Verlag 2007.

"A revised dendrochronological date for the fortress of Rusa II at Ayanis: Rusahinili Eiduru-kai," with M.W. Newton. In A. Sagona and A. Çilingiroğlu (eds.), *Anatolian Iron Ages 6: The Proceedings of the VIth Anatolian Iron Ages Colloquium held at Eskişehir, 16–20 August 2004. Ancient Near Eastern Studies, Suppl. 20*: 195–206. Leuven: Peeters Publishers 2007.

"Radiocarbon dating and the absolute chronology of the cemetery," with L.H. Barfield. In L.H. Barfield (ed.), *Excavations in the Riparo Valtenesi, Manerba 1976–1994*: 419–427. Firenze: Istituto Italiano di Preistoria e Protostoria 2007.

"Dendrochronology of the Byzantine World." In Elizabeth Jeffreys, (ed.), *Oxford Handbook of Byzantine Studies*: 182–192. Oxford University Press 2008.

"Dendrochemical analysis of a tree-ring growth anomaly associated with the Late Bronze Age eruption of Thera," with C.L. Pearson, D.S. Dale, P.W. Brewer, J. Lipton, and S.W. Manning. *Journal of Archaeological Science* 36 (2009) 1206–1214.

"Neutron Activation Analysis of Dendrochronologically-dated Trees," with K. Ünlü, D.K. Hauck, N.Ö. Cetiner, and J.J. Chiment. This volume.

"A 924-year oak regional tree-ring chronology for north central Turkey," with C.B. Griggs, M.W. Newton, J.D. Watkins, and S.W. Manning. This volume.

"Dendrochronology at Gordion," with M.W. Newton. In C.B. Rose and G. Darbyshire (eds.), *The Chronology of Early Iron Age Gordion*. University of Pennsylvania Museum. In press.

"Dating volcanic eruptions with tree-ring chemistry," with K. Ünlü, C.L. Pearson, and D.K. Hauck. *IEEE Potentials*. In press.

30 Chronologies posted on the International Tree-Ring Data Bank (ITRDB). See [http://hurricane.ncdc.noaa.gov/pls/paleo/fm_createpages.treering].

See also Aegean Dendrochronology Project Annual Progress Reports, 1979–2008.

Reviews

"The Greek Dark Ages by V. R. d'A. Desborough," *Anatolia* XVI (1972) 141–143.

"Tree-Ring Dating and Archaeology by M.G.L. Baillie," *American Journal of Archaeology* 87 (1983) 105–106.

"Trees and Timber in the Ancient Mediterranean World by Russell Meiggs," *Environment* 26:9 (November 1984) 27.

"Les cernes de croissance des arbres (La dendrochronologie). (Typologie des sources du moyen âge occidental, B.III-2, Fasc. 53.) by André-V. Munaut," *Speculum* (1992) 200–201.

"Radiocarbon 35 (1993) (Calibration and 14th International Radiocarbon Conference Issue)," *Society for Archaeological Science Bulletin* 19: 3/4 December 1996.

"Elmalı II, by Jayne Warner," *Bryn Mawr Classical Review* 7.3 (1996) 257–260.

"The Absolute Chronology of the Aegean Early Bronze Age: Archaeology, Radiocarbon and History, by Sturt Manning," *American Journal of Archaeology* 100:4 (1996) 784–785.

"A Slice Through Time: Dendrochronology and Precision Dating, by M.G.L. Baillie," *American Journal of Archaeology* 101:1 (1997) 192.

"Ancient Anatolia: Fifty Years' Work by the British Institute of Archaeology at Ankara, edited by Roger Mathews. London: British Institute of Archaeology at Ankara, 1998," *Bulletin of the American Schools of Oriental Research* 319 (August 2000) 87–88.

"Chronometric Dating in Archaeology: Advances in Archaeological and Museum Science, Vol. 2. R.E. Taylor and Martin J. Aitken, eds." *Society for Archaeological Sciences Bulletin* 22 (1999) 14–15.

Peter Kuniholm's Dendro Time

Fritz H. Schweingruber

In the '70s of the last century began the golden age of dendrochronology. Peter Kuniholm was one of the driving forces of this golden age. In focus was the development of centennial and millennial chronologies, which started with Andrew Elicott Douglass, Edmund Schulman, Bruno Huber, and Ernst Hollstein. Their research stimulated archaeologists, biologists, ecologists, and physicists of Peter's generation. The Ithaca conference clearly showed that Peter's Near Eastern chronology is the temporal backbone of the intensive and complicated human history at the beginning of written history. Parallel to his work, Europeans compiled millennial oak/pine chronologies for central and western Europe. They tell a precise yearly story of human migration and activities along lake shores in Neolithic and Bronze age times, and about the founding of big cities. In addition, Japanese scientists crossdated the Hinoki and Sugi chronologies for dating their famous temples.

Tree-rings of the past play a key role in one of the most important problems of the 21st century: the human induced climate change. Millennial chronologies contain information about short and long term climatic variations and tell us where we are in relation to the past. Peter's Near Eastern chronology and the western American conifer chronologies are important for reconstructing precipitation in the Mediterranean climate and the North American Pacific regions. Density and ring widths of samples in conifer chronologies from the northern and subalpine timberlines in Eurasia and North America contain information about temperature variations. Oak chronologies from western Europe and conifer chronologies from Japan and South America are still not fully understood climatologically.

Decoding the biological hieroglyphs was a main scientific challenge of the last 40 years. a) New statistical methods have been developed to distinguish biological aging phenomena from climatic influences. b) Pointer years attract world-wide attention to define the influence of volcanic explosions, cold periods and droughts, El Niño, and even human pollution events. c) Dating of scars allowed reconstruction of forest fires in arid and boreal zones and observing and dating callus helped to detect extreme frosts in the past. d) The development of hemispheric networks combined with the improvement of X-ray densitometry allowed a global climatic interpretation of tree rings. e) Tree-rings were and are still of crucial importance for the calibration of the radiocarbon scale. The Ithaca meeting, however, showed that it is still a problem to reach an annual resolution with radiocarbon dates. It seems to be clear that variations of the earth's magnetic field, the intensity of sunspots, and the burning of fossil fuel greatly influence the physical timescale. f) Recently intra- and inter-annual $^{12}C/^{13}C$ and $^{16}O/^{18}O$ ratio tree-ring chronologies are under debate. g) Powerful electronic devices allow the analysis and quantification of microscopic structures.

Where will the dendrochronological journey go? Tree-rings from different species are complicated anatomical structures, which react very sensitively to environmental influences and especially to stress. Therefore dendrochronologists have to understand the physiological and anatomical mechanisms behind the fundamental principles of cambial growth, which occurs in long-living trees as well in short-living shrubs and herbs. The dendrochronological "correlation-period" comes to an end and the "period of understanding" begins. The science of dendrochronology will no longer be limited to forested areas; annual rings occur in the subarctic tundra as well as in all open land of temperate regions. Hardly analyzed are the trees in the tropics despite our knowing that many form annual rings. Since tree-rings and growth layers occur in most countries of the world scientists have to communicate in different languages, have to bridge scientific borders, and senior scientists have to teach.

Peter, you are and were riding on two horses: archaeology and dendrochronology. You were always open for new developments, e.g. in isotopic physics and climatology. You always trained young students. We met so many of your "kids" in Switzerland. And, for most of us a miracle, you communicated in "dozens" of languages and your enthusiasm is of no limit. Slowly, slowly you will become a historical chronology, but a master chronology.

Article submitted April 2007

Perspective: Archaeology, History, and Chronology from Penn to the Present and Beyond

James Muhly

Introduction

"I came to Cornell in 1976 with half a suitcase of wood. Now we have an entire storeroom with some 40,000 archived pieces that cover some 7,500 years" (Kuniholm 2006: 2).

"So, you want to be a dendrochronologist?" That is actually the title of an article published in the *High Country News* for 24 January 2005, by Michelle Nijhuis. The article begins: "Sure, counting tree rings might sound like a cushy job" (Nijhuis 2005: 1). I had always suspected that this was the case, but did not expect to find it stated in print. The author, however, is cautious, stating that "before you set out into the bristlecone pines, make sure you know what you're in for." There are basic requirements for the job and she spells them out:

1. Strong Legs. The clearest records of climate are found in harsh environments.

2. Patience, patience, patience. As she puts it: "crossdating a set of tree-ring samples still takes a near pathological love of detail."

3. Love of statistical gymnastics, for isolating a single variable requires "a tolerance for big datasets, an affection for mind-bending statistics, and, last but not least, a very comfortable office chair" (Nijhuis 2005: 2-3).

I am not sure about that last one for it is my impression that, whenever Peter Kuniholm is in the office, he never sits down. He is always on the move.

Kuniholm seems to have met all these requirements. An article in a Greek newspaper, on the dendrochronology project, described him as "a husky classical archaeologist who minces no words" (Ritter 1993: 3) and the *Cornell Alumni Magazine* describes him as "Resembling sort of an academic younger brother to Colonel Sanders" (Saulnier 2002: 50—this article gives a wonderful description of Kuniholm's diplomatic skills, for which see below). Dendrochronology is what he decided to do, over thirty years ago, and dendrochronology is what he has done, with spectacular success. It might never have happened had Peter not gone out to Gordion as a graduate student, to work with Rodney Young. It was certainly the excavation of what was then identified as the tomb of Midas, King of Phrygia, that did the trick. Midas, as we will call the deceased for the time being, was buried inside what could be described as a log cabin, but can more accurately be described as "the world's oldest standing wooden building" (Manning et al. 2001: 2532).

Here was something that deserved serious study. Was there some scientific way of dating such a remarkable structure? Well, yes there was and, as Peter in the early 1970s had been appointed as the director of the Ankara branch of the American Research Institute in Turkey (or ARIT), after spending a year as a Fellow of the Institute (under Hans Güterbock), he soon found himself in the basement of the Ankara Museum measuring tree rings (continuing the work by Bryant Bannister and others, for which see below), making use of a measuring instrument provided by Elizabeth Ralph and a second-hand microscope provided by Froelich Rainey. This was not all entirely accidental. Peter Kuniholm was a student of Classical Archaeology at the University of Pennsylvania in a department chaired by Rodney Young, the director of excavations at Gordion. The University Museum, in the 1970s, was one of the great centers of archaeological science in the world, thanks to the establishment, in 1951, of the University of Pennsylvania's Museum Applied Science Center for Archaeology, known as MASCA. This was the brainchild of Froelich Rainey, the new director of the museum. He put in charge of MASCA Elizabeth Ralph from Chicago, a young physicist who had studied with Willard Libby, the "father of radiocarbon dating." Ralph's assistant was Henry Michael, one of the pioneers of dendrochronology in America (who just passed away, on 19 February 2006, at the age of 92: for the career of Michael see

Gidwitz 2001). In 1971 Michael and Ralph co-edited a book on *Dating Techniques for the Archaeologist*, for which Michael contributed a chapter on "Climates, Tree Rings, and Archaeology," pp. 49–56. Coming from such a background it is not surprising that Peter Kuniholm found himself involved in pioneering studies in the dendrochronology of Anatolia. Turkey was also a country that Peter knew very well because, before becoming a graduate student at Penn, he had taught at Robert College in Istanbul.

Research in the basement of the Ankara Museum eventually led to a PhD dissertation on Dendrochronology at Gordion and the Anatolia Plateau, completed in 1977 and published by University Microfilms (Ann Arbor MI) in 1979. I regret to say that this publication is not available in any of the libraries of Athens. I have my own copy, given to me by Peter in 1977, but the dissertation seems not to be otherwise available in Greece. Classical Archaeology is not a field known for being open to new techniques and new research strategies. No one quite knew what to do with Peter's conclusions in his dissertation. There was general agreement that the Phrygian capital of Gordion had been destroyed by Kimmerian invaders in the early 7th century BC (see Young 1978: 10; De-Vries 1980: 34; Mellink 1991: 629, 634; Sams 1997: 239–240, 243). Perhaps as a harbinger of things to come Keith DeVries, in his summary of work at Gordion 1969–1973, and beyond, refers to the destruction level but never mentions the Kimmerians (DeVries 1990; Voigt and Henrickson 2000: 51–52, are highly skeptical of the entire Kimmerian episode). Later Greek chronographers put this destruction at 696 BC. King Midas turned out to be a historical figure, for he was mentioned in Assyrian letters from Nimrud written during the reign of the Assyrian king Sargon II (721–705 BC), in apparent agreement with later Greek chronology. Special mention should be made of a letter from Nimrud found in 1952 (summarized by Mellink 1991: 622–623; new edition by Parpola 1987: 4–7, text no. 1). According to this letter both Mita of Muški and Urballa-Warpalawas of Tyana sent messengers to Sargon II, anxious to make peace with Assyria. But Mita betrayed the confidence of Urikki-Warikas, ruler of Que, a king known from the famous Luwian-Phoenician bilingual inscription from Karatepe, but now also from his own monumental Luwian-Phoenician inscription set up at the site of Çineköy, in the vicinity of Adana (Ipek et al. 2000; Lanfranchi 2005; Lebrun and De Vos 2006; Rollinger 2006: 284–286).

But Kuniholm's dates for the wood found in the destruction level (DL) at Gordion were in the early 9th century BC and, in the 1980s and 1990s, Anatolian archaeologists all concluded that these dates must reflect the use of old wood and that dendrochronology was not going to be of much use for absolute dating. Better to stick with imported Greek pottery, of which there is none in the DL, and later Greek historical traditions.

I do not wish to belabor this point, but I find it very curious that, in subsequent scholarship on Gordion, written by members of the Gordion excavation staff, this dissertation is virtually ignored. I cite just a few examples. Machteld Mellink, in her excellent discussion of "The Phrygian Kingdom," in the second edition of the *Cambridge Ancient History* (Mellink 1991), does cite the dissertation in her bibliography but never refers to it in her text. G. Kenneth Sams, in his article on "Gordion and the Kingdom of Phrygia," published in 1997 (Sams 1997), ignores the dissertation altogether, as does Keith DeVries in his publication of work at Gordion prior and subsequent to the death of Rodney Young in 1973; DeVries was the supervisor of Kuniholm's dissertation (DeVries 1990). In the final report on the Midas Mound Tumulus Kuniholm's work merits only a footnote (Young et al. 1981: 96, n. 17).

On the other hand, in the late 1970s and early 1980s, Russell Meiggs, the great Oxford historian of ancient Greece and Rome, was in the process of writing his superb book on *Trees and Timber in the Ancient Mediterranean World*, published by the Clarendon Press in 1982. Meiggs knew all about Kuniholm's research, had read his dissertation, and gave a magnificent account of the problems involved in dating the wood from the Midas Mound. The reason for this is very simple: Meiggs understood trees (Meiggs 1982: 458–461). For Meiggs it made perfectly good sense that juniper had been selected for the logs making up the outer casing of the tumulus, since juniper was a much stronger wood than pine. It was also not surprising, according to Meiggs, that the outer junipers were longer-lived than the pine beams of the burial chamber. Looking for suitable trees in the surrounding countryside, the available junipers, coming from slow-growing trees, were going to be older than the pine trees used for the burial chamber itself. But how much older? That, as Meiggs realized, was the decisive question.

It is certainly curious that Kuniholm's research was taken far more seriously by an ancient historian, working in Oxford, than by his archaeological colleagues working at Gordion itself. But Kuniholm persisted, for he knew that dendrochronology was a well-established field of research and, if it worked for the southwestern United States, for southern Germany and northern Ireland, then it was also going to work for Anatolia. The point I want to make here, one obvious to everyone in the field of dendrochronology, is that by the 1970s, dendrochronology was a new research technique only in the world of the eastern Mediterranean. Andrew Ellicott Douglass had already established accurate dates for 45 Pueblo monuments in the American southwest, back in the 1920s; the *Tree-Ring Bulletin*, one of the major journals in the field, had been founded in 1934, and, in 1937, Douglass had established the Laboratory of Tree-Ring Research at the University of Arizona (Douglass 1921; Robinson 1976). None of this had made any impact upon the worlds of Aegean and Anatolian archaeology.

Except, of course, for Rodney Young himself. Work on the Midas Mound began in 1957. In 1961 Young invited Dr. Bryant Bannister of the Laboratory of Tree-Ring Research at the University of Arizona, to come to Gordion in order to sample the juniper logs from the Midas Mound for possible dendrochronological dating. Bannister spent a week at Gordion, using his coring tool to take samples from the logs of the tomb's outer casing (Kuniholm 1977: xxxiv-xxxv). These samples were taken back to Tucson and measured by Bannister and his assistant Jeffrey Dean (now a senior professor at the lab and one of the participants at this conference). Very little was published on this work, but all the samples were included in Kuniholm's dissertation, there marked 'A' for Arizona (Kuniholm 1977: liv). One of these corings, A-TU-GOR-9, proved to be the then oldest log in the tumulus—806 years (Kuniholm 1977: 3). Since then a longer-lived log, with 918 rings, has been found in the tomb.

Even before Bannister and Dean, Rodney Young had already thought about the possibilities of doing dendrochronological dating of the wood from the Midas Mound. In his first report on the opening of the Midas Mound, published in 1958, Young states that cross-sections of three of the round juniper logs from the outer casing of the tumulus had been taken and sent to Prof. J. L. Giddings of Brown University, who had done the fundamental work on tree-rings in Alaska (Bering Straits), together with Froelich Rainey. Giddings reported that one of the cross-sections contained up to 700 growth rings. In describing this work Young goes on to say:

> "If our tomb is to be dated 725–700 BC, this tree would have started to grow around 1400 BC. The possibility of establishing a dendrochronological chart for Anatolia which might carry us right back to the time of the Hittite Empire depends on the availability of wood or charcoal samples from sites such as Boğhaz Kale and Kültepe as well as upon a sufficient uniformity of climate and rainfall over the Anatolian plateau at that remote period. The project is well worth pursuing"

(Young 1958: 148, n. 17; also quoted by Kuniholm 1977: xxxiv).

Apparently conditions at the University of Pennsylvania had to wait for a Kuniholm to come along before someone was willing to take up this challenge.

When I set out to write this paper for the dendrochronological conference at Cornell, in honor of Peter Kuniholm, I was convinced that the idea of doing tree ring dating, at Gordion and all of Anatolia, was the result of the establishment of MASCA at the University Museum. Now I realize how wrong I was. Rodney Young needed no one to tell him about the potential of dendrochronological dating; he was well aware of such possibilities already in 1957–1958. Developments in dendrochronological research over the past 50 years certainly would have come as no surprise to him.

What Peter Kuniholm set out to do turned out to be as much a project in public relations as it was a research project. He was determined to bring dendrochronology to the world of Aegean and Near Eastern archaeology and, in particular, to Greece and Turkey. He had to convince his cautious colleagues that this was something worth doing, and he had to raise his own money to support his own research. Even after coming to Cornell, it was only in 1989 that Peter was able to announce, in his Newsletter for December of 1989, after 13 years in the Department of Classics, that, on 1 July of that year, he had finally become associate professor in the Department of the History of Art and Archaeology.[1] Let this be fair warning to anyone foolish enough to set out to create a new field of research in a well-established academic discipline.

So, after some 30 years of intensive research, where do things now stand for what has become The Malcolm and Carolyn Wiener Laboratory for Aegean and Near Eastern Dendrochronology, and its new director, Prof. Sturt Manning? It is often said that, in the Humanities at American universities, scholars spend their entire careers rewriting their dissertations. What Peter Kuniholm has done is to expand continuously the research horizons of his dissertation. In 1996 Kuniholm published a paper in which he summarized dendrochronological research in the Aegean and the Near East down to 1995 (Kuniholm 1996). He calculated that the Cornell Lab had established chronological sequences covering 6,500 years of tree-ring chronologies, based upon over eight million microscope measurements made by some 400 project participants who had put in over 50,000 hours of lab time, all this in a period of some 20 years. Eight years later, in 2003,

---

[1] The actual quote was "I seem to have become a Permanent Denizen of the Department of the History of Art and Archaeology instead of an Adjunct Denizen of the Department of Classics." [eds.]

the chronological sequences covered up to 7,500 years spread over the last 9,000 years, based upon over ten million tree-ring measurements (Wilkie 2003). What has been created, for Anatolia, is a so-called "floating chronology," detailed in almost all of its aspects, but lacking firm anchors for the Bronze Age–Iron Age chronology.

It was this lack of absolute fixed points that limited the impact of dendrochronological work in Anatolia. In 1991 M. G. L. (Mike) Baillie, one of the leading scholars in this field (and one who took part in this Cornell conference), wrote a remarkable paper setting out, in a very straight-forward fashion, the reasons for his belief that "no-one in the scientific community has any doubt about the entire tree-ring system" (Baillie 1991: 19). This paper represents a lecture delivered before a very skeptical audience of scholars who firmly believed that the traditional Late Bronze and Early Iron Age chronology had to be shortened by some 400 years (James, Thorpe, Kokkinos, and Frankish 1987; James 1991). Baillie set out to show them that this simply was not possible. For him the absolute chronology of the Old World was based upon five long tree-ring chronologies (Baillie 1991: 15, with references):

1. the lower forest bristlecone pine chronology.

2. the upper tree-line bristlecone pine chronology.

3. the Irish bog oak chronology.

4. the German oak chronology.

5. the so-called Göttingen chronology.

Baillie makes no reference to work by Kuniholm in the Aegean or Anatolia, for that work lacked fixed anchors. He does state, however, in reference to the controversies regarding a fixed date for the eruption of Thera, that:

> "What is particularly interesting is the way in which the attempts to date the Theran eruption highlight the inherent weaknesses in archaeological interpretation, ancient history, volcanology, ice-core studies and routine radiocarbon analysis. Interestingly, the story merely serves to demonstrate the remarkable dating resolution offered by dendrochronology—in the whole debate the only fixed points are 1628 BC and 1627 BC, both derived from tree-ring studies; everything else is flexible" (Baillie 1991: 22-23; for those two points see the paper by Malcolm Wiener in this volume).

The persistent efforts of Peter Kuniholm to "close the gap" in the dendrochronological sequence for the Aegean and Anatolia are now starting to have an impact upon Anatolian and Classical archaeology. To appreciate what has happened we should first return to Gordion. In order to anchor at least part of the floating chronology, a series of 52 decade-long tree-ring samples from wood at Gordion was cut and sent to Heidelberg for high-precision radiocarbon measurements, making use of absolutely dated tree rings from southern Germany and northern Ireland. This work was reported on in *Science*, vol. 294 for 21 December 2001 (Manning et al. 2001). The construction timbers from the destruction level at the site of Gordion gave a date between 830 and 800 BC, while the trees used to build the great Midas Mound Tumulus gave a date of ca. 740 BC, much too early for any association of this tumulus with the historical Mita of Muški. This could only be seen as a confirmation of the results worked out in the 1970s. This work was elaborated upon in a paper published in the electronic version of *Antiquity*, vol. 77 for June 2003, written together with the late Keith DeVries, G. Kenneth Sams, and Mary Voigt, the director of new excavations at Gordion (DeVries et al. 2003). This study was based upon radiocarbon determinations by Bernd Kromer at the Heidelberg Akademie der Wissenschaften, made from short-lived seed samples found in jars resting on the floor of the destruction level at the site. These consisted of two barley samples, two of lentils, and one of flax, all from the destruction level of Terrace Building 2A. There were also three samples from the reeds used in the construction of the roof of Terrace Building 2A. The dates were in the fourth quarter of the 9th century, suggesting to the authors "that the destruction date may be around the lower end of the 827–803 range," about one century earlier than the accepted historical destruction at the hands of the Kimmerians (DeVries et al. 2003).

According to the new dendrochronological dates for construction on the city mound the felling of the timbers used in the construction of Clay Cut Building 3 was now put at ca. 916-906 BC while the timbers used in Terrace Building 2A were felled ca. 890-880 BC (DeVries et al. 2003). Peter Kuniholm has now informed me that a re-measured piece now brings this date down to 850 BC. Already, in 2001, it had been stated that Terrace Building 2A must have been destroyed in a conflagration between 830 and 800 BC (Manning et al. 2001: 2534). It seemed reasonable to assume, therefore, that construction work on the city mound began no later than the late 10th century BC. In work related to the conservation of the monumental Phrygian Gate, in the summer of 2003, a row of some twenty juniper logs was discovered under the east face of the wall of the gate. These had probably been laid as a leveling course to support the weight of the dry-stone masonry of the gate complex. It was possible to cut samples from four of the logs and the last existing ring (without bark) was 862 BC. On the basis of this evidence Kuniholm concluded that "The Phrygian palace enclosure wall and gate thus appear to be a ninth-century affair" (Kuniholm 2004). This chronology is actually quite close to that proposed by Oscar Muscarella back in 1995, in his survey of the early Iron Age history of the Phrygians (Muscarella 1995: 96):

"The first unambiguous archaeologically documented indication of organized political and social activity in Gordion's history was the initiation of the massive building program around 825–800 BC."

This is in reference to the Phase 6B master building plan at Gordion. Basically the same chronology is presented by Mary Voigt and Robert Henrickson in their study of the early Phrygian state (Voigt and Henrickson 2000: 48–50).

These early dates certainly did not come as a surprise to Peter Kuniholm. Already, back in 1977, Kuniholm had concluded, on the basis of his (preliminary) early dendro dates from the city mound at Gordion, that the Phrygians had been well established at Gordion by the middle of the 9th century, perhaps even earlier (Kuniholm 1977: 14–15). In a short review of his work at Gordion and other Phrygian sites, published in 1988, Kuniholm argued that the dendrochronological evidence indicated "a long history of human activity at Gordion" (Kuniholm 1988: 8). All of this has, of course, now been confirmed by the renewed archaeological work at the site of Gordion (for which see the paper by Mary Voigt in this volume; also Voigt and Cuyler Young, Jr. 1999: 201–203).

As for the juniper logs from the Midas Mound Tumulus the dates were now 743–741 BC for the construction of the tumulus (DeVries et al. 2003). It is important to realize that these new dates are all from the juniper logs used in constructing the outer casing of the great tumulus. To my knowledge no recent work has been done in dating the pine beams used in the construction of the burial chamber itself (and with good reason, as squared-off timbers can never give terminal dates and it would also now be impossible to get permission to take samples from the beams of the inner chamber). These high dates have been stoutly contested, especially by Oscar Muscarella of the Metropolitan Museum of Art (Muscarella 2003). What is at issue here is not just a series of radiocarbon and dendro dates from Gordion. The revised chronology has profound implications for the way scholars will come to see relations between the Greeks and the Phrygians in the 9th and 8th centuries BC. Traditionally things have been seen in terms of Phrygians borrowing from the Greeks. Will all of this scholarship now have to be turned upside down? At stake are such items as artistic styles and iconography, the art of making mosaics, even early alphabetic writing. The Gordion excavators have already concluded that:

"The chronological changes open up a severe gap between Phrygian material culture and the more modest culture of Geometric Greece and they put Phrygia more on a par with the flourishing Syro-Hittite region to the southeast" (DeVries et al. 2003).

We are in for some lively scholarship in the years ahead. Greek archaeologists certainly are not going to be happy with what will undoubtedly be seen as a rather backhanded description of the culture of Geometric Greece.

The other great interest of the Cornell Dendro Lab over the past 20 years has been the Middle Bronze Age in Central Anatolia and the possibility of obtaining accurate dendro dates from the sites of Kültepe and Acemhöyük. Central to success in getting wood samples from both sites has been Kuniholm's close personal relationship with their Turkish excavators, Tahsin and Nimet Özgüç. It is not just accidental that Peter Kuniholm comes from a background in the State Department. Once again we are dealing with a situation that is at once archaeological and deeply historical. Assyrian merchants were living at Kültepe, trading tin and textiles for gold and silver and making a good profit from doing business with the local Anatolian rulers. Most of the Assyrians lived in the merchant colony or kārum, and they were in constant communication with the main branch of the family business back in Assur. The details of this correspondence are contained in an estimated 35,000 Old Assyrian letters found in the kārum Kanish. They were written in the Old Assyrian dialect and are very difficult to read. Only a small handful of scholars work in Old Assyrian, most notably Klaus Veenhof of Holland. There are no tablets from the site of Acemhöyük, but we have bullae or sealings from the Sarıkaya palace belonging to Šamši-Adad I, king of Assyria, a contemporary of Hammurapi, the king of Babylonia, as well as of Aplahanda, king of Carchemish.

We still know very little about the history or chronology of the Old Assyrian period and accurate dates from Kültepe would be of enormous benefit to all scholars interested in the history and archaeology of Anatolia during the second millennium BC. The complex chronological and historical problems at both sites have been explicated in a superb paper by Maryanne Newton and Peter Kuniholm, "A Dendrochronological Framework for the Assyrian Colony Period in Asia Minor," published in 2004 in a volume in memory of Bob and Linda Braidwood (Newton and

Kuniholm 2004). Without going into details that I do not understand in any case, we are dealing with wood from two large buildings, both destroyed by violent conflagrations: the so-called Sarıkaya (or "Yellow Rock") palace at Acemhöyük and the Waršama palace, the palace of the local Anatolian ruler at Kültepe, from level 7 on the city mound, contemporary with level Ib at the kārum. The Sarıkaya palace now gives a dendro date of ca. 1774 BC, but with a roof repair at least as late as 1766 BC, and the Waršama palace a date of ca. 1832 BC, but with a repair at least as late as 1779 BC. Both of these dates present serious problems for our present understanding of Old Assyrian chronology.

What can be said about Old Assyrian chronology has already been said by Klaus Veenhof in a superb, small pamphlet published in 2003: *The Old Assyrian List of Year Eponyms from Karum Kanish and its Chronological Implications* (Veenhof 2003). Throughout their history the Assyrians maintained the practice of naming the year after an annual official known as a limmu, much like the eponymous archōn in Athens. On the basis of new discoveries of limmu-lists at Kültepe and at Mari, Veenhof has now worked out what must be a nearly complete list of limmu officials for kārum Kanish level II. They cover an interval of 138 years, from 1974 BC, the first year of the reign of King Irišum I, to 1836 BC, during the reign of Naram-Suen. For Veenhof these dates now represent the duration of level II at the kārum Kanish. In 1836 the kārum came to an end in a violent destruction.

There then followed a short hiatus prior to the refounding of the kārum in what is known as the kārum Kanish Ib. It cannot have been a long interval because some of the merchants who were active at the end of level II were still doing business during the early years of level Ib. What can we say about the dendrochronological dating of level II and level Ib in the kārum? Nothing whatsoever, for there is no wood with enough preserved rings to date it. All of the wood used for dendrochronology has come from the large public building from level 7 on the city mound at Kültepe, known as the palace of Waršama and from the large public building at Acemhöyük known as the Sarıkaya palace. Both of these buildings were contemporary with level Ib at the kārum, a period about which we still know very little. Veenhof estimates that it must have covered a period of ca. 60 years, from ca. 1800 to just after 1740 BC. We also know that level Ib was, to some degree, contemporary with the reign of Šamši-Adad I who, according to Veenhof, was king of Assyria from 1808 to his death in 1776, a period of 33 years.

Yet, according to the evidence from dendrochronology, the Sarıkaya palace was built in ca. 1774 BC, two years after the death of a ruler whose bullae were found in the burned remains of the palace. How these problems are going to be resolved only time will tell, but it is important to remember that all of Veenhof's dates are based upon the so-called "Middle Chronology," putting the reign of Hammurapi of Babylon at 1792–1750 BC and the Hittite raid upon Babylon, during the reign of king Muršiliš I, in 1595 BC. This chronology became canonical with the publication of a short 52-page pamphlet in 1940 by Sidney Smith, entitled Alalakh and Chronology (Smith 1940). If Sidney Smith turns out to be correct it has to be said that he will be correct for all the wrong reasons. Smith thought that the tablets from Alalakh level VII were contemporary with those from Mari. We now realize, thanks to the brilliant work of B. Landsberger (1954; see also Goetze 1957), that the Alalakh tablets date to the generation subsequent to the reign of Hammurapi.

The inscribed bullae from the Sarıkaya palace have played such a major role in evaluating the dendrochronological dates from this palace that it is important to understand the actual context of these sealings and their potential chronological significance. Fortunately we have a detailed account of this context from the excavator, Nimet Özgüç (1980). The important evidence comes from room 6 at the Sarıkaya palace. The clay bullae had been carefully removed from the bales and packages of the merchandise sent to the palace and then stored on wooden shelves built up against the walls of the room. One bulla had actually been found burned to a crisp, in the intense fire that destroyed the palace, still stuck to what survived of the wall of the room (N. Özgüç 1980: 62). This is the best evidence known to me for the Bronze Age storage of bullae on wooden shelves. What this means is that the shipments bearing these bullae came into the palace from outside and were then stored on wooden shelves within the rooms of the palace. For chronology this means that we have no idea exactly when these sealings entered the storage facilities of the palace. Nor do we have any idea how long they were stored, prior to the final destruction of the palace. These are very important chronological considerations that, it seems to me, have not received proper attention in all recent discussions on Middle Bronze Age dendrochronology for Anatolia.

It is also important to realize that none of the individuals mentioned in the Sarıkaya palace bullae actually lived in the palace itself nor were any of these sealings produced in the palace itself. The significance of this is obvious, as N. Özgüç realized: "So far, we have not found a single sealing which belonged to a person identified by an inscription as a resident of the palace, let alone the sealings of a local ruler" (N. Özgüç 1980: 62). The significance of this is that the sealings, of Šamši-Adad I of Assyria and Aplahanda of Carchemish, could have arrived (and subsequently been stored) at the Sarıkaya palace many years prior to the destruction of that palace at the end of kārum Kanish Ib, itself an event for which we have no convincing historical explanation (see T. Özgüç 2003).

Back in 1996 Kuniholm and his colleagues proposed that the dendrochronological dates for the Waršama and Sarıkaya palaces rendered the so-called High Chronology (dating Hammurapi 1848–1806 BC) "very unlikely"; supporting instead "either a Low or lower Middle chronology (or a new independent chronology in this range")" (Kuniholm et al. 1996: 782). In 2001 a somewhat revised version of this chronology was proposed:

"With the revised Anatolian tree ring dating, only a chronological solution close to the classic Middle Chronology, which places the reign of Šamši-Adad I between ca. 1832 +7/-1 BC and 1776 +7/-1 BC, is viable. The so-called low-Middle Chronology is also plausible" (Manning et al. 2001: 2534).

Kuniholm and his colleagues gave no indication as to exactly what might be represented by the so-called "low Middle Chronology," but exactly such a chronology has now been proposed by the Slavic scholar Boris Banjević (2005). Banjević deals with the evidence from solar and lunar eclipses, as recorded in Mesopotamian cuneiform texts. Such evidence has long been thought to support only the so-called High Chronology (Huber 1982; 2000). Banjević argues that "The Long Chronology is not supported by any evidence except the astronomical data and statistics" (Banjević 2005: 181) and is strongly opposed by all evidence from history and archaeology. In a detailed discussion that cannot possibly be repeated here Banjević proposes a new chronology, one that puts the reign of Šamši-Adad I at 1760–1728 BC, of Hammurapi at 1744–1702 BC, and of Samsuditana, the last ruler of the Ist Dynasty of Babylon, at 1577–1547 BC (Banjević 2005: 190). This chronology would be more in keeping with the available dendrochronological dates from Middle Bronze Age Anatolia.

In the years ahead there will be, I am sure, much discussion regarding the new Gordion chronology. There will probably be even more discussion regarding the new "Cornell" chronology for Early Iron Age Greece, based upon radiocarbon and dendro dates from the site of Assiros, in Macedonia (for excavation see Wardle 1997). It has long been recognized that the absolute chronology of the LBA/EIA transition in the Aegean, and especially the date for the beginning of Protogeometric pottery, was based upon little more than guesswork and "best estimates" of the available evidence. In a discussion with John Papadopoulos and Ian Morris, regarding the length of the Submycenaean period, James Whitley proposed that:

"Papadopoulos must realize that our knowledge of 'Dark Age' chronology will not be substantially improved until radiocarbon dating, and more importantly, dendrochronology are taken more seriously in Early Iron Age studies that they have been hitherto" (Whitley 1993: 224).

What Whitley called for has now been provided by the Aegean Dendrochronology Project at Cornell.

According to the traditional chronology, putting the beginning of the Protogeometric style of pottery at ca. 1050–1020 BC, the dates suggested for the relevant phases at Assiros were (Wardle 1997: 447):

- Phase 3: 1000–950 BC
- Phase 2: 950–900 BC.

On the basis of detailed radiocarbon and dendrochronological study of charred wood associated with these two phases the Cornell Lab now proposes a dating of ca. 1080 BC for Phase 3 and ca. 1070 BC for Phase 2 (Newton, Wardle, and Kuniholm 2005). These dates would in turn demand a date of ca. 1100 BC for the beginning of the Protogeometric period in Greece (Newton, Wardle, and Kuniholm 2005: 187–188).

What makes this new chronology of special interest is that, in stratigraphic association with Phase 3 at Assiros, Ken Wardle found a large fragment of a special type of Protogeometric pottery associated with sites in northern Greece, Thessaly, and Macedonia, but also found at Troy and in the fill of the Toumba at Lefkandi in Euboea (Wardle 1997: 448 and 455, Fig. 3:2; Newton, Wardle, and Kuniholm 2005: 176–177 and Pl. 2). All of these examples come from Richard Catling's Group I of this style of pottery (Catling 1998; Dickinson 2006: 207 and 208, Fig. 7.1). The large number of examples found at Troy, some seemingly of local manufacture (Mommsen et al. 2001: 194), have received special attention (Muhly 2003: 28) as they all come from Troy VII b3 (Lenz et al. 1998).

As their context at Troy cannot be later than the late 12th century BC, the revised dating for such pottery in terms of the chronology of early Protogeometric Greece is most welcome, but what of the presence of these amphorae with compass-drawn circle decoration in the fill of the Toumba at Lefkandi? This fill contained a total of 18,530 sherds, belonging to a late phase of Middle Protogeometric and, on the basis of the traditional chronology, dated to ca. 950 BC, or perhaps slightly earlier (Dickinson 2006: 22). To raise the dating for this Toumba deposit would involve raising the dates for the remarkable "Heroön," covered by the Toumba, and all its associated material (for background see Antonaccio 2002; Lemos 2002: 140–146; Dickinson 2006: 107–111; 114–116; 190–191). It is not surprising that the first reaction to this new chronology for "Dark Age" Greece has not been very supportive (Dickinson 2006: 20). My own guess is that we are going to come around to accepting the new dating for Gordion and for Protogeometric Greece, but not without a great deal of critical, often heated discussion. I, for one, would not want it any other way.

I look forward to many more years of research at the dendrochronology lab at Cornell and the radiocarbon Lab at the University of Heidelberg as well as elsewhere around the world. Many of my colleagues have, in recent years, become very pessimistic. The more we work the more complicated things become. Absolute truth still eludes us, and the hard facts of the present become the fictions of the future. In the world of the Aegean we have lived, for some time now, with a High Chronology for Late Bronze Age Greece, and a fierce opposition to that chronology, led by Manfred Bietak, Peter Warren, and Malcolm Wiener. With such opposition no one is going to get away with anything. Faith in the future kept Peter Kuniholm going for many years; we should all follow his example.

Acknowledgments

I want to thank Peter Kuniholm for reading a rough draft of this paper. He has saved me from many errors. Those that remain are entirely of my own making.

References

Antonaccio, C. 2002. Warriors, Traders, and Ancestors: the 'Heroes' of Lefkandi. In J. Munk Højte (ed.), *Images of Ancestors*: 13–42. Aarhus: Aarhus University Press.

Baillie, M.G.L. 1991. Dendrochronology and Thera. The Scientific Case. *Journal of the Ancient Chronology Forum* 4: 15–28.

Banjević, B. 2005. Ancient Eclipses and the Fall of Babylon. *Akkadica* 126: 169–193.

Catling, R.W.V. 1998. The Typology of the Protogeometric and Subprotogeometric Pottery from Troia and its Aegean Context. *Studia Troica* 8: 151–187.

DeVries, K. 1980. Greeks and Phrygians in the Early Iron Age. In K. DeVries (ed.), *From Athens to Gordion. The Papers of a Memorial Symposium for Rodney Young*: 33–49. Philadelphia: University Museum.

DeVries, K. 1990. The Gordion Excavation Seasons of 1969–1973 and Subsequent Research. *American Journal of Archaeology* 94: 371–406.

DeVries, K., Kuniholm, P.I., Sams, G.K., and Voigt, M.M. 2003. New dates from Iron Age Gordion. *Antiquity* 77(296): 1–3.

Dickinson, O. 2006. *The Aegean from Bronze Age to Iron Age. Continuity and change between the twelfth and eighth centuries BC.* London: Routledge.

Douglass, A.E. 1921. Dating our prehistoric ruins. *Natural History* 21: 2.

Gidwitz, T. 2001. Telling Time. *Archaeology* 54(2): 36–41.

Goetze, A. 1957. On the Chronology of the Second Millennium B.C. *Journal of Cuneiform Studies* 11: 53–61; 63–73.

Huber, P.J. 1982. Astronomical Dating of Babylon I and Ur III. *Monograph Journals of the Near East, Occasional Papers* I/4. Malibu: Undena Publications.

Huber, P.J. 2000. Astronomy and Ancient Chronology. In J.A. Armstrong and D.A. Warburton (eds.), *Just in Time: Proceedings of the International Colloquium on Ancient Near Eastern Chronology (2nd Millennium BC), Ghent 7–9 July 2000*: 159–176. *Akkadica* 119–120. Brussels: Fondation Assyriologique George Dossin.

Ipek, I., Lemaire, A., Tekoğlu, R., and Tosun, A. 2000. La bilingue royale louvite-phenicienne de Çineköy. *Comptes rendus des séances de l'Académe*: 961–1006.

James, P. 1991. *Centuries of Darkness. A challenge to the conventional chronology of Old World Archaeology.* London: Jonathan Cape.

James, P.J., Thorpe, I.J., Kokkinos, N., and Frankish, J.A. 1987. *Studies in Ancient Chronology Vol. 1: Bronze to Iron Age Chronology in the Old World: Time for a Reassessment?* London: Institute of Archaeology, University College.

Kuniholm, P.I. 1977. Dendrochronology at Gordion and on the Anatolian Plateau. Ph.D. Dissertation, University of Pennsylvania.

Kuniholm, P.I. 1988. Dendrochronology and Radiocarbon Dates for Gordion and Other Phrygian Sites. *Source. Notes in the History of Art* VII: 3/4: 5–8.

Kuniholm, P.I. 1996. The Prehistoric Aegean: Dendrochronological Progress as of 1995. *Acta Archaeologica* 67: 327–335.

Kuniholm, P.I. 2004. Aegean Dendrochronology Project December 2004 Progress Report. Cornell University: Aegean Dendrochronology Project.

Kuniholm, P.I. 2006. As quoted in *Science Daily Research News*; New Evidence Suggests the Need to Rewrite Bronze Age History: 29 April 2006.

Landsberger, B. 1954. Assyrische Königsliste und 'dunkles Zeitalter.' *Journal of Cuneiform Studies* 8: 31–45; 47–73; 106–133.

Lanfranchi, G.B. 2005. The Luwian-Phoenician Bilingual of Çineköy and the Annexation of Cilicia to the Assyrian Empire. In R. Rollinger (ed.), *Von Sumer bis Homer: Festschrift für Manfred Schretter zum 60. Geburtstag am 25 Februar 2004. Alter Orient und Altes Testament* 325: 481–496. Neukirchen-Vluyn: Verlag Butzon & Bercker Kevelaer.

Lebrun, R., and DeVos, J. 2006. A propos de l'inscription bilingue de l'ensemble sculptural de Çineköy. *Anatolia Antiqua* 14: 45–64.

Lemos, I.S. 2002. *The Protogeometric Aegean. The Archaeology of the Late Eleventh and Tenth Centuries BC.* Oxford: Oxford University Press.

Lenz, D., et al. Protogeometric Pottery at Troia. *Studia Troica* 8: 189–222.

Manning, S.W., Kromer, B., Kuniholm, P.I., and Newton, M.W. 2001. Anatolian Tree Rings and a New Chronology for the East Mediterranean Bronze-Iron Ages. *Science* 294: 2532–2535.

Meiggs, R. 1982. *Trees and Timber in the Ancient Mediterranean World.* Oxford: Clarendon Press.

Mellink, M. 1991. The Native Kingdoms of Anatolia, I. The Phrygian Kingdom. In J. Boardman et al. (eds.), *The Cambridge Ancient History*, 2nd ed., III/2: 622–643. Cambridge: Cambridge University Press.

Mommsen, H., Hertel, D., and Mountjoy, P.A. 2001. Neutron activation analysis of the pottery from Troy in the Berlin Schliemann Collection. *Archäologischer Anzeiger* 2001: 169–211.

Muhly, J.D. 2003. Greece and Anatolia in the Early Iron Age: The Archaeological Evidence and the Literary Tradition.

In W.G. Dever and S. Gitin (eds.), *Symbiosis, Symbolism, and the Power of the Past. Canaan, Ancient Israel, and their Neighbors, from the Late Bronze Age through Roman Palaestina. Proceedings of the Centennial Symposium W. F. Albright Institute of Archaeological Research and American Schools of Oriental Research, Jerusalem 29–30 May 2000*: 23–35. Winona Lake, Indiana: Eisenbrauns.

Muscarella, O.W. 1995. The Iron Age Background to the Formation of the Phrygian State. *Bulletin of the American Schools of Oriental Research* 299–300: 91–101.

Muscarella, O.W. 2003. The Date of the Destruction of the Early Phrygian Period at Gordion. *Ancient West & East* 2(2): 225–252.

Newton, M.W., and Kuniholm, P.I. 2004. A Dendrochronological Framework for the Assyrian Colony Period in Asia Minor. *Featuring Complex Societies in Prehistory: Studies in Memoriam of the Braidwoods*: 165–176. *TÜBA-AR* 7: Turkish Academy of Sciences Journal of Archaeology.

Newton, M.W., Wardle, K.A., and Kuniholm, P.I. 2005. Dendrochronology and Radiocarbon Determinations from Assiros and the Beginning of the Greek Iron Age. *To Arkhaiologiko Ergo sti Makedonia kai Thraki* 17: 173–189.

Nijhuis, M. 2005. So you want to be a dendrochronologist? *High County News* (Paonia, CO): 24 Jan. 2005.

Özgüç, N. 1980. Seal Impressions from the Palaces at Acemhöyük. In E. Porada (ed.), *Ancient Art in Seals*: 61–99. Princeton: Princeton University Press.

Özgüç, T. 2003. *Kültepe Kaniš/Neša: The earliest international trade center and the oldest capital city of the Hittites.* Tokyo: The Middle East Culture Center in Japan.

Parpola, S. 1987. *The Correspondence of Sargon II, Part I: Letters from Assyria and the West. State Archives of Assyria*, I. Helsinki: Helsinki University Press.

Ritter, M. 1998. Collecting wood for the seeds of ancient history. *Kathimerini* 29 Sept. 1998: 3.

Robinson, W.R. 1976. Tree-ring dating and archaeology in the American Southwest. *Tree-Ring Bulletin* 36: 9–20.

Rollinger, R. 2006. The Terms 'Assyria' and 'Syria' Again. *Journal of Near Eastern Studies* 65: 283–287.

Sams, G.K. 1997. Gordion and the Kingdom of Phrygia. In R. Gusmani, M. Salvini, and P. Vannicelli (eds.), *Frigi e Frigio. Atti del 1 Simposio Internationale, Roma 16–17 ottobre 1995*: 239–248. Rome: Consiglio Nazionale delle Ricerche.

Saulnier, B. 2002. Life Lines. *Cornell Alumni Magazine* 104(5): 48–55.

Smith, S. 1940. *Alalakh and Chronology.* London.

Veenhof, K.R. 2003. *The Old Assyrian List of Year Eponyms from Karum Kanish and its Chronological Implications.* Ankara: Turkish Historical Society, VI/64.

Voigt, M.M., and Young, T.C., Jr. 1999. From Phrygian Capital to Achaemenid Entrepot: Middle and Late Phrygian Gordion. *Iranica Antiqua* 34: 191–241.

Voigt, M.M., and Henrickson, R.C. 2000. Formation of the Phrygian state: the Early Iron Age at Gordion. *Anatolian Studies* 50: 37–54.

Wardle, K.A. 1997. Change or Continuity: Assiros Toumba at the Transition from Bronze to Iron Age. *AEMΘ* 10 A: 443–460.

Wilkie, N.C. 2003. Citation for P.I. Kuniholm's Receipt of Pomerance Award for Scientific Contributions to Archaeology from Archaeological Institute of America, *American Journal of Archaeology* 107: 279.

Whitley, J. 1993. Woods, Trees and Leaves in the Early Iron Age of Greece. *Journal of Mediterranean Archaeology* 6(2): 223–229.

Young, R.S. 1958. The Gordion Campaign of 1957: Preliminary Report. *American Journal of Archaeology* 62: 139–154.

Young, R.S. 1978. The Phrygian Contribution. In *Proceedings of the Xth International Congress of Classical Archaeology. Ankara-Izmir 1973, Vol. I*: 9–24. Ankara: Turkish Historical Society.

Young, R.S., et al. 1981. *Three Great Tumuli: The Gordion Excavations Final Reports, Vol. I.* Philadelphia: University Museum.

Article submitted May 2007

Excursions into Absolute Chronology

M. G. L. Baillie

Abstract: *Recent experience suggests that, with the exception of geologically obvious extinction-level events, there are essentially no short-term global environmental events in the historical record. If we take the example of 1816—The Year Without a Summer, that was apparently triggered by the eruption of Tambora—as far as is known, this event seems to have had environmental significance only in the north Atlantic region (Harington 1992). This in itself could be due either to the effects being localized or to inadequate reporting for most of the globe in the early 19th century. Irrespective of whether 1816 was a global environmental event, or not, it is typical of the problems encountered when trying to establish global events in the past. The situation exists because of the spatially-patchy historical record combined with the apparent failure of ancient writers to document environmental events, even those taking place around them, in any consistent manner. In this paper several events will be documented to provide a focus on the issue. It goes without saying that absolute chronology is fundamental in attempts to understand past events.*

Global events

The issue of abrupt or short-term global environmental events was highlighted in 1994, when it was realized that tree-ring chronologies from a wide spread of geographical locations around the world all showed reduced growth towards the middle of the sixth century AD (Baillie 1994). This so-called "540 event" event was interesting because a dry fog, interpreted as the effects of a major volcano, had been noted in 536–537 (Stothers and Rampino 1983a), and it was known that there were widespread famines from China (Weisburd 1985) to the Mediterranean (at least), and the outbreak of a great plague in the early 540s. The widespread tree-ring effects showed that the environmental downturn followed on for at least a decade after 536, and appeared to be two-stage in character. This 540 package acted as a catalyst for the identification of other events wherein information from differing sources—including history, ice-core records, astronomy, alternative histories, and myths—gave hints of catastrophic undertones.

Once a global environmental event has been identified it is fundamental that the underlying cause should be defined. This is where a multi-disciplinary approach is essential. Various records can be interrogated to look for hints of just what had taken place at the time, and immediately one is thrown into the issue of chronology. The tree-ring records make it clear that the first abrupt downturn is in 536, though in several records—e.g. European oak (Baillie 1994), Mongolian pine (D'Arrigo et al. 2001), and Scandinavian temperature-sensitive pines (Zetterberg et al. 1994; Briffa 2000)—there is what appears to be a clear second stage in the early 540s. Immediately a question can be raised, namely, was the entire decade-long environmental event due to a single cause in 536 or was there a primary event in 536 and a second, re-enforcing, event in or around 540? Clearly it would be important to know. The answer, with respect to cause, should lie in the ice-core records from Greenland; and here, immediately, we embark on a chronological saga.

Ice acidity and AD 536

In the definitive listing of the dates of volcanic acid in the Dye-3 and GRIP ice cores there were identifiable, but unremarkable, acid layers at 527 and 532 in Dye-3 and 529 and 534 in GRIP respectively (Clausen et al. 1997). Given the exceptional extent and duration of the 536–545 environmental downturn, the fact that the ice-core researchers did not highlight these acidities as being significant left open a raft of questions. Indeed, as late as 2002 Larsen et al. were willing to state that the nearest *big* volcanic acidity in the ice lay at 527 ±1 and that "the AD 527 volcanic eruption is the only eruption in the period." On this basis it seemed fair to test the proposition that there was no evidence for an exceptional volcano at the time of the 536–545 environmental event, and it was this testing that led to exploration of the possibility that the event might have an extraterrestrial cause (Baillie 1994, 1995, 2007). Interestingly, both Celtic and Japanese myths involving dates around AD 540 relate to sky gods such as Arthur/Mongan in the British Isles (McCafferty and Baillie 2005) and Benzaiten in Japan (Aston 1956), and carry with them the notion of a "wasteland" and "stones falling from the sky" respectively. Initially, therefore, an extraterrestrial cause for the global downturn could not be ruled out.

Chronologically, looking at the available ice acidity and tree-ring evidence for the 6th century, there had to be a possibility that the 532/534 acid layer in the ice cores actually related to the historical 536 dry-fog record. However, that possibility would have left no clue as to the cause of the second "540" stage of the environmental downturn. So a second possibility had to be considered; one that would involve the ice-core dating being less secure than is normally suggested. That second possibility is that the 527–529 acidity be moved forward to 536, and the 532–534 acidity by the same amount to around 541. It costs a dendrochronologist nothing to make such a suggestion because the ice core chronologies undoubtedly have errors (see below) and have not been independently proven to be correct in the first millennium AD, or before. However, irrespective of the implications of having to move the European ice core chronology by a few years, the suggestion—that could be generated by anyone looking at the tree-ring and ice-core records across the period—would make sense because it would explain the two-stage nature of the 536–545 downturn.

This was the background against which new ice-core information has to be judged. In 2008 Larsen et al. published the results of a re-analysis of the acidities in the 6th century in three ice cores, namely Dye-3, GRIP, and NGRIP; the results were surprising. It transpired that there was a large northern eruption at AD 529 ±2 and a "hitherto underestimated" eruption of "unparalleled magnitude in the last two millennia" at AD 533–534 ±2. Larsen et al. proposed that this latter eruption was the cause of the environmental downturn in 536 and thereafter, in their view to *at least* 550. This showed that the ice-core dates were at least slightly flexible, and could be moved forward by a few years, though Larsen et al. would not countenance the second possibility, i.e. of moving their 529 acidity to 536 and their 533–534 acidity to 540–541; a move of seven years from the original Clausen et al. (1997) dates. As a response to the Larsen et al. paper this latter 7-year ice-core move was proposed (Baillie 2008). Supporting evidence for the 7-year move was a consistent offset between all four ice acidities in the 6th century AD and notable tree-ring effects in precisely dated tree-ring series (see Table 1).

Table 1 represents the current situation with volcanoes and their effects in the 6th century. The ice-core scenario (Table 1,b) would place volcanoes at 517.5, 531.5, 536, and 570, with only the AD 536 volcano showing clear evidence of severe environmental effects (Larsen et al. 2008). The tree-ring scenario (Table 1,c) would have volcanoes and their effects in each of the years 522, 536, 541, and 574 (Baillie 2008), see also Figure 1. In the case of the latter scenario the world would possibly make a little more sense. We must, however, deal with the reason for the reluctance on the part of the ice-core workers to allow their dates to move to conform to the proposed tree-ring scenario. The principal sticking point is the claim that the European ice-core chronology is anchored at AD 79 with both acid and tephra from Vesuvius being observed in the ice at that time. Unfortunately the critical evidence for the tephra identification is no more than "C. Barbante pers. comm. 2005" (Vinther et al. 2006), and the results are neither published nor available; the Vesuvius anchor point therefore must remain the subject of doubt, and indeed, even when published, potentially the subject of debate (given the known debates over the identification of tephra from Thera in the 17th century BC).

It is implicit in the tree-ring scenario in Table 1 that the errors in the ice-core chronology *must* be greater than those suggested by the ice-core workers. Such a challenge requires elaboration, and this, in turn, requires an excursion into the world of ice-core chronology to look for possible causes of error. Fortunately we are told about the ice-core dating procedures in some detail. Vinther et al. (2006) tell us that their chronology was constructed by identifying marker volcanic horizons in the various cores and independently counting the annual layers between the markers. For example, AD 934 is a marker horizon with acid attributed to the eruption of the Icelandic volcano Eldgjá; they also note that:

> ...1283 years are counted in between A.D. 79 (Vesuvius) and A.D. 1362 (Öraefajokull). This is only one year more than the 1282 years known from historical records (Vinther et al. 2006; [27]).

It seems that this chronological procedure has been devised and used in-house by the ice-core workers. Of particular interest is the use of the idea of *independent* counting. Dendrochronologists use this idea to underpin the ultimate replication of their chronologies when they compare the chronologies constructed independently by independent workers in different laboratories, often in different countries. Vinther et al. may

a	515	529	533.5	567.5	Raw ice-core dates (Clausen et al. 1997; Larsen et al. 2008)
b	517.5	531.5	536	570	Ice dates moved 2.5 years (Larsen et al. 2008)
c	522	536	540.5	574.5	Raw ice dates moved 7 years (Baillie 2008)
d	522	536	542	575	Severe narrow rings in Swedish pines (H. Grudd, pers. comm.)
e	522	536	541	574	Frost rings in Bristlecone pines (Salzer and Hughes 2007)

Table 1: The ice acid dates **a)** as provided by layer counting; **b)** moved forward 2.5 years to make AD 533–534 equivalent to 536; **c)** same dates as (a) moved forward by 7 years to bring them into line with tree-ring evidence; **d)** notable reduced growth in temperature-sensitive Swedish pines; **e)** frost rings in American bristlecone pines.

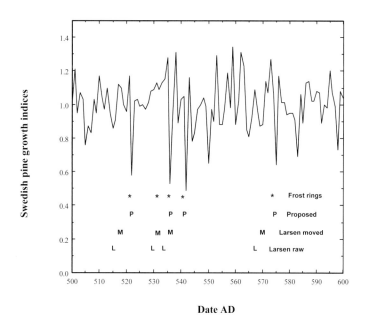

Figure 1: Swedish pine ring-width indices (courtesy of Håkan Grudd) showing extreme negative growth at AD 522, 536, 542, and 575. Coincident bristlecone pine frost rings are plotted (Salzer and Hughes 2007). Ice core acidities both raw **(L)** and moved 2.5 years **(M)** are plotted (Larsen et al. 2008). **(P)** represents proposed ice core dates, moved forward by 7 years, to conform to the tree-ring effects (Baillie 2008).

claim that their ice layers are counted independently, but that claim only has validity if the layers in each core are counted by truly independent workers, without the constraints of imposed zero-error points such as AD 1362 or AD 79. Put simply, Baillie (2008) is challenging the robustness of this ice core dating procedure, while the various ice-core workers (Vinther et al. 2006; Larsen et al. 2008) are challenging the scientific and antiquarian worlds to accept the correctness of their layer counting and the correct identification of their marker horizons.

Probing the ice-core dating procedures

It is possible to argue that there may be grounds for doubt about the European ice-core chronology. Dye 3 was the first long ice core to be analysed in detail. With earlier cores, Crête ran back only to around AD 550, while Camp Century was only used from AD 40 back in time (Hammer et al. 1980). Vinther et al. (2006) tell us how they generate their error estimates and it can be assumed that these statements apply to the Dye-3 chronology errors. First it is important to recognise that there have to be error estimates for ice cores because there are some layers that:

> Can neither be dismissed nor confirmed as annual layers...[These] are recorded as uncertain years. Half of the number of uncertain years is subsequently included in the final timescale, while the other half is discarded. The maximum counting error is then defined as half of the number of uncertain years. This is to say, that uncertain years enter into the dating as 0.5 +/- 0.5 years.

This is a revelation, especially to a chronologist. We know that by 50 BC in the Dye-3 timescale the error is ±4 years (Hammer 1984). So this implies that the ice-core workers had encountered eight "uncertain years" on the way down to 50 BC. Of that eight, they had included four in their dating. And here we can see where they may have gone wrong. This is the high accumulation Dye-3 core, wherein it is stated that it is very unlikely that there would be any "complete years without precipitation" (Vinther et al. 2006). This self-cancelling idea—with half the number of uncertain years included and half discarded—would, in the opinion of this author, work only in a system which had *missing* layers and *double* layers. We are told Dye3 in all probability does not have any missing years, thus the procedure described by Vinther et al. *will tend only to add years* to the chronology, never to take them away. This may be the reason why the Dye-3 dates and subsequently all the European ice-core dates *tend to be too old*. It is why 50 BC should move to 44 BC, why AD 529 should move to AD 536, and AD 533.5 to AD 540.5 (Baillie 2008; and Table 1). On this logic the acid layer that the ice-core workers refer to as Vesuvius actually lies at around AD 86 and has nothing to do with Vesuvius. Put another way, if there really is Vesuvius tephra in AD 79–80 ice, then it is vitally important that the results be published for scrutiny.

Other extreme environmental events

Let us now return to the theme of the dating of extreme events. In the remainder of this article it is intended to look at two other environmental events that have attracted attention in recent years, namely the arrival of the Black Death into Europe in 1348 (Baillie 2006), and the Eldgjá volcanic eruption normally dated to AD 934 (Clausen et al. 1997) or the later AD 930s (Zielinski et al. 1995). In both cases, in honour of the chronology-building work carried out by Peter Kuniholm and his team, it is intended to involve information from Aegean tree-ring chronologies. The intention is to show how, in many cases, there is much relevant scientific information that can be interrogated in trying to obtain a handle on the causes of the events. Of necessity data will be presented and interpreted to show just how complicated such an apparently simple activity can become.

The Black Death period

While the initiation of the Justinianic plague coincides remarkably with the two-stage global tree-ring downturn across the period 536–550, it was not initially realized that the arrival of the Black Death also coincided with another, albeit less extreme, global tree-ring downturn (Baillie 2006). The availability of robust tree-ring chronologies from Europe, Fennoscandia, the Polar Urals, the Aegean, North and South America, New Zealand, and Tasmania allowed the construction of an annual resolution "world tree-ring master chronology" back to before AD 1200. When this data was first being interrogated it was assumed that there would be little agreement between distant chronologies for differing tree species. As a result groups of four chronologies were meaned to represent an Old World and a New World (the Americas plus New Zealand and Tasmania) chronology; there was no prior reason to imagine that there would be any similarity between such diverse master chronologies. In fact, as Figure 2 shows, there is a surprising level of agreement implying that most of the time the primary forcing factor on tree-growth must be temperature related; it being impossible to imagine that rainfall could be so coherent.

The notable thing about Figure 2 is that while the two chronologies are in agreement most of the time, in the fourteenth century there are dramatic departures between the two records in the periods 1314–1320 and 1325–1335. It is clear that something was affecting the conditions for global tree-growth in an unusual fashion in the decades before the Black Death. Even more surprising was the observation of the remarkable similarity between the two major chronologies following 1350 and indeed lasting until around 1400. Dendrochronologists are conditioned to think in regional terms and this observation of coherent behaviour in a global tree-ring dataset raised some interesting questions discussed elsewhere (Baillie 2006).

This exercise took no notice of individual chronologies concentrating as it did on the idea of global and hemispheric signal. However it is possible to de-convolve the data and look at any one regional chronology against the mean of all the other datasets. For the purposes of this article the Aegean chronology, which is itself the mean of three separate species chronologies from the Aegean region, namely pine, juniper, and oak (data courtesy Peter Kuniholm), was plotted against the mean of the other seven world chronologies, Figure 3. The data in Figure 3 has been 5-point smoothed to show the main trends. It is apparent that for the period plotted, 1150 to 1400, there is relatively good agreement between the two completely independent datasets. There are however two points of departure, one around 1280 and another between 1335 and the late 1340s. In the latter case the world signal shows a downturn across the 1340s, consistent with Figure 2. However, the Aegean growth is opposite for most of the decade of the 1340s declining only as it approaches 1350.

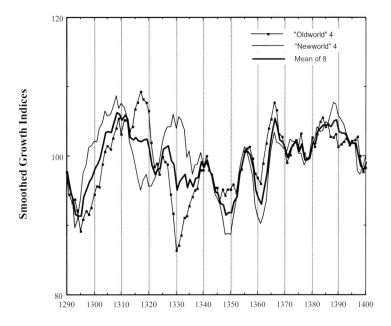

Figure 2: Two tree-ring chronologies representative of the northern and southern hemispheres, showing the notable growth depression across the 1340s and indicating that the Black Death has a clear environmental context.

We can imagine that tree growth in the Aegean region might be negatively affected by summer drought. Thus, if the assumption is that global trees exhibit reduced growth due to reduced temperature, this might well be consistent with enhanced growth in the Aegean region; reduced temperature limiting evapotranspiration in that region and allowing enhanced growth. What is also apparent is that in the critical years of 1348–1350 when the Black Death arrived into Europe the Aegean tree growth also declines despite the fact that there are suggestions of these years having cooler and wetter conditions. Overall, reviewing Figure 3 it seems that tree-ring records highlight the decade of the 1340s no matter how they are plotted. One reason for labouring this issue is that it is increasingly unclear just what the Black Death was. The rejection of bubonic plague as the principal vector for the Black Death was initiated by Twigg (1984), and has been carried on by Hoyle and Wickramasinghe (1993), Cohn (2002), Scott and Duncan (2005) and Baillie (2006). The fact that the so-called "plague" has a clear environmental context should cause both scientists and historians to look at all their evidence with a fresh eye. Overall, the Aegean chronologies provide a slightly different view of this important event in human history, but serve to highlight the generalized nature of the global growth decline just at 1348, the year of the arrival of the Black Death into Europe.

The AD 930s

Progressing back in time from the 14th century, the tree-ring/ice story begins to take on increased complexity. First, the Aegean tree-ring record drops to a single record: the pine chronology stops in 1148, the juniper in 1037, leaving only the single oak record. Then there are problems between the European GRIP and American GISP2 ice records. It has been shown elsewhere (Baillie 2006) that the two records agree well from the 1640s (where the GRIP record begins) back to around 1030. Figure 4 shows the ammonium records from the two cores (5-point smoothed for clarity) for the period AD 1000–1200. It is clear that back to 1030 the records agree well, however, the large signal centred on 1014 in the GRIP record is offset in the GISP2 record where it occurs around 1020. Although there is no *a priori* reason to doubt either one of these datasets, the disagreement pre-1030 shows that one of them is probably wrongly dated compared with the other. However, there are other datasets with which the ice results can be compared. In 1984 LaMarche and Hirschboeck suggested a link between frost rings in bristlecone pines and the environmental effects of large explosive volcanic eruptions. Salzer and Hughes (2007) have published a comprehensive list of bristlecone frost-ring dates from the bristlecone pine chronologies. This is significant because dates for volcanic acid in ice cores (Clausen et al. 1997; Zielinski et al. 1994) can now be compared directly with

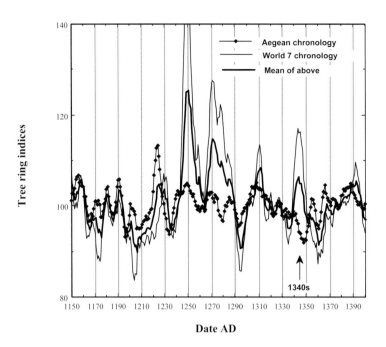

Figure 3: An Aegean oak/juniper/pine chronology plotted with the other seven world chronologies from Figure 1 showing an opposite response in the Aegean trees in the 1340s but agreement at the time of the Black Death 1348–1350.

bristlecone frost-ring dates. Recently McCormick et al. (2007) showed some tentative linkages between the dates of volcanic acid layers in the GISP2 core and historical cold winters. All of these sources of chronological data can now be compared. In Tables 2 and 3 the dating has been approached as follows. Historical and tree-ring dates are fixed in time (with the sole proviso that the date of the frost ring could represent the date of an eruption or possibly one or even two years after).

All the ice-core dates automatically have small dating errors and it would be fair to suggest the error on every quoted date should be thought of as spanning one or two years. But that is simply an error related to dealing with reading dates from ice. There are other systematic errors. For example there is the offset between GRIP and GISP2 mentioned above (Baillie 2006). Then there is the suggestion that back in the 6th century the European Dye-3 and GRIP ice-core dates are probably about 7 years too old (Baillie 2008). In Tables 2 and 3 all of these sources of information are applied and a few simple rules adhered to. These are:

a) from AD 950 back to AD 789 the Dye-3 and GRIP dates are unchanged;

b) from AD 950 back to AD 789 all the GISP2 dates are moved back five years;

c) from AD 789 back to AD 500 all the Dye-3 and GRIP dates are moved *forward in time* by seven years;

d) from AD 789 back to AD 500 all the GISP2 dates are *moved back* seven years.

These dating revisions are simple and reflect the suggestions made by Baillie (2006 and 2008). Synchronisms suddenly appear throughout the tables, not just between historical/frost-ring dates and ice acidities, but also between previously mis-matched acidities in the European and American ice cores. Table 3 starts to make sense where no sense existed before between the European (Dye-3 and GRIP) and American (GISP2) ice cores.

Let us assume for the purposes of this discussion that the GISP2 core dates need to move back in time by around 5 years by 936–939, i.e. that layer should be moved back to 931–934 (as in Table 2). This then fits with the extensively discussed Eldgjá eruption in Iceland (Stothers 1998). The eruption is important because it seems to have happened just at the end of an initial 60-year settlement period in Iceland and may well mark the transition from initial settlement to a more stable regime. In terms of dating, Stothers (1998), who looked at the event in detail, concluded that "convergent lines of evidence point to 934 as the year of the eruption." However Stothers did point out that much of the dating evidence is open to interpre-

History	Frost-rings	Dye3	GRIP	GISP2	GISP2 back 5
			943		
939–940				936–939	
934–935*	**934**	**932**	**934**		**931–934**
913				913–915	
	909				**908–910**
				902	
	899	**895**	**898**	**900**	**897**
					895
	889	889			
	884				
873–874		**871**	**871**	**875–876**	**870–871**
859–860				856–858	851–853
855–856				853–854	848–849
	835				
823–824				827	
821–822	**822**			**822–825**	**822**
	816				817–820
	789				

Table 2: Historic cold winters (McCormac et al. 2007) [* cold winter (Stothers 1998)]; Frost ring dates (Salzer and Hughes 2007); Dye-3 and GRIP dates from Clausen et al. (1997) GISP dates Zielinski et al. (1994). Last column GISP2 dates moved back 5 years. Bold indicates consistent agreement.

History	Frost-rings	Dye3	GRIP	GISP2	GISP-7	Dye3+7	GRIP+7
				767			
763–764		755	757	757	**760**	**762**	**764**
		743	744		**750**	**750**	**751**
	715						
	694			702	**695**		
	692			695–969			
	687				**688–689**		
	684			691	**684**		
	681					**681**	**682**
	674	674	675				
			645	639			652
626	**627–628**				**632**		**629**
			622				
	574	567*	568*			**574**	**575**
	541	533*	534*			**540**	**541**
536	**536**	529*	529*	529		**536**	**536**
	531						
	522	514*	516*		**522**	**521**	**523**

Table 3: Frost ring dates (Salzer and Hughes 2007) are fixed. Dye-3 and GRIP dates from Clausen et al. (1997) except with asterisks (Larsen et al. 2008); GISP dates Zielinski et al. (1994). Last three columns are GISP2 dates moved *back* 7 years; Dye-3 and GRIP dates moved *forward* 7 years. Bold indicates consistent agreement.

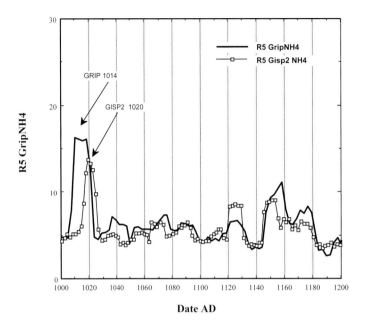

Figure 4: Smoothed ammonium records from the GRIP and GISP2 ice cores showing the breakdown of agreement before 1030.

tation as to the exact year and he left open the door for a slightly more flexible dating, i.e. 933 would not be out of the question and would still fit well with the bristlecone frost ring in 934. This flexibility is well exemplified by the following statement in the Irish *Annals of the Four Masters* (O'Donovan 1848):

> 932 Fire from heaven burned the mountains of Connaught this year and the lakes and streams dried up; and many persons were also burned by it.

Recent research by McCarthy (2005) suggests that the correct date for this entry may well be 933. Given the proximity of western Ireland to Iceland, and the dates, it seems highly likely that this reference is to some phenomenon associated with the Eldgjá eruption. For example, it might relate to the dumping of fluorine and even possible tectonic movements affecting river flow.

The GISP2 particulate record

The GISP2 core was analysed for particulates in terms of numbers of particles and their size (Zielinski and Mershon 1997). If we take the acid peak in GISP2 as occurring between 271.542 to 272.879m, the dust occurs between depths 272.897 to 273.769m. It would be reasonable to state on the basis of this depth information that the two "acid and dust events" are separated by the order of a few years, i.e. if the acid were deposited 931–934, the particulate matter is most likely deposited between 928 and 931. It seems highly unlikely that the dust from the Icelandic volcano Eldgjá would arrive on Greenland so long *before* the acid from the same volcano; so the dust must relate to some other environmental cause. However, now that we have two consecutive events involving dust and acid across 928–931 and 931–934 we can add in Kuniholm's Aegean oak chronology. The relevant section is plotted in Figure 5.

It is immediately apparent that the effects of *something* can be seen in the Aegean oak chronology as notably depressed growth in the years 930–934; possibly starting as early as 929. It looks in this case as if the particulates may be a signature for some environmental driver that affected trees in the Mediterranean area. It should be noted that there is no equivalent growth change in other chronologies from Europe or Fennoscandia. This of course raises a question about the Eldgjá eruption. Was Eldgjá the principal mover or was there a more complex package of environmental happenings that have merely been attributed to this eruption? Zielinski et al. (1995) noted that the tephra evidence from the assumed Eldgjá layer in GISP2 contained more than one population of tephra, suggesting another volcano somewhere.

Returning to Stothers (1998), despite concluding that the Eldgjá eruption most likely occurred in 934, he does at one point in his text state, on the basis of the death of the "lawspeaker" of the first Icelandic parliament possibly being due to the eruption, "...If

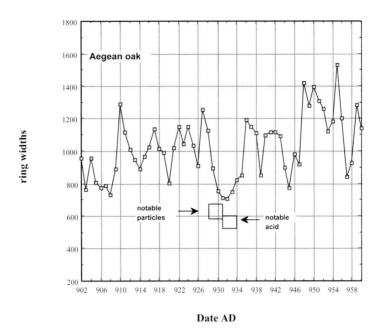

Figure 5: Kuniholm's Aegean oak chronology exhibiting notable reduced growth from 930 to 935 coincident with the particulate and acid evidence from the re-dated GISP2 ice core.

so, the eruption can be formally dated at either 930 or 934...."

This discussion, when combined with the early 930s growth pause in the Aegean oaks, raises the question why the year 934 was decided upon in the first place? Overall, the reaction of the Aegean trees may be the best guide we have that this whole package may have been initiated just around 930. It has however to be mentioned that Clausen et al. (1997) date the main acid peak in the GRIP core to 934–935 although they also show that fluorine begins to rise as early as 933, consistent with Herron's (1982) observations on Dye-3. It is also worth noting that relevant evidence for events such as that in the 930s can come from unusual sources. Figure 6 shows a plot of all the Irish archaeological sites that have been dated by dendrochronology in the last four decades. The sudden change in site frequency at AD 930 may be a hint that Eldgjá, or the package of which Eldgjá was a part, had some real effects on people on the ground.

Conclusion

These examples serve to show that precise chronology, and the interpretation of precise chronology, is, and will remain, a difficult area. New ice-core analyses and new interpretations are refining the original ice dating framework, making it more compatible with the tree-ring and historical chronologies. However, we may need to step back from any ice acidities other than those that demonstrate environmental effects in tree-rings. While we have discussed Eldgjá and the problems of its dating, one simple set of facts shows why it is probably not a good example for testing hypotheses. Salzer and Hughes (2007) record a frost ring in 934. D'Arrigo et al. (2001) note frost rings in Siberian pines in Mongolia in 938. Scuderi (1990) records a tree-ring minimum in foxtail pines from the Sierra Nevada in 936. The Aegean trees show notably reduced growth from 930 to 934. It seems that in comparison with a profound environmental event, like the one seen at AD 536 and thereafter, the goings on in the 930s lack the focus that one might expect from a single massive impulse. Despite that caution, new datasets are providing new perspectives on existing questions and can demand their re-examination. Peter Kuniholm's progress, in the elaboration of the Aegean tree-ring grid, could not be better demonstrated when we see Aegean chronologies providing new perspectives on known environmental events.

Acknowledgments

The author wishes to thank the many people in the tree-ring community who have generously supplied or exchanged data to allow the construction of global scale time series. The article is part of ongoing collab-

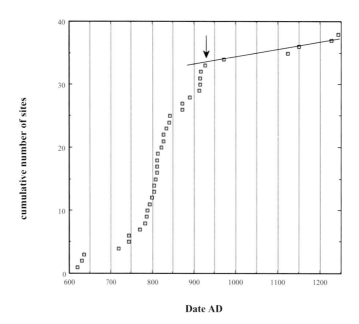

Figure 6: A cumulative plot of all the Irish archaeological sites dated by dendrochronology between AD 600 and 1250 showing the abrupt change in frequency of site construction using oak at AD 930.

orative research with ice-core workers partly funded by 14-CHRONO.

References

Aston W.G. 1956 *Nihongi: Chronicles of Japan from the earliest times to* AD *697*. London: George Allen and Unwin.

Baillie, M.G.L. 1994. Dendrochronology raises questions about the nature of the AD 536 dust-veil event. *The Holocene* 4(2): 212–217.

Baillie, M.G.L. 1995. *A Slice Through Time: dendrochronology and precision dating*. London: Routledge.

Baillie, M.G.L. 2006. *New Light on the Black Death: the cosmic connection*. Stroud: Tempus.

Baillie, M.G.L. 2007. The case for significant numbers of extraterrestrial impacts through the late Holocene. *Journal of Quaternary Science* 22(2): 101–109.

Baillie, M.G.L. 2008. Proposed re-dating of the European ice core chronology by seven years prior to the 7th century AD. *Geophysical Research Letters* 35, L15813, doi:10.1029/2008GL034755.

Briffa, K.R. 2000. Annual climate variability in the Holocene: interpreting the message of ancient trees. *Quaternary Science Reviews* 19: 87–105.

Britton, C.E. 1937. A Meteorological Chronology to AD 1450. *Geophysical Memoirs* No. 70, HMSO, London, 39.

Clausen, H.B., Hammer, C.U., Hvidberg, C.S., Dahl-Jensen, D., and Steffensen, J.P. 1997. A comparison of the volcanic records over the past 4000 years from the Greenland Ice Core Project and Dye 3 Greenland ice cores. *Journal of Geophysical Research* 102, No. C12, 26, 707–726,723.

Cohn, S.K. 2002. *The Black Death Transformed: Disease and Culture in Early Renaissance Europe*. London: Edward Arnold.

D'Arrigo, R., Frank, D., Jacoby, G., and Pederson, N. 2001. Spatial Response to Major Volcanic Events in or about AD 536, 934 and 1258: frost rings and other dendrochronological evidence from Mongolia and Northern Siberia. *Climatic Change* 49: 239–246.

Hammer, C.U. 1984. Traces of Icelandic Eruptions in the Greenland Ice Sheet, *Jökull* 34: 51–65.

Hammer, C.U., Clausen, H.B., and Dansgaard, W. 1980, Greenland Ice Sheet Evidence of Post-Glacial Volcanism and its Climatic Impact, *Nature* 288: 230–235.

Harington, C.R., ed. 1992. *The Year Without a Summer? World Climate in 1816*. Ottawa: Canadian Museum of Nature.

Herron, M.M. 1982. Impurity Sources of F, Cl, NO and SO in Greenland and Antarctic Precipitation. *Journal of Geophysical Research* 87: 3052–3060.

Hoyle, F., and Wickramasinghe, C. 1993. *Our Place in the Cosmos*. London: Phoenix.

LaMarche, V.C.,Jr., and Hirschboeck, K.K. 1984. Frost Rings in Trees as Records of Major Volcanic Eruptions. *Nature* 307: 121–126.

Larsen, L.B., Siggaard-Andersen, M.-L., and Clausen, H.B. 2002. The sixth century climatic catastrophe told by ice cores. Abstract from the 2002 Brunel University Conference Environmental Catastrophes and Recoveries in the Holocene 29 Aug.–2 Sept. 2002 (available at Atlas Conferences Inc. Document caiq-21).

Larsen, L.B., Vinther, B.M., Briffa, K.R., Melvin, T.M., Clausen, H.B., Jones, P.D., Siggaard-Andersen, M.-L., Hammer, C.U., Eronen, M., Grudd, H., Gunnarson, B.E., Hantemirov, R.M., Naurbaev, M.M., and Nicolussi, K. 2008. New ice core evidence for a volcanic cause of the A.D. 536 dust veil. *Geophysical Research Letters* 35, L04708, doi:10.1029/2007GL032450.

McCafferty, P., and Baillie, M. 2005. The Celtic Gods: comets in Irish mythology. London: Tempus.

McCarthy, D. 2005 Chronological Synchronisation of the Irish Annals available at https://www.cs.tcd.ie/Dan.McCarthy/chronology/synchronisms/annals-chron.htm (fourth edition).

McCormick, M., Dutton, P.E., and Mayewski, P.A. 2007 Volcanoes and the Climate Forcing of Carolingian Europe, A.D. 750–950. *Speculum* 82: 865–895.

O'Donovan J. 1848. *Annals of the Kingdom of Ireland by the four masters*. Dublin: Hodges and Smith.

Salzer, M.W., and Hughes, M.K. 2007. Bristlecone pine tree rings and volcanic eruptions over the last 5000 yr. *Quaternary Research* 67: 57–68.

Scott, S., and Duncan, C. 2005. *Return of the Black Death: the World's Greatest Serial Killer*. Chichester: Wiley.

Stothers, R.B. 1998. Far reach of the tenth century Eldgjá eruption, Iceland. *Climatic Change* 39: 715–726.

Stothers, R.B., and Rampino, M.R. 1983a. Volcanic Eruptions in the Mediterranean Before AD 630 From Written and Archaeological Sources, *Journal of Geophysical Research* 88: 6357–6371.

Stothers, R.B., and Rampino, M.R. 1983b. Historic Volcanism, European Dry Fogs, and Greenland Acid precipitation, 1500 B.C. to A.D. 1500. *Science* 222: 411–413.

Twigg, G. 1984. *The Black death: a biological reappraisal*. London: Batsford.

Vinther, B.M., Clausen, H.B., Johnsen, S.J., Rasmussen, S.O., Andersen, K.K., Buchardt, S.L., Dahl-Jensen, D., Seierstad, I.K., Siggaard-Andersen, M.-L., Steffensen, J.P., Svensson, A. Olsen, J., and Heinemeier, J. 2006. A synchronized dating of three Greenland ice cores throughout the Holocene. *Journal of Geophysical Research*, 111, 22(2): 101–109.

Weisburd, S. 1985. Excavating words: a geological tool. *Science News* 127: 91–96.

Zetterberg, P., Eronen, M., and Briffa, K.R. 1994. Evidence on Climatic Variability and Prehistoric Human Activities Between 165 B.C. and A.D. 1400 Derived from Subfossil Scots Pines (*Pinus sylvestris* L.) Found in a Lake in Utsjoki, Northernmost Finland. *Bulletin of the Geological Society of Finland* 66: 107–124.

Zielinski, G.A., Germani, M.S., Larsen, G., Baillie, M.G.L., Whitlow, S., Twickler, M.S., and Taylor, K. 1995. Evidence of the Eldgjá (Iceland) eruption in the GISP2 Greenland ice core: relationship to eruption processes and climatic conditions in the tenth century. *The Holocene* 5, 2: 129–140.

Zielinski, G.A., Mayewski, P.A., Meeker, L.D., Whitlow, S., Twickler, M.S., Morrison, M., Meese, D.A., Gow, A.J., and Alley, R.B. 1994. Record of volcanism since 7000 B.C. from the GISP2 Greenland ice core and implications for the volcano-climate system. *Science* 264: 948–952.

Zielinski, G.A., and Mershon, G.R. 1997. Paleoenvironmental implications of the insoluble microparticle record in the GISP2 (Greenland) ice core during the rapidly changing climate of the Pleistocene-Holocene transition. *Geological Society of America Bulletin* 109: 547–559.

Article submitted February 2009

One Hundred Years of Dendroarchaeology: Dating, Human Behavior, and Past Climate

Jeffrey S. Dean

Abstract: *Dendroarchaeology, the use of tree-ring data for archaeological dating and for the study of past human behavior and human-environment interactions, originated in the North American Southwest at the beginning of the twentieth century with the work of Andrew Ellicott Douglass, the founder of modern dendrochronology. Douglass's initial success stimulated a spate of archaeological tree-ring work that, although marked by sporadic advances and retreats, has, at the beginning of the new millennium, spread all aspects of dendroarchaeology throughout the world. Peter Ian Kuniholm's quest to bring tree-ring dating to the eastern Mediterranean was a major driver of this expansion. While still in the exploratory stage in many regions, the subdiscipline has achieved full integration into archaeological research in others. Despite some looming problems, dendroarchaeology faces a bright future that will consolidate past accomplishments and generate new topical and geographic applications.*

"Dendroarchaeology" (Sass-Klaassen 2002; Towner 2002, 2007) is a neologism now used to encompass the constantly ramifying aspects of what used to be called "archaeological tree-ring dating." The new term recognizes that tree-ring applications in archaeology have expanded well beyond dating to incorporate the analysis of human behavior and the study of past environmental conditions. Although dating remains primary, the chronological aspect of dendroarchaeology now also includes characterizing and evaluating tree-ring dates within their archaeological contexts, the comparative study of dates derived by various chronometric techniques, and estimating tree-felling dates from samples lacking exterior rings. Aspects of human behavior that can be abstracted from archaeological tree-ring sample collections reveal how people acquired wood from the natural environment and put it to a wide variety of uses. Environmental analyses include using archaeological tree-species assemblages to identify vegetation changes since the sites were occupied and using archaeological tree-ring chronologies to reconstruct aspects of past climate variability.

Background

Early speculations about and investigations of tree rings as dating tools and climate proxies notwithstanding, modern dendrochronology was founded in the early 20th century by the astronomer Andrew Ellicott Douglass in Flagstaff, Arizona, as an offshoot of his research into the effects of sunspots on terrestrial climate. Lacking meteorological records long enough to be compared with sunspot cycles, he turned to tree-ring width sequences as longer measures of climate variability. Investigation of local ponderosa pines led him to independently rediscover crossdating (the existence of identical patterns of ring-width variation in different trees over a wide area), to develop a 400-year chronology of ring-width variability common to those trees, and to establish a statistical relationship between ring widths and winter precipitation. These accomplishments impelled the subsequent development of modern dendrochronology, all of whose ramifications are firmly and exclusively grounded in crossdating.

Although early dendrochronologists would not have recognized the term, dendroarchaeology began with the first use of crossdating to assign a calendar date to a wood sample of unknown age. This defining event likely occurred in 1904 when Douglass crossdated a stump with his ponderosa pine ring chronology and astounded a local farmer by correctly identifying 1894 as the year in which the tree was cut. From this inauspicious beginning, archaeological tree-ring dating evolved into a global phenomenon fundamental to understanding general and specific aspects of human behavior.

In 1920, Douglass was plunged into systematic archaeological tree-ring dating when crossdating between wood from two Anasazi sites in New Mexico established that one (Pueblo Bonito) was under construction fifty years before the other (the Aztec Ruin). The astonishing implications of Douglass's heretofore unimagined ability to separate prehistoric human events in terms of terrestrial years galvanized Southwestern archaeologists. Their desire to attain absolute calendar dates for prehistoric sites sucked Douglass into a prolonged effort to join the living-tree and archaeological sequences into a single continuous ring-width chronology that could be used to date archaeological wood. Hundreds of samples and nine years later this quest culminated in the celebrated "bridging of the gap" episode at the Whipple Ruin in Show Low, Arizona. The renowned sample HH-39 linked the 500+-year "floating" archaeological sequence with the dated chronology anchored in time by the Flagstaff living trees. This connection allowed Douglass to assign calendar dates instantly to the scores of sites included in the prehistoric sequence and established dendrochronology as the foundation of Southwestern archaeological chronology.

Although Douglass scaled back his commitment to archaeological dating after 1929, other institutions and individuals established complementary tree-ring dating programs. In 1937, the University of Arizona created the Laboratory of Tree-Ring Research (LTRR) to commemorate and perpetuate the science of dendrochronology, including archaeological dating. These actions created a network of institutions and a cadre of practitioners that provided a foundation for maintaining and expanding archaeological tree-ring dating. Douglass's success fired the imaginations of scholars in other parts of the world as well, and dating programs were initiated in the North American Great Plains, the U.S. Midwest, Alaska, Scandinavia, Germany, and elsewhere. This flurry of activity produced crossdated tree-ring chronologies for many different regions, refined tree-ring methods, developed statistical crossdating techniques, derived numerous archaeological dates, and laid the groundwork for propagating dendrochronological research throughout the world.

This nascent growth was nipped in the bud by the outbreak of World War II. Most of the European operations were severely curtailed or terminated. The redirection of resources to the war effort impacted U.S. tree-ring efforts as well, with most labs being significantly cut back for the duration or shut down and many trained dendrochronologists being diverted into civilian and military aspects of the conflict. Upon their postwar return to civilian life, most of these individuals took up other professions thus depriving dendrochronology of the critical mass of adept practitioners necessary for future growth. The monetary and workforce demands of peace-time recovery further impeded the revitalization and expansion of global dendrochronology. Only the LTRR and a few dedicated individuals carried archaeological tree-ring dating through this dendrochronological Dark Age.

As "normality" returned after the war, dendrochronology slowly emerged from the doldrums, and archaeological tree-ring activities intensified. LTRR in the Southwest and individuals in the Great Plains, the eastern U.S., and Alaska responded vigorously to the growing demand for dates. In the 1950s, LTRR combined the tree-ring collections of institutions that had abandoned dendrochronology with its own samples into a reanalysis that consolidated all Southwestern archaeological tree-ring dates and converted them to a standard format. This project launched LTRR's current archaeological tree-ring dating program, which engages in all aspects of dendroarchaeology. In Europe, renewed postwar activity triggered a robust recovery of archaeological tree-ring dating in Scandinavia and the Soviet Union, and Bruno Huber's resumption of activity in southern Germany inspired and trained a succession of scholars who helped expand archaeological tree-ring dating into other institutions and regions.

A massive dendrochronological growth spurt that began in the 1960s was motivated in great measure by the drive to apply dendroclimatology in new areas. Long, high-resolution (annual) dendroclimatic reconstructions captured aspects of climate variability not represented in the short historical accounts, clarified how representative meteorological data were of longer periods, and promoted the investigation of human-environment interactions over long time spans. Growth in the archaeological aspects of dendrochronology was stimulated by convergences of many factors that varied from place to place. In the U.S., the explosive expansion of government mandated "cultural resource management" created a huge demand for dates, produced large numbers of samples, and generated unprecedented funding levels. Furthermore, investigation of behavioral and environmental issues created a wider appreciation of dendrochronology's value to archaeology. In Europe, the desire for archaeological dates was augmented by misgivings about the bristlecone pine calibration of radiocarbon determinations, which produced dates inconsistent with prevailing ideas about the chronology of northern Europe. This concern sparked the construction of multimillennial tree-ring chronologies to independently evaluate the bristlecone pine adjustments. These long chronologies allowed the dating of hundreds of archaeological sites throughout Europe, confirmed the major features of the bristlecone pine calibration, and provided the basis for dendroclimatic reconstructions that, among other things, documented the reality of global warming.

The emergence of dendroarchaeology

The modern (post-1960) era in dendrochronology has seen the evolution of a wide-ranging dendroarchaeology (Towner 2002, 2007) that goes well beyond the simple dating of wood and charcoal samples from archaeological contexts. Given dendrochronology's accuracy, precision, and high temporal resolution, chronological interests remain paramount, driving the establishment and/or expansion of archaeological dating programs wherever favorable conditions of tree growth and archaeological preservation exist. In the Western Hemisphere, LTRR has expanded its efforts into all of western North America from Alaska to central Mexico; the Lamont-Doherty Earth Observatory, Cornell University, and University of Tennessee laboratories are working in the eastern U.S. and Mexico, and numerous individuals are dating archaeological materials in other areas. The tree-ring laboratory in Mendoza, Argentina, has initiated archaeological dating at high elevation sites in the Andes. In Europe, scores of archaeological dating programs have sprung up so that routine dating is available from the Urals to Great Britain and from Scandinavia to the Mediterranean Sea. Cornell's Aegean Dendrochronology Project does dating throughout the eastern Mediterranean region, and LTRR and others have begun investigating archaeological tree-ring dating potential along the Mediterranean rim of Africa. Russian scholars have long been involved in dating archaeological materials, particularly in southern Siberia where local programs and cooperative activities with other countries are building chronologies and deriving dates. Dendrochronologists at Lamont-Doherty Earth Observatory and elsewhere are beginning to explore archaeological tree-ring dating in the Himalayas and China.

Dendroclimatology has played an enormous role in the environmental component of dendroarchaeology since the early 20th century, probably beginning with ideas about the Southwestern "Great Drought" of AD 1275-1299. Dendroclimatic reconstructions of a variety of climatic variables (precipitation, temperature, drought indices, streamflow, freezes, and others) are widely used to investigate the effects of environmental variability on human populations and the impact of human activity on the physical environment over time spans ranging from single years to millennia. In addition, comparing archaeological species assemblages with the modern arboreal compositions of their localities identifies changes in vegetation communities since the sites were occupied.

No other aspect of dendroarchaeology has experienced as meteoric an expansion as the behavioral component. This development has been driven by the growing appreciation of the multitude of ways in which humans use wood and by increasing interest in wood-use behavior as an index of the human-environment interface. Most archaeological tree-ring samples come from wooden elements that are artifacts just like pots, projectile points, structures, and other components of the archaeological record. Treating archaeological tree-ring sample collections as populations of artifacts illuminates past human attitudes and practices toward trees as a natural resource and wood as a raw material, important components of human groups' perceptions of their physical and built environments. Examples of human wood-use behaviors that can be extracted from archaeological tree-ring collections are virtually endless, limited only by people's creativity in using wood to satisfy various needs and desires.

Finally, because human actions determine the presence, datable properties, and distributions of wood in sites, information on wood-use behavior is crucial to relating tree-ring dates to their archaeological contexts. The behavioral component of dendroarchaeology is fundamental to devising efficient ways to sample sites and to perfecting assumptions, principles, and procedures for evaluating archaeological tree-ring dates, in other words, to building the method and theory of archaeological tree-ring dating (Ahlstrom 1985, 1997; Baillie 1991a, 1991b; Bannister 1962; Dean 1969, 1978, 1996b). Because different societies treat trees and wood in different ways, tree-ring dates from archaeological sites will be differentially distributed through time and among the sites' components. Therefore, information on the wood-use behavior involved in each case is vital to understanding tree-ring dates from the archaeological sites. This information is most conveniently encoded in conceptual models of the wood-use behavior specific to each group involved. These models include information on species selection, tools used, wood modification objectives and techniques, use of dead wood, reuse of timbers, structure repair, the structural contexts in which wood occurs, and many others. Such models, coupled with the archaeological contexts and physical attributes of the wooden elements, the nature of tree-ring dates, and evaluative assumptions, principles, and procedures, provide the conceptual framework for applying archaeological tree-ring dates to past human actions and events. Finally, the unequalled accuracy, precision, and resolution of tree-ring dates provide a baseline for characterizing the products of other dating techniques and combining the whole into general theories of archaeological chronometry.

Peter Ian Kuniholm and the Aegean Dendrochronology Project

Kuniholm's observation (2002:63) that "each new [dendrochronological] venture into an unknown area is really a form of pioneering" neatly sums up his own groundbreaking forays into the dendrochronological unknown of the eastern Mediterranean region. His tireless campaign to bring the accuracy and precision of tree-ring dating to the Near East was a primary driver of the dendroarchaeological revolution. His dendrochronological career was triggered by the demonstration that *Juniperus* logs in Tumulus MM at Gordion in Anatolia crossdated with one another to produce an 806-year floating ring chronology (Bannister 1970) that, according to archaeological estimates, ended in the late eighth century BC. Building on this discovery, Kuniholm launched a 40-year tree-ring chronology-building effort throughout the eastern Mediterranean region ranging from Turkey to Greece and extending into the Balkans and Italy. His intense curiosity, infectious enthusiasm, boundless energy, and linguistic gifts allowed him to transcend national, political, and institutional boundaries, circumvent bureaucratic impediments, and surmount scientific barriers in pursuing these goals. Major outcomes of the ADP research are a long Turkish tree-ring chronology, which now extends from the present back into the third millennium BC with only some small gaps in the "Roman" period, and many other dated and floating tree-ring sequences. The ADP chronologies produced hundreds of dates for archaeological sites throughout the eastern Mediterranean that helped resolve numerous chronological problems and created a few others. Along the way, Kuniholm became involved in other contentious chronological issues such as dating the eruption of Thera, the wiggle-match dating of dendrochronologically positioned sequential radiocarbon determinations (Manning et al. 2001), archaeological tree-ring dating methods, and the evaluation of archaeological tree-ring dates. Another important achievement is the ADP itself, which provides an institutional foundation and a formal mechanism for propagating the research beyond the accomplishments of one person.

It cannot be said that Kuniholm made these achievements single handedly, for he proved adept at involving numerous other people and institutions in the research. He is renowned for engaging legions of students to labor industriously in the basement of Goldwin Smith Hall on the Cornell campus and to accompany him on collecting adventures to odd corners of the world. Similarly, he exhibited an aptitude for enlisting the help of local residents, leaders, archaeologists, and politicians without whose aid the research would have been much more difficult or impossible. In addition, he was able to persuade numerous governmental and private institutions to provide financial support for the ADP. Equally noteworthy are the thousands of individual contributions to the Project motivated by Kuniholm's unceasing fundraising efforts. Most notable is the long-term financial backing that sustains the Malcolm and Carolyn Wiener Laboratory of Aegean and Near Eastern Dendrochronology at Cornell University. Finally, continuing and increased institutional recognition and support from the University, represented by the appointment of Sturt Manning to succeed Kuniholm as director of the laboratory, will help secure and expand Kuniholm's dendroarchaeological legacy.

The future of dendroarchaeology

Recent assessments of the past, present, and future (Baillie 2002; Dean 1996a, 1997: Kuniholm 2001, 2002; Nash 1999, 2002; Sass-Klaassen 2002; Towner 2002, 2007) of dendroarchaeology (not always called by that name) identify several important issues concerning the future of the subdiscipline. One of the most promising current trends is the expansion of the discipline beyond pure chronology (dating) into the realms of human behavior and the environment. Given the vast number of human activities that involve the procurement, modification, and use of wood, the potential for future expansion of the behavioral domain of dendroarchaeology is limited only by our ability to conceptualize such behaviors and devise means for recognizing them in the archaeological record. Apart from the obvious realized and future potential of dendroclimatic reconstruction for understanding the climate component of human adaptive behavior, other environmental aspects of dendroarchaeology have much to contribute. In particular, this component undoubtedly will continue to illuminate human-environment interactions, especially human impact on local environments. Dendroarchaeology will be equally involved in integrating different measures of past environmental variability such as alluvial dendroclimatology, geomorphology, palynology, paleohydrology, and packrat midden analysis. Because each of these paleoenvironmental techniques is sensitive to and informs on different environmental variables and different spatial and temporal scales of variation, no one of them encompasses the full spectrum of relevant variability. Integrated paleoenvironmental reconstructions have been successfully used in agent-based models of Anasazi subsistence, settlement, and village organization in the northern Southwest (Kohler et al. 2005). Much future effort undoubtedly will be devoted to this kind of research.

Another trend is the increasing attention being given to formalizing procedures for evaluating tree-ring dates from archaeological contexts. Based on expanding knowledge of human wood-use behavior, the nature of tree-ring dates, and the subtleties of archaeological context, dendroarchaeology leads in the effort to develop a body of assumptions, principles, and procedures to better evaluate archaeological tree-ring dates and strengthen their applicability to past human events. Because tree-ring dates are accurate to the calendar year and lack the uncertainty ($\pm$) associated with statistical dating techniques, they are ideally suited to illuminating the potential and pitfalls of independent dating in archaeology. These admirable attributes also can be something of a curse when trying to relate highly accurate and precise tree-ring dates to archaeological sequences necessarily afflicted with low temporal resolution; archaeologists sometimes find it easier to deal with the looser date ranges associated with radiocarbon and other techniques than with "points" in time. Nonetheless, dendroarchaeology is likely to remain in the vanguard of efforts to build method-specific chronometric theories and a general theory of archaeological dating that allow the characterization and comparison of dates produced by different chronometric techniques and the structured application of different kinds of dates to archaeological situations.

Dendrochronologists and archaeologists are beginning to recognize megascale aspects of tree-ring data and past human behavior that deserve further exploration. These revelations are manifest in discoveries that what were thought to be distinct local tree-ring sequences can be crossdated into regional-scale chronologies (Kuniholm 2001; 2002), that anomalies in tree growth (suppressions, surges, frost damage, etc.) correspond across wide areas (Cook et al. 1999; Stahle et al. 2007), and that many such anomalies appear to be associated with other natural occurrences such as volcanic eruptions (Salzer 2000) and human events such as depopulations and migrations, epidemics (Acuna-Soto et al. 2005), and settlement shifts (Benson et al. 2007). Pursuit of these intriguing possibilities likely will loom large in future research and stimulate larger-scale perspectives on dendrochronology that have heretofore prevailed. Extraordinary connections such as these will require especially rigorous empirical validation to forestall extrapolating these inferences beyond the bounds of hard evidence and credibility.

Such large-scale considerations are a principal reason for maintaining and strengthening communication, cooperation, and data sharing among dendrochronological programs world-wide. As noted by Kuniholm (2002:66), dendrochronology has been blessed with an unusually high degree of interaction, an openness that greatly facilitated the explosive growth of the discipline. Therefore, it behooves the discipline as a whole to encourage the widespread sharing of ideas, data, and results and to resist the apparently growing tendency to sequester information through proprietary practices that can only impede further progress.

A problem noted in nearly all assessments of dendrochronology is that of the proper management of the colossal number of tree-ring samples that have been accumulated over the last century (Baillie 2002; Dean 1996a). LTRR alone harbors approximately 400,000 archaeological samples and at least a million additional specimens from living trees and dead tree remnants, and an equal or greater number probably are housed elsewhere. These collections occupy an enormous amount of space that many university and corporate administrators would prefer to devote to more lucrative purposes. Yet scientific ethics require that these collections be maintained in accessible fashion to validate and, if necessary, check the dates and measurements derived from the samples. Furthermore, these collections provide the raw material for future research endeavors, many of which were not anticipated or even possible when the samples were collected. For example, the LTRR archaeological collections recently have been used for additional dating (Street 2001), for dendroclimatic reconstructions (Dean and Funkhouser 1995), to identify distant sources of prehistoric construction timbers (Reynolds et al. 2005), to characterize prehistoric wood-use behavior (Windes and McKenna 2001), to differentiate cottonwood (*Populus* spp.) from aspen (*Populus tremuloides*) beams (Tennessen et al. 2002), and to identify growing-season freezes that could have damaged prehistoric crops (Arnott and Adams 2006). A major part of managing these collections and making them available for further study is the adequate documentation of their contents. Digital catalogs, for example, would simplify and facilitate access to these repositories, but funding to create electronic inventories of large extant collections is extremely difficult to acquire. Given this state of affairs, adequate collections management will remain a concern of dendroarchaeology for some time to come.

An issue touched on by Baillie (2002) and Kuniholm (2002) is one that will change over the years as dendrochronology matures as a science, a process that will proceed at different rates in different places. In much of the world, dendrochronology is in a dynamic growth mode with new lands and species to explore, new chronologies to build and old ones to extend, new research opportunities to exploit, and emergent data bases to expand and refine. In other regions, however (the American Southwest is a good example), these fundamental tasks have already been accomplished. While Southwestern tree-ring chronology building proceeds apace for specific research purposes, there is no need to announce most of these chronologies to the discipline as a whole beyond releasing the data to the ITRDB. Southwestern dendroarchaeology has passed from the developmental stage into a mature phase characterized in part by the large-scale production of dates and other information used in exploring important archaeological issues (social organization, behavioral adaptation, intergroup interaction, demography, chronometric theory and method, etc.) that cannot be addressed without refined chronological controls or high-quality paleoenvironmental information. The problem with these innovative and important pursuits is that they produce few dazzling breakthroughs that excite university administrators, the press, and the public, a situation that can negatively impact institutional appreciation and support. This problem is not unique to dendroarchaeology; it also afflicts other chronometric sciences as reflected in the widespread termination of radiocarbon, archaeomagnetism, and luminescence dating programs. The predicament is so acute that the United States currently has only one program each that routinely provides archaeological tree-ring (LTRR), archaeomagnetic (Illinois State Museum), or luminescence (University of Washington) dating services.

The considerations enumerated above increase the "fragility of the discipline" outlined by Baillie (2002; see also Kuniholm 2002). The prevalence of one-person operations makes dendroarchaeology particularly vulnerable to the death, retirement, or career change of individual practitioners and highly susceptible to changing institutional priorities. Many programs have disappeared upon the departure of the individual responsible for the research. Such transience also endangers the sample collections that are the backbone of the discipline (Baillie 2002). Related to the reliance on individual researchers is the unusual dependence of dendroarchaeology on highly skilled, trained, and experienced technicians who have devoted years to achieving high professional proficiency. Permanent professional staff specialists are essential for long-term program stability and continuity. Such individuals are difficult to find and replace and yet are favorite targets of institutional budget cutters. In the U.S., the situation is so dire that staff losses can seriously diminish investigators' ability to seek research grants and take on new projects.

Conclusion

Despite some threats to the discipline, there are many reasons to suppose that dendroarchaeology has a bright future. Geographical expansion is booming, with the dendrochronological potential of new territories (northern Africa, Central and South America, the Himalayas, China, southeastern Asia, etc.) being actively explored. The development of new topical applications is rampant and limited only by the imagination applied to the effort. New laboratories and individual practitioners are spreading through both the public and private sectors, and dendrochronologists are appearing in a greater variety of academic departments than ever before. The sharing of theory, methods, and data achieved by embedding dendroarchaeology in comprehensive dendrochronological programs benefits all the associated subdisciplines. Similarly, integrating dendroarchaeology into multidisciplinary programs in archaeological science materially broadens the scope of the discipline and strengthens its ties with other natural and social sciences. Creative new mechanisms for broadening the financing of tree-ring research, pioneered by Kuniholm's fundraising strategy, are mitigating the impact of diminishing governmental and institutional support. Given these positive developments, it can be safely predicted that dendroarchaeology's second century will be as productive as its first.

References

Acuna-Soto, R., Stahle, D.W., Therrell, M.D., Gomez-Chavez, S., and Cleaveland, M.K. 2005. Drought, epidemic disease, and the fall of Classic Period cultures in Mesoamerica (AD 750-950), hemorrhagic fevers as a cause of massive population loss. *Medical Hypotheses* 65: 405–409.

Ahlstrom, R.V.N. 1985. The interpretation of archaeological tree-ring dates. Ph.D. dissertation, The University of Arizona, Tucson, Ann Arbor: University Microfilms International.

Ahlstrom, R.V.N. 1997. Sources of variation in the Southwestern tree-ring record. *Kiva* 62: 321–348.

Arnott, H.J., and Adams, R. 2006. Frost rings in timber cores from Spring House, Mesa Verde, Colorado. *Texas Journal of Microscopy* 37: 56–57.

Baillie, M.G.L. 1991a. Suck-in and smear: two related chronological problems for the 90s. *Journal of Theoretical Archaeology* 2: 12–16.

Baillie, M.G.L. 1991b. Marking and marker dates: archaeology with historical precision. *World Archaeology* 23: 233–243.

Baillie, M. 2002. Future of dendrochronology with respect to archaeology. *Dendrochronologia* 20: 69–85.

Bannister, B. 1962. The interpretation of tree-ring dates. *American Antiquity* 27: 508–514.

Bannister B. 1970. Dendrochronology in the Near East: current research and future potentialities. *Proceedings of the 7th International Congress of Anthropological and Ethnological Sciences, vol. 5*: 336–340.

Benson, L.V., Berry, M.S., Jolie, E.A., Spangler, J.D., Stahle, D.W., and Hattori, E.M. 2007. Possible impacts of early-

11th-, middle-12th-, and late-13th-century droughts on western Native Americans and the Mississippian Cahokians. *Quaternary Science Reviews* 26: 336–350.

Cook, E.R., Meko, D.M., Stahle, D.W., and Cleaveland, M.K. 1999. Drought reconstructions for the continental United States. *Journal of Climate* 12: 1145–1162.

Dean, J.S. 1969. Chronological analysis of Tsegi Phase sites in northeastern Arizona. *Papers of the Laboratory of Tree-Ring Research* No. 3. Tucson: The University of Arizona Press.

Dean, J.S. 1978. Independent dating in archaeological analysis. In M.B. Schiffer (ed.), *Advances in archaeological method and theory, vol. 1*: 223–255. New York: Academic Press.

Dean, J.S. 1996a. Dendrochronology and the study of human behavior. In J.S. Dean, D.M. Meko, and T.W. Swetnam (eds.), *Tree rings environment and humanity: proceedings of the international conference, Tucson, Arizona, 17–21 May 1994*: 461–469. Tucson: Radiocarbon.

Dean, J.S. 1996b. Behavioral sources of error in archaeological tree-ring dating: Navajo and Pueblo wood use. In J.S. Dean, D.M. Meko, and T.W. Swetnam (eds.), *Tree rings environment and humanity: proceedings of the international conference, Tucson, Arizona, 17–21 May 1994*: 497–503. Tucson: Radiocarbon.

Dean, J.S. 1997. Dendrochronology. In R.E. Taylor and M.J. Aitken (eds.), *Advances in archaeological and museum science*: 31–64. New York and London: Plenum.

Dean, J.S., and Funkhouser, G.F. 1995. Dendroclimatic reconstructions for the southern Colorado Plateau. In W.J. Waugh (ed.), *Climate change in the Four Corners and adjacent regions: implications for environmental restoration and land-use planning*: 85–104. Grand Junction, Colorado: U.S. Department of Energy, Grand Junction Projects Office.

Kohler, T.A., Gumerman, G.J., and Reynolds, R.G. 2005. Simulating ancient societies. *Scientific American* 293(1): 75–84.

Kuniholm, P.I. 2001. Dendrochronology and other applications of tree-ring studies in archaeology. In D.R. Brothwell and A.M. Pollard (eds.), *Handbook of Archaeological Sciences*: 35–46. New York: John Wiley and Sons.

Kuniholm, P.I. 2002. Archaeological dendrochronology. *Dendrochronologia* 20: 63–68.

Manning, S.W., Kromer, B., Kuniholm, P.I., and Newton, M.W. 2001. Anatolian tree rings and a new chronology for the east Mediterranean Bronze-Iron Ages. *Science* 294: 2532–2535.

Nash, S.E. 1999. *Time, trees, and prehistory: tree-ring dating and the development of North American archaeology, 1914–1950*. Salt Lake City: The University of Utah Press.

Nash, S.E. 2002. Archaeological tree-ring dating at the millennium. *Journal of Archaeological Research* 10: 243–275.

Reynolds, A.C., Betancourt, J.L., Quade, J., Patchett, P.J., Dean, J.S., and Stein, J. 2005. ^{87}Sr/^{86}Sr sourcing of ponderosa pine used in Anasazi great house construction at Chaco Canyon, New Mexico. *Journal of Archaeological Science* 32: 1061–1075.

Salzer, M.W. 2000. Dendroclimatology in the San Francisco Peaks region of northern Arizona, USA. Ph.D. dissertation, The University of Arizona, Tucson. Ann Arbor: University Microfilms International.

Sass-Klaassen, U. 2002. Dendroarchaeology: successes in the past and challenges for the future. *Dendrochronologia* 20: 87–94.

Stahle, D.W., Fye, F.K., Cook, E.R., and Griffin, R.D. 2007. Tree-ring reconstructed megadroughts over North America since A.D. 1300. *Climatic Change* 83: 133–149.

Street, D.G. 2001. How fast is a kiva? The dendroarchaeology of Long House, Mesa Verde National Park, Colorado. *Kiva* 67: 137–165.

Tennessen, D., Blanchette, R.A., and Windes, T.C. 2002. Differentiating aspen and cottonwood in prehistoric wood from Chacoan great house ruins. *Journal of Archaeological Science*. 29: 521–527.

Towner, R.H. 2002. Archaeological dendrochronology in the southwestern United States. *Evolutionary Anthropology* 11: 68–84.

Towner, R.H. 2007. Dendroarchaeology. In S.A. Elias (ed.), *Encyclopedia of Quaternary Science*: 2307–2315. Amsterdam: Elsevier.

Windes, T.C., and McKenna, P.J. 2001. Going against the grain: wood production in Chacoan society. *American Antiquity* 66: 119–140.

Article submitted October 2007

The Absolute Dating of Wasserburg Buchau: A Long Story of Tree-ring Research

A. Billamboz

Abstract: *Along with its palisade system of bog pine (*Pinus rotundata *Link), the Wasserburg Buchau at Federsee was the first site dedicated to tree-ring investigations on the northalpine range. On the basis of a large sampling made by H. Reinerth in 1939, B. Huber was able to construct two tree-ring chronologies for both the inner and outer palisades, showing a short time of construction within a few years, respectively. The missing cross-dating implied a time shift of a least 100 years between both building activities. Since new excavations during the last 20 years allowed a resampling (Baden-Württemberg Office for the Protection of Ancient Monuments and Pfahlbaumuseum Unteruhldingen), the missing link between Huber's two pine chronologies was found. Together with larger heteroconnections, this link led to the absolute dating of the site along with its whole history of construction. Furthermore, the evaluation of tree-ring data from a paleodendroecological perspective provided consistent information about the evolution of climate and environmental changes as well as their consequences for the settlement development during the course of the Late Bronze Age.*

Wasserburg Buchau in the pioneer phase of European dendrochronology

Apart from initial attempts at tree-ring analyses on timber from archaeological sites in northeastern Europe in the early 1930s (e.g., in Biskupin, Poland), B. Huber's systematic investigations at Wasserburg Buchau are generally considered the first application of dendrochronology in Central European archaeology (for particulars of the history of dendrochronology and pile dwelling research, see Billamboz 2004). This well-known fortified site, located in the southern part of the Federsee bog, had just been excavated a few years previously within the framework of a large-scale research program on settlement archaeology headed by R. R. Schmidt, founder of the Institute of Prehistoric Research (Urgeschichtliches Forschungsinstitut) at the University in Tübingen. His assistant, H. Reinerth, believed he had found two phases of occupation, both related to the Late Bronze Age (Reinerth 1928). Reinerth was in charge of the post-excavations conducted in 1937, and shortly afterward, he—meanwhile the head of the Reich Office for German Prehistory (Reichbund für Deutsche Vorgeschichte)—got in touch with B. Huber, the director of the Institute of Forestry Botany in Tharandt (Technical High School, Dresden). Along with his students W. Wittke and J. Zittwitz, Huber had just initiated preliminary tree-ring studies to test the possibility of applying dendrochronology to native tree species.

After the initial investigations of archaeological timber at Dümmersee in 1939, Reinerth's office and the laboratory at Tharandt signed an agreement in July 1940 to collaborate on an extensive analysis of the material from the Wasserburg site at Federsee. It stipulated the following main conditions, among others: first, all tree-ring curve plots were to be produced in duplicate, one for each partner; second, the archeologists would be in charge of evaluating the dating results and their archaeological interpretation, as well as establishing a tree-ring chronology for the prehistoric period, while tree-ring data would be made available to the researchers at Tharandt for evaluation in regard to climate change and woodland development. On this basis, extensive sampling of the palisade zone at the eastern edge of Wasserburg Buchau was conducted late in the same year (Figure 1).

Since these constructions were made principally of soft wood, it was thought they would lend themselves better to comparison with the tree-ring investigations in the American Southwest, which had until then been based solely on long-lived coniferous tree species such as yellow pine and sequoia. The northwestern gate of the settlement was included in the sampling zone. Ad-

Figure 1: Wood sampling during the winter at the end of 1940 in the north-eastern part of the palisade system at Wasserburg Buchau (view from the North towards the gate in the foreground). Note the pile samples at the edge of the excavation trenches (courtesy of Dr. G. Schöbel, Pfahlbaumuseum Unteruhldingen).

ditional samples, mostly of ash, were collected along the northern edge of the central built-up area. Together with W. Holdheide, who was responsible for the identification of the tree species by means of wood anatomy, Huber completed a report on his investigations a year later (Huber and Holdheide 1942). For the most part, logs of bog pine (*Pinus rotundata* Link) were used in the construction of the palisades.[1] Bog pine is a robust species belonging to the mountain pine (*Pinus mugo* Turra) family that thrives in tree form on the fringes of raised bogs; nowadays it is found north of the Alps between the Vosges and the Beskides (for more on the ecology and development of this tree species at Federsee and in the collinean region of Upper Swabia, see Bertsch 1931/32, 1950).

The tree-ring series show a high degree of variation in radial growth, with some very narrow and even incomplete rings. Measurements were taken using thin sections to allow the identification of ring boundaries under incident light with the aid of a bright-field microscope (Figure 2, note the consideration of the late wood proportion!). In this way, Huber and his co-workers were able to establish two bog-pine chronologies, one for the inner palisade extending over 125 years, and another of 152 years for the outer palisade (Figure 3). Due to the lack of suitable references at the time, it was not possible to provide absolute dates. Since there was evidently no overlap between the two chronologies, Huber concluded that the two palisades were not built at the same time, but rather a gap of at least 100 years must have separated the two phases of building activity. That the waney edge was present in most round timbers used in the palisades proved useful in determining felling dates on a relative scale. This made it possible to settle a heated debate between Hans Reinerth and Oscar Paret, the then-head of the Office for the Protection of Monuments in Stuttgart: whereas the former argued that the palisades had been erected in a very short time, the latter believed that the density of piles was the result of a combination of repeated building activity and later repairs (Paret 1941). Considering the high concentration of felling dates within a few years (not including the outer palisade), the question appeared to be resolved in Reinerth's favor.

The distribution of logs according to their cambial age led Huber to the conclusion that they had not been culled from natural woodland, but from a bog stand already modified by human activity. At the time, it was not possible to obtain results from the analyses of other tree species. A year later, Huber obtained samples from the lakeshore site of Unteruhldingen on Lake

---

[1] According to the results obtained by E. Neuweiler (unpublished report of 1928 on the analyses of timber from the excavations carried out during the 1920s), bog pine was also the tree species used most frequently in the central area of the settlement (about 50%).

Constance and established an oak chronology with a depth of 33 samples. Using a new statistical procedure, Huber presented an initial synchronization between the oak chronology there and the bog pine IP (inner palisade) chronology at Wasserburg, and thus was the first to establish a link between two wetland settlement phases at Lake Constance and the Federsee on a relative scale (Huber 1943).[2]

Sixty years later: the missing ring and absolute dating

In the 1960s, a new attempt was made to obtain a precise date for the construction of the palisades at Wasserburg Buchau based on the radiocarbon method, for which six wood samples were taken in the southwestern corner of the settlement (Wall 1998). By averaging the results obtained from five of these samples, a date of 2895 ±40 BP was obtained for the erection of the inner palisade (IP), while the sixth sample produced a date of 2640 ±85 BP for the construction of the outer palisade (AP). Thus Reinerth's hypothesis that the outer palisade had been built before the inner one was proved false. Parallel to the radiocarbon dating, the wood samples were analyzed by the tree-ring laboratory in Munich, and synchronizing the tree-ring series with both of the pine chronologies that had been established by that time proved to be no problem (Huber and Giertz-Siebenlist 1998).

A further step toward establishing absolute dates was taken in 1982 with the founding of the Landesdenkmalamt Baden-Württemberg's tree-ring laboratory at Hemmenhofen. As part of the "Bodensee-Oberschwaben" project, further samples were taken from selected segments distributed along the entire perimeter of the palisades (Billamboz and Schlichtherle 1982). Again, the tree-ring analysis concentrated primarily on bog pine. The previously obtained data could be supplemented with several chronologies and dendro-groups. We were able to add a new chronology to the two first pine chronologies of the palisade system and relate it to a thin palisade closer to the center of the settlement (chronologies A-B-C, Billamboz 1996). The low sample depth of the new chronology, A, did not allow us to confirm a possible synchronization between A and B. In regard to chronology B, it was possible to identify three cutting phases for the IP sequence (B1-3).

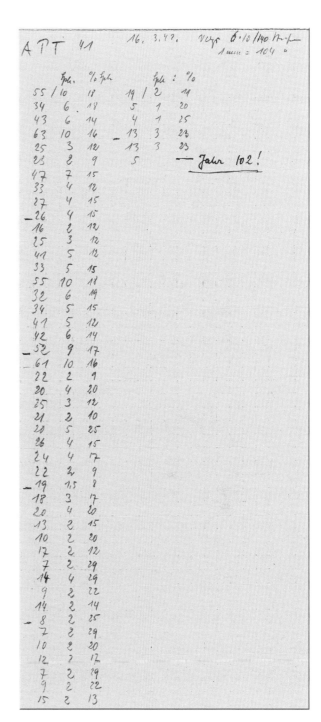

Figure 2: Copy of an original measurement sheet as filled out by B. Huber and his co-workers. Header line: sample name (sample taken in the gate area of the outer palisade), date of measurement, scale factor of microscope. Measurements: col. 1 = ring width, col. 2 = width of late wood, col 3 = percentage of late wood); "Jahr 102!": end year of relative dating on the AP chronology.

---

[2] The synchronization between two tree-ring series was based on the percentage of cases of year-to-year variations showing a different interval trend of growth (tree ring larger or smaller than in the previous year). In the 1970s, this method was replaced by one taking into account the coincident variations of growth (Gleichläufigkeit = percentage of cases of agreement). A definitive absolute date for the Unteruhldingen sequence was later obtained in our laboratory (Schöbel/Billamboz 1996).

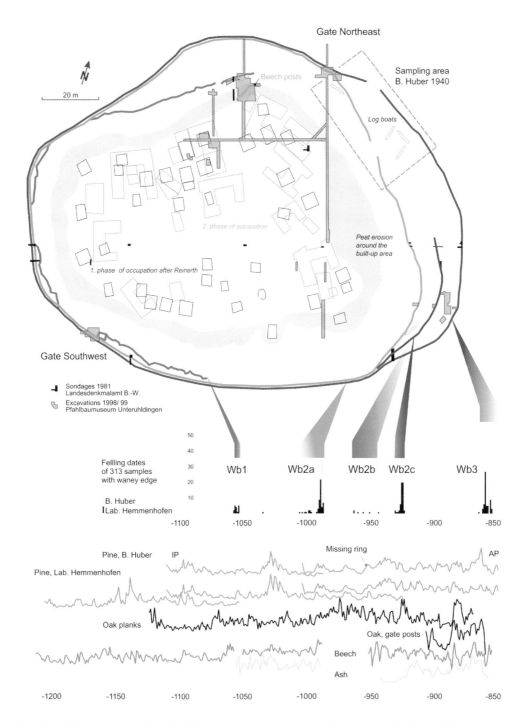

Figure 3: Plan of the Wasserburg Buchau highlighted by dendrochronology. The synchronization of different tree-species, incorporating the first tree-ring data compiled by Huber, resulted in a clear picture of the settlement's development. Several phases of construction can be inferred on the basis of the frequency diagram of felling dates. The dated palisade segments demonstrate a steady extension of the settlement eastward over the entire period of occupation. To facilitate further comparison, both settlement phases as defined by H. Reinerth have been incorporated in the core area of the settlement.

Chronology C increased the sample depth of the AP sequence, though not its length. Thus there was no overlap and the three chronologies still remained separate entities on a relative scale.

Once absolute dates had been obtained from several oak planks that had been stored in the Federsee museum ever since they were excavated by Reinerth (Billamboz 1992), post-excavations were conducted by Landesdenkmalamt Baden-Württemberg and Pfahlbaumuseum Unteruhldingen (Schöbel 1998/99, 1999a-b, 2000) that presented an opportunity for targeted sampling—and thus held out the prospect of establishing an absolute chronology for the entire settlement enclosure.

This new material allowed us to extend chronology A and cross-date it with B. Then, after a missing ring was detected and replaced in the first section of Huber's AP sequence, we were able to link chronologies A and B to C. The joining of the three sequences produced a 356-year chronology, which could be assigned to the time span between 1207 and 852 BC by means of synchronization with the South German oak master and other regional non-oak chronologies. In addition, analysis of other tree species allowed a number of dendro-groups to be assembled, for some of which it was also possible to determine absolute dates. This was the case for oak (here probably *Quercus robur* L.), beech (*Fagus silvatica* L.), and ash (*Fraxinus excelsior* L.; for the correlations between local chronologies, see Billamboz 2003, 2006).

History of construction and wooden floor structures

Now that cross-dating had been achieved, it was possible to assign calendar dates to the previously established construction phases and integrate the relative dates published by Huber. The data on the excavated areas and associated wood samples were entered into a CAD-modeled space to facilitate their evaluation. In this way, we could match up the subsistent structures detected during the post-excavations with Reinerth's site plan more or less satisfactorily. The outlines of the three palisades, in particular, appear to correspond, but certain divergences exist for some structures of the central area of the settlement that have yet to be sufficiently explained.

If we restrict the focus to the felling dates associated with the palisades (Figure 3), we can reconstruct the development of the settlement as follows: Starting from the innermost, very sparse palisade erected in 1058–1054 BC (phase Wb1), we can discern a steady enlargement of the settlement enclosure eastwards, as already pointed out in an earlier paper (Billamboz 1996). This holds true for the inner palisade segments,

which were erected in the years 1006–988 (Wb2a), 964–945 (Wb2b), and 932–925 BC (Wb2c), respectively, as well as for the outermost palisade, which shows evidence of repeated repairs along the entire perimeter of the settlement, conducted between 867 and 852 BC. The southwestern and northeastern gates of the settlement are associated with this last phase of building activity. For their construction, young oak trees were felled in the spring of 862 BC, which, however, of course does not exclude the possibility of building activity in other years.

Figure 4: A bog pine stand (Pfrunger-Burgweiler bog in Upper Swabia). A member of the mountain pine family, bog pine (*Pinus rotundata* Link) is distributed from Galicia to the Carpathian basin. Its tree form is called "Spirke" in South Germany. This hardy species makes few demands on its environment and thrives on the poor soils of raised bogs, where it faces little or no competition. Its high resin content makes its wood resistant to humidity and well suited to building purposes. Because of the sensitivity with which its growth patterns respond to changing ecological conditions, bog pine tree-ring series constitute valuable archives for the study of natural changes in bog ecosystems, especially for the reconstruction of water-table variations.

If we return for a moment to the question debated by Reinerth and Paret—whether the palisades were erected at one time or are, instead, the result of repeated construction measures—it turns out, ironically enough, that on the basis of the dendro-dates obtained, both were right to a certain extent. Indeed, the respective palisade segments on the eastern edge of the settlement can be interpreted as the product of a single phase of building activity, while the more massive western palisade represents the sum of various

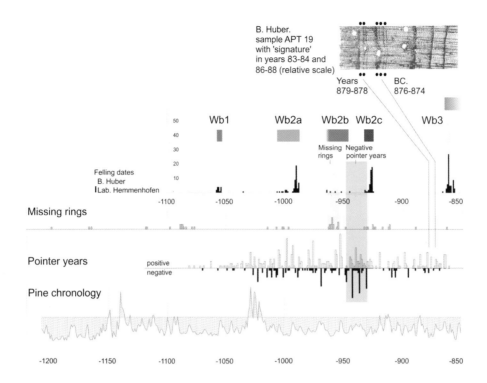

Figure 5: With its highly sensitive tree-ring series, bog pine yields information on environmental changes during the course of occupation, particularly with regard to extreme years of growth. Pointer years are calculated according to the departure from the mean in units of standard deviation (index positive 1 or negative –1). A particular concentration of negative pointer years can be seen from 950 to 920 BC, between the construction phases Wb2b and 2c. Even the tree-ring series of the pine logs used for the palisade extension Wb2c were characterized by incomplete and missing rings. Such disruptions in growth patterns can probably be linked to a rise in the water table in the southern part of the Federsee bog. For an example of this type of growth pattern, which was already recognized by B. Huber, see above (Huber/Holheide 1942, part of Figure 1). An absolute time scale can be now assigned to this characteristic signature.

measures conducted in the course of several settlement phases over a period of 250 years.

In regard to the structures in the inner area of the settlement, it has proved difficult to verify Reinerth's interpretation of the evidence as consistent with two settlement phases. The complete desiccation of the timber structures remaining after the extensive excavations conducted by Reinerth greatly reduces the prospects of obtaining dendrochronological dates from them compared to the vertical posts of the palisades. Despite these obstacles, we were able to date some house structures and assign them to the phases Wb1 and Wb2a. These results are inconsistent with Reinerth's interpretation. As has been pointed out by W. Kimmig (1992), it seems more likely that Reinerth's reconstruction of the first settlement reflects an amalgamation of asynchronous structures that could not be separated due to the difficulties posed by the stratigraphic situation and unavailability of dendrochronological dates at the time. The same holds true for the more elaborate floor plans of Reinerth's second phase, which can also be explained as assemblages of asynchronous elements.

Since the possibility of drastic erosion directly following the settlement phase cannot be excluded, as the example of the neighboring Early Bronze Age site Siedlung Forschner shows (Billamboz 2006), it is conceivable that the structures associated with the final building phase Wb3 had already disappeared by the time Reinerth conducted his campaign. The documentation of his excavations is currently being re-examined, with particular attention to the excellent photographs taken by Reinerth and his co-workers at the time. Incorporating them into the CAD plan could allow certain problems concerning the location of some structures to be solved and facilitate analysis of the various findings (e.g. separation of superimposed floor constructions by means of identification of particular timbers according to macroscopic criteria, such as the characteristic branching pattern of bog pine).

The development of the settlement in relation to environmental changes

Another issue of major interest to settlement archaeology is that of settlement continuity. In the case of Wasserburg Buchau, we can shed light on this question with the help of dendrochronology, in particular the evaluation of tree-ring data from a paleoecological perspective. Our observations so far (Billamboz 2003, 2006) suggest that, because the radial growth of wetland tree species is largely determined by groundwater fluctuations, tree-ring variations are good indicators of ecological changes in wetland areas and thus can provide valuable information on settlement conditions. At Wasserburg Buchau, we can draw on numerous bog-pine series for such an evaluation. This robust species (*Pinus rotundata* Link, commonly called "Spirke" in southern Germany) is found in tree form especially on the slopes of raised bogs, where there tends to be good natural drainage (Figure 4). As recent studies (Freléchoux 1997, Freléchoux et al. 2000; Sobania 2002) have shown, this species exhibits a high rate of growth in phases in which the vegetation period is warm and dry. In contrast, wet summers that lead to a rise in the water table inhibit the formation of tree rings, and in particular, of late wood. Based on these considerations, the high degree of variation in ring width and late-wood density observed in the pine series from Wasserburg Buchau would appear to indicate that growth conditions in the exploited bog pine stands were unstable and that the lateral development of the bog toward the Federsee did not follow a steady, consistent course during the Late Bronze Age.

To better illustrate this phenomenon, we calculated the frequency of "pointer years" (years in which radial growth deviates significantly from the mean). For this purpose, single series were indexed within a five-year moving window according to the "skeleton plot" method (Cropper 1979; Index = central value - mean value / standard deviation). The distribution of positive and negative pointer years corresponds to the occurrence of index values exceeding 1 or -1 (Figure 5, below). Of particular interest in this respect are negative pointer years, which are associated with a stronger impact of factors inhibiting growth, in this case, a rising water table, in particular.

A major concentration of negative pointer years is recorded between 950 and 930 BC, which corresponds precisely to the interval between two phases in the construction of the inner palisades. Thus the tree-ring analysis underscores the strong relationship between environmental changes and building activity in the settlement. Furthermore, missing and incomplete rings were observed in the individual series. The severe reduction, or even absence, of radial growth must similarly be interpreted as stress-related; with their deeper roots, older pine trees seem to be more severely affected by a rise in the water table. The recorded data are incorporated into Figure 5. A particular concentration emerges around 960 BC, corresponding to the extension of the palisade system in Wb2b. These measures might be interpreted as a communal effort by the settlers to protect their home against the rising water. These observations reinforce the impression that the settlement developed in several waves. Thus the history of construction at Wasserburg Buchau makes it appear likely that the site was not inhabited continuously between 1050 and 850 BC, but rather that the settlers took advantage of periods of favorable climate to occupy it. That considered, it appears plausible that this fortified site functioned as a "bridgehead" to the lake, particularly in times of receding water levels and attendant drying out of the bog.

Notification and acknowledgments

This paper is a more concentrated version of an original publication in German, presented in the review *Plattform* (Zeitschrift des Vereins für Pfahlbau und Heimatkunde E. V. Unteruhldingen. Original title: A. Billamboz, Die Wasserburg Buchau im Jahrringkalender. *Plattform* 13/14, 2004/2005, 97–105); translation by A. Barrett, Berlin, with many thanks from the author.

I would especially like to thank G. Schöbel, the director of the Pfahlbaumuseum in Unteruhldingen, for placing documents and photographs from the museum's archives at my disposal, particularly those related to B. Huber and the early years of European dendrochronology. In addition, my thanks go to my son, C. Billamboz, who was commissioned by the Pfahlbaumuseum to digitize the old excavation plans and incorporate them into the CAD-model space; to S. Buckow, N. Bleicher, J. Kempe, and I. Sobania, who assisted in conducting the necessary tree-ring measurements; and to A. Kalkowski, who was responsible for the final layout of the figures.

References

Bertsch, K. 1931/32. Paläobotanische Monographie des Federseeriedes. *Bibl. bot. 103*. Stuttgart: Schweizerbart'sche Verlagsbuchhandlung.

Bertsch, K. 1950. Nachträge zur vorgeschichtlichen Botanik des Federseerieds. *Veröff. der Württembergischen Landesstelle für Naturschutz u. Landschaftspflege* 19 (Ludwigsburg 1950), 88–128.

Billamboz, A. 1992. Bausteine einer lokalen Jahrringchronologie des Federseegebietes. *Fundber. Baden-Württemberg* 17:1 293–306.

Billamboz, A. 1996. Tree-rings and pile-dwellings in southwestern Germany. Following in the footsteps of Bruno Huber. In J.S. Dean, D.M. Meko, and T.W. Swetnam (eds.), *Tree Rings, Environment and Humanity, Proceedings of the International Conference, Tucson, Arizona, 17–21 May, 1994*: 471–483. Tucson: University of Arizona Press.

Billamboz, A. 2003. Tree rings and wetland occupation in Southwest Germany between 2000 and 500 BC: Dendroarchaeology beyond dating in tribute to F. H. Schweingruber. *Tree-Ring Research* 59(1): 37–49.

Billamboz, A. 2004. Dendrochronology in lake-dwelling research. In F. Menotti (ed.), *Living on the Lake in Prehistoric Europe*: 117–131. London: Routledge.

Billamboz, A. 2006. Jahrringuntersuchungen in der Siedlung Forschner und weiteren bronze- und metallzeitlichen Feuchtbodensiedlungen Südwestdeutschlands. Aussagen der angewandten Dendrochronologie in der Feuchtbodenarchäologie. Erscheint in: Die früh- und mittelbronzezeitliche "Siedlung Forschner" im Federseemoor. Befunde und Dendrochronologie. Regierungspräsidium Stuttgart, Landesamt für Denkmalpflege, Esslingen. Siedlungsarchäologie im Alpenvorland, Forschungen und Berichte zur Vor- und Frühgeschichte in Baden-Württemberg. [Preprint auf CD-Rom, 2006].

Billamboz, A., and Schlichtherle, H. 1982. Moor- und Seeufersiedlungen. Die Sondagen 1981 des "Projekts Bodensee-Oberschwaben." *Archäologische Ausgrabungen in Baden-Württemberg* 1981: 36–50.

Billamboz, A., and Schöbel, G. 1996. Dendrochronologische Untersuchungen in den spätbronzezeitlichen Pfahlbausiedlungen am nördlichen Ufer des Bodensees. In Siedlungsarchäologie im Alpenvorland IV. *Forsch. u. Ber. Vor- u. Frühgesch. Baden-Württemberg* 47: 203–221. Stuttgart: Theiss.

Cropper, J. Ph. 1979. Tree-ring skeleton plotting by computer. *Tree-Ring Bulletin* 39: 47–54.

Fréléchoux, F. 1997. Étude du boisement des tourbières hautes de la chaîne jurassienne: Typologie et dynamique de la végétation—approche dendroécologique et dendrodynamique des peuplements arborescents. Thèse, Univ. Neuchâtel.

Fréléchoux, F., Buttler, A., and Gillet, F. 2000. Dynamics of bog-pine-dominated mires in the Jura Mountains, Switzerland: A tentative scheme based on synusial phytosociology. *Folia Geobotanica* 35: 273–288.

Huber, B., and Holdeide, W. 1942. Jahrringchronologische Untersuchungen an Hölzern der bronzezeitlichen Wasserburg Buchau am Federsee. *Berichte der Deutschen Botanischen Gesellschaft* Band LX, H. 5: 261–283.

Huber, B., and Giertz-Siebenlist, V. 1998. Nachträge zur Dendrochronologie der "Wasserburg Buchau." In Siedlungsarchäologie im Alpenvorland V. Forschungen und Berichte zur Vor- und Frühgeschichte in Baden-Württemberg 68: 87–89: Stuttgart: Theiss.

Huber, B. 1943. Über die Sicherheit jahrringchronologischer Datierung. *Holz als Roh- und Werkstoff* 6: 263–268.

Kimmig, W. 1992. Die "Wasserburg Buchau" eine spätbronzezeitliche Siedlung. *Materialhefte zur Vor- und Frühgeschichte in Baden-Württemberg* 16. Stuttgart: Theiss.

Paret, O. 1941. Der Untergang der Wasserburg Buchau. Zur Vorgeschichtsforschung am Federsee. *Fundberichte Schwaben Neue Folge* 10: 1–50.

Reinerth, H. 1928. Die Wasserburg Buchau. Eine befestigte Inselsiedlung aus der Zeit 1100–800 v. Chr. Führer zur Urgeschichte 6, Augsburg: Benno Filser.

Schöbel, G. 1998/99. Wiederausgrabungen in der "Wasserburg Buchau," einer spätbronzezeitlichen Ufersiedlung im Federseemoor. *Plattform* 7/8: 130–131.

Schöbel, G. 1999a. Wiederausgrabungen in der spätbronzezeitlichen Ufersiedlung "Wasserburg Buchau" im Federseemoor bei Bad Buchau, Kreis Biberach. Arch. Ausg. Baden-Württemberg 1998: 74–77.

Schöbel, G. 1999b. Nachuntersuchung in der spätbronzezeitlichen Ufersiedlung "Wasserburg-Buchau" bei Bad Buchau, Kreis Biberach. In Köninger et al. (eds.) *Moor- und Taucharchäologie im Federseeried und am Bodensee. Die Unternehmungen 1999 des Landesdenkmalamtes Baden-Württemberg, Referat 27*. NAU (Nachrichtenblatt Arbeitskreis Unterwasserarchäologie) 6: 45–46.

Schöbel, G. 2000. Die spätbronzezeitliche Ufersiedlung "Wasserburg Buchau," Kreis Biberach. In: (Bay. Ges. für Unterwasserarchäologie Hrsg.) Inseln in der Archäologie. *Archäologie unter Wasser* 3: 85–100. München: Janus.

Sobania, I. 2002. Jahrringökologische Studien an Moorkiefern (*Pinus rotundata* Link) entlang eines ökologischen Gradienten im Pfrunger-Burgweiler Ried (Oberschwaben, Südwestdeutschland). Diplomarbeit, Universität Stuttgart.

Wall, E. 1998. Archäologische Federseestudien. Untersuchungen zu Topographie, Stratigraphie, Hydrologie und Chronologie der vorgeschichtlichen Siedlungen am Federsee. In: Siedlungsarchäologie im Alpenvorland V (Hrsg. Landesdenkmalamt Baden-Württemberg). *Forschungen und Berichte zur Vor- und Frühgeschichte in Baden-Württemberg* 68: 11–76. Stuttgart: Theiss.

Article submitted May 2007

Is there a Separate Tree-ring Pattern for Mediterranean Oak?

Tomasz Wazny

Abstract: *East Mediterranean trees show relatively homogeneous growth patterns. Tree-ring study has confirmed in general a common response to the large scale climatic anomalies that influenced ring growth in this region. However, no correlation to central and north European patterns has been found. Discovery of such a relationship could provide the absolute dating of the now-floating Aegean BC chronology. To answer the question of the possible existence of a pan-European tree-ring pattern for oak, tree-ring width chronologies representing three different time-windows were compared. The results confirmed two separate oak tree-ring patterns with the Alps as a distinct boundary. However, in the East, the Carpathian Mountains—in contrast to the Alps—seem to act as a bridge between North and South climate systems. Romania reveals itself to be the key region connecting both parts of Europe.*

Since 1973 the Aegean Dendrochronology Project (ADP) at Cornell University has worked toward the development of long-term tree-ring chronologies for the Mediterranean and Balkan regions. After 34 years of work, Peter Ian Kuniholm and his co-workers and students established numerous chronologies covering the last 9000 years (i.e. Hughes et al. 2001). His area of interest covers the broad geographical zone from northwestern part of Italy in the West to the Caucasus Mountains in the East and from the southern slopes of Alps in the North to Egypt in the South (Figure 1). The region is about 3000km in latitude by 1500km in longitude. The ADP chronologies have provided datings of numerous historic timbers and contributed to the reconstruction of the history of particular sites such as a Bronze Age settlement near Sozopol in Bulgaria (Kuniholm et al. 1998), numerous historic buildings and Ottoman monuments (Kuniholm, 2000), and archaeological sites in Greece and Anatolia (i.e. Newton and Kuniholm 2004). They have also been used for the study of paleoclimate (Griggs et al. 2007).

Oak is one of the most important species creating the forest stands of Europe. The range of its natural distribution covers almost the entire continent except part of Fennoscandia, where the conifers dominate, and the high mountainous areas. Due to its strong physical and mechanical properties, oak has been used by humans for all forms of building activity. Its high natural durability and resistance against major degradation are factors contributing to oak's importance in almost all cultural layers.

The genus Quercus is represented by 22 native species in Europe (Tutin et al. 1993). Most of them grow in the southern part of the continent. Trees included in the chronologies developed by the Cornell Tree-Ring Laboratory belong to the four species: *Q. robur* L. (pedunculate oak); *Q. petraea* (Mattuschka) Lieb. (sessile oak); *Q. frainetto* Ten. (Italian or Hungarian oak); and *Q. hartwissiana* Steven. (no English common name). *Q. robur* and *Q. petraea* are the most widely spread oak species. They grow under a broad variety of ecological conditions from Turkey and Greece to the central part of Sweden and coast of Norway. The range of natural distribution of both trees is shown in Figures 2 and 3. Two other species of oak are found only in warmer climatic conditions: *Q. frainetto* is native to SE Europe and Turkey, while *Q. hartwissiana* occurs only in eastern Bulgaria, north and eastern Turkey, and in the West Transcaucasus. Most of the historic timbers used for chronology building in the Mediterranean represent these species (Kuniholm and Striker, 1987; Griggs et al 2007); however, a few other species may also be taken into consideration. *Q. pubescens* (downy oak), native to southern Europe and central part of the continent, is one of them. Its heavy and resistant wood has applications in carpentry and furniture production. Three native oak species of northern, central and western Europe

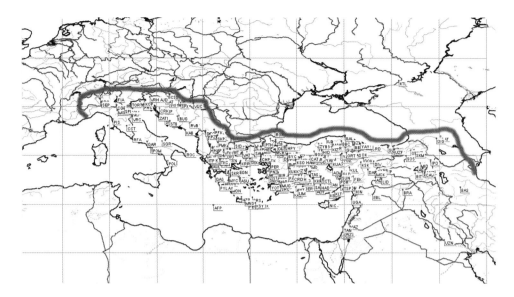

Figure 1: Map of Eastern Mediterranean showing location of sites investigated by the Cornell Tree-Ring Laboratory and northern boundary of the study area.

(*Q. robur*, *Q. petraea*, and *Q. pubescens*) have been often successfully crossdated (Haneca 2005). The North European long-term chronologies were built with tree-ring series of the first two species.

All the above-mentioned oaks are deciduous and they belong to the subgenus *Quercus*. Oaks from this group form distinct rings with sharp ring boundaries (Figure 4). Their wood-anatomical structures are almost identical; therefore, the different species within the subgenus *Quercus* cannot be distinguished on this basis (Schweingruber 1990). Huber et al (1941) noticed, however, that for penduculate and sessile oak there existed slight differences in the structure of earlywood vessels and in the area of transition between earlywood and latewood. These differences occur when tree-ring widths exceed 1.5mm, but they are not sufficient for identification of species. Grosser (1977) pointed out a 50% probability of erroneous identification in such a case.

Tree-ring patterns of different European oak species growing in a particular geographical region are very similar. Changes in tree-ring widths demonstrate identical responses to the changing meteorological conditions (Cedro 2007). Differences in soil and climatic requirements between species are generally much less than differences between sites (Ufnalski 2006). There are also natural hybrids between oak species, but their form and extent have been questioned (i.e. Aas 1990, Kremer et al. 2002). As a consequence ring-width series of different species are always combined into one chronology and simply treated as "oak." The two most widely distributed species—*Quercus robur* and *Quercus petraea*—are represented in most of the European tree-ring chronologies.

East Mediterranean trees, whether oaks or conifers, show relatively homogeneous tree-ring patterns. Kuniholm and Striker (1987) noticed significant interspecies correlation and in general a common response to climate variation over long distances. Study of signature years confirmed the existence of patterns showing large scale climatic anomalies that influenced ring-growth in the Aegean (Hughes et al. 2001). Despite this, one could find it astonishing that no correlation to central- and north-European patterns has been observed. Discovery of such a correlation could provide the absolute dating of the now-floating Aegean BC chronology. An increasing number of recently created tree-ring chronologies in Europe, especially oak chronologies, motivated me to make an attempt to "cross the Alps" and to connect existing Mediterranean oak chronologies with absolute northern European master chronologies. Another important question which arose concerns territorial limits/boundaries of these chronologies and the reasons for their distinct placements.

In 1983 Mike Baillie asked: "Is there a single British Isles oak tree-ring signal?" (Baillie 1983). Further study revealed that not only the British Isles but all of western Europe has the same tree-ring signal (Pilcher et al. 1984). The existence of a pan-European tree-ring pattern seems to be much more questionable, and I have decided to use the chance given by the possibility of simultaneous work in both parts of Europe to find the answer.

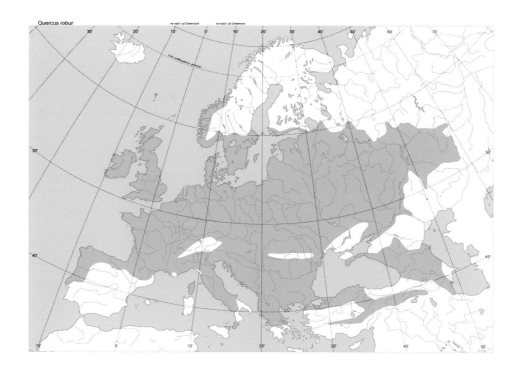

Figure 2: Distribution range of *Quercus robur* (based on Boratynski (1995). http://pl.wikipedia.org/wiki/Grafika:QuercusRobur_ZasiegGatunku01.png

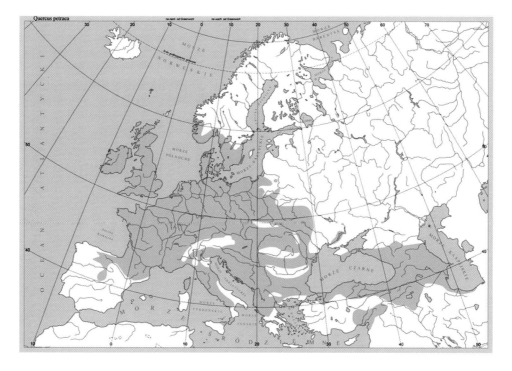

Figure 3: Distribution range of *Quercus petraea* (based on Boratynski (1995). http://pl.wikipedia.org/wiki/Grafika:QuercusPetraea001.png

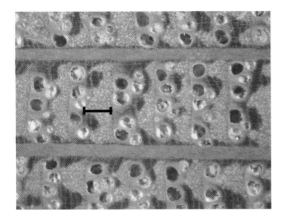

Figure 4: Cross section of oak, subgenus *Quercus*. Six distinct annual ring boundaries are visible (left to right). Every annual ring consists of earlywood formed in spring, with large vessels and more compact latewood formed in the second part of the growing season. The scale bar represents 1mm.

Material and methods

More than 100 tree-ring width chronologies representing three chosen time-windows have been compared in order to answer the question of whether we can link the two parts of Europe and to define the possible borders and the reasons behind where they exist. The following time-windows have been selected with respect to the availability of chronologies:

1. from the present back to AD 1500
2. between 400 BC and AD 400 (including the so-called Roman Gap in the Mediterranean chronologies)
3. European Late Bronze Age followed by the Hallstatt period

The methodology was based on the availability of absolutely dated oak chronologies representing regions not far away from the expected boundary. In addition, and as a type of control, millennia-long chronologies from Germany have been brought into the calculation. Chronologies have been selected both from the resources of the International Tree-ring Data Bank[1] and from data obtained from collaborating laboratories. To enhance the common signal, regional chronologies have been joined into the composite chronologies.

The relationship between time-series of the ring-width data has been measured by the *t*-score calculated from the correlation. "*T*-statistics" were introduced to dendrochronology by Baillie and Pilcher (1973) and in the following years the algorithm has been modified many times. It is implemented in most software used for cross-dating; however, different formulas used for calculation cause slight differences in the results obtained for the same time-series (Sander and Levanic 1996). Independent of some critical remarks, which we hear mainly from the statisticians, the results are reliable and compatible with results obtained by different laboratories. The most important justification for the choice of this method is: It works!

In practice, a *t*-value of 4 is the minimum level of accepted match, while a *t*-score exceeding 10 suggests that both compared tree-ring series represent the same tree. Dendrochronologists should not rely only on *t*-scores, of course. Statistics can produce artifacts; misinterpreted results can lead to false conclusions and false dating results. Therefore every positive result of statistical correlation must be confirmed visually. Visual verification of cross-dated series is traditionally done by comparison of graphs reflecting the growth rhythms of the trees.

Results and discussion

The first level of information about tree-ring patterns of European oaks is obtained from statistical parameters describing the ring-width series. Descriptive statistics of the main oak master chronologies are summarized in Table 1. Chronologies have been selected according to the geographical criteria to represent different regions of both parts of Europe.

The main parameters of north European chronologies are relatively uniform regardless of climatic conditions. Mean ring width varies from 1.29mm to 1.50mm except for the Burgundy chronology, which has a mean ring width of 1.69mm. High autocorrelation (statistics describing dependence of values on preceding values) of 0.70–0.83 shows that annual growth is dependent on conditions from the previous year. These signals from the previous year are transmitted mainly by materials stored in wood structure, buds, hormones, etc. Mean sensitivity is another important parameter describing tree-ring series. Sensitivity, calculated as a percentage change from each measured yearly ring value to the next, informs how strong the tree-ring series reflect the environmental factors. Low mean sensitivity indicates regular growth of oak trees and reduced year-by-year variability. The mean sensitivity of European oaks amounts to only 0.11 to 0.13.

The South European pattern looks similar to the North European oak tree-ring patterns in general. Only the mean ring width is smaller, which means that the average Mediterranean tree grew more slowly.

We have done further comparison of tree-ring series in three selected time-windows. The first period taken into consideration represents the Late Bronze

---

[1] The ITRDB is maintained by the National Oceanic and Atmospheric Administration (NOAA) Paleoclimatology Program and World Data Center for Paleoclimatology. This professional and non-profit organization has been established to provide a permanent location for the storage of high-quality dendrochronological data from around the world. Tree-ring data are freely available to the scientific community.

	Region	Mean ring width [mm]	Standard deviation	Auto-correlation	Mean sensitivity
1	Southern Sweden	1.50	0.44	0.82	0.124
2	North Germany, Hamburg	1.29	0.33	0.77	0.134
3	South Germany	1.35	0.33	0.79	0.121
4	Northern Poland	1.37	0.38	0.83	0.123
5	Netherlands	1.40	0.30	0.80	0.112
6	Eastern France, Burgundy	1.69	0.36	0.80	0.115
7	England, East Midland	1.38	0.30	0.70	0.130
8	Romania, North	1.33	0.31	0.73	0.131
9	**Northern Turkey**	**0.88**	**0.19**	**0.74**	**0.120**
10	**Northern Greece**	**0.83**	**0.21**	**0.78**	**0.127**
11	**Central-Western Greece**	**1.16**	**0.36**	**0.87**	**0.133**

Table 1: Descriptive statistics of selected European oak ring width series. All chronologies are truncated from AD 1451 to 1950. Chronologies 1–8 represent the northern part of the continent, while chronologies 9–12 (in bold) are the Mediterranean representatives.

Age and Hallstatt period. The number of chronologies available for this period is limited, but the Hallstatt period was very promising for this study because of rapid climate change and preceding climate disturbances.

This period of unstable weather conditions over a broad geographical range was reflected by the similar reactions of trees growing in different geographical regions and resulted in distinct correlations over the entire distance. In this way it was possible to achieve absolute dating of a settlement called Biskupin in Poland by means of North-German chronologies based on material from Lower Saxony (Wazny 1994). The Hallstatt period's transition between a warm, dry subboreal period and a wet, cool subatlantic influenced not only trees but also caused migration and expansion of cultures such as the Scythian culture (van Geel et al., 2004). Climate change of the early first millennium BC was caused by decline of solar activity and had a global range (Chambers et al. 2007). Unfortunately such long-distance correlations between tree-ring chronologies in the time of climate deterioration seem not to work over the different climate zones and across the Alps in particular (Figure 5).

If we go back deeper into the Bronze Age, we find it remarkable that the archaeological site of Lavagnone—located near the southern shore of Lake Garda, close to the southern slope of the Alps, with hundreds of preserved timbers—is datable only by radiocarbon. The 297 year-long tree-ring sequence based on 107 samples does not cross-date with any of the South German chronologies and is dated by radiocarbon wiggle-matching to 2213–1917 ± 10 BC (Griggs et al. 2002). Similar negative results were found in a compilation of tree-ring series obtained from 13 palafitte settlements located in the area of Lake Garda and spanning 700 years: 2233–1533 cal BC (Martinelli 2007). (This regional chronology includes series dated by Griggs et al. 2002). Although the distance between northern Italy and Switzerland or southern Germany is not even 300km, no link between both parts of the Alps can be found.

The second time-window encompasses the first centuries BC and AD—a time of both expansion and fall of the Roman Empire, of an enormous increase in building activity and the development of large cities. Public and private buildings became larger and more ambitious, and these required timbers of larger size and better quality (Meiggs 1982). Simple roof constructions evolved into large-format roofing systems carrying ceramic tiles. Unfortunately for dendrochronologists, Roman architects invented many new structural novelties, technologies, and materials like concrete, stone arches, and concrete vaults. Fired bricks and easily accessible volcanic stones replaced flammable wooden beams. Wood was used in huge amounts for charcoal production, as a fuel for firing of bricks and tiles, metal production, for heating the famous Roman baths and many other purposes. Despite the preference given to stone architecture, many buildings retained traditional ceilings and wooden roofs. Now, only empty beam holes, post holes, and sometimes carbonized traces remain (Figure 6). The existence of beam holes does not mean that all of them were filled with high quality oak. However, the occurrence of oak in preserved building structures indicates that it was one of the most used wood species in many parts of ancient Mediterranean world. Despite the huge wood utilization, the Roman period has become the most problematic period in the development of the East Mediterranean dendrochronology because of lack of suitable timbers (Kuniholm 2002).

Remote northern provinces of the Roman Empire relied on local material resources delivered mainly

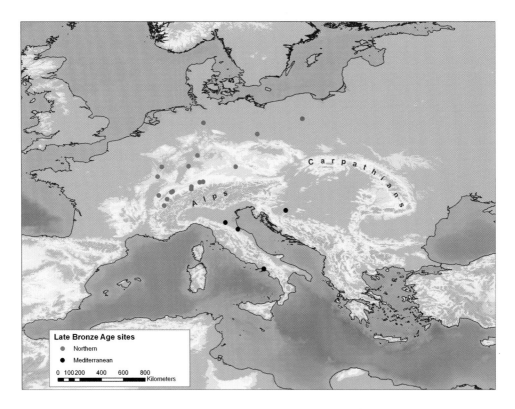

Figure 5: Location of oak chronologies representing Late Bronze Age and Hallstatt period.

Figure 6: The Forum Romanum. Brickwork with empty beam-holes.

from the forest. Numerous wooden Roman wells, bridge pilings, trackways, remnants of Roman and native villages, villas, forts, waterfront structures, and shipwrecks have been found in France, Germany, England, and the Netherlands. Ships made of oak suggest adaptation of local woodworking traditions. In the Mediterranean Basin light fir was a preferable shipbuilding material because of speed requirements for the ships (Meiggs 1982). Although rich findings have enabled the construction of numerous oak chronologies for regions in the west and the north Mediterranean countries (Figure 7), they unfortunately do not show any significant correlation with the South European oak pattern.

The third, modern time-window is filled by chronologies based on growing trees and extended by means of timbers from standing buildings. Tree-ring patterns reflect the present forest structure, which in most parts of Europe differs from the natural forest in the sense of species composition and the age structure of the trees. Distribution of chronologies over the continent is more even, but this should be expected given the denser network of available tree-ring series. In this time-window we notice the first success: one significant link between Poland/Germany and Macedonia/Kosovo has been affirmed—via Romania (Figure 8). Romanian oak chronologies representing the regions of Maramures and Transylvania tend to lean more to the North European tree-ring pattern, but a combined Romanian chronology with the tentative name DRACULA2 shows significant correlation with the Macedonian chronology YUGOLATE, which represents the Mediterranean pattern.

The results unambiguously show the existence of two different tree-ring patterns for oak growing in the northern and southern parts of Europe. The Alps create a sharp border, which disperses as it moves east, and finally the two patterns meld somewhere on the flat plains of the Carpathian Basin. The search for similarities and connections between tree-ring patterns was motivated by the wide distribution area of the most important oak species, *Quercus robur* and *Quercus petraea*, which cover both parts of Europe, and by the possible global occurrence of similar patterns of tree growth suggested by Schmidt and Gruhle (2006). In the case of the North European oak master chronologies a significantly high correlation exists over distances of about 400km (Hollstein 1980, Wazny and Eckstein 1991). Furthermore in some years the dendrochronological signal reaches far beyond this range, especially during the continent-wide signature years.

The shortest distance between chronologies representing the southern and northern parts of Europe is no more then 250km in the Alps. Despite this short distance the two sets of tree-ring patterns do not show any significant correlation. The same distance is between sites situated inside the former Roman province of Pannonia, where Roman timbers from Slovenian Celje and Croatian Sisak do not match with a Hungarian oak chronology from the same period. This lack of correlation is surprising in this instance because there are no natural boundaries between the sites. The same Hungarian chronology representing the vicinity of Budapest crossdates with Austrian and German master chronologies and definitely represents the northern pattern. Theoretically, looking at the map of Europe and its relief (Figure 7), we see that this region provides the greatest opportunity to link both parts of Europe, so if the relief does not prevent the correlation, what could be the reason? Climate conditions and the weaker influence of Mediterranean climate in the present Republic of Hungary are the most logical explanations, while the insufficient number of chronologies no doubt contributes to the problem.

In terms of climate, the two parts of the continent belong to different climate regimes. Mediterranean plants must face two different types of stress – winter cold and summer drought (Cherubini et al. 2002). Low winter temperatures are distinctive especially in northern and eastern parts of Mediterranean region and at the higher altitudes. The annual rings of oaks growing in North Aegean are limited mainly by May-June precipitation (Griggs et al 2007); however, oaks possess deep root systems that can easily reach water stored deep in the soil, and therefore they should be less affected by lack of precipitation. Oak's dependence on summer precipitation has been observed also in northern Europe, where the trees find almost optimal conditions in a climate balanced between maritime and continental influences. The Carpathian Mountains surrounding the Carpathian Basin from the North, East, and South are a unique geographical unit located between the mild-wet Atlantic, the winter-cold continental, and the summer-dry Mediterranean climates. This mountain range with its characteristic 1500km long arch seems to be a bridge connecting different vegetation zones and the junction of the main European climate systems. The Mediterranean "bridgehead," found in mountainous Macedonia and Kosovo, is climatically closer to that in the Carpathians because of higher altitude and the resulting higher precipitation.

This study should be treated as preliminary because the network of East European oak chronologies is full of holes and therefore a better selection of data was not possible. There are also problems with the accessibility of already existing tree-ring series.

The International Tree-Ring Data Bank is the only source of data for tree-ring research open for scientists.

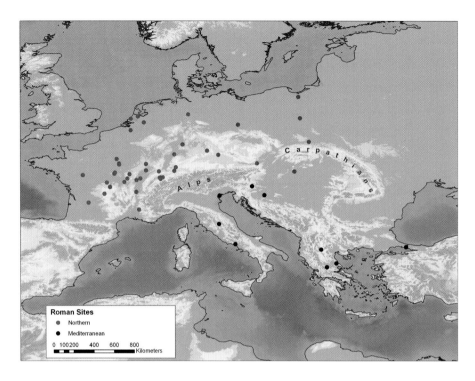

Figure 7: Location of oak tree-ring chronologies representing period 400 BC–AD 400.

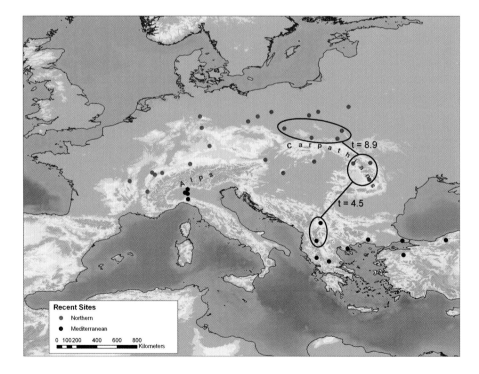

Figure 8: Location of oak chronologies representing the last 500 years with visualization of regions covered by composite chronologies. *T*-scores showing correlation between the composite chronologies are displayed.

Figure 9: Oak constructions in the prehistoric salt exploitation site in Figa, Romania. Preliminary results of radiocarbon dating are around 1000 BC.

From more than 3300 chronologies representing over 50 countries there are 34 chronologies representing the period 400 BC–AD 400 (as of 2008). Only nine of them concern European oak; all have been created for the Netherlands by Esther Jansma and her coworkers from the RING Foundation, Utrecht/Netherlands. The period before 500 BC is represented in the ITRDB resources by 24 chronologies. Only 11 oak chronologies from Europe are among them, all submitted by Andre Billamboz, Landesamt für Denkmalpflege, Hemmenhofen/Germany.

Tree-ring data from Europe are not published for many reasons. The most important of them is the activity of European commercial laboratories. Another reason is the practice of keeping data for the scientists' own needs so they may have an advantage in getting funds for new projects. "Publication" of tree-ring chronologies has meant publishing only information about the existence of the chronology, but not publication of the data itself. Tree-ring chronologies can have many "owners," and all of them must agree to make data available to the asking laboratory. As it turns out there are many ways to make the problem of tree-ring data policy even more complicated: one particular millennia-long chronology belongs partly to Institution A and is available in the ITRDB, while the second part of the chronology is the property of the Institution B and is not published. In such situations the only way to obtain data is to exchange data based on collaboration and contacts between laboratories. Therefore, I would like to express my special gratitude to colleagues who have given me access to their tree-ring data.

The existence of separate tree-ring patterns for Mediterranean and North European oak along with ill-defined borders of the North and South patterns (except the Alps), have forced the development of numerous new oak chronologies, especially for the transition area east of the Alps. Results of this research reveal that Romania is the most promising region, and not only because of the "Carpathian bridge." This country, rich with forest resources, has a long tradition of wood-working lasting to the present day[2] Transylvanian Bronze-Age salt mines with wooden shafts and facilities for salt exploitation (Figure 9) are a testimony of the highly developed wood-dependent culture in ancient times. Roman engineers and builders, whose normal passion was to use stone for building, were forced to use wood to build new *castra* in part of the province of Dacia between the Danube river and the mountains. For dendrochronology, the potential of the Balkan countries is almost untapped and it is there that the greatest chance exists to tie the Aegean tree-ring chronologies with the tree-ring master chronologies of northern and western Europe.

Conclusions

1. Comparison of two groups of European oak chronologies representing the southern and central-northern part of the continent confirms the existence of two separate tree-ring patterns. The Alps create a distinct boundary.

2. A potential and promising direction to find links between both parts of the continent is Pannonia because of the lack of natural boundaries. Serbia, Croatia, and part of Slovenia belong to the Mediterranean climate zone, while Hungary represents the northern pattern.

3. The Carpathian Mountains, in contrast to the Alps, seem to act as a bridge between North and South climate systems. Romania situated in this key region is a huge and almost untouched reservoir of historic timbers representing the last 4000 years. Additionally, deposits from the Danube and Suceava rivers contain subfossil oak logs for the potential creation of millennia-long chronologies to trace variability of different climatic influences.

Acknowledgments

Part of this research has been realized during my Fulbright Scholarship at Cornell University. I would

---

[2] In the program "Wooden Culture through Europe," organized by the Council of Europe in 2000-2002, Romanian carpenters taught their Scandinavian colleagues how to build a log cabin using traditional wood-working techniques!

like to thank all researchers who made their unpublished tree-ring data available: Andre Billamboz, Istvan Botar, Dieter Eckstein, Heinz Egger, Olafur Eggertson, Michael Friedrich, Michael Grabner, Carol Griggs, Andras Grynaeus, Jutta Hoffmann, Marek Krapiec, Peter Kuniholm, Catherine Lavier, Hans Hubert Leuschner, Hans Linderson, Burghart Schmidt, Willy Tegel. I wish also to thank Peter Kuniholm and Mary Jaye Bruce for reading this text and for their helpful comments.

References

Aas, G. 1990. Kreuzbarkeit und Unterscheidung von Stiel und Traubeneiche. *Allg. Forstzeitschr.* 45: 219–221.

Baillie, M.G.L. 1983. Is there a single British Isles oak tree-ring signal? In A. Aspinall and S.E. Warren (eds.), *Proceedings of the 22nd Symposium on Archaeometry*: 73–82. University of Bradford.

Baillie, M.G.L., and Pilcher, J.R., 1973. A simple crossdating program for tree-ring research. *Tree Ring Bulletin* 3: 7–14.

Boratynski, A. 1995. Podstawy systematyki debow. *Quercus*, Wydanie specjalne z okazji otwarcia wystawy „Dab" (Fundamentals of oak systematics. *Quercus*. A special volume on the occasion of "Oak" exhibition). Osrodek Kultury Lesnej, Goluchow: 27–37 (in Polish).

Cedro, A. 2007. Tree-ring chronologies of downy oak (*Quercus pubescens*), pedunculate oak (*Q. robur*) and sessile oak (*Q. Petraea*) in the Bielinek Nature Reserve: Comparison of the climatic determinants of tree-ring width. *Geochronometria* 26: 39–45.

Chambers, F.M., Mauquoy, D., Brain, S.A., Blaauw, M., and Daniell, J.R.G. 2007. Globally synchronous climate change 2800 years ago: Proxy data from peat in South America. *Earth and Planetary Science Letters* 253: 439–444.

Cherubini, P., Gartner, B.L., Tognetti, R., Bräker, O.U., Schoch, W., and Innes, J.L. 2003. Identification, measurement and interpretation of tree rings in woody species from mediterranean climates. *Biological Reviews* 78: 119–148.

van Geel, B., Bokovenko, N.A., Burova, N.D., Chugunov, K.V., Dergachev, V.A., Dirksen, V.G., Kulkova, M., Nagler, A., Parzinger, H., van der Plicht, J., Vasiliev, S.S., and Zaitseva, G.I. 2004. Climate change and the expansion of the Scythian culture after 850 BC: a hypothesis. *Journal of Archaeological Science* 31: 1735–1742.

Griggs, C.B., Kuniholm, P.I., and Newton, M.W. 2002. Lavagnone di Brescia in the Early Bronze Age: Dendrochronological Report. In R.C. de Marinis (ed.), *Studi sul'Abitato dell'Eta del Bronzo del Lavagnone Desenzano del Garda*: 19–34.

Griggs, C.B., DeGaetano, A.T., Kuniholm, P.I., and Newton, M.W. 2007. A regional reconstruction of May-June precipitation in the north Aegean from oak tree-rings, AD 1089-1989. *International Journal of Climatology* 27: 1075–1089.

Grosser, D., 1977. *Die Hölzer Mitteleuropas.* Berlin, Heidelberg, New York: Springer Verlag.

Haneca, K. 2005. Tree-ring analyses of European oak: Implementation and relevance in (pre-) historical research in Flanders. Ph.D. dissertation, Ghent University.

Hollstein, E. 1980. Mitteleuropäische Eichenchronologie. Trierer dendrochronologische Forschungen zur Archäologie und Kunstgeschichte. *Trier Grabungen und Forschungen* 11. Mainz am Rhein: Verlag Philipp von Zabern.

Huber, B., Holdheide, W., and Raack, K. 1941. Zur Frage der Unterscheidbarkeit des Holzes von Stiel- und Traubeneiche. *Holz als Roh- und Werkstoff* 4: 373–380.

Hughes, M.K., Kuniholm, P.I., Eischeid, J., Garfin, G., Griggs, C.B., and Latini, C. 2001. Aegean tree-ring signature years explained. *Tree-Ring Research* 57: 67–73.

Kremer, A., Dupouey, J.L., Deans, J.D., Cottrell, J., Csaikl, U., Finkeldey, U., Espinel, S., Jensen, J., Kleinschmit, J., Van Dam, B., Ducousso, A., Forrest, I., de Heredia, U.L., Lowe, A.J., Tutkova, M., Munro, R.C., Steinhoff, S., and Badeau, V. 2002. Leaf morphological differentiation between *Quercus robur* and *Quercus petraea* is stable across western European mixed oak stands. *Annals of Forest Science* 59: 777–787.

Kuniholm, P.I. 2000. Dendrochronologically Dated Ottoman Monuments. In U. Baram and L. Carroll (eds.), *A Historical Archaeology of the Ottoman Empire: Breaking New Ground*: 93–136. New York: Plenum Press.

Kuniholm, P.I. 2002. Dendrochronological Investigations at Herculaneum and Pompeii. In W.F. Jashemski and F.G. Meyer (eds.), *The Natural History of Pompeii*: 235–239. Cambridge University Press.

Kuniholm, P.I., Kromer, B., Tarter, S.L., and Griggs, C.B. 1998. An Early Bronze Age Settlement at Sozopol, near Burgas, Bulgaria. In M. Stefanovich, H. Todorova, and H. Hauptmann (eds.), *James Harvey Gaul: In Memoriam*. 399–409. Sofia: The James Harvey Gaul Foundation.

Kuniholm, P.I., and Striker, C.L. 1987. Dendrochronological Investigations in the Aegean and Neighboring Regions, 1983–1986. *Journal of Field Archaeology*, 14: 385–398.

Meiggs, R. 1982. *Trees and timber in the ancient Mediterranean world*. Oxford: Clarendon Press.

Martinelli, N. 2007. Dendrocronologia delle palafitte dell'area gardesana: situazione delle ricerche e prospettive. In *Contributi di archeologia in memoria di mario Mirabella Roberti, Atti del XVI Convegno Archeologico Benacense, Cavriana 15-16 ottobre 2005,* "Annali Benacensi," XIII–XIV, 103–120.

Newton, M.W., and Kuniholm, P.I. 2004. A Dendrochronological Framework for the Assyrian Colony Period in Asia Minor. *Türkiye Bilimler Akademisi Arkeoloji Dergisi*, 7: 165–176.

Pilcher, J.R., Baillie, M.G.L., Schmidt, B., Becker, B. 1984. A 7272-year tree-ring chronology for western Europe. *Nature* 312 (5990): 150–152.

Sander, C., Levanic, T. 1996. Comparison of t-values calculated in different dendrochronological programmes. *Dendrochronologia* 14: 269–272.

Schmidt, B., Gruhle, W. 2006. Globales Auftreten ähnlicher Wuchsmuster von Bäumen—Homogenitätsanalyse als neues Verfahren für die Dendrochronologie und Klimaforschung. *Germania* 84: 1–35.

Schweingruber, F.H. 1990. *Anatomie europäischer Hölzer— Anatomy of European Woods*. Swiss Federal Institute for Forest, Snow, and Landscape Research, Birmensdorf. Bern: Paul Haupt Verl.

Tutin, T.G., Heywood, V.H., Burges, N.A., Valentine, D.H., Walters, S.M., Webb, D.A. 1993. *Flora Europaea*. Cambridge University Press.

Ufnalski, K. 2006. Teleconnection of 23 modern chronologies of *Quercus robur* and *Q. petraea* from Poland. *Dendrobiology* 55: 51–56.

Wazny, T. 1994. Dendrochronology of Biskupin—absolute dating of the early iron-age settlement. *Bulletin of the Polish Academy of Sciences, Biological Sciences*, 42: 283–289.

Wazny, T., and Eckstein, D. 1991. The dendrochronological signal of oak (*Quercus* spp.) in Poland. *Dendrochronologia* 9: 181-191.

Article submitted March 2008

Dendrochronological Research at Rosslauf (Bressanone, Italy)

Maria Ivana Pezzo

Abstract: *The Sovrintendenza ai Beni Culturali of the Province of Bolzano carried out excavations which revealed in Rosslauf the presence of dwellings constructed by the autochthonous Rhaetian population. The building was destroyed by fire and, as a consequence of slow combustion, the wooden material was preserved. The excavations uncovered a number of barrel staves stacked at one side of a room of the basement, and pieces from a large vat. The particular type of combustion that occurred within the basement of the dwelling preserved the carbonized remains of the wooden objects in a manner such that it was possible to conduct dendrochronological analysis, determine the ligneous species, and construct a 117-year chronology.*

The town of Bressanone is situated in the center of the Isarco Valley, where the valley bottom broadens at the confluence of the Isarco and Rienza watercourses. The area's climate is particularly clement because the morenic plateau to the north protects it against cold northerly winds, while eastwardly sloping uplands give constant exposure to the sun, even during the winter months. These features combine with stable hydrological conditions to create an optimal location for human settlement. Indeed, there are traces evident of practically uninterrupted human habitation in Bressanone from the Neolithic Age (Relazioni–Beni archeologici 1997) until the present day.[1] Rosslauf extends along the right bank of the River Isarco, behind the historic center of Bressanone and between the present-day Via Brennero and Via Dante.

From April to June 2002, on commission by Dr. Umberto Tecchiati of the Sovrintendenza ai Beni Culturali of the Province of Bolzano, the Società Ricerche Archeologiche directed by Gianni Rizzi carried out excavations which revealed the presence beneath the medieval and Roman layers of dwellings constructed by the autochthonous Rhaetian population. The main building consisted of two storeys, with a basement created by a stone perimeter wall clad with timber and insulated with sandy gravel which functioned as an interspace. The floors of the structure were supported by cross-beams resting on alignments embedded in the walls. The technique used for the construction was that known as "Blochbau," or the log-cabin system with interlocking timbers—which demonstrates the considerable construction skill possessed the inhabitants at the time. The building was destroyed by fire, but the ceiling supported by wooden beams and composed of gravel material collapsed onto the objects in the basement, so that combustion was slow and took place in an anaerobic (i.e. oxygen-free) environment. As a consequence of this slow combustion, the wooden material was carbonized and perfectly preserved. The excavations uncovered a number of barrel staves stacked at one side of the room, and pieces of a large vat (Figure 1). Of outstanding interest was part of a six-spoke wheel. Pottery finds of note included two ribbed earthenware jars with curled brims, and pottery shards ascribable to Fritzens-Sanzeno culture (Perini's Rhaetian A), steep-sided cups decorated with pine-needle motifs, small ansated jugs decorated with "metopes" containing imprints delimited by vertical bands of three parallel lines. The pottery finds probably date to the first half of the 5th century BC.[2]

The dendrochronological analysis

The particular type of combustion that occurred within the basement of the dwelling preserved the carbonized remains of the wooden objects in a manner

---

[1] For a description of pre-Roman finds in the Bressanone area, see L. Allavena Silverio 2002: 444–451.

[2] This description of the excavation and the finds (still unpublished) has been made possible by information kindly provided by Gianni Rizzi and Alessandro Manincor of the Società Ricerche Archeologiche.

Figure 1: The reconstruction (by Gianni Rizzi) of the room interior shows the barrels, wheels and large containers along the wall, with the vat in the background. To be noted is the complete absence of nails and metal elements in the barrels and the vat, with wooden hoops instead being used to bind the staves together. The room was completely lined with planks and beams.

such that it was possible to conduct dendrochronological analysis, determine the ligneous species, and construct a 117-year chronology. Dendrochronology is the study of the annual growth of trees or timber sensitive to climate changes, and it enables the dating of wooden objects or their carbonized remains. Used as the reference curve for the Rosslauf material was the curve plotted for southern Bavaria by B. Becker,[3] which spans from AD 1985 to 546 BC. The arboreal species for the Becker curve is oak. Recent studies (Pindur 2001: 62–75) have successfully compared trees of different species which have grown in similar climatic conditions and in locations at a more than 250-kilometre distance from each other. Comparisons between the Rosslauf curve and Becker's curve yielded particularly significant and noteworthy data.

The charcoal analysed for the present study originated from the beams of the building, and from the barrels and the vat. Fifty samples were taken, and 30 were analysed. The samples' species were determined and, when possible, also the year of the last ring present. The species (Schweingruber 1990) were identified by Stefano Marconi and Maurizio Battisti of the Museo Civico of Rovereto. In some instances the species could be narrowed to two choices at best, due to small sample size.

---

[3] The curve has not yet been published. See Kuniholm 2002: 66.

Samples measured

ROSS-1 Sample from post 54, south-west corner (C3 502/6). Measurements: max. length 8cm; width 3.5cm; thickness 4.5cm. Rings: 15. Species: larch (*Larix decidua* Mill.). The sample exhibited an annular sequence with only a few measurable rings; it was consequently not included in the chronology.

ROSS-2 Sample from beam 5 (portion of post), ancient hut, north-east interior. Measurements: max. diam. 10cm; max. thickness 5cm. Rings: 30. Species: larch (*Larix decidua* Mill.) cf. spruce (*Picea abies* (L.) Karst.). The sample exhibited an annular sequence with regular growth and quite thick measurable rings; because this was a roundel, the pith was present but not the bark.

ROSS-3 Sample from beam 11 (portion), ancient hut, centre-south interior. Measurements: max. diam. 14cm; max. thickness 8cm. Rings: 12. Species: larch (*Larix decidua* Mill.) cf. spruce (*Picea abies* (L.) Karst.). The sample exhibited an annular sequence with few measurable rings but of notable thickness.

Barrels and vat

The samples below were used to construct a curve for each individual barrel; the year of the last ring present is given for the dated samples.

ROSS-20 Sample from barrel N. Measurements: max. length 3cm; max. width 3cm; max. thickness 5.6cm. Rings: 32. Species: larch (*Larix decidua* Mill.). The sample was taken from one of the staves of barrel N and exhibited an annular sequence of very thin measurable rings. The last ring present was dated to 505 BC.

ROSS-21 Sample from barrel N. Measurements: max. length 3.5cm; max. width 2cm; max. thickness 3.5cm. Rings: 38. Species: larch (*Larix decidua* Mill.). The sample was taken from one of the staves of barrel N and exhibited an annular sequence of rather thin measurable rings. The last ring present was dated to 485 BC.

ROSS-22 Sample from barrel N. Measurements: max. length 3cm; max. width 4cm; max. thickness 3.5cm. Rings: 22. Species: larch (*Larix decidua* Mill.). The sample was taken from one of the staves of barrel N and exhibited an annular sequence of rather thin measurable rings. The last ring present was dated to 490 BC.

ROSS-23 Sample from barrel N. Measurements: max. length 3cm; max. width 4cm; max. thickness 5cm. Rings: 21. Species: larch (*Larix decidua* Mill.). The sample was taken from one of the staves of barrel N and exhibited a perfectly preserved annular sequence of rather thin measurable rings. The last ring present was dated to 498 BC.

ROSS-24 Sample from barrel N. Measurements: max. length 1.6cm; max. width 1.6cm; max. thickness 3cm. Rings: 23. Species: larch (*Larix decidua* Mill.). The small sample was taken from one of the staves of barrel N and exhibited an annular sequence of very thin measurable rings. The last ring present was dated to 500 BC.

ROSS-25 Sample from barrel N. Measurements: max. length 2cm; max. width 1.2cm; max. thickness 1.7cm. Rings: 25. Species: larch (*Larix decidua* Mill.). The small sample was taken from one of the staves of barrel N and exhibited an annular sequence of very thin measurable rings. The last ring present was dated to 489 BC.

ROSS-26 Sample from barrel N. Measurements: max. length 2.3cm; max. width 1.1cm; max. thickness 2.6cm. Rings: 24. Species: larch (*Larix decidua* Mill.). The small sample was taken from one of the staves of barrel N and exhibited an annular sequence of very thin measurable rings. The last ring present was dated to 515 BC.

ROSS-27 Sample from ligneous fragments found on barrel N, north side. Measurements: max. length 1.6cm; max. width 1.6cm; max. thickness 3cm. Rings: 33. Species: spruce (*Picea abies* (L.) Karst.). The very small sample was taken from stave fragments on the exterior of barrel N, north side. The annular sequence was characterized by extremely thin measurable rings. The last ring present was dated to 492 BC.

ROSS-28 Sample from ligneous fragments found on barrel N, north side. Measurements: max. length 2.2cm; max. width 0.8cm; max. thickness 2.8cm. Rings: 32. Species: larch (*Larix decidua* Mill.). The very small sample was taken from stave fragments on the exterior of barrel N, north side. The annular sequence was characterized by extremely thin measurable rings. The last ring present was dated to 496 BC.

ROSS-29 Sample from ligneous fragments found on barrel N, north side. Measurements: max. length 2.6cm; max. width 0.8cm; max. thickness 2.5cm. Rings: 33. Species: larch (*Larix decidua* Mill.). The very small sample was taken from stave fragments on the exterior of barrel N, north side. The annular sequence was characterized by extremely thin measurable rings. The last ring present was dated to 519 BC.

ROSS-30 Sample from staves of barrel B 1. Measurements: max. length 4.2cm; max. width 2cm; max. thickness 4.3cm. Rings: 31. Species: larch (*Larix decidua* Mill.). The sample was taken from one of the staves of barrel B 1. The annular sequence was characterized by rather thin measurable rings.

ROSS-31 Sample from the inside bottom of barrel C. Measurements: max. length 3.9cm; max. width 2.2cm; max. thickness 4.3cm. Rings: 84. Species: larch (*Larix decidua* Mill.). The sample was taken from a stave bearing evident traces of working. To be noted are three parallel grooves cut at a distance of 0.7cm from each other on the outer surface. The stave was part of the outer cladding but was found inside the barrel at the bottom. The annular sequence was characterized by extremely thin measurable rings. The last ring was dated to 494 BC.

ROSS-37 Sample from ligneous fragments found at barrel N, north side. Measurements: max. length 6cm; max. width 1.6cm; max. thickness 3.5cm. Rings: 67. Species: larch (*Larix decidua* Mill.). The sample was taken from a stave of barrel N, north side. The annular sequence was characterized by extremely thin measurable rings. The last ring was dated to 539 BC.

ROSS-41 Sample from barrel B 1. Measurements: max. length 2.3cm; max. width 1.5cm; max. thickness 3.2cm. Rings: 50. Species: larch (*Larix decidua* Mill.). The small sample was taken from staves of barrel B 1. The annular sequence was characterized by extremely thin measurable rings.

ROSS-42 Sample from barrel B 1. Measurements: max. length 3.5cm; max. width 1.4cm; max. thickness 4cm. Rings: 33. Species: larch (*Larix decidua* Mill.). The sample was taken from staves of barrel B 1. The annular sequence was characterized by extremely thin measurable rings.

ROSS-43 Sample from barrel B 1. Measurements: max. length 2cm; max. width 0.9cm; max. thickness 1.3cm. Rings: 56. Species: larch (*Larix decidua* Mill.). The small sample was taken from staves of barrel B 1. The annular sequence was characterized by extremely thin measurable rings.

ROSS-44 Sample from vat D. Measurements: max. length 3.1cm; max. width 1.3cm; max. thickness 3cm. Rings: 59. Species: larch (*Larix decidua* Mill.). The small sample was taken from one of the staves of vat D. The annular sequence was characterized by extremely thin measurable rings.

ROSS-45 Sample from barrel L. Measurements: max. length 2.2cm; max. width 2.3cm; max. thickness 2.1cm. Rings: 58. Species: larch (*Larix decidua* Mill.). The small sample was taken from one of the staves of barrel L. The annular sequence was characterized by extremely thin measurable rings. The last ring present was dated to 485 BC.

ROSS-46 Sample from barrel L. Measurements: max. length 2.3cm; max. width 1.1cm; max. thickness 2.6cm. Rings: 37. Species: larch (*Larix decidua* Mill.). The very small sample was taken from one of the staves of barrel L. The annular sequence was characterized by extremely thin measurable rings. The last ring present was dated to 496 BC.

ROSS-49 Sample from barrel C. Measurements: max. length 4.4cm; max. width 1.3cm; max. thickness 5.6cm. Rings: 59. Species: larch (*Larix decidua* Mill.). The sample was taken from one of the staves of barrel C. The annular sequence was characterized by rather thin measurable rings. The last ring present was dated to 543 BC.

ROSS-50 Sample from barrel U. Measurements: max. length 4.7cm; max. width 1.5cm; max. thickness 2.8cm. Rings: 47. Species: larch (*Larix decidua* Mill.). The sample bore evident signs of working on the external surface and was taken from a stave of barrel U. The annular sequence was characterized by extremely thin measurable rings. The last ring present was dated to 490 BC.

ROSS-51 Sample from barrel U. Measurements: max. length 3.5cm; max. width 1.5cm; max. thickness 2.1cm. Rings: 43. Species: larch (*Larix decidua* Mill.). The sample bore evident signs of working on the external surface and was taken from a stave of barrel U. The annular sequence was characterized by extremely thin measurable rings.

ROSS-52 Sample from barrel U. Measurements: max. length 2cm; max. width 1.4cm; max. thickness 1.8cm. Rings: 33. Species: larch (*Larix decidua* Mill.). The small sample was taken from a stave of barrel U. The annular sequence was characterized by extremely thin measurable rings.

ROSS-53 Sample from barrel U. Measurements: max. length 2.3cm; max. width 1.2cm; max. thickness 2cm. Rings: 28. Species: larch (*Larix decidua* Mill.). The small sample was taken from a stave of barrel U. The annular sequence was characterized by extremely thin measurable rings.

ROSS-54 Sample from barrel U. Measurements: max. length 1.5cm; max. width 1.5cm; max. thickness 1.2cm. Rings: 38. Species: larch (*Larix decidua* Mill.). The very small sample and was taken from a stave of barrel U. The annular sequence was characterized by extremely thin measurable rings.

ROSS-55 Sample from barrel U. Measurements: max. length 3cm; max. width 3.4cm; max. thickness 2.8cm. Rings: 33. Species: larch (*Larix decidua* Mill.). The sample was taken from a stave of barrel U. The annular sequence was characterized by extremely thin measurable rings. The last ring present was dated to 508 BC.

ROSS-56 Sample from barrel U. Measurements: max. length 2.5cm; max. width 4cm; max. thickness 4cm. Rings: 34. Species: larch (*Larix decidua* Mill.). The sample was taken from one of the collapsed staves on the bottom of barrel U. The annular sequence was characterized by extremely thin measurable rings.

ROSS-57 Sample from barrel L. Measurements: max. length 5cm; max. width 1.7cm; max. thickness 5.5cm. Rings: 42. Species: larch (*Larix decidua* Mill.). The sample had a smooth surface and was taken from a thin stave of barrel L. The annular sequence was characterized by extremely thin measurable rings. The last ring present was dated to 509 BC.

Conclusions

Larch (*Larix decidua* Mill.) was the most frequently identified arboreal species, in the cases of both the barrel and vat staves, and the beams. Other samples proved to be spruce (*Picea abies* (L.) Karst.). Although the beam samples were larger in size than the fragments from the barrel staves, they yielded scant data because the rings, of considerable width, were of insufficient number for the last ring present to be determined. Planks from the outer part of the log were used to make the staves; this interpretation was confirmed by the fact that the rings were extremely narrow, almost parallel and without curvature, which indicates a position close to the exterior of the log.

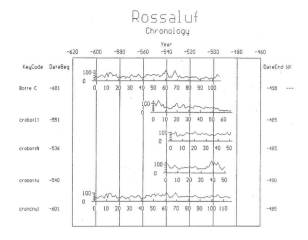

Figure 2: Graphs of the barrel chronologies with the final Rosslauf master chronology.

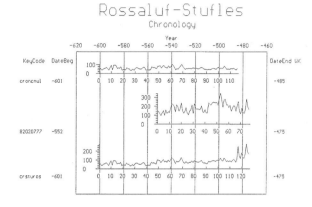

Figure 3: Graphs of the chronologies for Rosslauf and Stufles.

Analysis of the samples determined the *terminus post quem* of barrels C, L, N, and U. Barrel C, with only two samples (ROSS-31 and ROSS-49), furnished the longest chronology: 601–494 BC. Comparison of this chronology with Becker's chronology yielded good results which were subsequently confirmed by the data obtained by analysis of the other samples.

Barrel L, for which three samples were dated (ROSS-45, ROSS-46, ROSS-57), had a chronology covering the time span 551–485 BC. Barrel N had nine dated samples (ROSS-20, ROSS-21, ROSS-22, ROSS-23, ROSS-24, ROSS-25, ROSS-26, ROSS-27, ROSS-28), and the chronology obtained ranged from 536 to 485 BC.

Two samples from barrel U were dated (ROSS-50 and ROSS-55), and its chronology was from 540 to 490 BC.

The Rosslauf charcoal yielded a 117-year chronology (601–485 BC). Combination of this chronology with that for Stufles (STU-777) recently constructed at the Laboratorio di Dendrocronologia of Rovereto produced a 127-year chronology (601–475 BC).

Acknowledgments

This study has been made possible by the constant support and valuable advice of Dr. Umberto Tecchiati. I am grateful to Gianni Rizzi for the information about the excavation and the Rosslauf material. I am also indebted to Dr. Franco Finotti and Dr. Lorenzo Dal Rì. Also indispensable has been the assistance provided by the Malcolm and Carolyn Wiener Laboratory for Aegean and Near Eastern Dendrochronology, Cornell University, Ithaca, NY, in particular by Prof. Peter Ian Kuniholm, thanks to whose teaching my passion for dendrochronology was born.

References

AA.VV., 1997. Relazioni-Beni archeologici, Denkmalpflege in Südtirol 1991/1995 Tutela dei Beni Culturali in Alto Adige, Bolzano, 14.

Allavena Silverio L. 2002. Un piccolo edificio di epoca romana, in L. Dal Rì and S. Di Stefano (eds.), *Archeologia romana in Alto Adige, Bolzano*: 444–451.

Hillam J. 2003. *Dendrochronology*. London: English Heritage.

Kaennel, M., and Schweingruber, F.H. (eds.) 1995. *Multilingual Glossary of Dendrochronology*. Swiss Federal Institute for Forest, Snow and Landscape Research. Berne, Stuttgart, Vienna: Haupt.

Kuniholm, P.I. 2002. Archaeological dendrochronology, *Dendrochronologia*, 20(1): 63–68.

Pezzo, M.I. 2003a. Verkohlte Holzproben aus einem raetischen Haus. *Der Schlern*, 77, April, Heft 4, 4–9.

Pezzo, M.I. 2003b. Neue dendrochronologische Untersuchungen in Brixen/Stufels, *Der Schlern*, 77, Juli, Heft 7, 44–48.

Schweingruber, F.H. 1990. *Anatomie europaeische Hoelzer: ein Atlas zur Bestimmung europaeische Baum-, Strauch- und Zwergstrauchhoelzer*, Eidgenoess. Forschungsanst. für Wald Schnee u. Landschaft, Birmensdorf: Bern, Stuttgart.

Pindur, P. 2001. Dendrochronologische Untersuchungen an Zirben aus dem Waldgrenzbereich der Zillertaler Alpen, *Innsbrucker Geographische Gesellschaft, Innsbrucker Jahresbericht* 1999/2000, 62–75.

Article submitted April 2007

The Development of the Regional Oak Tree-ring Chronology from the Roman Sites in Celje (Slovenia) and Sisak (Croatia)

Aleksandar Durman, Andrej Gaspari, Tom Levanič, Matjaz Novšak

Abstract: *Prof. Peter Ian Kuniholm began extensively collecting samples for dendrochronological analysis from archaeological sites on the territory of former Yugoslavia (especially from Croatia, Slovenia, and Bosnia-Hercegovina) in 1986. It was then that scholars there first began thinking about this new method for absolute dating. This work of Kuniholm's stimulated, among other things, the founding of the Laboratory of Dendrochronology at the Slovenian Forestry Institute in Ljubljana, Slovenia. Notable among the numerous analyses carried out by the dendrochronology laboratory at Cornell are those of the wooden posts in the Kupa riverbed in the environs of today's Sisak in Croatia. Here in ancient times more than a hundred posts were driven at a place that the modern inhabitants call Kovnica (The Mint); these posts reinforced a harbor for unloading iron and the shore on which metalworking shops stood.*

Introduction

Colonia Flavia Siscia was for some time the capital city of the province of Pannonia Savia and the consolidated administrative center of iron production for the provinces of Illyricum and Pannonia (Figure 1). For several centuries a mint for coins (including golden ones) also existed here. Siscia handled nearly a million tons of iron, either administratively or literally, in its metalworking shops (more than four million tons of slag from Roman times remain in the administrative area of this ancient city). The city was the main supplier of iron to the Pannonian limes on the Danube– the most dangerous and most heavily fortified Roman border of all, stretching about 700km from Vindobona (Vienna) to Singidunum (Belgrade). Roman samples that the dendrochronology lab at Cornell has analyzed include numerous piles, the remains of the wooden construction of an aqueduct across the bed of the Kupa River, and the remains of a flat-bottomed barge, and several chronologies have emerged (Figures 2, 3). Piles from the Older Iron Age (Hallstatt) and the Later Iron Age (La Tène) have also been extracted from the riverbed and analyzed. Roman era pile-work was renewed every seven years, as Kuniholm has demonstrated.

Celje excavation

The rescue excavations in the area of the new motorway underpass along Mariborska Avenue in Celje, Slovenia, conducted by the Regional Office of the Institute for the Protection of the Cultural Heritage of Slovenia and the archaeological excavation company Arhej in 2003–2004, revealed the remains of a Late Roman wooden bridge as well as a short section of via publica Celeia-Poetovio. Located some 400m north of the city centre of Municipium Claudium Celeia (the province of Noricum), the bridge crossed one of the northern branches of the Savinja River. A burial ground from the 4th century AD was documented along the southern approach to the bridge, while a continuation of the road towards the northeast reaches out of the excavation area (Figures 4, 5). After a catastrophic flood the approximately 30m wide river branch turned into an oxbow, and the former channel was subsequently filled with silt and clay.

Construction

The distribution of 204 piles indicates the existence of at least four piers, while the abutments were swept away with a significant part of the banks (Figure 6). Piers measured approximately 6m in length, corre-

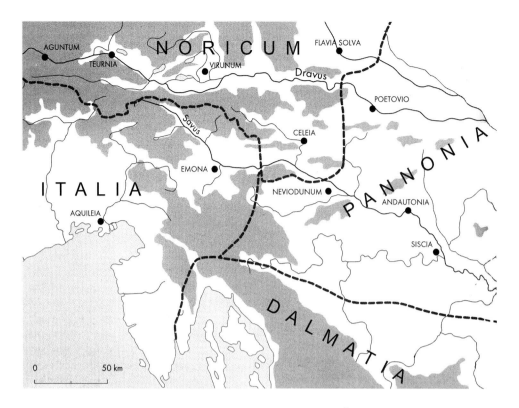

Figure 1: Map of the region in the Early Imperial period (after Šašel Kos, Scherrer 2002, 8,9).

Figure 2: Sisak. The site Kovnica ("the Mint") in the Kupa river-bed during low water in August 2003 (photo: Andrej Gaspari).

Figure 3: Sisak-Kovnica. Excavations 1985. The remains of a flat-bottomed barge of a Celtic-Roman type, ^{14}C dated to the mid of the 3rd century AD, were re-used to relieve the pressure of the water flow to wooden construction (photo: K. Kiš).

Figure 4: Celje. Mariborska Avenue. Northern part of the excavation area with the remains of a Late Celtic and Early Imperial sanctuary complex (late 1st century BC–ca. 50 AD) and residential-artisanal architecture from the 2nd century AD. The Late Roman bridge crossed the depression next to the present-day railway bridge (photo: Rafko Urankar, Jure Krajšek).

Figure 5: Celje. Mariborska Avenue. Excavation of the Roman bridge (piles are marked with sticks) was conducted under extremely difficult conditions with the ground water constantly rising. Simple stratigraphic sequence of the channel infill and the 19th century stone bridge are visible next to the excavator (photo: Srečko Foršt).

sponding to the recorded width of the road. The density of the piles increases from the first (14) to the fourth pier (70), which might represent the response to the main axis of the river current. Closely spaced piers with intervals measuring ca. 4m (I-II), 4.3m (II-III), and 2.5 (1.8m) (III-IV) make the construction more like a particularly strong causeway than a bridge. Piles, made almost exclusively from oak trees, were rammed vertically into a relatively flat river-bottom consisting of pebbles and sand (Figures 7, 8). The piles preserved were between 2.5 and 4.3m in height. Round and oval as well as shaped (square, rectangular, triangular, hexagonal, and octagonal) cross-sections from 10 x 12 to 30 x 30cm occur. The latter were obviously used to secure the basic ground-plan of each pier, as they are aligned in more or less straight lines parallel and perpendicular to the current. The distribution of the piles with round cross-sections, represented primarily within groups with shorter lengths (up to 2.5m), shows a more irregular pattern, perhaps a result of subsequent strengthening or/and renovation of the piers. All preserved piles have a pointed tip from 0.3 to 1.5m in height and bear no traces of iron shoes. On the other side no abrasion marks, which would point to the original river-bottom, were observed on the upper parts of the piles.

Dendrochronological analysis, methodology, and results

Three chronologies based on samples collected during archaeological excavations of the Late Roman bridge in Celje have been built. Tree species were identified and tree-ring widths were measured on a LINTAB measuring table to a 0.01mm precision. Raw data were further processed, error checked and crossdated in the PAST-4 program. Out of 170 samples collected from the piles of the Roman bridge, 111 samples had more than 40 tree rings, which was the lower limit for further processing of the samples. Out of these 111 samples we could successfully crossdate 39 samples (35% of samples that meet minimum standard for a synchronization or 18% of all samples received in the laboratory). Altogether we built three chronologies with different sample depths and lengths. The longest, a 316 year-long chronology, is called CE-8002cr; the other two—CE-8001cr and CE-8004cr—are 139 years long (Table 1). Dating of Slovene chronologies was done with ^{14}C, and the precision of this dating was increased by using wiggle-matching in the OxCal v.3.10 program. The calibrated period based on two radiocarbon dated samples of the same pile after wiggle-matching was 280–329 Cal AD; for further processing of our chronologies we decided to take the average value for this period, 305 Cal AD.

Keycode	Length	Min. Value	Max. Value	Mean Value	Std. Dev.	Mean Sens.	From*	To*
CE-8001cr	139	46.0	373.8	128.6	66.70	0.132	167 AD	305 AD
CE-8002cr	316	67.8	337.9	149.0	53.49	0.134	7 AD	322 AD
CE-8004cr	139	46.0	364.6	137.0	61.82	0.138	167 AD	305 AD

Table 1: Basic data regarding chronologies from the Roman bridge in Celje, Slovenia. All analyzed samples belong to oak (*Quercus* sp.) * No absolute date; dated with ^{14}C.

Keycode	Length	Min. Value	Max. Value	Mean Value	Std. Dev.	Mean Sens.	From*	To*
Sisak-999	253	43.0	192.0	103.40	28.60	0.159	14 AD	266 AD
Sisak-666	147	27.0	185.0	96.74	31.72	0.196	120 AD	266 AD
Sisak-000	240	47.0	249.0	99.20	26.85	0.195	39 BC	200 AD

Table 2: Basic data regarding chronologies from Sisak, Croatia. All analyzed samples belong to oak (*Quercus* sp.) * No absolute date; dated with ^{14}C.

		CE-8001cr	CE-8002cr	CE-8004cr	Sisak-999	Sisak-666
CE-8002cr	t_{BP} / Glk%	14.30 / 85.30				
CE-8004cr	t_{BP} / Glk%	36.90 / 88.50	15.00 / 89.60			
Sisak-999	t_{BP} / Glk%	4.38 / 68.00	3.99 / 58.70	4.64 / 72.00		
Sisak-666	t_{BP} / Glk%	**5.12** / **69.50**	**5.97** / **69.70**	**5.34** / **69.50**	16.10 / 81.00	
Sisak-000	t_{BP} / Glk%	3.09 / 67.60	2.21 / 57.00	2.98 / 76.50	7.30 / 70.30	4.47 / 62.30

Bold Significant matches between chronologies of the Roman bridge from Celje, Slovenia and those of the Roman platform from Sisak, Croatia

▨ Comparison between chronologies from the same site

Table 3: Comparison of all oak site chronologies from Celje and Sisak.

		CE-8001cr	CE-8002cr	CE-8004cr
HOLLSTEIN4	t_{BP} / Glk%	0.55 / 60.80	0.81 / 52.10	0.55 / 52.20
LOTHRINGEN3	t_{BP} / Glk%	1.37 / 55.00	1.88 / 54.70	1.61 / 50.70
HOH-ROM1	t_{BP} / Glk%	-0.26 / 56.10	1.53 / 54.70	0.17 / 54.70

Table 4: Comparison of Celje site chronologies with German Roman chronologies for oak (source of German oak chronologies: ITRDB NOAA).

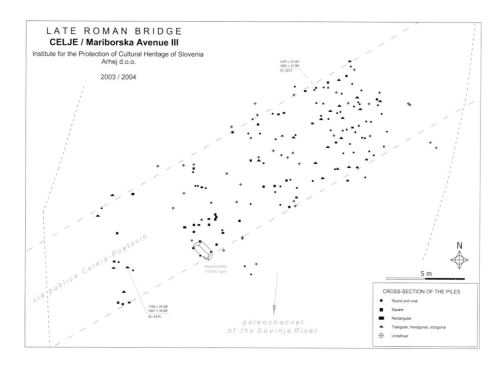

Figure 6: Celje Mariborska Avenue. Ground plan of the bridge (produced by: Andrej Gaspari).

Figure 7: Celje. Mariborska Avenue. Closely spaced piles in the third pier (photo: Andrej Gaspari).

Figure 8: Celje. Mariborska Avenue. An example of a pile with square cross-section and long centered point (photo: Srečko Firšt)

Comparison of oak chronologies from Celje and Sisak with already existing oak chronologies from the Roman site in the River Kupa (Sisak, Croatia), built by Kuniholm and Durman (Table 2) gives the best match between CE-8002cr and Sisak-666 (tBP= 5.97; Glk%= 69.70; overlap=253 years; Figure 9). Based on a high similarity between the Celje and Sisak chronologies we tried to join both chronologies into a single regional chronology; however the final result did not meet our expectations. Comparison between oak chronologies of the Roman bridge from Celje and three Roman oak chronologies from Germany (data obtained from the ITRDB database) gave no significant relationships. This indicates that oak chronologies north of the Alps are very hard if not impossible to use for synchronizing and dating oak chronologies south of the Alps. However, material from Sisak was significantly crossdated with material from Slovenia. This gives us an enormous potential to crossdate oak material from different parts of the Balkan Peninsula and also encourages further research in this region.

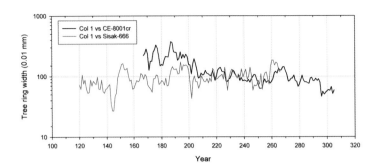

Figure 9: Comparison of two short and most similar chronologies CE-8001cr and Sisak-666 (tBP = 5.12; Glk%= 69.50; overlap=100 years).

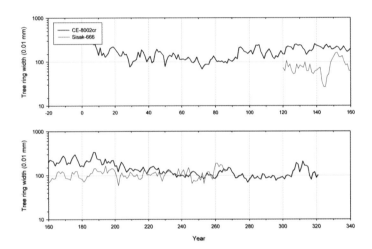

Figure 10: Comparison of two chronologies with the highest crossmatch CE-8002cr and Sisak-666 (tBP= 5.97; Glk%= 69.70; overlap 139 years).

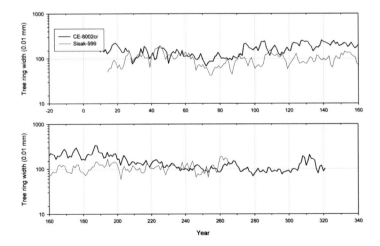

Figure 11: Comparison of two site chronologies with second best crossmatch CE-8002cr and Sisak-999 (tBP= 3.99; Glk%= 58.70; overlap 253 years)

References

Durman, A. 2002. Iron resources and production for Roman frontier in Pannonia. *Historical Metallurgy* 36: 24–32.

Gaspari, A., Erič, M., and Šmalcelj, M. 2006. A Roman river barge from Sisak (Siscia), Croatia. In L. Blue, F. Hocker, and A. Englert (eds.), *Connected by the Sea: Proceedings of the Tenth International Symposium on Boat and Ship Archaeology, Roskilde, Denmark, 2003*: 283-289. Oxford: Oxbow Books.

Šašel Kos, M., and Scherrer, P. (eds.) 2002. The Autonomous Towns of Noricum and Pannonia. *Situla* 42, Ljubljana: Narodni muzej Slovenije.

Article submitted March 2007

Dendroclimatology in the Near East and Eastern Mediterranean Region

Ramzi Touchan and Malcolm K. Hughes

Abstract: *Dendroclimatology is the study of the interrelationship between tree-ring growth and climate variability and its use to reveal patterns of climate variation in the past. Indirect evidence of climatic variability such as long time series of tree-ring growth measurements spanning several centuries serve as proxy records of past conditions. We introduce some key aspects of the history of dendrochronology in the region as necessary background to our discussion of the development of dendroclimatology there. The current status of dendroclimatology in the Near East is summarized briefly, and its potential application periods of interest to archaeologists discussed.*

Introduction

Given that dendrochronology, and its applications to both archaeology and climatology, had its origins in the semi-arid Southwest of the United States, it is not surprising that there is a long history of tree-ring research in the similarly dry eastern Mediterranean region. Such mid-latitude regions, especially in their mountains, contain many local environments that are likely to yield highly variable tree rings showing strong common patterns of growth, the ideal raw material for dendrochronology. As in other regions, most of the early history of dendrochronology in the Near East concerns the quest to check for the existence of crossdating, followed by investigations of the biology of tree-ring formation, in turn providing the basis for the extraction of information on past climates from tree rings.

Early dendrochronology in the Near East

In the early 1920s, Andrew Ellicott Douglass of the University of Arizona was approaching the problem of dating archaeological materials in the Southwest, eventually meeting with major success in 1929 when he was able to link series from living trees with those from southwestern archaeological timbers. As early as 1924, and probably earlier, Douglass was also interested in establishing an Egyptian and Near East tree-ring project (see, for example, January 7, 1924, letter to Douglass from J. George Allen, Secretary of the Oriental Institute, University of Chicago, concerning coffin planks), something he developed further in the 1930s in the form of a small feasibility study (Bryant Bannister, personal communication, February 2007). His aim was to demonstrate crossdating in Egyptian tree-ring samples, particularly from coffins held in American museums. He met with mixed success in gaining access to these samples, and was prevented from completing the feasibility study by the onset of World War II (see Bannister letter to P.I. Kuniholm on May 13, 1988).

Bryant Bannister—a former student of Emil Haury and research assistant of A.E. Douglass—inspired by Douglass's interest in Near East dendrochronology, was the first dendrochronologist to attempt systematic tree-ring dating of near eastern archaeological sites (Bannister 1970). He collected and analyzed tree-ring specimens from an eighth century BC tomb in Gordion, Turkey, carried out preliminary examination of wood samples from Egyptian coffins and pyramids, and collected and crossdated samples of living Cedar of Lebanon (*Cedrus libani*) in Lebanon.

Bannister laid out the necessary conditions for the successful application of the methods of archaeological dendrochronology in this region. First, he wrote that "the prehistoric inhabitants of the region...must have made fairly extensive use of wood, preferably for construction purposes, and this wood or its derivative charcoal must still be preserved so that both cellular and ring structure remain evident." His second basic requirement was that "...the specimens crossdate with each other. Crossdating is fundamental to the use of the tree-ring dating method and can be defined as the precise identification in different wood samples of similar and synchronous ring patterns formed by the succession of variable growth ring thicknesses...." Bannister (1970) further emphasized the absolute necessity for replication of tree-ring patterns by collection of as many samples as possible for each structure under investigation, and also by establishment of collections at multiple locations. He pointed out the need to study the environmental causes of the tree-ring variability on which the dating is based, and discussed the idea of developing long dating chronologies for regions in the eastern Mediterranean, specifically Turkey.

Recent times: building chronologies and networks, understanding records

Bryant Bannister (1970) described his Gordion work in these terms: "A beginning has now been made in the Near East, and it is hoped that the results obtained through tree-ring analysis will ultimately contribute to a better understanding of the ancient world." A few years later Peter Ian Kuniholm took up this challenge, embarking on a remarkable campaign of chronology building and application, not only using materials from an extraordinary range of archaeological and architectural sources, but also exploring the basis for dendrochronology in the region on a broad scale by building chronologies of living trees from several dozen forests reaching from southern Italy to eastern Turkey. He has made great progress toward a multimillennial dating chronology capable of providing independent calendar dates for the felling of timber used in ancient times (Kuniholm and Striker 1987; Kuniholm 1990; 1994). He pioneered the connection of extreme tree-ring events (dry/wet) to historical events and was able to add an additional dimension to the study of environmental history and its relationship to human society and population dynamics. His achievements are celebrated in this volume.

At about the same time as Kuniholm was commencing his journey back in tree-ring time, several Israeli scientists used tree-ring series to date archaeological sites (Lev-Yadun et al. 1984; Liphschitz and Biger 1988; Lev-Yadun 1992; and Liphschitz 1992) in the southern part of the region. Other reports concerned the basic biology of tree-ring formation in such regions, examining the cycle of cambial activity in a given species of tree to determine if a ring represents one year of growth (Liphschitz et al. 1981, 1984, and 1985; Liphschitz and Lev-Yadun 1986; and Fahn et al. 1986).

Dendroclimatology in the Near East and eastern Mediterranean

Where there is crossdating, there is the possibility of detecting a climate signal in collections of tree-ring data. Gassner and Christiansen-Weniger (1942) demonstrated that tree growth is significantly influenced by precipitation in parts of the north-central Anatolian plateau. Their work was followed, some decades later, by a number of dendroclimatic studies in various parts of the region (Liphschitz and Waisel 1967; Shanan et al. 1967; Chalabi and Serre-Bachet 1981; Chalabi and Martini 1989; Munaut 1982; and Parsapajouh et al. 1986) and by an acceleration of publication in the late 1990s and early 2000s (Touchan et al. 1999; Touchan and Hughes 1999; Akkemik 2000; Hughes et al. 2001; D'Arrigo and Cullen 2001; Touchan et al. 2003; Akkemik and Aras 2005; Akkemik et al. 2005; Touchan et al. 2005 a & b; Griggs et al. 2007, and Touchan et al. 2007). Touchan et al. (1999) developed the first dendroclimatic reconstruction in the Near East for southern Jordan, a 396-year-long reconstruction of October–May precipitation based on two chronologies of *Juniperus phoenicia*. They showed that the longest reconstructed drought, as defined by consecutive years below a threshold of 80% of the 1946–1995 mean observed October–May precipitation, lasted four years. The longest drought recorded in the 1946–1995 instrumental data lasted three years. Based on the results of the reconstruction, seven droughts of three or more years have occurred during the past 400 years. The chronology from southern Jordan covers 527 years (1469–1995). To ensure the reliability of the reconstruction, analysis was limited to the period (1600–1995) for which the chronology is well replicated. Touchan and Hughes (1999) were able to build the first tree-ring chronologies in northern Jordan from *Pinus halepensis* and *Quercus aegilops* and one chronology of *Pinus halepensis* from Carmel Mountain in Israel. They demonstrated a significant correlation between the northern site chronologies (northern Jordan and northwestern Israel), but no correlation between the northern site chronologies and the sites from southern Jordan.

More recently in Turkey, Akkemik (2000) investigated the response of a *Pinus pinea* tree-ring chronology from the Istanbul region to temperature and precipitation. In 2001, two dendroclimatological research manuscripts were published (Hughes et al. 2001 and D'Arrigo and Cullen 2001). D'Arrigo and Cullen (2001) presented the first 350-year dendroclimatic reconstruction of February–August precipitation for central Turkey (Sivas), also using Peter Kuniholm's materials. A subsequent reconstruction was developed by Akkemik et al. (2005) for a March–June precipitation season from oak trees in the western Black Sea region of Turkey. They found that during the past four centuries drought events in this region persisted for no more than two years, and extreme dry and wet events occurred generally in one-year intervals. These various tree ring studies in Turkey suggest that runs of dry years generally extend for one or two years and rarely for more than three years. *Pinus nigra* tree rings were used by Akkemik and Aras (2005) to reconstruct April–August precipitation (1689–1994) for the southern part of central Turkey, whereas oak tree rings from a number of locations were used by Griggs et al. (2006 and 2007, respectively) to reconstruct May–June precipitation in the North Aegean for the period AD 1089–1989.

Hughes et al. (2001) worked with the network of living tree chronologies developed by Kuniholm. By demonstrating that crossdating over large distances in Greece and Turkey has a clear climatological basis, with signature years consistently being associated with specific, persistent atmospheric circulation anomalies (Hughes, Kuniholm et al. 2001), a basis was established for the building of a specifically dendroclimatic network in the region. Such a network would differ from one collected for the purpose of checking for crossdating, or anchoring a long chronology to the present, in a number of important respects. In particular, species, sites, and trees must be selected with the dendroclimatic purpose in mind; strong replication of numbers of trees per site and number of sites per region is required, as is a geographic distribution of chronologies capable of capturing geographic variations of the relevant aspects of climate.

Touchan and Hughes set out to develop precisely such a regional network of chronologies specifically developed for use as proxy records of climate and to publish climatic reconstructions based on them (Figure 1). At the time of writing this network includes 36 chronologies from 42 sites, the longest of which dates back to AD 1017. Touchan et al. (2003) used tree-ring data from living and dead trees in southwestern Turkey to reconstruct spring (May–June) precipitation several centuries back in time. Their reconstructions show clear evidence of multiyear to decadal variations in spring precipitation. The longest period of spring drought was four years (1476–1479). Spring droughts of three years in length have occurred only from 1700 to the present. The longest reconstructed wet periods were found during the 16th and 17th centuries. They also found that spring drought (versus wetness) is connected with warm (cool) conditions and southwesterly (continental) circulation over the eastern Mediterranean. Touchan et al. (2005a) were the first to develop a Standardized Precipitation Index (drought index) reconstruction for most of Turkey and some adjoining regions from tree-rings for the period AD 1251–1998. This index has become a widely accepted drought index that provides an objective method for determining drought conditions at multiple time scales. Their study provided important regional information concerning hydroclimatic variability in southwestern and south-central Turkey. Touchan et al. (2005b) continued their investigations of the relationships between large scale atmospheric circulation and regional reconstructed May–August precipitation for the eastern Mediterranean region (Turkey, Syria, Lebanon, Cyprus, and Greece). They reported the first large-scale systematic dendroclimatic sampling for this region from different species. Six May–August precipitation reconstructions ranging in length from 115 to 600 years were reported. No long term trends were seen during the last few centuries. Large-scale atmospheric circulation influences on regional May–August precipitation were identified over the eastern Mediterranean region. For example, precipitation in this season is driven by anomalous below (versus above) normal pressure at all atmospheric levels and by convection (subsidence) and small pressure gradients at sea level.

Tree rings and other records of past climate and its impacts

A pioneering comparison of tree-ring-derived data and independent reconstructions of large-scale sea-level pressure (SLP) and surface air temperature based on primarily documentary evidence showed that large-scale climatic patterns associated with precipitation and tree-ring growth in this region have been substantially stable for the last 237 years (Touchan et al. 2005a). This, to some degree, provides mutual validation for both sources of information on past climate, and also suggests that the reconstructions have some physical validity.

Investigations linking dendroclimatic reconstructions to other proxy records of past climate, specifically to historical documents, are reported elsewhere (D'Arrigo and Cullen 2001; Akkemik 2005; Touchan et al. 2005 a & b; and Touchan et al. 2007). For example, all five studies identified the year 1660 as a dry summer while Purgstall (1983) reported that catastrophic fires and famine in Anatolia occurred in the same year. Touchan et al. (2007) reported that running a 70-year moving average on the longest reconstruction showed that the years AD 1518–1587 comprised the most humid 70-year period in the AD 1097–2000 May–June precipitation reconstruction. Pfister et al. (1999) reported that the 16th century marks a significant shift in the climate in Europe, a period of generally cooler conditions that had profound effects

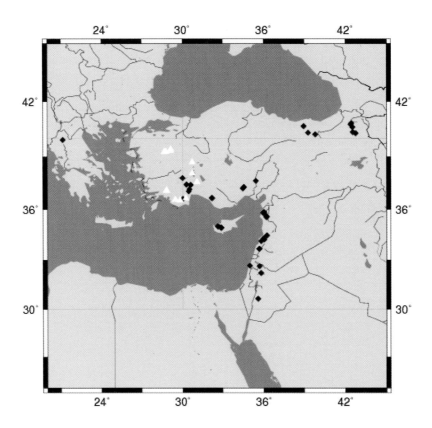

Figure 1: Locations of 47 developed (♦) and 12 in process (▲) tree-ring sites (Touchan and Hughes network from 1995-2004).

on environment and society. Touchan et al. (2007) also indicated that the period 1591–1660 represents the second driest in their reconstruction, corresponding with the findings of Kuniholm (1990) and Griswold (1977, 1983, 1993). These last authors reported that the late 16th and early 17th centuries in Anatolia were characterized as an unstable period politically, socially, and climatically, which caused a large-scale change in land use and large sudden fluctuations in urban populations.

What do these developments mean for the study of climate in ancient times?

Crossdating on large geographic scales provides evidence of a climate imprint on tree rings. So, the distribution of years in which trees over large regions showed common variation may be used to infer something about climate conditions in those years, using the analysis of such events in conjunction with observed climate conditions in the modern period as a guide. If timbers for multiple trees are present at a number of locations contemporaneously, such inferences may be defended on the basis of the patterns we see today. Given that robust common features in regional networks of tree rings most often reflect some kind of extreme climatic condition (a shift of rain-bearing storm tracks north or south of the region of interest, resulting in drought, for example) even such limited inferences may be of value to those seeking to interpret the interactions of the natural environment and ancient societies.

How much further is it possible to go in the direction of quantitative reconstructions of climate hundreds and thousands of years ago, analogous to those that have been made for recent centuries? If sufficient material is available to construct crossdated site or regional chronologies from species and places known to be good recorders from modern studies, and if there is reason to believe that the relevant ecological conditions of tree growth of the ancient material resemble those of the living trees, it may be possible to glean useful information on interannual to multidecadal variability in the past. This applies even if the archaeological tree-ring materials are not anchored to the present—that is, if they are "floating"—but they must be strongly dated against one another. It would be possible to quantify some of the uncertainties in such a "floating dendroclimatic reconstruction" by mimicking it using subsets of the modern data set and checking the results against the instrumental data.

Why do this? The climate that people experience holds much more meaning than averages or frequency distributions of conditions that are expected in the various seasons of the year. Experience on various time scales—e.g., memories of last year's dry spring, or the perception that it never rains in May because it hasn't for several years—may well affect behavior. Climate variation from year to year is not simply random. It has structure in time, in part determined by very large scale patterns in the circulation of the atmosphere and the oceans. The year-to-year or decade-to-decade sequence of seasonal climate conditions may well be a particularly informative feature of climate in the context of archaeology, and it is precisely this that quantitative reconstructions could reveal.

Acknowledgments

We are deeply indebted to Professor Bryant Bannister for drawing our attention to several important sources, and for his encouragement and inspiration. We also thank Professor Jeffrey S. Dean for his suggestions and providing us with a file that contains Douglass and Bannister communications since 1924 concerning tree rings in Egyptian materials. Our work reported here has been supported by the following grant from the US National Science Foundation Earth System History Program (Grant No. 0075956) and part from EU funded FP6 project 017008 "Millennium."

References

Akkemik, Ü. 2000. Dendroclimatology of umbrella pine (*Pinus pinea* L.) in Istanbul, Turkey. *Tree-Ring Bulletin* 56: 17–23.

Akkemik, Ü., and Aras, A. 2005. Reconstructed April–August precipitation (1689–1994) for southern part of central Turkey by using *Pinus nigra* tree rings. *International Journal of Climatology* 25(4): 537–548.

Akkemik, Ü., Dağdeviren, N., and Aras, A. 2005. A preliminary reconstruction (A.D. 1635–2000) of spring precipitation using oak tree rings in the western Black Sea region of Turkey. *International Journal of Biometeorology* 49: 297–302.

Bannister, B. 1970. Dendrochronology in the Near East: current research and future potentialities. *Proceedings of the seventh International Congress of Anthropological and Ethnological Sciences, Moscow, 1964* 5: 336–340.

Chalabi, M.N., and Serre-Bachet. F. 1981. Analyse dendroclimatologique de deux stations syriennes de *Quercus cerris* ssp *pseudocerris*. *Ecologia Mediterranea* 7: 3–21.

Chalabi, M.N., and Martini, G. 1989. Etude dendroclimatologique de l'*Abies cilicica* (Ant. et Ky) de Syre. *Research Journal of Aleppo University* 12: 57–89.

D'Arrigo, R., and Cullen, H. 2001. A 350-yr reconstruction of Turkish precipitation. *Dendrochronologia* 19: 169–177.

Fahn, A., Werker, E., and Baas, P. 1986. *Wood Antomy and Identification of Trees and Shrubs from Israel and Adjacent Regions*. Jerusalem: The Israel Academy of Sciences and Humanities.

Gassner, G., and Christiansen-Weniger, F. 1942. Dendroklimatologische Untersuchungen über die Jahresringentwicklung der Kiefern in Anatolien. *Nova Acta Leopoldina*: Abhandlung der Kaiserlich Leopoldinisch-Carolinisch deutschen Akademie der Naturforscher NF, Band 12, Nr 80.

Griggs, C.B. 2006. A tale of two: Reconstructing climate from tree-rings of the North Aegean, AD 1089–1989, and Late Pleistocene to present: Dendrochronology in upstate New York. Ph.D. Dissertation, Cornell University, Ithaca, NY.

Griggs, C.B., DeGaetano, A.T., Kuniholm, P.I., and Newton, M.W. 2007. A regional high-frequency reconstruction of May-June precipitation in the north Aegean from oak tree rings, A.D. 1089–1989. *International Journal of Climatology* 27 (8): 1075–89. DOI 10:1002/joc.1459.

Griswold, W.J. 1977. The Little Ice Age: its effect on Ottoman history, 1585–1625. Paper presented at the Middle East Studies Association Meeting, New York.

Griswold, W.J. 1983. *The great Anatolian rebellion 1000–1020/1591–1611*. Berlin: Klaus Schwartz Verlag.

Griswold, W.J. 1993. Climatic change: a possible factor in social unrest of seventeenth century Anatolia. In H.W. Lowry and D.J. Quataert (eds.), *Humanist and Scholar: Essays in honor of Andreas Tietze*: 37–57. Istanbul: Isis Press.

Hughes, M.K., Kuniholm, P.I., Eischeid, J.K., Garfin, G., Griggs, C.B., and Latini, C. 2001. Aegean tree-ring signature years explained. *Tree-Ring Research* 57: 67–73.

Kuniholm, P.I, and Striker, C.L. 1987. Dendrochronological investigations in the Aegean and neighboring regions, 1983-1986. *Journal of Field Archaeology* 14: 385–398.

Kuniholm, P.I. 1996. Long tree-ring chronologies for the Eastern Mediterranean. In Ş. Demirci, A.M. Özer, and G.D. Summers (eds.), *Archaeometry '94: The Proceedings of the 29th International Symposium on Archaeometry*: 401-409. Ankara: TÜBİTAK.

Kuniholm, P.I. 1990. The archaeological record: evidence and non-evidence for climate change. In S.K. Runcorn and J.-C. Pecker (eds.), The Earth's Climate and Variability of the Sun Over Recent Millennia, *Phil. Trans. R. Soc. Lond. A*: 645–655.

Livezey R.E, and Smith, T.M. 1999. Considerations for use of the Barnett and Preisendorfer (1987) algorithm for canonical correlation analysis of climate variations. *Journal of Climate* 12: 303–305.

Lev-Yadun, S., Liphschitz, N., and Waisel, Y. 1984. Ring analysis of *Cedrus libani* logs from the roof of El-Aqsa mosque. *Eretz-Israel* 17: 92–96, 4*–5* (Hebrew and English summary).

Lev-Yadun, S. 1992. The origin of the cedar beams from Al-Aqsa Mosque: Botanical, historical and archaeological evidence. *Levant XXIV*: 201–208.

Liphschitz, N., and Waisel, Y. 1967. Dendrochronological studies in Israel. *Quercus boisseri* of Mt. Meron region. *La-Yaaran* 17: 78–91, 111–115 (in Hebrew).

Liphschitz, N., Lev-Yadun, S., and Waisel, Y. 1981. The annual rhythm of activity of the lateral meristems (cambium and phellogen) in *Cupressus sempervirens* L. *Annals of Botany* 47: 485–496.

Liphschitz, N., Lev-Yadun, S., Rosen, E., and Waisel, Y. 1984. The annual rhythm of activity of the lateral meristems (cambium and phellogen) in *Pinus halepensis* Mill. and *Pinus pinea* L. *IAWA Bulletin* n.s., 5(4): 263–274.

Liphschitz, N., Lev-Yadun, S., and Y. Waisel. 1985. The annual rhythm of activity of the lateral meristems (cambium and phellogen) in *Pistacia lentiscus* L. *IAWA Bulletin* n.s., 6(3): 239–244.

Liphschitz, N., and Lev-Yadun, S. 1986. Cambial activity of evergreen and seasonal dimorphics around the Mediterranean. *IAWA Bulletin* n.s., 7(2): 145–153.

Liphschitz, N., and Bieger, G. 1988. Identification of timber and dating of historical buildings in Eretz-Israel (Palestine) by dendrohistorical and dendrochronological methods. *Dendrochronologia* 6: 99–110.

Liphschitz, N. 1992. Building in Israel throughout the Ages: One cause for the destruction of the Cedar forest of the Near East. *GeoJournal* 27.4: 345–352.

Munaut, A.V. 1982. The Mediterranean area. In M.K. Hughes, M.P. Kelly, J.R. Pilcher, V.C. LaMarche (eds.), *Climate from Tree Rings*: 151-154. Cambridge University Press.

Nash, S. 1999. *Time, Trees, and Prehistory: Tree-Ring Dating and the Development of North American Archaeology, 1914–1950*. Salt Lake City: University of Utah Press.

Parsapajouh, D., Bräker, O.U., and Schär, E. 1986. Étude dendroclimatologique du bois de *Taxus baccata* du nord de l'Iran. *Schweizerische Zeitschrift für Forstwesen* 137: 853–868.

Pfister, C., Brázdil, R. 1999. Climatic variability in sixteenth-century Europe and its social dimension: a synthesis. *Climatic Change* 43: 5–53.

Shanan, L., Evenari, M., and Tadmor, N.H. 1967. Rainfall patterns in the central Negev Desert. *Israel Exploration Journal* 17: 163–184.

Touchan, R., and Hughes, M.K. 1999. Dendrochronology in Jordan. *Journal of Arid Environments* 42: 291–303.

Touchan, R., Meko, D.M., and Hughes, M.K. 1999. A 396-year reconstruction of precipitation in Southern Jordan. *Journal of the American Water Resources Asscociation* 35: 45–55.

Touchan, R., Garfin, G.M., Meko, D.M., Funkhouser, G., Erkan, N., Hughes, M.K., and Wallin, B.S. 2003. Preliminary reconstructions of spring precipitation in southwestern Turkey from tree-ring width. *International Journal of Climatology* 23: 157–171.

Touchan, R., Xoplaki, E., Funkhouser, G., Luterbacher, J., Hughes, M.K., Erkan, N., Akkemik, Ü., and Stephan, J. 2005a. Reconstructions of Spring/Summer Precipitation for the Eastern Mediterranean from Tree-Ring Widths and its Connection to Large-Scale Atmospheric Circulation. *Climate Dynamics* 25: 75–98.

Touchan, R., Funkhouser, G., Hughes, M.K., and Erkan, N. 2005b. Standardized precipitation indices reconstructed from tree-ring width for the Turkish region. *Climatic Change* 72 (3): 339–353.

Touchan, R., Akkemik, Ü., Hughes, M.K., and Erkan, N. 2007. May-June Precipitation of southwestern Anatolia, Turkey during the last 900 years from tree rings. *Quaternary Research* 68(2): 196–202.

Article submitted March 2007

A 924-year Regional Oak Tree-ring Chronology for North Central Turkey

Carol B. Griggs, Peter I. Kuniholm, Maryanne W. Newton,
Jennifer D. Watkins, and Sturt W. Manning

Abstract: *We present a 924-year (AD 1081 to AD 2004) regional oak tree-ring chronology for north central Turkey built from 154 samples from three forest sites and nine historic buildings. We also provide an estimate for the sapwood count of the oaks across northern Turkey—22 sapwood rings, plus 9 or minus 6 at 1 standard deviation—and show that the age of the tree is a significant factor in determining those values.*

Introduction

We have combined 154 oak samples from three forest sites and nine historic buildings into a regional oak tree-ring chronology for north central Turkey (east of Istanbul to Ordu, approximately 29–38°E, 40–42°N) (Figure 1). These samples comprise a small percentage of the hundreds of wood samples collected by the Malcolm and Carolyn Wiener Laboratory for Aegean and Near Eastern Dendrochronology for dating the construction and building phases of historic buildings across Turkey for the 2nd millennium AD, but they contain the best regional oak growth patterns. Their ring-width patterns securely crossdate across the area, indicating that their parent trees responded similarly to similar climate parameters, and that the origins of the historic building timbers were most likely within north central Turkey. This work also indicates that regionalism can affect Anatolian dendrochronology, especially in particular time periods. The robust north central Turkey regional oak chronology dates from AD 1081 to 2004 (all dates hereafter are AD = CE). Most of the samples' measurements were also used in a larger regional chronology to reconstruct May-June precipitation for the north Aegean from 1089 to 1989 (Griggs et al. 2007).

Methods

Basic dendrochronological methods were used to collect, prepare, and measure the samples for tree-ring dating (Fritts 1976; Baillie 1982; Cook and Kairiukstis 1989). The time series of each sample's measurements was usually detrended with a negative exponential curve to remove the normal growth trend of wider to narrower rings spanning the life of a tree. A floating point average or cubic spline curve was used when there was significant low-frequency variability in a sample that was not apparent in the ring patterns of other samples from that site.

Three statistical tests plus visual comparisons were used in "crossdating" to determine the secure relative placement of each site's samples to each other, then to crossdate the site chronologies to build a securely-dated regional chronology. One test is an adaptation of the Student's t-test, calculated from the correlation coefficients between two samples in running standardized short sequences over time, which is a standard measure for the dendrochronological crossdating of oaks (Baillie and Pilcher 1973). The use of the standardized short sequences allows for low-frequency variations that are slightly different in two time series, variations that lower the value of the strictly linear correlation coefficient. The second test is the Pearson correlation coefficient (r score), used as a measure of quality in crossdating (Cook and Kairiukstis 1989). The third statistical test is the trend coefficient, a percentage of how many times the ring-widths of two sequences both increase or decrease from year-to-year along the period of their overlap.

Several decades of tree-ring research around the world have shown that while the tree-ring series' values are not independent—as is needed to use the established levels of significance in the Student's t-score or in correlation coefficients—the utility of these tests to locate possible crossdates is viable when the numer-

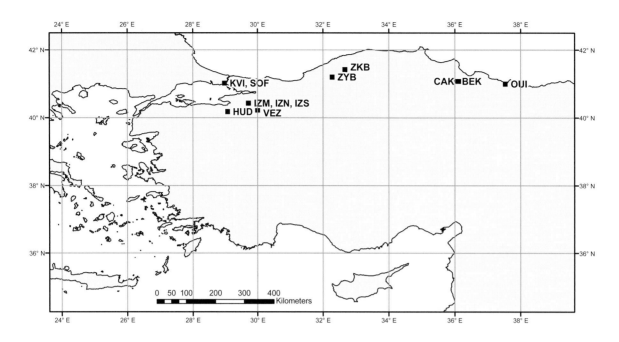

Figure 1: Map of the Anatolian region indicating the sampling locations of the included forest and building sites. The three-letter codes are listed in Table 1. (Map courtesy of P. W. Brewer)

ical values are treated with caution and the indicated date is considered in the context of the site and a complete assessment of the wood. For the present research, based on our experience of Aegean-region oaks, we set the requirement of at least one of the following to consider a crossdate: ≥ 4.00 for the Student's t-score, or ≥ 0.350 for the correlation coefficient. However, the real decision on the security level of the crossdating is based not on the statistical values but on a visual comparison. Very close matches over short sequences within the overlap can give highly significant, but false, statistical values. It takes a close visual examination of the patterns to determine whether there is a secure match at any possible position that is indicated by the statistics. If a visual match is good, but the numbers are not highly significant at one placement, a sequence is compared with all securely-dated samples and chronologies at that place in time before it gets assigned to that date, and a less-than-significant relationship is duly noted for future inspection with any new samples. Table 1 contains a list of the averaged descriptive statistics for each historic and forest site chronology.

For an estimate of the number of sapwood rings normal for the oaks from this region and climate, we used the sapwood counts of the 71 included samples that have bark or a waney edge. The average, median, standard deviation, and skewness of the sapwood count were calculated. The data set was also broken down into smaller groups and examined for variations over time, space, and also according to the lifespans of the individual trees. The median sapwood ring count for the 71 samples is 22; one standard deviation is plus 9, or minus 6, rings.

To calculate the most probable building date for any site chronology that does not contain samples with bark or a waney edge, we look at the dates of the beginning of sapwood rings in the samples, the lengths of the samples, and the end date of the chronology. For chronologies with just one sample with sapwood, we generally take its beginning sapwood date and add add 21 to that value to get the most likely building date (21 = 22 minus 1, since the beginning date is the first year of sapwood). If the chronology's end date is significantly later than the end date of the sample with sapwood, we use the chronology's end date as the sapwood boundary date. For chronologies with a range of beginning sapwood dates, we generally take the middle number and add 21 (+9/-6) with the same consideration of sample length and the end date of the chronology.

Discussion and Results

The North Turkey regional chronology is built of 16 site chronologies listed in Table 1 and illustrated in Figure 2A. The chronology itself and its sample depth over time are shown in Figure 2B. The analyses used

Lat °N	Long °E	Region or City	Site Name	Site ID	Subdivision or *Quercus* species	Chronology Begins	Ends	Length (years)	No of samples	Average overlap	Average *t*-scores	Average correlation
41.02	28.97	İstanbul	Beyoğlu, Karaköy Vapur İskelesi	KVI	Black Sea	1602	1852	251	4	108	4.54	0.368
			Hagia Sophia	SOFB	Bannister	1394	1581	188	6	73	4.22	0.445
				SOFT	Türbe	1356	1615	260	6	129	4.64	0.374
				SOFN	NW Buttress	1188	1332	145	6	72	3.55	0.391
40.18	29.07	Bursa	I. Murat Hüdavendigâr Camii	HUD		1111	1384	274	6	161	5.05	0.370
40.43	29.72	Bilecik	İznik, Elbeyli, Mara Camii	IZM		1398	1554	157	6	66	5.17	0.543
		Bilecik	İznik, Nilüfer İmareti	IZN		1136	1375	240	6	76	4.28	0.413
		Bilecik	İznik, Hagia Sophia	IZS		1081	1241	161	4	105	4.22	0.382*
40.23	30.00	Bilecik	Vezirhan	VEZ		1526	1657	132	4	85	4.09	0.383
41.20	32.28	Zonguldak	Yenice, Bakraz Forest	ZYBH	*Q hartwissiana*	1624	1984	361	11	168	4.78	0.354*
41.20	32.28	Zonguldak	Yenice, Bakraz Forest	ZYBP	*Q petraea*	1712	2004	293	8	171	7.37	0.476
41.42	32.67	Zonguldak	Karabük, Büyükdüz Forest	ZKB	*Q petraea*	1699	2004	306	20	159	6.25	0.445*
41.08	36.05	Samsun	Kavak, Çakallı Forest	ÇAK	*Quercus* spp.	1835	1989	155	17	80	4.50	0.442
41.08	36.15	Samsun	Kavak, Bekdemirköy, Camii	BEK1	1st Story	1089	1484	396	26	167	6.89	0.450
				BEK2	2nd Story	1483	1875	393	14	135	5.06	0.384
41.00	37.53	Ordu	Ünye, İkizce, Eski Camii	OUI		1266	1493	228	10	106	4.28	0.370*

*Because of individual sample's growth idiosyncrasies beyond the normal reduction in ring-width over time, a floating point average was used to detrend one or more of the included samples.

Table 1: The historic and forest site locations plus the dates of the chronologies, their lengths and number of samples, the average Student's *t*-score values and the average correlation coefficient values (Pearson's *r*) between the samples.

for crossdating and constructing the regional chronology raise several other important points about the oaks, their sapwood, and the history of the region.

The oaks and their sources in the study region

The area covered by this chronology consists of two well-defined ecoregions, the Black Sea and the Marmara Transitional ecoregions (Atalay, 2002: map). Both are at the southern limit of the ranges of the included oak species (Davis 1982; Griggs 2006: chapter 1). The forest sites are located within the Black Sea ecoregion, generally right on the boundary of two subregions in the humid temperate coastal region and the semi-arid cooler (altitude 500+m) region on the southern, rainshadow, side of the Black Sea coastal mountain range.

The source of the wood used in each of the included historic buildings is not known (except for the few that sit within present-day forests), but oaks currently grow across both ecoregions. The western boundary of the included area is not the geographical boundary set by the locations of the historic buildings; rather it was determined by examining the correlations between each site chronology and other forest and historic building chronologies from across Turkey, Greece, and surrounding countries, to establish the source of the wood. For most of the study region, the transportation of oak timbers from very far was not needed due to the abundance of oak forests, but for the more populated region of northwestern Turkey, particularly Istanbul, importation was probably necessary, especially during peaks of building construction. Ease and cost effectiveness of transport makes it likely that the source of those timbers would have been from around either the Sea of Marmara or the Black Sea. The samples from the included 17th to 19th century buildings crossdate better with the Black Sea forest oaks than with the oaks from two forests in the Marmara Transitional ecoregion which we have sampled, the Devecikonağı forest south of the Sea of Marmara (DEV), and the Belgrade forest to the northwest of Istanbul (IBO). Neither of the latter two forests is included in the North Turkey chronology since their chronologies crossdate better with site chronologies to the west and with other chronologies that do not belong in the north central Turkey group. This indicates a significant amount of influence from the Sea of Marmara and the Aegean Sea at both sites, and also from the low elevation at the Belgrade Forest. Their chronologies exhibit a regional pattern unique to the different climatic-environmental zone in the Sea of Marmara Transitional ecoregion. The chronologies do crossdate significantly with the regional north central Turkey chronology presented here (DEV at $t =$ 6.14, $r = 0.38$, $n = 232$; IBO at $t = 4.09$, $r = 0.27$, $n = 217$) and similarly with the north Greece oak chronology (in preparation) enabling us to build a wider north Aegean regional oak chronology (Griggs et al. 2007).

For the earlier historic buildings, each included site chronology has better crossdating with site chronologies from other sites in the eastern half of the region, rather than with sites around the Aegean Sea, which confirms that the timber source was most likely from the Black Sea ecoregion. The western boundary of the study region is thus loosely defined as the division between the continental climate regime found in north central Turkey and the region with a significant maritime influence on the climate regime, the Marmara Transitional region.

We cannot identify oak at the species level from the wood samples alone (Huber and von Jazewitsch 1956: 29; W. Schoch and D. Eckstein, pers. comm. 2002), but the included samples are all of the subgeneric class *Quercus* Oersted (Schweingruber 1990: 403; Griggs 2006: chapter 1). Again, the secure crossdatings among samples and among site chronologies indicate a similar climate response; and from a regional perspective, the species of oak is not important (Kelly et al. 1989; 2002).

Most of the oak species of forest samples were identified by the foresters. Our analysis of those species-specific oaks indicates that there is more variance between individual trees' ring growth for certain species. In particular, there is a significant difference in the correlations between the *Quercus petraea* samples (average $t = 7.37$, average $r = 0.476$) and the *Q. hartwissiana* samples ($t = 4.78$, $r = 0.354$). The correlation between each species' chronology and other site chronologies, however, is at the same level of significance (*Q. petraea* and *Q. hartwissiana* chronologies correlate with the ÇAK chronology at 4.45 and 4.10, and with the BEK 2nd story chronology at 4.65 and 5.07 t-scores, respectively). We see the same differences in the statistics between samples in our historic site chronologies, with samples from some sites correlating very well, and others with more difficult crossdating. These variations could be site-specific or period-specific rather than species-specific, but our experience with the forest data sets indicates the latter.

The additional oak samples collected since 1986, a large number of which contain their complete sapwood, have also given us the opportunity to examine the average number and range of the sapwood ring count in oaks for various smaller areas within the eastern Mediterranean region. The values of 25.6 ± 9 noted in Kuniholm and Striker (1987) are based on samples from the whole region and limited to the longer ring sequences collected for dendrochronological analysis at that time. For northern Turkey, the analysis of

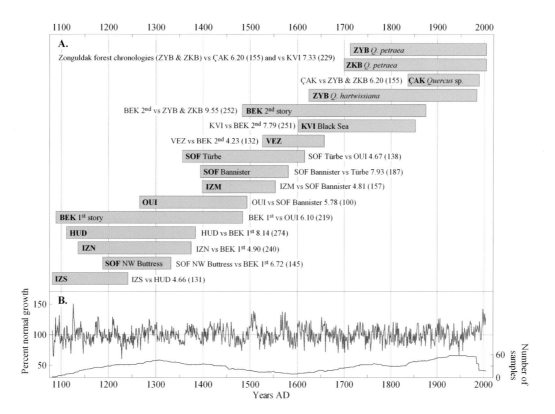

Figure 2: A. Graphic representation of the components of the north central Turkey oak chronology, with averaged best t-score crossdate values among each of the elements indicated. **B.** The regional chronology, calculated from the average of the samples' de-trended measurements, and its sample depth over time.

the sapwood count in the 71 included samples with bark or waney edge give a sapwood estimate of 22 +9/-6 rings (the median value plus or minus one standard deviation; Figure 3A). Over the eastern Mediterranean region, there is no significant geographic variation in the sapwood count, or any variation over the 900 years covered by the oaks, but the variation in sapwood count due to the lifespan of the individual trees is significant, especially in this region (Figure 3B). The trees with fewer than 125 rings generally have fewer than 22 sapwood rings and the older trees have a higher sapwood count, with those above 150 rings having an average of around 26 sapwood rings. The sapwood estimates are used to evaluate the most probable building date and possible range of dates for each historic building chronology that contains no samples with bark or waney edge

Site chronologies from forests and buildings in the Black Sea ecoregion

All three forest sites are within the Black Sea ecoregion (Figure 1). The chronologies of these sites were used to date the historical oak chronologies securely to calendar years. In particular, the three chronologies of two sites near Zonguldak have long chronologies beginning from 1629 to 1712, and an excellent regional climate signal due to the sites' high altitudes. The samples' species were recorded during their collection, enabling us to divide the samples from the Yenice site (ZYB) into two chronologies, one of *Quercus hartwissiana* and one of *Quercus petraea*. The Karabuk (ZKB) samples are all of the latter species. The fourth forest chronology, from the Çakallı samples (ÇAK) from Kavak in the Samsun province, has a shorter sequence, but the timber source is on the eastern boundary of the study region. The ÇAK samples are sections from logs that had been stolen, but then found and confiscated by the Kavak Forestry District along with the tractors on which they were being transported. The exact (and local) source of these samples was determined when the logs' stumps were subsequently identified; this confirmed the assumption that transport on tractors was a good indicator of wood from a nearby forest. The excellent crossdating of the ÇAK chronology with the ZYB and ZKB chronologies clearly indicates that all the chronologies contain the same climate signal.

Of the historic chronologies from the Black Sea ecoregion, the combined chronology from two building phases of a mosque in Bekdemir (BEK), a village south of the provincial capital of Samsun and near

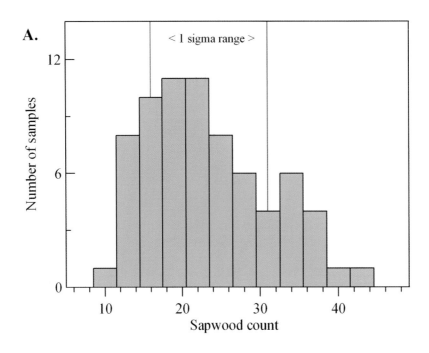

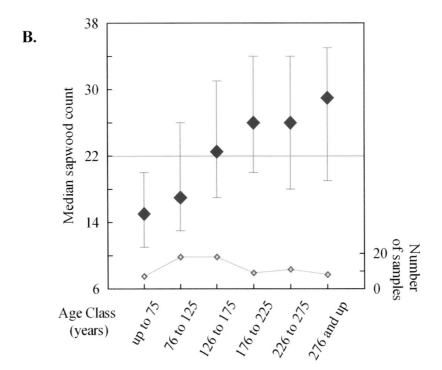

Figure 3: A. Sapwood counts in the 71 samples with bark or waney edge range between 11 and 42 rings, with the 1σ range 15 to 31. **B.** The relationship of the median sapwood count to the age of the oaks from this region. The number of samples in each group is indicated at the bottom of the graph. The average number of sapwood rings in the trees with more than 150 rings is the same as was reported in Kuniholm and Striker (1987).

Kavak, covers an extraordinary 787 years from 1089 to 1875. The building itself is a small two-story mosque with no detailed history (Kuniholm 2000: 97–99). The early chronology, built of the wood from the first story, dates from 1089-1484 with no sapwood, and the second story chronology from 1483-1875, a bark date, with the two phase chronologies overlapping by only two years (Figure 2). The first story's samples contain no sapwood, so its building date was most probably ca. 1506 or later, with the second story added in 1875–76. The second story chronology securely crossdates with both the Zonguldak forest (ZYB, ZKB) chronologies, and the first story chronology crossdates securely with all other North Turkey building chronologies from the same period.

The easternmost site chronology is composed of the ring-widths of 10 samples cut from timbers forming the walls of the İkizce Eski Camii (OUI), a historic mosque located directly in the Black Sea humid temperate region in Ordu (Kuniholm 2000: 115–116). The chronology included here ends in 1493, a modification of the chronology reported in 2000 due only to the removal of one sample with less-than-secure crossdating that had extended the 1493 date to 1522. The remaining ten samples' end dates have a long range from the late 14th to the late 15th centuries, and some timbers are clearly re-used (see Kuniholm 2000: 116). No sapwood is present, and the most likely final building date of the mosque is 1515 or later. The chronology crossdates securely with the first floor chronology from Bekdemir, the site closest to Ordu, and also with two of the chronologies from the Hagia Sophia in Istanbul, an indication that the wood for those Hagia Sophia building phases had been transported from somewhere closer to Ordu than Istanbul.

Site chronologies from buildings in the Marmara Transitional ecoregion

The historic buildings from this region include three collections of oak timbers from buildings in the town of İznik (ancient Nicaea), east of both İznik Lake and the Sea of Marmara (Figure 1 and Table 1). The timbers used in the buildings at İznik could have been harvested from either ecoregion. Four samples from the church of Haghia Sophia are combined into the IZS tree-ring chronology. Three are from tie-beams in the drum of the prothesis, and the other sample is from a tie-beam in the drum of the diakonikon. The dendrochronological results indicate that the church was built in a single construction, with the last ring of the chronology dating to 1241 with a possible sapwood boundary at 1232. The sapwood ring count of 22 indicates ca. 1253 or later as the most probable building date.

Six samples were cored from the complex known as the Nilüfer İmareti (IZN) in İznik that today houses the city archaeological museum. Four samples came from tie-beams in the east porch, one from a lintel in the lower window in the exterior central main bay, and one from a lintel in the lower window in the west wall of the exterior south wing. The inscriptional date of 1388–1389 is corroborated by the outer preserved ring of this chronology, 1375, with the sapwood beginning at 1352, 1362, and 1363 on three samples—which, allowing for the typical sapwood value of 22 years, would yield approximate dates around 1374, 1384 or 1385. A hypothetical actual cutting date of 1387 (likely from the inscription, given observed Aegean practice of cutting timber the winter before use) is well within the +9/-6 range.

The third tree-ring collection at İznik was from a ruined unnamed mosque in Elbeyli Kasabası, at Mara Mevkisi (IZM). The last preserved ring in the IZM chronology, made up of six timbers from stretchers and header beams, dates securely to 1554; the lower average of 17 sapwood rings indicates that the 1554 is very likely the felling year, and the felling date is probably within two years of the building date. The samples with sapwood rings are all relatively short (n=52, 91, and 112), and their sapwood begins in 1543, 1538, and 1537, respectively. The lower average value of 20 sapwood rings for shorter samples (Figure 3) and the dendrochronological date gives a most likely building date of 1556; 1554 is well within the range of possible building dates.

Southeast of İznik, in the area of Bilecik, on the old silk caravan road from İznik-Lefke to Söğüt and Eskişehir are the remains of a large building called the Vezirhan (the name of both the building and the town) (Kuniholm 2000: 111–112). This building was a caravansaray; of the four samples from this building included in the north central Turkey chronology, three are from framing timbers in the niches on the south end of the building, and one is from a stretcher in the southeast wall. The Vezirhan chronology (VEZ) puts the building's construction date at 1657 with a possible waney edge in one sample. Sapwood begins in three samples at 1622, 1625, and 1632: these dates support the "possible waney edge" date of 1657 as the most probable building date: the reported 1659/1660 date referred to by Kâtip Çelebi is well within the possible construction range (Kuniholm 2000: 111—who notes that this date is however a year or two after he died!).

The I. Murat Hüdavendigâr Mosque is located in Bursa, just southeast of the Sea of Marmara. Samples were collected from tie beams throughout the mosque, but for many samples the ring count was not enough to crossdate securely. The HUD chronology is composed of only six long sequences and ends in 1384 with an additional incomplete ring for 1385 in a sample with a possible waney edge. Sapwood begins in 1353, 1365, and 1369, supporting the 1385 building date. The 1385 date also corroborates with the historic record (*vakfiye*, or foundation document) of the lengthy construction period for this building, which was authorized for building in 1366 but was not completed until 1385 when it was officially dedicated (Gabriel 1958; Kuniholm 2000: 122).

The wood from the the Karaköy Vapur İskelesi (Kuniholm 2000: 99–100) was excavated during an archaeological project in Beyoğlu in the fall of 1995 by the Istanbul Archaeological Museum. The 15 oak samples from this site were broken into three groups due to sample size and their crossdating. Only four longer-lived samples are contained in the KVI Black Sea chronology and included in the north central Turkey chronology; the other samples crossdate better with forest chronologies from west of Istanbul (and these samples—not included here—enable a dating from bark of the wood cutting at 1858 as in Kuniholm 2000: 100). The KVI Black Sea chronology's outer ring dates to 1853, with sapwood beginning at 1842 in the one sample that ends in 1853.

The original construction of the Hagia Sophia in Istanbul is dated to around 532–527, but there have been many additions and alterations to the original structure. We have collected more than 174 samples from multiple contexts in this building: only three components appear to have been built with wood that came from the study region (during the last 900 years). The chronology of one of the components, the northwest buttress (SOF NW Buttress), consists of eight timbers with an end date of 1332 but with no sapwood, thus a construction date of 1354 or later. Another chronology (SOF Bannister) includes eight samples from bannisters on both sides of the south ramp leading from the ground floor to the gallery, ending in 1580. There was no sapwood in any of the eight samples, thus the date of the addition of the bannisters would be 1602 or later. The third chronology (SOF Türbe) consists of seven samples from the east porch of the türbe of Mustafa I and Ibrahim, and ends in 1615, again with no sapwood. The most probable construction date of the türbe porch is 1637 or later: Mustafa I died in 1639. The excellent crossdating of these Hagia Sophia chronologies with the Bekdemir first story chronology and the OUI site chronology implies that the timbers used in the construction of these two Istanbul buildings was transported from the southern Black Sea coast.

The north central Turkey regional chronology is made up of these 16 site chronologies (Figure 2). We have been able to build a separate regional chronology for the Aegean basin that covers about the same calendar dates: a similar report is in preparation. The measurements of the samples included (and ancillary information) in the north central Turkey regional chronology are available at the International Tree-Ring Data Bank website and from the Malcolm and Carolyn Wiener Laboratory for Aegean and Near Eastern Dendrochronology website: http://dendro.cornell.edu/.

Conclusion

The Malcolm and Carolyn Wiener Laboratory for Aegean and Near Eastern Dendrochronology at Cornell is working to build long regional tree-ring chronologies for various species in the eastern Mediterranean region. The north central Turkey regional oak chronology, AD 1081–2004, provides a key basic resource for archaeometric work in the secure dating of built structures, timber and wood objects from northern Turkey, and for environmental analyses (e.g. Griggs et al. 2007), and also a resource to test and identify at least the approximate origin of oak wood employed across a wide area of northern to northwestern Anatolia. It forms the first element of a wider supra-regional oak chronology in development. The present north central Turkey chronology provides a more complete history of environment and material culture for this large and historically important region for most of the 2nd millennium AD.

Acknowledgments

The Malcolm and Carolyn Wiener Laboratory for Aegean and Near Eastern Dendrochronology gratefully acknowledges support for this research from the National Science Foundation, the Malcolm H. Wiener Foundation, individual Patrons of the Aegean Dendrochronology project, and Cornell University. We thank the many past Laboratory members and students who worked on wood included in the chronology presented here.

References

Atalay, İ. 2002. *Türkiye'nin Ekolojik Bölgeleri (Ecoregions of Turkey)*. T.C. Orman Bakanlığı Yayınları No: 163. İzmir: Meta Basımevi.

Baillie, M.G.L. 1982. *Tree-ring Dating and Archaeology*. Chicago: University of Chicago Press.

Baillie, M.G.L., and Pilcher, J.R. 1973. A simple cross-dating program for tree-ring research. *Tree-Ring Bulletin* 33: 7–14.

Cook, E.R., and Kairiukstis, L.A. (eds.). 1989. *Methods of Dendrochronology: applications in the environmental sciences*. Dordrecht: Kluwer Academic Publishers.

Fritts, H.C. 1976. *Tree Rings and Climate*. London and New York: Academic Press.

Gabriel, A. 1958. *Une capitale turque: Brousse, Bursa.* Paris: E. De Boccard.

Griggs, C.B. 2006. Tale of Two: Reconstructing Climate from Tree-Rings of the North Aegean, AD 1089-1989, and Late Pleistocene to Present Dendrochronology in Upstate New York. Dissertation, Cornell University.

Griggs, C.B., DeGaetano, A.T., Kuniholm, P.I., and Newton, M.W. 2007. A regional high-frequency reconstruction of May-June precipitation in the north Aegean from oak tree rings, A.D. 1089-1989. *International Journal of Climatology* 27 (8): 1075–89. DOI 10:1002/joc.1459.

Goodwin, G. 1971. *A History of Ottoman Architecture.* Baltimore: Johns Hopkins Press.

Huber, B., and von Jazewitsch, W. 1956. Tree-ring studies of the forestry-botany Institutes of Tharandt and Munich. *Tree-Ring Bulletin* 21: 28–30.

Kelly, P.M., Leuschner, H.H., Briffa, K.R., and Harris, I.C. 2002. The climatic interpretation of pan-European signature years in oak ring-width series. *The Holocene* 12: 689–694.

Kelly, P.M., Munro, M.A.R., Hughes, M.K., and Goodess, C.M. 1989. Climate and signature in West European oaks. *Nature* 340: 57–60.

Kuniholm, P.I. 2000. Dendrochronologically Dated Ottoman Monuments. In U. Baram and L. Carroll (eds.), *A Historical Archaeology of the Ottoman Empire: Breaking New Ground*: 93–136. New York: Kluwer Academic/Plenum Publishers.

Kuniholm, P.I., and Striker, C.L. 1987. Dendrochronological investigations in the Aegean and neighboring regions, 1983–1986. *Journal of Field Archaeology* 14: 385–398.

Kuniholm, P.I., and Striker, C.L. 1983. Dendrochronological investigations in the Aegean and neighboring regions, 1977–1982. *Journal of Field Archaeology* 10: 411–420.

Schweingruber, F.H. 1990. *Anatomie europäischer Hölzer.* Berne: Verlag Paul Haupt.

Article submitted September 2008

Dendrochronology on *Pinus nigra* in the Taygetos Mountains, Southern Peloponnisos

Robert Brandes

Abstract: *Widespread fir die-back and a decline of the timberline in some high mountains of the Hellenic peninsula appear to have a causal connection with climatic dryness in the second half of the 20th century. But, is it correct to see this as a consequence of climate change? Dendrochronological methods are applied to find out about the "normality" of drought years and the fluctuations of precipitation in the past. A Pinus nigra chronology from southern Peloponnisos provides some interesting answers, including new perspectives on* Abies *species and timberline dynamics in Greece.*

Archaeology and geography: meeting via tree rings

Some archaeologists and ecologists share a common interest: the information about the past stored in tree rings. It was living trees, growing in the mountains of southern Greece, which made me meet Peter Ian Kuniholm and his team of students in July 2000. A joint field trip into the Taygetos mountains to sample *Pinus nigra* built the foundation for a cooperation between the Cornell Aegean Dendrochronology Project and my doctoral thesis project at the Geographical Institute of the University Erlangen-Nürnberg, Germany. The following contribution provides an ecological perspective on Aegean tree rings connected with questions of possible climate change impacts.

The background: timberlines and fir die-back

A study of 11 high mountain areas on the Hellenic peninsula (Figure 1) revealed that the die-back of firs (*Abies borisii-regis* and *A. cephalonica*) is the most important dynamic process at the Greek timberlines (Brandes 2007). The dying of older firs is accompanied by a general insufficiency of regeneration at the upper margin of the timberline-ecotone (i.e., the particularly sensitive ecological transition zone between the oromediterranean mountain forest and the treeless altomediterranean vegetation; see Figure 2). In some areas, among them the Taygetos, this already leads to a local treeline-decline at 50–200m altitude (Figure 3).

In central Greece and in the northern Peloponnisos the selective die-back of firs in mixed stands with *Juniperus foetidissima* leaves only the very drought resistant junipers remaining at the treeline (Figure 4).

Fir die-back, occurring not only at the timberlines but also inside the mountain forests, is caused by a process in which drought plays the central role. Indeed, a close temporal and causal connection between drought years and wide-spread fir die-back has been documented in Greece (Brofas and Economidou 1994; Markalas 1992). The lack of water causes stress to the trees and thus also favors the attack of biotic factors (Figure 5).

Mistletoes increase the water deficit of the trees since they continue to waste water even in times of drought (Sinclair et al. 1987). Unfavorable site conditions—shallow or altogether missing soil cover on rocky, steep limestone slopes—do not allow storage of water for a longer period of time. Even though Greece is not heavily industrialized, air pollutants from eastern Europe reach the country in summer, causing high ozone-levels which can surpass the values achieved in central Europe by 10% (Lelieveld et al. 2002; Larsen et al. 1990). Anthropogenic influence sometimes plays a role since it contributed locally to a deterioration of the site conditions. More often, pasturing and woodcutting increased the spread of pathogenic fungi by causing injuries on living trees or leaving behind stumps; these exposed, bark-free wood-surfaces greatly favor the attack of the fungi. Once being present in a tree stand they spread by root contacts from tree/stump to tree (Woodward et

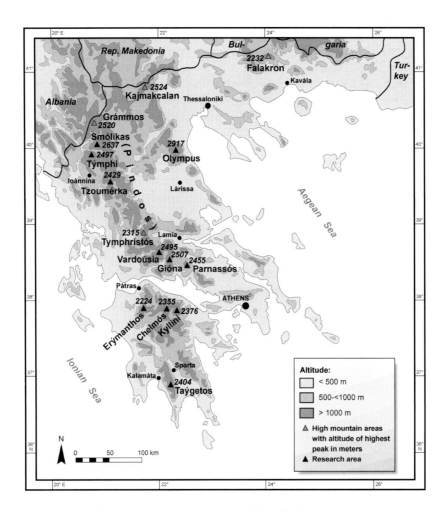

Figure 1: The high mountains of the Greek mainland.

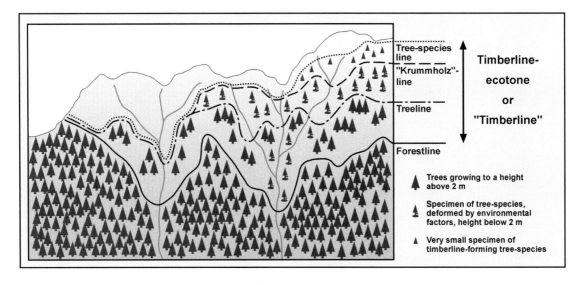

Figure 2: The upper timberline. The ecological boundary of the forest.

Figure 3: Fir die-back at the timberline in Taygetos, causing a decline of the treeline.

Figure 4: At least 600 year-old *Juniperus foetidissima*, about 6m high, at 1750m a.s.l. on mount Kyllini, northern Peloponnisos. Firs can not colonize this dry SW slope of the mountain.

al. 1998; Sinclair et al. 1987). The stress caused by the other factors of the process-framework greatly facilitates a successful fungal attack. Only *Armillaria* spp. kill single trees rather quickly. The more widely spread fungus in the Greek fir forests is *Heterobasidion annosum* (root rot), which slowly destroys the root system of a tree, leading to further difficulties in water uptake. Thus *H. annosum* strongly contributes to the "sick-phase" of the firs that may last for many years or even decades (Tsopelas 1999; Tsopelas and Korhonen 1996). The outbreak of bark-beetle calamities stands in very close connection with drought-years, mainly if the lack of rainfall occurs in spring and early summer (Kailidis and Markalas 1988). The insects profit from the weak immune defence of the stressed firs and as well from higher temperatures going along with drought. Episodic, regional, or nation-wide bark beetle calamities have been documented by the Greek forest authorities since 1929.

How "normal" are drought years and their ecological impacts on Greek fir forests and timberlines?

The widespread "sickness" and die-back of firs in Greece corresponds very well with the results of climatological studies revealing decreasing trends of precipitation in the eastern Mediterranean during the second half of the 20th century (Maheras and Anagnostopoulou 2003; Luterbacher and Xoplaki 2001; Potsdam Institut für Klimafolgenforschung 1998; Amanatidis et al. 1993). Kutiel et al. (1996: 89) called the drought years of the decade 1980–1990 the most severe in 100 years. Indeed, between 1987 and 1990 fir forests all over Greece suffered great losses (up to 40% of the trees in some stands) by bark-beetle calamities (Athanasaki 1999; Brofas and Econimidou 1994; Markalas 1992).

Against the background of the 20th century "global warming" (Kasang and Kubasch 2000), the regressive timberline dynamics seem a bit astounding at first glance since this stands in contrast to observations at other high mountain and nordic timberlines of the northern hemisphere (Holtmeier 1974 and 2003; Arno 1984), including the central Balkans (Meshinev et al. 2000). There, an advance of tree-growth into the open zone beyond the existing tree stands has been observed in several places. It is caused by the often hard to differentiate influences of periods of warmer winter climate (climate change?) and the abandonment of former pastures.

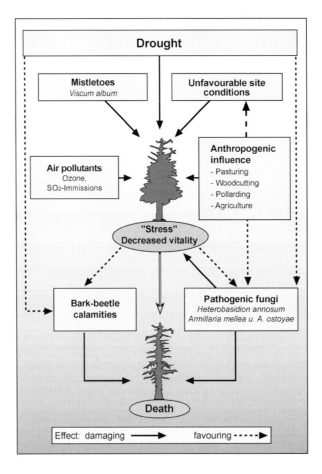

Figure 5: Process framework for "fir die-back."

The treeline-decline in some Greek high mountains cannot be generally ascribed to human interference either, because the intensity of pasturing has strongly decreased there since the late 1950s. Moreover, fir die-back and insufficient regeneration also occur at inaccessible sites on very steep, rocky slopes. Thus, the explanation of this phenomenon must be seen mainly in the ecology of Greek timberlines, differing for example from the Alps or even from the mountains of Bulgaria. In Greece, timberlines are caused not only by the harshness of the winter climate, but by a "two-sided," seasonally changing climatic impact, i.e. by snow and frost on the one hand plus by dryness, strong insolation, and overheating of soil surfaces on the other hand (Brandes 2007). Especially for successful regeneration (germination of seeds and establishment of young trees) a series of years with favorable weather conditions in both winter and summer is essential. Thus, even though pasturing has decreased in recent decades, a too-often occurring lack of moisture during the vegetation period could have contributed greatly to insufficient regeneration at timberline.

Both fir die-back and regressive timberline dynamics in Greece correspond well with the expected scenario of adaptational processes in vegetation due to changing climatic conditions (Brandes and Ise 2007). Earlier and heavier forest decay in cool mountain areas, increased mortality, and failing reproduction are some of them (Seuffert 2000: 67, Gates 1993: 246, Walker 1991: 305). That bark-beetles and pathogenic fungi play an important role within the process framework of fir die-back also meets expectations perfectly (Paine and Baker 1993: 73). The same holds true for the assumption that global warming is most likely to exacerbate tree stress near the margins of a species' geographical ranges. The range boundaries of Mediterranean Abies-species in general are at the same time the southernmost distribution boundary of the entire genus in this part of the world. Since their comparatively small areas are confined to isolated mountains regions, a major part of their habitat shows the sensibility of an ecological boundary.

However, before readily applying the term "fingerprints of climate change" (Walther et al. 2001) to the observations on Greek firs one fundamental question should be answered first: are severe drought years or repeated dry periods a recent phenomenon of the 20th century?

Jan	Feb	Mar	April	May	June	July	Aug	Sept	Oct	Nov	Dec	Year
1.4	1.3	3.7	7.9	13.9	16.1	19.6	19.7	16.0	11.1	8.0	3.6	10.2

Table 1: Mean monthly temperatures (in °C), Taygetos, calculated for 1600m a.s.l. (Brandes 2007)

In this case meteorological records can not provide answers since they date back in Greece only to 1860 (Athens). Still today climatological data for the high mountains remain scarce. Therefore, a different scientific tool must be applied in order to learn more about the occurrence of drought and about precipitation fluctuations in the past: dendrochronology. The analysis of tree rings has long become a valuable method enabling us to put events of our own times in perspective and to develop a picture of what is "normal" (Schweingruber 1988: 168).

Study area: high mountains in the southern Peloponnisos

The Taygetos is the southernmost high mountain area on the Greek mainland (Figure 1). Geologically limestones (including dolomite and marble) with rather shallow soils prevail. A dense mountain forest with *Abies cephalonica* and *Pinus nigra* covers most of the oromediterranean vegetation-zone, beginning at approximately 1000m above sea level (a.s.l.). The timberline-ecotone lies between 1600m and 2000m and is formed by both species, with *Abies* often locally dominating in number. The climate of the area is characterized by the contrast of cool, humid winters and dry summers. Figure 6, a climatic diagram of a village east of Kalamata, gives a data-sketch of this "Mediterranean mountain climate."

Between November/December and March the zone above approximately 1600m a.s.l. is covered by snow. Calculated mean monthly temperatures also reveal that the vegetation period at this altitude lasts from April to November (Table 1). Even at timberline summer temperatures are not cool enough to limit or interrupt the growth of trees.

Precipitation is characterized by an enormous variability in all months of the year. In summer there are only a few days with rainfall, usually brief thunderstorms. In some years though there is no summer rain at all for several weeks.

Methods: straightforward dendro

Even though fir die-back formed the background of the dendrochronological research in the Taygetos, the analyses were carried out on *Pinus nigra* Arnold ssp. pallasiana (Lamb.) Holmboe. The reason lies in the fact that it was impossible to find old and healthy *Abies cephalonica* growing on dry sites. Even on apparently healthy firs an infestation with pathogenic fungi might already have "disturbed" the year-ring pattern, obscuring the wanted climatic information (Kiennen and Schuck 1983, Fritts 1976).

The drought resistant black pine does not get affected by pathogenic fungi or mistletoes. Even on very dry, rocky sites this species grows to healthy, centuries-old specimens, providing excellent dendro material for a view into the past. In other regions of the Mediterranean *Pinus nigra* has already proved to be very suitable for dendrochronological and dendroclimatological research (see for example studies of Genova and Fernandez 1999 or Eckstein and Vogel 1992).

The sampling took place between 15 July and 1 October 2000. Altogether 88 living trees were sampled with a 40cm long Pressler increment corer. Additionally, on request of the author, two trees were cut down by the forest department of Sparti, contributing very valuable tree-discs. The usually only 5–13m high trees were sampled at 13 different locations, lying mostly between 1400m and 1700m a.s.l. (min. 1220m, max. 1855m) on the southern and eastern slopes of the Taygetos. In order to obtain a precipitation-sensitive chronology, the sampling was based on a careful single-tree-selection according to ecological criteria (c.f. Cook and Kairiukstis 1990, Fritts 1976). Figure 7 shows the prevailing site conditions (i.e. dry, rocky, limestone substrate) and an example of the preferentially sampled growth form (i.e. tree architecture indicating high age and very little competition of neighboring trees).

As a first step in the process of revealing statistical climate-growth relationships, the statistical agreement of the year-ring curves was tested with XCORINA and then COFECHA (Holmes 1983). This led to a selection of curves from only 45 trees covering the time-span 1729–1999 (samples and period of best statistical homogeneity). The raw data of these year-ring curves were standardized with ARSTAN (Cook and Holmes 1999) in order to remove their autocorrelation. Subsequently they were used to build the "residual" chronology. The year-ring indices of this chronology became the dependent variables in the calculation of "response functions" (Bräuning 1999, Kaennel and Schweingruber 1995, Schweingruber 1988, Strumia et al. 1997), using the program PRECON (Fritts 1999). Monthly values of temperature and precipitation from Kalamata for the period 1952–1999 were the indepen-

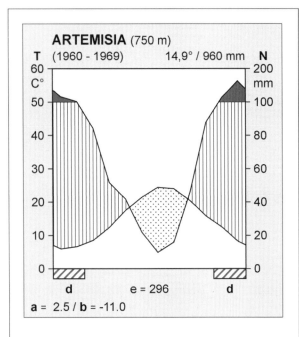

Figure 6: Mediterranean mountain climate: a diagram from Artemisia.

Figure 7: Example of sampled *Pinus nigra* growth forms and site conditions.

dent variables of the response function model (calculated from July of the previous year to August of the current vegetation period).

Results: A precipitation-sensitive pine chronology and good Aegean crossdating

Observations during sampling, year-ring measuring, and crossdating

In the field it already became apparent that, despite the relatively high altitude, water supply and not temperature is the main factor controlling the growth of the sampled black pines. Only on a few sites with deep soil and therefore better moisture conditions, as in little valleys and dolines (selected for comparison with the usual dry sites), did trees reach a height of up to 18m, a circumference at breast-height of up to 4m, and an age of 300–400 years. For comparison, the two cut-down *Pinus nigra*, growing on rocky sites, reached 10m and 11m height, circumferences of 2.35m and 2.55m, and ages of 421 and 462 years respectively. The study by Heinze (1996) clearly revealed that stronger growth of *Pinus nigra* on deeper soils is not caused by better nutrient supply but by better water supply. Also, on the cores, great differences in the width of the year-rings from trees sampled at dry and moist sites were easily recognizable with the naked eye.

Common to all samples was a sensitive year-ring pattern, clearly indicating strong climatic impacts and the capability of the trees to react to this climatic variability (Figure 8). A further characteristic was the frequent occurrence of micro-rings, which are only a few rows of cells wide. At first glance these might be mistaken as false rings. However, the comparison of the individual year-ring curves with an earlier Taygetos black pine chronology existing at Cornell and the Aegean pine master chronology quickly revealed that all visible rings were genuine year rings.

The crossdating of the curves also showed that rings were "missing" on various cores. The analysis of the complete tree discs delivered the explanation for this feature: so-called "wedging," i.e. partially absent rings (Figure 8). They are caused by environmental stress, in our case by dry-warm years with low net production of food and growth-controlling substances (c.f. Fritts 1966: 973). The fact that rings were frequently missing on cores from very dry, rocky sites, but never from the sites with better soil conditions, was a further indication for moisture being the ecological key factor for the sampled *Pinus nigra*.

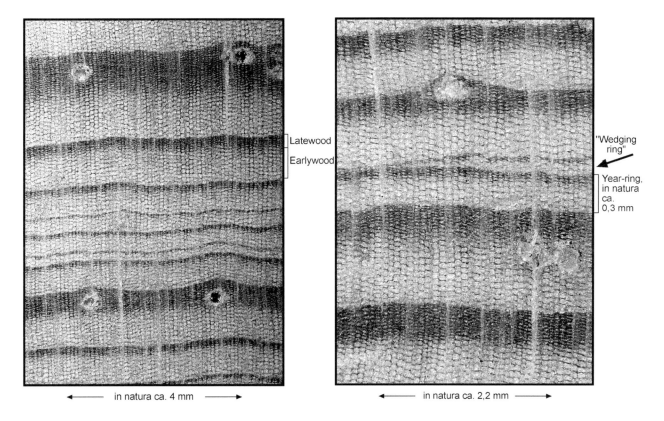

Figure 8: Sensitive year-ring pattern and wedging ring.

Chronology 1538–1999

The year-ring curves of all 90 sampled trees were similar enough to combine them with XCORINA into one chronology, termed the "AllTay" chronology (Figure 9). The number of samples is given below the index-curve. Data-points marked with year-dates render those years prominent in which the average growth reached less than 80% of the mean value (100% = year-ring index 1000). This was done only for the time after 1640, since the period of 1538 to 1640 is covered by less than 8 samples. Such a small sample-density and the fact that the sampled trees were still young at that time (with their growth therefore being subject more to non-climatic influences) reduces the reliability of the chronology.

An interesting observation related to this chronology is the comparison of the mean index-curve of the two stem-discs (10 measured radii each) from the cut down trees, growing on dry sites with southern exposure in 1520m and 1420m a.s.l., and the AllTay chronology (without the data of these trees). In order to guarantee a high number of samples in the All-Tay chronology both data-sets had been shortened to the period of 1700–1999. As Figure 10 shows there is a remarkable visual and statistical conformity. This example reveals the paramount influence of climatic factors on the year-ring width of the sampled trees (c.f. Fritts 1976: 23). Considering the rather huge altitudinal range of the sampling sites, this conformity is a further indication for precipitation being the main factor controlling the growth of the sampled trees.

Climate-growth relationships

Figure 11 shows the results of the response-function analysis. It draws the following picture of climate-growth relationships:

- Significant positive correlation coefficients indicate that the growth of the sampled *Pinus nigra* depends substantially on precipitation in May and June.

- A significant negative correlation coefficient also exists for precipitation in March. This does not reflect a growth limitation by too much moisture, though. In Taygetos a lot of precipitation in March means huge amounts of snow above approx. 1300m a.s.l.. This causes low soil temperatures and a delay of the beginning of tree growth in spring (c.f. model at Fritts 1976: 236). A dry March, on the other hand, causes earlier snow-melt, a warming of the substrate, and consequently an earlier beginning of year-ring width formation.

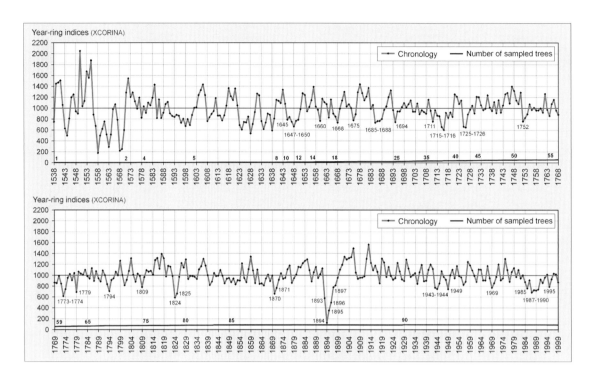

Figure 9: Year-ring chronology, *Pinus nigra*, Taygetos, 1538–1999.

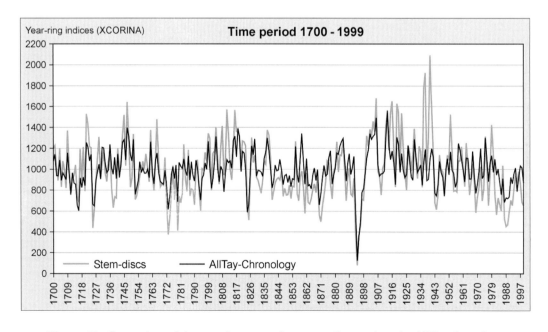

Figure 10: Comparison of the year-ring curve of two stem-discs against the AllTay chronology.

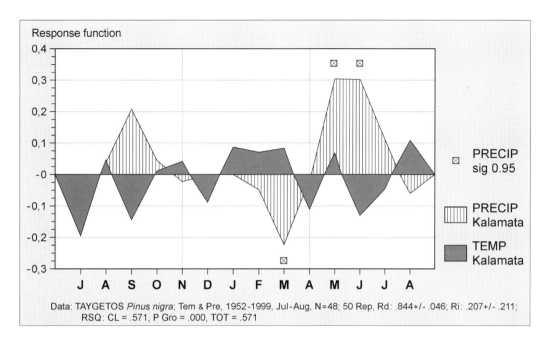

Figure 11: Results of the response function model for *Pinus nigra*, 1952–1999, 14 months.

- Even though the level of significance is not reached, the coefficient for the previous September at least strongly indicates that high autumn precipitation, going along with mild temperatures, favors the formation and accumulation of food-reserves within the tree; this influences the growth in the following year positively (c.f. model at Fritts 1976: 234). The September of the same vegetation period was not included in the response function model because test calculations done before applying this method did not reveal any strong correlation for this month. In part, this could be explained by the fact that dryness can shorten the growth-period, i.e. in dry years a normal or even moist September does not lead to increased late-wood growth (c.f. model at Fritts 1976: 233).

Against the background of the climate-ecological and site conditions the results of the response function analysis can be very well substantiated. In all, these results fulfill the "central demand for the determination of climate-growth relationships" (Bräuning 1999: 59), i.e. the ecological plausibility of the statistical relations.

The marked summer dryness and the lack of moisture reserves in the dry substrate usually limit cell-growth of the trees during July and August (correspondingly no significant correlation in the response function calculations). Only an unusually wet summer probably leads to rare exceptions and contributes positively to a greater year-ring width (for details see Brandes 2007). The winter months, November to February, are too cold and snowy for tree-growth in the upper margin of the mountain forest (no correlation with year-ring width either). Therefore, the weather conditions during the transitional seasons play the key role. Snow-melt and the soil temperatures decide the start of tree-growth in March/April. In still cool April there is usually enough moisture available (originating from snow-melt and usually enough rainfall). The much warmer months May and June (table 1) provide even better growth conditions, providing that there is enough rain.

Comparison with other dendrochronological studies

The climate-growth relationships of *Pinus nigra* from the mountains of southern Greece fit into the picture drawn by other dendrochronological studies from the eastern Mediterranean; Peter Kuniholm's intensive dendrochronological research in the Aegean region revealed the existence of signature years, allowing the dating of wood within a distance of up to 2000km and a time span reaching back to about 7000 BC. Correspondingly, there is also a remarkable, clearly climatically caused similarity between the AllTay chronology and other chronologies built at Cornell (Table 2).

This frequent existence of signature years in the Aegean chronologies (even from different tree genera, among them *Abies, Cedrus, Juniperus, Pinus, Picea, Cedrus*) made Hughes et al. (2001) ask about the causes of this large-scale "signal" in the year-rings of the eastern Mediterranean. In a first step

Location	Tree species	Distance (km)	Overlap (years)	Gleichläufigkeit	t-value	D-value
Drama, Zagradeniye, GR	*Pinus nigra*	560	275	66.4%	6.94	114.0
Çanakkale, Kalkım, TR	*Pinus nigra*	520	343	63.2%	6.80	89.5
Adana, Pos, TR	*Pinus nigra*	1130	319	62.9%	5.46	70.4
Feke, Düşmüş Ormanı, TR	*Pinus nigra*	1210	245	65.6%	4.28	66.6
Grevena, Pindos, GR	*P. heldreichii*	360	323	60.6%	6.70	70.7
Katara Pass, Pindos, GR	*P. heldreichii*	320	463	60.4%	5.64	58.6
Monte Pollino, Calabria, IT	*P. heldreichii*	590	318	60.99%	5.62	61.2
Antalya, Elmalı, TR	*Cedrus libani*	680	332	61.0%	3.41	37.6

Table 2: Comparison between the Taygetos chronology and other chronologies from the Aegean region (Kuniholm, pers. comm. 2004).

they revealed the climate-growth relationships of 23 chronologies, using the response function method. As a result they defined the total April through June precipitation as the common climate signal for the region. In order to explain this, they looked at data of atmospheric circulation and discovered recurring seasonal patterns causing either unusual moist or dry weather conditions during spring and early summer.

Touchan et al. (2003) identified only the amount of May and June precipitation as most important for the year-ring width of the trees growing in the mountains of southwestern Turkey. The relative "insignificance" of April rain in their study as well as in the results of *Pinus nigra* from Taygetos, might be based on the fact that in April trees growing in high mountains are usually provided with sufficient moisture from snowmelt. Also, even with little rainfall, the rather low temperatures in April (compared to May) keep water losses of the trees still small.

One more interesting study shall be mentioned: Waldner and Schweingruber (1996), working with conifers from subalpine and boreal timberline sites all over Europe, observed only in southeastern Europe and in Spain a simultaneous decrease in year-ring width and an increase in late wood density. This indicated the occurrence of extreme drought years. Moreover, it proved that lack of water inhibits tree growth there even at the timberline.

Discussion: What do we learn about the climate of the past?

Climatic interpretation of the chronologies

The response function calculations also yielded the result that nearly 60% of the variation in the residual-chronology can be explained by climatic influence (RSQ, i.e. $r^2 = 57.1\%$; Figure 11). This may be called very satisfying, since tree growth is influenced by factors other than climate as well. There is no doubt that data from a meteorological station at high altitude (instead of the ones from Kalamata, at only 11m a.s.l.) would lead to an even higher degree of climatically explained variance. If high altitude data were available, temperature probably might also reach significant correlation levels, for example in March. Still, there is no reason to assume that the achieved results would be fundamentally different. The data from Kalamata reflect the regional weather conditions sufficiently, because periods of drought or heavy rainfall, caused by high pressure areas or cyclones, are reflected in the precipitation data of all altitudes. Better climate data would hardly diminish the revealed paramount significance of moisture to tree growth because in southern Greece there is a very close inverse relationship between temperature and precipitation, caused by high insolation (i.e. dry weather goes along with high temperatures and vice versa).

Even though the hygric conditions are the most important factor for the year-ring width of *Pinus nigra*, some aspects speak against the attempt to reconstruct precipitation data from the Taygetos chronology. The most important ones are: the lack of high altitude climatic data reaching back several decades; the varying influence of a second climatic component, i.e. the temperatures in March; plus the generally high variability of precipitation (for more details see Brandes 2007).

Thus, the chronologies from the Taygetos do not tell about any certain amount of rainfall. Instead, they reflect the more complex "growth climate" for the sampled *Pinus nigra*, i.e. the thermal and hygric conditions for the growth of black pines during their vegetation period. Therefore, the chronologies do not allow statements about winter climate.

Since the year-ring width mainly depends on water supply, we may still regard the chronology as an indicator of moisture conditions during the vegetation period. The paramount, ecologically highly plausible significance of rainfall in late spring/early summer makes it likely that very low index values in the chronology are caused by dry weather during that time of the year. Close to average year-ring indices tell less about the amount of rainfall in late spring and early summer. As Brandes (2007) demonstrated, exceptional weather conditions in other months can either accentuate or compensate the influence of May/June on year-ring widths. For example in 1974 an unusually wet August (99.2mm, i.e. 542% of the long-year mean value) surely contributed to a rather average year-ring index, despite May/June having been relatively dry.

The high interannual variability of climate, mainly of rainfall, causes strong and abrupt variations in the course of the chronology. Looking at the 20th century, the chronology confirms that a number of years with unfavorable growth-climate occurred, especially during the 1980s. But it also clearly shows that, in general, single drought years or series of them are not unusual in southern Greece at all (for detailed discussion see Brandes 2007).

Somewhat astonishing is the fact that the years 1893–1896 have the lowest index values of the last four centuries. This requires some explanation. Undoubtedly these years were characterised by low precipitation in the whole Aegean, since many other chronologies from this region also show index-values being 25-50% below the average (Kuniholm, pers. comm.). In Taygetos, the optical impression of drastically reduced growth is increased by the fact that a very high number of samples had "missing rings" in the years 1894/1895. To compensate the missing part of the year-ring sequence, these years were given the value "1" in the raw data files, thus "artificially" lowering the index-value.

The high number of missing rings clearly reveals a time of intense climatic stress to the trees in Taygetos. However, it seems very likely that a biotic factor could have contributed to the very low growth in these years: *Thaumetopoea pityocampa*, the "processionary moth." An exceptional mass outbreak of this insect, which is the most important insect pest in Greek pine forests (Markalas 1987, Avtzis 1983), might have led to an intense defoliation of the pines, since in spring these insects feed on pine needles. Usually the trees can produce new needles in May-June, compensating for the damage. Maybe, for whatever reason, this was not the case in those years.

Low-frequency curves in the range of decades have often proved to be better suited for a climate-historical interpretation of year-ring chronologies than the high-frequency curves themselves (Bräuning 1999, Eckstein and Vogel 1992, Richter and Eckstein 1990). Therefore, the chronology was filtered by a 10-year lowpass-filter (Figure 12). Again, the end of the 20th century appears as a rather long-lasting unfavorable period for tree-growth, but similar periods did occur in the past, too. It must be added again, that the strong oscillations during the 16th and early 17th century are biased by a very low number of samples, the period 1893–1896 by the above mentioned problem correcting the "missing rings."

Verification by historical sources

In order to confirm that the Taygetos chronologies may be seen as a proxy climatic record for the precipitation conditions during the vegetation period, a verification with historical sources informing about drought years in the Aegean region was meaningful (Brandes 2007). Some examples are given in the following.

For the 16th century the chronology is based on only very few samples. Nevertheless, we can identify the drought years which occurred in Turkey in 1564, 1585, and 1595–1599 (c.f. Kuniholm 1990) as single years of low growth in the Taygetos chronology too. For Crete, which lies only 250km to the Southeast, the time 1556–1570 is documented as a pronounced dry-period (Grove and Conterio 1995: 232 ff.). The AllTay chronology and the low-pass filtered version both show clearly that the southern Peloponnisos had been affected by lack of rain as well.

For the 17th century Grove and Conterio (1995: 238) cite Venetian sources reporting in 1626 about the total drying-up of 140 cisterns in Candia, Crete. This suggests that even the previous years had been rather dry—an assumption which is strongly supported by the year-ring chronologies from Taygetos. Summing up their research on weather reports from 16th and 17th century Crete, Grove and Conterio (1995: 223) also give an interesting statement, which we should bear in mind, whenever we hear about extreme weather in the eastern Mediterranean: "In some years, and groups of years, weather conditions occurred which were apparently anomalous by twentieth century standards, especially winter and spring droughts, exceptionally severe winters, and summer rain."

For the 18th and 19th century there is the example of 1715, according to Xoplaki et al. (2001: 600) a "year of great famine" in central and southern Greece, following dryness all over the country from November 1712 to summer 1714. This situation is reflected in the high- and low-frequency chronologies as well. In 1873/74 thousands of people and animals died of starvation in Turkey. According to Kuniholm (1990: 252) year-ring chronologies from the Aegean show subnormal growth for 1873 and 1874, following four other years of significantly subnormal growth. The Taygetos chronologies also identify 1861–1874 as a period of

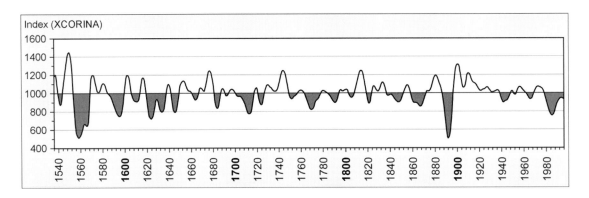

Figure 12: Year ring chronology, 1538–1998, smoothed by 10-year low-pass filter.

unfavorable rainfall conditions, 1870 probably being the worst single year in this region.

These few examples demonstrate the great potential for a further cooperation between historians and dendrochronologists. It should be possible to fully reconstruct the (regional) climate-history of Greece for the recent centuries.

For the 20th century there is an excellent coincidence between the years of mass fir die-back all over Greece in 1987–1990 (see above), caused by drought between 1985 and 1990, and strongly subnormal *Pinus nigra* growth in these years.

Conclusions: what will the future bring?

The high- and low-frequency chronologies reflect the climate-ecological conditions for trees growing in the mountains of southern Greece. Therefore, Brandes (2007) draws the conclusion that "drought-years" in the *Pinus nigra*-chronologies indicate extreme environmental stress on the much less drought resistant *Abies cephalonica*. In the Taygetos we can indeed observe that black pines outlive two or even three generations of firs.

The last example mentioned under Verification by Historical Sources above reveals that, at least in some years with strongly subnormal pine growth, bark-beetle calamities occurred, killing many firs. This assumption is supported by the results of Kailidis and Markalas (1988) who found out that 50% below average rainfall in spring and early summer (i.e. in the time also influencing pine growth most) leads to low vitality of firs and serious bark-beetle infestations in the *Abies*-forests of Greece.

Periods of subnormal pine-growth, as shown in the low-frequency chronology, most likely represent times of increased fir die-back and/or weak *Abies*-regeneration—as we observed them in most recent decades. Concerning the timberline, it must be emphasized again that regeneration there depends on a series of years with favorable climatic conditions in winter and during the vegetation period. Correspondingly, strong regeneration in the upper part of the ecotone (which is characterized by particular adverse site and climate-ecological conditions) is likely to happen only in intervals of decades, as it is often the case at timberlines in general (Holtmeier 2003). These relations give a plausible explanation for the observed local advance or decline of the treeline. Of course pasturing is an important timberline-forming factor in Greece too. But only in heavily grazed, easily accessible areas it becomes the decisive factor. In all, given the rather unfavorable hygric conditions at the end of the 20th century, the regressive timberline dynamics are nothing that requires hitting the "climate change" alarm bottom yet.

Having revealed the "normality" of drought-years and precipitation fluctuations, we can hardly evaluate the processes of fir die-back and treeline-decline as extraordinary and should therefore be careful to ascribe them to a 20th century climatic change (Brandes and Ise 2007). Probably, in Greece the genus *Abies* has suffered from the consequences of hygric stress for a very long time already. Information from archaeology and history indicates that drought-years as well as decennial and centennial precipitation fluctuations have been essential parts in the spectrum of natural climatic variability in the Aegean for many centuries. This relative constancy of climate also manifests itself in the good agreement between ancient descriptions and today's plant cover and wind trajectories (Kuniholm 1990, Meiggs 1982, Thirgood 1981: 23ff., Philippson 1948: 168, Lamb 1977: 146, 270, 263; Hempel 1999 and 2000).

These results suggest that already in antiquity seasonal water deficits should have had negative influences on the physical condition of *Abies*. Indeed we find indications of this in statements by Theophrastos (370–285 BC) and Pliny (first century AD), saying that the best timber used in Greece came from Macedonia and the now Turkish Black Sea coast, with the worst coming from Parnassos, Euboea, and Arcadia. There is good reason to assume that they were referring to *Abies*-species, since in antiquity firs were the most important timber for the making of ships as well as for the construction of buildings (cf. Meiggs 1982: 17, 26, 56, 118). It seems likely that in those times it was not only because of unfavorable site conditions or species-specific characteristics that *Abies cephalonica*-trunks (coming from Parnassos, Euboea, Arcadia) delivered shorter lengths and inferior quality compared to *A. alba* from rainy Macedonia or *A. bornmuelleriana* and *A. nordmanniana* from the wet slopes of the Pontic mountains. In much dryer southern and central Greece, bark-beetle epidemics and pathogenic fungi (probably very early spread well above their natural rate by human activities in forests) have most likely always played a major role, making the wood of relatively tall old firs useless—just like it is today. Thus, together with difficulties of access and transport existing until the 1950s (McNeill 1992: 73, Rothmaler 1943), insufficient wood quality could very well have been one more reason that kept timber cutters from venturing into the mountain regions.

This consideration is most prominently confirmed by the example of Parnes (today national park "Párnitha"), a mountain of 1413m altitude, only 20km north of the Greek capital. During the 5th century BC, when Athens already depended on timber imports from Macedonia in order to realize huge building programs, the timber resources of Mount Parnes were not drawn upon. The import by sea was preferred not only because of transport difficulties by land (which would have required large wagon-building programs first) but also because of the firs' poor timber-quality (c.f. Meiggs 1982: 16, 123, 125, 202). In the further course of antiquity not even charcoal works seemed to have caused a total destruction of the forest on Mount Parnes. This can be concluded from the fact that the area was used for hunting bears and wild boar during the time of Pausanias (late 2nd century AD; Meiggs 1982: 381). Reports from the early 19th century even speak of "inexhaustible supply of timber" there (McNeill 1992: 72). Still today vast but unhealthy fir stands cover the mountain—i.e., currently timber-cutting would not pay either.

At present nobody can say with certainty if the climatic development of the most recent decades is a fluctuation within the natural spectrum or the beginning of true climatic changes. As Bolle (2003b: 7) elucidates, in the Mediterranean basin any prognosis for the future climatic development will generally stay particularly difficult. This is due to the complexity of its nature, caused by the interplay of orography, maritime influence, large-scale as well as regional circulation patterns and manifold ecosystems.

Moreover, the utility of scenario-based climate prognoses or of models for assessing the impact of future climate on ecosystems in the Mediterranean is strongly reduced by two weak points. Firstly, the models do not unveil changes in the occurrence of extreme events which play a very important ecological role. Secondly, they don't allow reliable simulations on a regional scale which would be very important regarding the complex structure of Mediterranean landscapes (Bolle 2003a: 2).

In any case the widespread sickness and die-back of *Abies* and the dynamic processes related to it clearly reveal a high susceptibility of the Greek fir forests and timberlines to drought. This calls for a forestal practice considering much more than the threats by climate change scenarios (Brandes and Ise 2007). From a scientific point of view Greek firs and the mountain forests of the Hellenic peninsula in general will surely remain interesting study objects in the 21st century.

Acknowledgments

Many cordial thanks go to Mary Jaye Bruce for editing this contribution. For their help in working with the tree-ring data I am grateful to Peter Ian Kuniholm and the Cornell Aegean Dendrochronology Project team, to Dr. Christoph Dittmar (FH Weihenstephan) and Prof. Dr. Dieter Eckstein (University Hamburg). For their support in Greece I thank Dr. Panagiotis Tsopelas (NAGREF, Athens) and the forest departments of Sparti and Kalamata. Last but not least I wish to express my sincere gratitude to the German and Greek branches of BP for generously promoting my research in Greece with the necessary fuel and to Superfast Ferries (Piraeus) for kindly providing transport from Italy to Greece.

References

Amanatidis, G.T., Paliatsos, A.G., Repapis, C.C., and Bartzis, J.G. 1993. Decreasing precipitation trend in the Marathon area, Greece. *Intern. Journal of Climatology* 13: 191–201.

Arno, S.F. 1984. *Timberline: Mountain and Arctic Forest Frontiers*. Seattle: Mountaineers Books.

Athanasaki, S. 1999. Relationships between the change in bioclimatic type and the phenomenon of intensive dying of fir (Greek with English summary). *Dasiki Erevna* (Forest Research; Athens), 12: 3–8.

Avtzis, N. 1983. Das Auftreten von Thaume-topoea pityocampa Schiff. in Griechenland (Griechisch mit deutscher Zusammenfassung). *Dasiki Erevna* (Forest Research; Athens), 4 (2): 137–144.

Bolle, H.-J., ed. 2003a. *Mediterranean Climate: variability and trends*. Berlin: Springer.
Bolle, H.-J. 2003b. Climate, Climate Variability, and Impacts in the Mediterranean Area: An Overview. In H.-J. Bolle (ed.), *Mediterranean Climate: variability and trends*: 5–86. Berlin: Springer.
Bräuning, A. 1999. Zur Dendroklimatologie Hochtibets während des letzten Jahrtausends. Dissertationes Botanicae, Stuttgart, 132.
Brandes, R. 2007. Waldgrenzen griechischer Hochgebirge. Unter besonderer Berücksichtigung des Taygetos, Südpeloponnes (Walddynamik, Tannensterben, Dendrochronologie). Dissertation Universität Erlangen-Nürnberg. Erlanger Geographische Arbeiten, Sonderband 36.
Brandes, R., and Ise, M. 2007. Fingerprints of climate change in Mediterranean mountain forests? Observations on Mediterranean fir-species threatened by climate change. *Geo-Oeko*, 1–2, Vol. 28: 1–26.
Brofas, G., and Economidou, E. 1994. Le dépérissement du Sapin du Mont Parnasse (Grèce): Le rôle des conditions climatiques et écologiques. *Ecologia Mediterranea* 20, 1/2: 1–8.
Cook, E.R., and Holmes, R.L. 1999. Users manual for Program ARSTAN. Laboratory of Tree-Ring Research, Univ. of Arizona, Tucson (adapted from Users Manual for Program ARSTAN in Holmes et al. 1986, 50–65).
Cook, E.R., and Kairiukstis, L.A., eds. 1990. *Methods of dendrochronology: applications in the environmental science*. Boston: Kluwer.
Eckstein, D., and Vogel, J. 1992. Dendroklimatologische und dendroökologische Untersuchungen an *Pinus brutia* auf Zypern. Unveröff. Untersuchungsbericht, Ordinariat für Holzbiologie, Universität Hamburg.
Fritts, H.C. 1966. Growth-rings of trees: their correlation with climate. *Science* 154: 973–979.
Fritts, H.C. 1976. *Tree rings and climate*. New York: Academic Press.
Fritts, H.C. 1999. PRECON: A statistical model for analysing the tree-ring response to variations in climate. University of Arizona, Tucson. See: http://www.ltrr.arizona.edu/webhome/hal/precon.html.
Gates, D.M. 1993. *Climate change and its biological consequences*. Sunderland, MA: Sinauer Associates.
Génova, M., and Fernández, A. 1999. Tree rings and climate of *Pinus nigra* subsp. *salzmannii* in Central Spain. *Dendrochronologia* 16/17: 75–85.
Hagedorn, J. 1977. Probleme der periglazialen Höhenstufung in Griechenland. In H. Poser (ed.), *Formen, Formengesellschaften und Untergrenzen in den heutigen periglazialen Höhenstufen der Hochgebirge Europas und Afrikas zwischen Arktis und Äquator*: 227–237, Göttingen: Vandenhoeck und Rupert.
Harris, K. 2001. XCORINA: The Cornell Tree-Ring Analysis System. Cornell University, Ithaca.
Hempel, L. 1999. Stürmische Nordwinde im östlichen Mittelmeer innerhalb des Systems der "Grosswetterlagen" Europas. *Geoökodynamik* 20: 323–341.
Hempel, L. 2000. Stürmische Südwinde im östlichen Mittelmeer innerhalb des Systems der "Grosswetterlagen" Europas. *Geoökodynamik* 21: 285–303.
Holmes, R.L. 1983. Computer-assisted quality control in tree-ring dating and measurement. *Tree-Ring Bulletin* 43: 69-78. For COFECHA see "Dendrochronology Program Library," http://www.ltrr.arizona.edu/software.
Holtmeier, F.-K. 1974. Geoökologische Beobachtungen und Studien an der subarktischen und alpinen Waldgrenze in vergleichender Sicht. *Erdwiss. Forschg.* Bd. VIII.
Holtmeier, F.-K. 2003. Mountain timberlines. Ecology, Patchiness, and Dynamics. *Advances in Global Change Research* 14.
Hughes, M.K., Kuniholm, P.I., Eischeid, J., Garfin, G.M., Griggs, C.B., and Latini, C. 2001. Aegean tree-ring signature years explained. *Tree-Ring Research* 57 (1): 67–73.
Kailidis, D., and Markalas, S. 1988. Dürreperioden in Zusammenhang mit sekundärem Absterben und Massenvermehrungen rindenbrütender Insekten in den Wäldern Griechenlands. *Anz. Schädlingskunde, Pflanzenschutz, Umweltschutz* 61: 25–30.
Kasang, D., and Cubasch, U. 2000. Die globale Mitteltemperatur im 20. und 21. Jahrhundert. *Peterm. Geogr. Mitt.* 144, 4: 54–55.
Kiennen, L., and Schuck, J. 1983. Untersuchungen über die Zuwachsentwicklung bei erkrankten Tannen. *Eur. J. For. Path.* 13: 289–295.
Kuniholm, P.I. 1990. Archaeological evidence and non-evidence for climatic change. *Phil. Trans. R. Soc. Lond. A* 330: 645–655.
Kutiel, H., Maheras, P., and Guika, S. 1996. Circulation and extreme rainfall conditions in the eastern Mediterranean during the last century. *International Journal of Climatology* 16: 73–92.
Lamb, H.H. 1977. *Climate. Present, past and future. Vol. 2*. London: Methuen.
Larsen, J.B., Yang, W., and Tiedemann, A. 1990. Effects of ozone on gas exchange, frost resistance, flushing and growth of different provenances of European silver fir (*Abies alba* Mill.). *Eur. J. For. Path.* 20: 211–218.
Lelieveld, J., Berresheim, H., Borrmann, S., et al. 2002. Global Air Pollution Crossroads over the Mediterranean. *Science* 298: 794–799.
Luterbacher, J., and Xoplaki, E. 2001. 500-year Winter Temperature and Precipitation Variability over the Mediterranean Area and its Connection to the Large-scale Atmospheric Circulation. In H.-J. Bolle (ed.), *Mediterranean Climate: variability and trends*: 133–153. Berlin: Springer.
Maheras, P., and Anagnostopoulou, C. 2003 Circulation Types and Their Influence on the Interannual Variability and Precipitation Changes in Greece. In H.-J. Bolle (ed.), *Mediterranean Climate: variability and trends*: 215–240. Berlin: Springer.
Markalas, S. 1987. Der Befall von (auf Terrassen angelegten) Kiefernaufforstungen durch den Prozessionsspinner (*Thaumetopoea pityocampa* Schiff.). *Forstarchiv* 58(5): 205–207.
Markalas, S. 1992. Site and stand factors related to mortality rate in a fir forest after combined incidence of drought and insect attack. *Forest Ecology and Management* 47: 367–374.
McNeill, J.R. 1992. *The mountains of the Mediterranean world: an environmental history*. Cambridge: Cambridge University Press.
Meiggs, R. 1982. *Trees and timber in the ancient Mediterranean world*. Oxford University Press.
Meshinev, T., Apostolova, I., and Koleva, E. 2000. Influence of warming on timberline rising: a case study on *Pinus peuce* Griseb. in Bulgaria. *Phytoceonologia* 30: 3–4: 431–438.
Paine, T.D., and Baker, F.A. 1993. Abiotic and biotic predisposition. In T.D. Schowalter and G.M. Filip (eds.), *Beetle-pathogen interactions in conifer forests*: 61–79. London: Academic Press.
Philippson, A. 1948. *Das Klima Griechenlands*. Bonn: Ferd. Dümmlers Verlag.
Potsdam Institut für Klimafolgen-Forschung (P.I.K.) 1998. *Zweijahresbericht für 1996/97*. Potsdam.
Richter, K., and Eckstein, D. 1990. A proxy summer rainfall record for southeast Spain derived from living and historic pine trees. *Dendrochronologia* 8: 67–82.

Rothmaler, W. 1943. Die Waldverhältnisse im Peloponnes. *Intersylva (Zeitschrift der internationalen Forstzentrale)*, 3. Jahrgang: 329–342.

Schweingruber, F.H. 1988. *Tree rings: Basics and Applications of Dendrochronology.* Dordrecht: Reidel.

Seuffert, O. 2000. Klimawandel - Erkenntnisse, Defizite und Erfordernisse bei Erfassung und Prognose. *Peterm. Geogr. Mitt.* 144: 66–71.

Sinclair, W.A., Lyon, H.H., and Johnson, W.T. 1987. *Diseases of trees and shrubs.* Ithaca: Cornell University Press.

Thirgood, J.V. 1981. *Man and the Mediterranean forest.* London: Academic Press.

Touchan, R., Garfin, G.M, Meko, D.M., et al. 2003. Preliminary reconstructions of spring precipitation in Southwestern Turkey from tree-ring width. *Int. J. Climatol.* 23: 157–171.

Tsopelas, P. 1999. Distribution and ecology of *Armillaria* species in Greece. *Eur. J. For. Path.* 29: 103–116.

Tsopelas, P., and Korhonen, K. 1996. Hosts and distribution of the intersterility groups of *Heterobasidion annosum* in the highlands of Greece. *Eur. J. For. Path.* 26: 4–11.

Waldner, P.O., and Schweingruber, F.H. 1996. Temperature influence on decennial tree-ring width and density fluctuations of sub-alpine and boreal conifers in Western Europe since 1850 A.D. *Dendrochronologia* 14: 127–151.

Walker, B.H. 1991. Ecological consequences of atmospheric and climate change. *Climatic Change* 18: 301–316.

Walther, G.-R., Burga, C.A., and Edwards, P.J., eds. 2001. *"Fingerprints" of Climate Change: Adapted behaviour and shifting species ranges.* New York: Plenum Press.

Woodward, S., Stenlid, J., Karjalainen, R., and Hüttermann, A., eds. 1998. *Heterobasidion annosum: Biology, Ecology, Impact and Control.* Cambridge: Cambridge University Press.

Article submitted January 2007

Could Absolutely Dated Tree-ring Chemistry Provide a Means to Dating the Major Volcanic Eruptions of the Holocene?

Charlotte L. Pearson and Sturt W. Manning

Abstract: *This paper reviews current evidence to support the possibility that some aspect of the eruption chemistry of a particular volcanic event may be trapped and preserved in annual growth rings accumulated by trees growing through the eruption period. It considers the problems in applying standard dendrochemical principles to date the key undated volcanic events of the Holocene.*

Introduction

Discussion of the significance of volcanically induced impacts on climate, human history, and the natural environment of the last 10,000 years has frequently proven controversial (Sadler and Grattan 1999; Buckland 1997; Brunstein 1996; Pyle 1989; Parker 1985). A key underlying reason for this is a failure to obtain precise and accurate dates for certain key volcanic events. Whilst recent advances in the detection of cryptotephra horizons and high resolution analysis of the Greenland ice cores (e.g. Vinther et al. 2006; Davies et al. 2004)—plus Salzer and Hughes's (2007) latest approach to Bristlecone pine minima—present encouraging prospects for an eventual long term, annual resolution record of volcanism, in general, dates beyond the last few hundred years are not absolute and can be rather problematic (Bronk Ramsey et al. 2004; Southon 2002).

There is growing interest in investigating the potential offered by well-replicated, absolutely dated long tree-ring chronologies (e.g. Leuschner et al. 2002; Grudd et al. 2002; Eronen et al. 2002) from around the globe for dating volcanic eruptions via tree-ring chemistry (Pearson 2006; Pearson et al. 2005; Pearson et al. 2006; Ünlü et al. 2005). Over the past few decades there has been much speculation over possible causes of sudden, short-term growth anomalies in such chronologies, where certain precisely dated annual growth increments were extraordinarily enhanced for a short time or severely stunted so that the annual ring or rings produced are only a few cells thick. Where such replicated anomalous growth cannot be explained by specific or local factors (e.g. fires, canopy clearance, insect attack, etc.) it is thought to represent short term perturbations in climate, attributed as, or suggested to be, the effects of volcanism on the earth's atmosphere (Salzer and Hughes 2007; Manning et al. 2001; Gervais and MacDonald 2001; Grudd et al. 2000; D'Arrigo 1999; Baillie and Munro 1988; Briffa et al. 1998; Kuniholm et al. 1996; Scuderi 1990; Baillie 1995; LaMarche and Hirschboeck 1984).

The mechanism evoked to explain these climatic shifts is that the sulfur dioxide-based stratospheric aerosol generated by a major volcanic eruption backscatters incoming solar radiation and light, lowering ground temperatures for c. one to three years by small but significant amounts. The degree of impact varies with the size, type, and geographical location of the eruption, as well as with the time of year and other coinciding climatic conditions (Adams et al. 2003; Oppenheimer 2003; Robock 2002; Chenoweth 2001; Pyle et al. 1996; McCormick et al. 1995; Sear et al. 1987; Rampino and Self 1982). The argument for a causal connection between growth anomalies and eruptions is based on apparent correlations between the dates of specific anomalies with historically attested volcanism in the last few hundred years (eg. Briffa et al. 1998), and further back in time, with volcanically induced acidity spikes represented in the polar ice-core records (Vinther et al. 2006; Cole-Dai et al. 2000; Clausen et al. 1997; Cole-Dai et al. 1997; Legrand et al. 1997; Zielinski et al. 1997; Stothers and Rampino 1983; Hammer et al. 1980). Although improved chronologies are continuously being produced (e.g. Salzer and Hughes 2007), over time—especially beyond the last few centuries—the statistical correlation is less than conclusive in many cases, and the exact nature of the volcano-climate-tree-growth linkage is by no means universally accepted (Sadler and Grattan 1999; Buckland et al. 1997; Brunstein 1996; Yamaguchi and Grissino-Mayer 1993; Pyle 1989; Parker 1985).

If tree rings have the potential to act as annually dated time capsules of environmental geochemistry, this may present a new approach to proving a causal connection between absolutely dated tree-ring anomalies and specific volcanic eruptions. By linking elemental signatures such as those obtained from ice layers and/or tephra shards with absolutely dated growth increments, master chronologies from around the globe could be employed to reconstruct an absolute history of volcanism for the Holocene and beyond. This would have wide geographical coverage and could be used to further refine and cross-calibrate the ice-core records and tephra chronologies.

This paper will review the prospects for such an approach, first proposed by various researchers in the 1980s, e.g. Hughes (1988), for finding a dendrochemical resolution for the absolute dating of past volcanism.

The Hypothesis

The basic principle of dendrochemistry is that the chemical composition of the annually produced woody increment can act, in part, as an archive of the chemistry of the growth environment at the time of formation (Cutter and Guyette 1993, Amato 1988). Based on this premise, multi-elemental analyses of tree rings hold the potential to produce dated sequences of palaeoenvironmental elemental change. This potential has been proven in numerous dendrochemical studies over the last few decades as a tool for reconstructing patterns of anthropogenic pollution (Tommasini et al. 2000; Watmough and Hutchinson 1999; Shortle et al. 1997; Watmough et al. 1997; Martin et al. 1997; Latimer et al. 1996; Eklund 1995; DeWalle et al. 1991; Guyette et al. 1991; Stewart et al. 1991; Baes and McLaughlin 1989; Bondietti et al. 1989; Long and Davis 1989; Frelich et al. 1988; Ragsdale and Berish 1988; Maclauchlan et al. 1987; Legge et al. 1984; Robitaille 1981; Symeonides 1979; Ward et al. 1974).

If we regard a volcanic eruption as a pollution event dispersing a particular chemical signature into the environment at a specific time, then previous dendrochemical studies suggest that under the right conditions it may well be possible to detect some part of that signature in a contemporary tree-ring sequence.

From existing research, two sub-hypotheses can be drawn as to the form of elemental signature one might hope to find in response to a volcanic eruption.

The first hypothesis is that a direct sample of some unique part of the eruption chemistry (either from the gaseous cloud or as part of the physical loading of tephra) may be identified in individual tree rings. By linking the occurrence of unusual elements, combinations, and/or higher concentrations of particular elements (associated with certain eruptions), an actual chemical fingerprint could be derived. This could then be used to directly link absolutely dated tree rings with volcanic eruptions, ice-core acidity profiles, and stratified tephra horizons. The second hypothesis is that a volcanogenic increase in global or local acidity levels could chemically alter the environment in which a particular tree was growing and result in changes in the uptake of elements naturally available in that environment.

There is a growing body of evidence to support both of these hypotheses, from older pollution studies and from more recent investigations targeting this specific research question. Before going on to evaluate this supporting evidence, however, consideration must be given to the problems inherent within basic dendrochemical principles.

Problems with dendrochemistry

The concept of a tree ring as a time capsule of environmental chemistry is clearly a highly simplified postulation of an incredibly complex and variable system. Trees do not passively record their environmental chemistry; rather they filter preferentially according to specific biological needs from an ever-changing range of site-specific inputs (Pearson 2006; Pearson et al. 2005; Smith and Shortle 1996; Cutter and Guyette 1993).

The biological factors which mitigate the formation of any chemical record in a tree are at best poorly understood and vary not only from species to species, but in all likelihood from tree to tree. In addition, the various adsorbed elements all have certain characteristics which determine the way in which the tree is able to utilize them. A key issue here is the degree of element mobility, and consequently the dispersal of that particular element in the xylem. The lower the mobility, the greater the reliability of a given element in the creation of a chronology of environmental chemistry. The solubility of the various ions—determined not only by the characteristics of individual elements but also by factors such as the pH of the sap—governs their concentration in the xylem solution. The more soluble an element, the more mobile it is. The more mobile, the less likely it is to remain in the outermost tree ring and accurately record chemical changes as they occur.

Other factors influencing the potential chemical record are the ionic radius of an element, which determines how strongly the ions will be held to exchange sites on the cell walls, and whether or not an element is essential to the growth of the tree. Essential elements such as copper, calcium, zinc, magnesium, iron, and potassium may be preferentially translocated around the tree, with increased mobility before being distributed according to purpose in the xylem. Such elements must therefore be approached with caution in dendrochemical studies, but can also prove very useful in that trees will often take up as much as is available, and thus variations may accurately reflect external chemical change.

Another important factor determining the internal distribution of elements within wood is the transition zone between the sapwood and the heartwood. Here differences in the equilibrium of specific element concentrations can govern whether new heartwood is enriched by or depleted in a particular element. Changes in concentration at the boundary between the heartwood and sapwood for various elements have been noted by Watmough and Hutchinson (2002, 1999), Penninck et al. (2001), Shortle et al. (1997), Okada et al. (1993), Chun and Hui-yi (1992), Tomita et al. (1990), McClenahen et al. (1989), Wardell and Hart (1973). Other physiologically derived patterns have been observed in terms of the radial distribution of certain elements which naturally decline or increase with age (Aoki et al. 1998; Prohaska et al. 1998; Jonsson et al. 1997; Oliveira et al. 1997; Haas and Muller 1995; Myre and Camiré 1994; Bailley and Reeve 1993; Hagemeyer 1993; Bondietti et al. 1989; Frelich et al. 1988) and concentrations of certain elements in particular anatomical components of a single tree ring (Sunden et al. 2000; Berglund et al. 1999; Lövestam et al. 1990; McClenahen et al. 1989; Saka and Goring 1983). Such patterns do not appear to be consistent for a particular element, but rather vary for the same element from species to species and are an essential consideration when interpreting any dendrochemical data set. Pearson (2006) databased more than 651 observations from 79 different species of tree collated from 208 dendrochemistry papers in order to produce a resource whereby previously observed elemental behaviors can be searched by tree species, pattern type, anatomical association, etc., as a research tool to aid with subsequent data interpretation.

The environment in which a tree grows supplies a further range of external factors which can also have a profound influence on tree-ring chemistry. The nature of the underlying bedrock and soil chemistry, substrate depth, water table level, local wind directions, climate, aspect, and slope can all determine not only background levels of elements available for uptake, but also exposure and response to various changes in environmental chemistry. Soil chemistry and depth in particular can be fundamental in determining not only the availability of certain elements, but also the main uptake path via which a tree absorbs nutrients. A deep, fertile soil is likely to reduce sensitivity for recording short-term atmospheric elemental change by delaying uptake through complex soil chemical interactions, and by encouraging dominant uptake of nutrients via the roots, potentially resulting in transportation of elements all around the xylem. In contrast, uptake through the leaves can result in rapid deposition in the most recently formed tree ring (Watmough 1999; Lin et al. 1995). Therefore, it has been concluded (Pearson et al. 2006, 2005) that the most suitably located species for dendrochemical studies will be those growing in more marginal environments, i.e., in poor, shallow, fast-draining soils. It is hypothesized that such conditions would minimise the background contribution from the original soil chemistry and time lags caused by chemical interaction in the soil environment. Also, that this would lead to a dominant uptake of loaded precipitation through the bark and leaves, routes more likely to result in a rapid deposition of any multi-elemental signature in newly forming cells.

Supporting evidence

Irrespective of the complexities inherent to dendrochemical studies, there are numerous publications which confirm that given the right tree and the right conditions, good correlations can be made between the onset and cessation of known anthropogenic pollution events and contemporary tree rings (e.g. Penninckx et al. 2001; Martin et al. 1997; DeWalle et al. 1991; Bondietti et al. 1990, 1989). There are also an increasing number of publications from which further evidence can be derived in support of the two hypotheses proposed by which dendrochemistry could be used to date volcanic eruptions.

The strongest evidence to support the first hypothesis was found by Hall et al. (1990). In this early study two anomalously high peaks in concentrations of rare earth elements (cerium, neodymium, lanthanum, samarium, gadolinium, lutetium, and thulium) were found in tree rings corresponding to eruptions of the Mount St. Helens volcano. The *Pseudotsuga menziesii* they sampled was growing 15km northeast of the volcano and had received direct fallout of 15cm of volcanic ash. More recently, Sheppard et al. (in press) found an increase in sulfur and phosphorus in tree rings sampled near Parícutin volcano, reflecting a sustained increase in these elements from the onset of the 1943 eruption onwards. Again the trees had received a direct fallout of tephra, with uptake direct from the eruption chemistry the most likely scenario, although the authors do not rule out the impact of rising volcanogenically derived acidity on pre-existing soil sulfur and phosphorus as an additional source. A similar effect (though in this case from an anthropogenic source) was recorded by Tendel and Wolf (1998). Here again trees were found to directly record an increase in sulfur levels in response to a known increase in sulfur dioxide. In this study the effect was found to apply to different tree species over a wide geographical area.

Conversely, it should be noted that a study by Watt et al. (2007) specifically to detect volcanic cation deposition from known events in trees growing on the flanks of Mount Etna, Sicily, found no correlations between dendrochemical patterns and volcanic activity. Nevertheless, the previous examples would seem to provide sufficient evidence to suggest that there are prospects for identifying some kind of directly linkable volcanic signature in tree rings. However, the second hypothesis, which relates to a somewhat less provable link with volcanism, has even greater support.

The impact of anthropogenically created acid rain on forest ecosystems has been the subject of numerous dendrochemical studies (e.g. Penninckx et al. 2001; Watmough, and Hutchinson 1999; Frelich et al. 1998; Hutchinson 1998; Prohaska et al. 1998; Shortle et al. 1997; Freer-Smith and Read 1995; Lin et al. 1995; Hoffmann et al. 1994; Markrt 1994; DeWalle et al. 1991; Guyette and Cutter 1991; Bondietti et al. 1989, 1990; Guyette et al. 1989; Häsänen and Huttunen 1989; Legge et al. 1984; Symeonides 1979). Rising soil acidity, due to increased precipitation of sulfuric or nitric acid from anthropogenic pollution sources, can lead to alterations in the relative availability of nutrients and ions in the soil, and preferential uptake and translocation within trees. Acid precipitation has also been shown to enhance susceptibility to adsorption through the bark and leaves (Lin et al. 1995; Percy and Baker 1988; Huttunen et al. 1983) leading to a rapid response from trees where such uptake paths are dominant. All this transfers easily to a volcanic hypothesis, as work currently underway by Ünlü et al. (2005) demonstrates specifically in relation to the apparent uptake of gold. Evidence to support the hypothesis has been provided by Padilla and Anderson (2002) who observed a series of rises or peaks in concentrations of barium, copper, and zinc in tree rings formed around the time of the Laki (1783), Tambora (1815), and Krakatau (1883) eruptions. They attribute these results to a decrease in soil pH resulting from volcanically induced acid rain, with the most convincing of the associations for barium, copper, and zinc for Tambora and copper for Krakatau. Furthermore, Pearson (2006) suggests there is potential that such responses can be provable for a wider range of elements in sequences over several hundred years at annual resolution. Using ICP-MS for multi-elemental analysis, the approach was to investigate the prospects for proving hypothesis one, but the majority of the results provided supporting evidence for hypothesis two.

Figure 1 shows a composite of elemental concentration patterns from two pine trees—one growing in Cyprus, one in Turkey, around the 1815 Tambora eruption. Both trees show unusual elemental responses around the time of the eruption. Whilst these data are just a small subset from a larger study, they typify the responses found from different trees around the world for the same major volcanic eruption. The large error bars on some of the data show an increased heterogeneity in the tree rings around the time of the eruption—hypothesized to be due to increased uptake/surplus of particular elements resulting in "hotspot" concentrations (e.g. Sunden et al. 2000), possibly associated with anatomical features. Of the elemental responses found by Pearson 2006, most fell into either this category, or the type of response reported by Padilla and Anderson 2002: a short sustained increase in concentrations (Figure 2).

There is therefore good evidence to suggest that at the very least, dendrochemical studies of long tree-ring sequences may serve to provide a further proxy indicator to connect absolutely dated tree rings with specific volcanic eruptions. However, in order for this to be really useful it needs to be applicable to a range of older wood obtained from the world's dendrochronological records.

Pilot Study

To explore this further, two samples of *Juniperus* sp. from the Aegean Dendrochronology Project archive from Porsuk, Turkey, were provided by Peter Kuniholm (see Figure 3). Both samples were dated as part of the Aegean Dendrochronology Project and cover an exceptional growth spike around relative year 854 (1650 BC +4/-7). The anomaly, indicative of a sudden, short-term change to a cooler, moister growth environment, is displayed by 61 trees of different species, retrieved from the foundations of a Hittite wall at the site of Porsuk, in south-central Turkey (Kuniholm et al. 2005). The trees are thought to have grown in the Taurus Mountains approximately 840 kilometers downwind of Thera (Peter Kuniholm pers. comm.), and it has been speculated that the anomaly must relate to the impact of the Minoan eruption (most recently dated between 1627–1600 BC according to Friedrich et al. 2006; see also Manning et al. 2006).

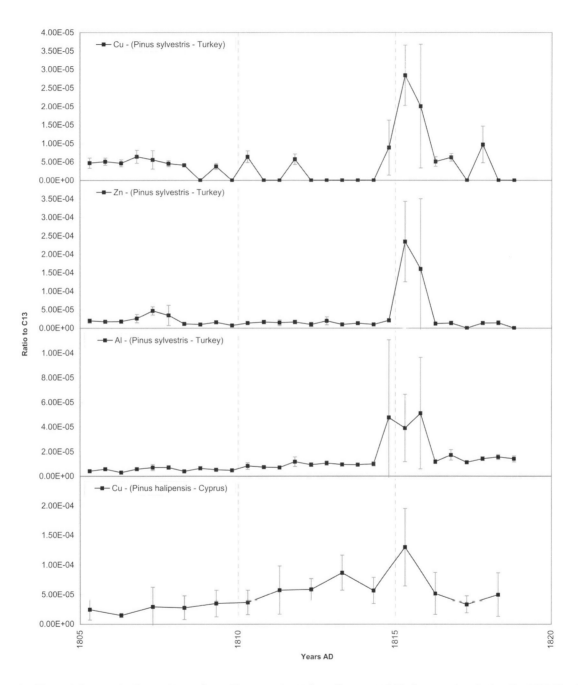

Figure 1: Elemental concentration patterns from *Pinus* sp. trees from Cyprus and Turkey growing during the 1815 Tambora eruption. Both trees in disparate locations show unusual elemental responses reflecting an increase in the heterogeneity of the tree-ring chemistry around the time of the eruption.

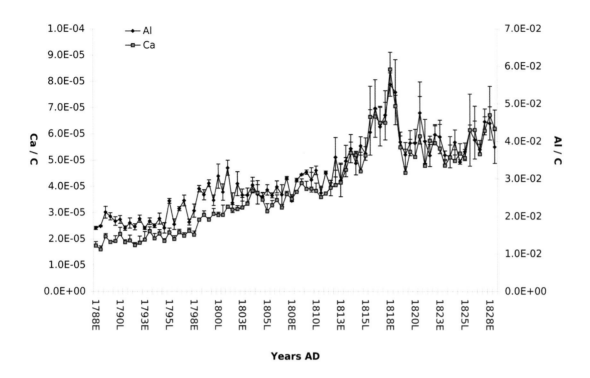

Figure 2: A example of a short sustained increase in elemental concentrations around the time of the Tambora eruption. From Pearson et. al (2005).

The two samples were sub-sampled at annual resolution with a steel blade, under x20 magnification, and 0.25gm of each ring was then prepared and analyzed by solution ICP-AES using the methodology described by Pearson (2006). Due to very narrow ring size/absence of material, it was not possible to sample the same length of sequence at this resolution for both C-TU-POR-26A and C-TU-POR-3A. The resulting sequences comprised relative years 840–870 for C-TU-POR-26A and 825–882 for C-TU-POR-3A. Figure 4 provides an example of the data collected for sulfur and zinc for each of the trees. These data typify the two main types of elemental patterns to be derived from the samples. The pattern for zinc is one of occasional spikes, also displayed by aluminum, silver, and zirconium. Of these, only zinc displayed any arguable correlation between the two trees, and as can be observed from Figure 4, this is not an exact correlation. There is an offset of one year between the largest of the spikes in each sequence. If these zinc spikes—the two largest in each of the sequences—could be argued the result of ion mobilization due to increased environmental acidity, the delayed correlation in C-TU-POR-3A may relate to a difference in the soil depth and/or chemistry at the growth site. C-TU-POR-3A displays slightly wider, more regular rings than C-TU-POR-26A and it could be hypothesized that this may indicate that the former comes from a more stable (less marginal) local growth environment where a slight time lag in the soil might be anticipated. Alternatively, Hall et al. (1990) attributed a similar one-year elemental discrepancy, relating to a date of a known event, to elemental mobility in the xylem which could indeed plausibly explain a time lag of several years. Whatever the explanation in this case, both of the zinc spikes correlate to within a year of the ring 854 growth anomaly. This is particularly interesting when considered against the data for sulfur. C-TU-POR-3A shows a clear increase in this element the year following the onset of the growth anomaly, correlating with the zinc. C-TU-POR-26A also displays a sulfur increase in the years following the growth ring anomaly; however, it is not such a convincing response.

Other elements displaying an increase at the same time as the sulfur in the two samples were barium, strontium, magnesium, sodium, and calcium. In the absence of well replicated agreement between the two trees (plus additional replicates that would need to be added to substantiate these initial findings) it is perhaps inappropriate to speculate as to the causal mechanisms behind the observed patterns; however, there are other aspects of the dataset which beg discussion.

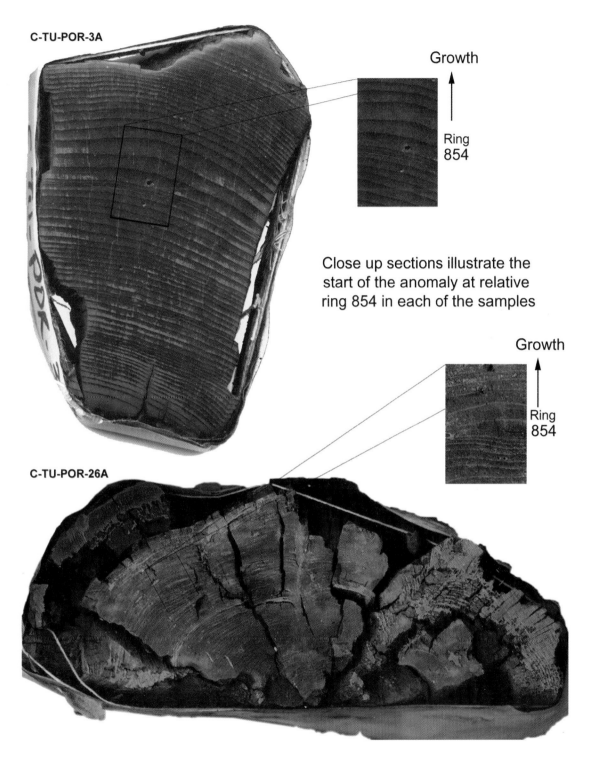

Figure 3: Samples C-TU-POR-26A and C-TU-POR-3A from the Aegean and Near Eastern Dendrochronology. *Juniperus* sp. from Porsuk, Turkey, showing the relative year 854 growth anomaly previously suggested to be attributed to the Minoan eruption of Thera.

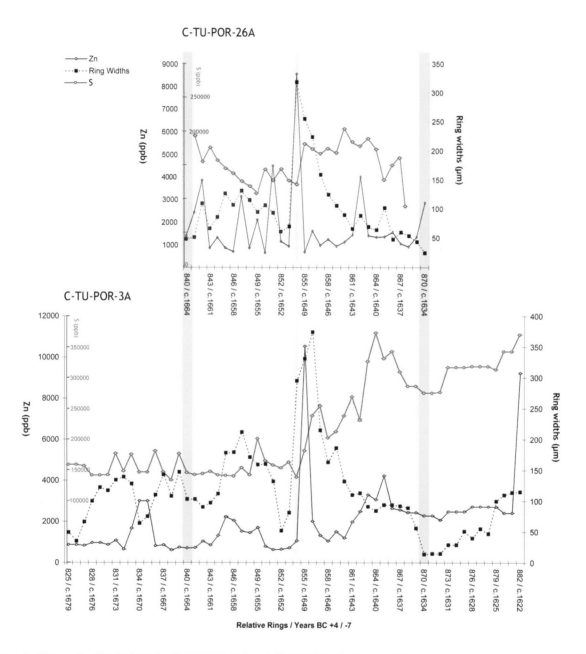

Figure 4: Zinc and sulfur in Samples C-TU-POR-26A and C-TU-POR-3A showing different but distinct responses around ring 854.

A key consideration is that in these two samples, the concentrations for all elements are high for wood. In particular the strontium levels are around 200ppm, an order of magnitude higher than those found in a piece of modern juniper analyzed at the same time. Initial conclusions based on the high concentrations and shared patterns of these elements were that the most likely explanation for the previously observed patterns was a coincidental result of contamination from groundwater permeation. Yet, as has been pointed out (Peter Kuniholm pers. comm 2008) the coincidence of the elemental response with the tree-ring anomaly in C-TU-POR-3A is so clear that it is perhaps more far-fetched to suggest coincidental saturation to that specific point in the wood than to insinuate that what we are observing in both samples is a response, influenced by numerous unknown site specific and physiological controls, to some sort of environmental event. Indeed, more recent studies have reported similarly high concentrations in other species (in proximity to a known depositional source), (e.g. Sheppard et al. in press, report sulfur concentrations in *Pinus* sp. from 50 up to 1000 ppm). If further work can replicate the initial datasets with corresponding sequences from other trees at Porsuk over longer series of tree rings, at annual or at least sub-annual resolution, it may be that these preliminary results present the first step to connecting a multi-elemental anomaly (either derived from a direct deposition of tephra to the growing tree or increased acidification from volcanic output) to the most major growth-ring anomaly in the last 6000 years of the Aegean chronology.

The ultimate goal would be to eventually tease out an actual multi-elemental fingerprint linking directly to the environmental event which triggered that anomaly. This would be especially relevant as there now seems to be a considerable amount of evidence to suggest that there were two major volcanic events between 1600 and 1660 BC: that around 1627 BC (e.g. Salzer and Hughes 2007; Friedrich et al. 2006; Manning et al. 2006; Grudd et al. 2000; Clausen et al. 1997; Baillie and Munro 1988; LaMarche and Hirschboeck 1984) and that around 1642 BC (e.g. Crowley et al. in prep; Salzer and Hughes 2007; Vinther et al. 2006; Manning et al. 2006; Hammer et al. 1987, 2003; Hantemirov and Shiyatov 2002; Eronen et al. 2002). The Porsuk anomaly is currently dated to 1650 +4/-7 BC, and given the likely close geographical proximity of the Porsuk trees to the eruption of Thera, it would seem highly relevant to test out the possibilities for isolating an elemental trace for this event in these samples.

In terms of the data so far, however, the discrepancy between the elemental patterns from the two trees underlines the complexity involved in any dendrochemical study, but particularly the extra layer of unknowns attached to any study using unprovenanced material with no growth site specific data. Unfortunately, in order to attempt to trace volcanic signatures for many key volcanic eruptions of the Holocene these are likely to be the main types of sample available for study.

Conclusions

There are now a substantial number of studies proving that annual concentrations of specific elements in tree rings can directly reflect changes in external environmental chemistry. Furthermore there seems to be good evidence to suggest that this may be applied to the detection of volcanically induced changes in environmental chemistry.

Further defining, quantifying, and substantiating such associations is likely to be an extremely complex and difficult process. Whether or not (and how) a particular eruption might be represented in the chemistry of a tree ring relies upon a massive number of variables, many of which are as yet poorly understood. Nevertheless, with a rigorous program of research using carefully selected samples and continuously improving analytical techniques, it now appears that there are prospects that tree-ring chemistry may one day reveal at the very least another proxy for absolute dates for volcanic eruptions.

Future work should focus on very specifically designed studies, which take into full consideration all aspects of the history of the sample in question and previously observed physiologically derived patterns of elements and ring width. Additional work should be carried out in order to gain a better understanding of the physiological processes associated with elemental response, and all data should be substantiated by replication from several sampling points in the same tree and the same sequence from a number of trees.

When dealing with wood from the world's master chronologies, the main factors to consider should be, firstly, how was the wood preserved?—i.e., was it retrieved from an environment such as a bog in which it may have been contaminated with post-depositional chemistry, rendering it impossible to detect the more subtle pre-depositional elemental patterns? Secondly, is anything known about the original growth environment of the tree so that site specific factors influencing the uptake of particular elements can be considered? Thirdly, is there going to be sufficient sample material to allow for the replications needed to produce a robust, reproducible data set? As most analytical processes are destructive, rare samples covering key events may not be available, or ring widths may be too narrow to provide sufficient annually resolved material.

It seems the best possible prospects in terms of responsiveness to environmental elemental change lie with samples known to have grown in marginal, dry environments, and which have been preserved in cold or dry conditions where the potential for contamination is low (for example, from the North American Bristlecone Pine chronology). It is hoped that by focusing on samples such as these and using and developing the latest in analytical techniques, it may one day be possible to unlock yet more information from the tree-ring archive and attribute absolute dates to some of the key volcanic eruptions of prehistory.

Acknowledgments

This review is based on work which took place as part of a three year PhD project at The University of Reading, UK. It was supervised by Professors Sturt Manning and Max Coleman with the help of Dr. Kym Jarvis and the staff of the National Environmental Research Council–Inductively Coupled Plasma facility at Kingston-Upon-Thames, UK. It was funded by a University of Reading studentship and the UK Natural Environment Research Council. Professors Martin Bell and Paul Buckland were examiners. During the course of the project a large cross-section of the dendrochronological community were kind enough to contribute wood samples and advice. In particular Peter Kuniholm and Maryanne Newton of The Malcolm and Carolyn Wiener Laboratory for Aegean and Near Eastern Dendrochronology, but also Mike Baillie and David Brown of the School of Archaeology and Palaeoecology, Queen's University Belfast; Christer Karlsson and Håkan Grudd of the Swedish University of Agricultural Sciences, Siljansfors Experimental Forest; Gretel Boswijk and John Ogden of the University of Auckland; Pröstur Eysteinsson and Hrafn Oskarsson of the Icelandic Forestry Commission; Mike Barbetti of the Australian Key Centre for Microscopy and Microanalysis, The University of Sydney; Rosanne D'Arrigo and David Frank of the Tree-Ring Laboratory of Lamont-Doherty Earth Observatory, Columbia University; and Martin Bridge of University College London.

References

Adams, J.B., Mann, M.E., and Ammann, C.M. 2003. Proxy evidence for an El Nino-like response to volcanic forcing. *Nature* 426: 274–278.

Amato, I. 1988. Tapping tree rings for the environmental tales they tell. *Analytical Chemistry* 60: 1103A–1107A.

Aoki, T., Katayama, Y., Kagawa, A., Koh, S. Yoshida, K. 1998. Measurement of trace elements in tree rings using the PIXE method. *Nuclear Instruments and Methods in Physics Research B* 136: 919–922.

Baes C.F. III, and McLaughlin, S.B. 1984. Trace elements in tree rings: evidence of recent and historical air pollution. *Science* 224: 494–497.

Bailey, J.H.E., and Reeve, D.W. 1993. Determination of the spatial distribution of trace elements in black spruce, *Picea mariana* Mill., by imaging microprobe secondary ion mass spectrometry. *International Symposium on Wood and Pulping Chemistry* 2: 848–857.

Baillie, M.G.L. 1995. *A Slice Through Time*. London: Batsford.

Baillie, M.G.L., and Munro, M.A.R. 1988. Irish tree rings, Santorini and volcanic dust veils. *Nature* 332: 344–346.

Berglund, A., Brelid, H., Rindby, A., and Engstrom, P. 1999. Spatial distribution of metal ions in spruce wood by synchrotron radiation microbeam X-ray fluorescence analysis. *Holzforschung* 53: 474–480.

Bondietti, E.A., Baes, C.F. III, and McLaughlin, S.B. 1989. Radial trends in cation ratios in tree rings as indicators of the impact of atmospheric deposition on forests. *Canadian Journal of Forest Research* 19: 586–594.

Bondietti, E.A., Momoshima, N., Shortle, W.C., and Smith, K.T. 1990. A historical perspective on divalent cation trends in red spruce stemwood and the hypothetical relationship to acidic deposition. *Canadian Journal of Forest Research* 20: 1850–1858.

Bronk Ramsey, C., Manning, S.W., and Galimberti, M. 2004. Dating the Volcanic Eruption at Thera. *Radiocarbon* 46: 325–344.

Brunstein, F.C. 1996. Climatic significance of the Bristlecone Pine latewood frost-ring record at Almagre Mountain, Colorado, U.S.A. *Artic and Alpine Research* 28: 65–76.

Buckland, P.C., Dugmore, A.J., and Edwards, K.J. 1997. Bronze age myths? Volcanic activity and human response in the Mediterranean and North Atlantic regions. *Antiquity* 71: 581–593.

Chenoweth, M. 2001. Two major volcanic cooling episodes derived from global marine air temperature, AD 1807–1827. *Geophysical Research letters* 28: 2963–2966.

Chun, L., and Hui-yi, H. 1992. Tree-ring element analysis of Korean pine (*Pinus koraiensis* Sieb. et Zuc.) and Mongolian oak (*Quercus mongolica* Fisch. Ex Turcz.) from Changbai Mountain, north east China. *Trees* 6: 103–108.

Clausen, H.B., Hammer, C.U., Hvidberg, C.S., Dahl-Jensen, D., Steffensen, J.P., Kipfstuhl, J., and Legrand, M. 1997. A comparison of the volcanic records over the past 4000 years from the Greenland Ice Core Project and Dye 3 Greenland Ice Cores. *Journal of Geophysical Research* 102: 26707–26723.

Cole-Dai, J., Mosley-Thompson E., Wight, S.P., and Thompson, L.G. 2000. A 4100–year record of explosive volcanism from an East Antarctica ice core. *Journal of Geophysical Research* 105: 24431–24442.

Cole-Dai, J., Mosley-Thompson, E., and Thompson, L.G. 1997. Annually resolved southern hemisphere volcanic history from two Antarctic ice cores. *Journal of Geophysical Research* 102: 16761–16771.

Cutter, B.E., and Guyette, R.P. 1993. Anatomical, chemical, and ecological factors affecting tree species choice in dendrochemistry studies. *Journal of Environmental Quality* 22: 611–619.

D'Arrigo, R.D., and Jacoby, G.C. 1999. Northern North American tree-ring evidence for regional temperature changes after major volcanic events. *Climatic Change* 41: 1–15.

DeWalle, D.R., Swistock, B.R., Sayre, R.G., and Sharpe, W.E. 1991. Spatial variations of sapwood chemistry with soil acidity in Appalachian forests. *Journal of Environmental Quality* 20: 486–491.

Eklund, M. 1995. Cadmium and lead deposition around a Swedish battery plant as recorded in oak tree rings. *Journal of Environmental Quality* 24: 126–131.

Eronen, M., Zetterman, P., Briffa, K.R., Lindholm, M., Merilainen, J., and Timonen, M. 2002. The supra-long Scots pine tree-ring record for Finnish Lapland: Part 1, chronology construction and initial inferences. *The Holocene* 12: 673–680.

Freer-Smith, P.H., and Read, D.B. 1995. The relationship between crown condition and soil solution chemistry in oak and sitka spruce in England and Wales. *Forest Ecology and Management* 79: 185–196.

Frelich, L.E., Bockheim, J.G., and Leide, J.E. 1988. Historical trends in treering growth and chemistry across an air quality gradient in Wisconsin. *Canadian Journal of Forest Research* 19: 112–121.

Friedrich, W.L., Kromer, B., Friedrich, M., Heinemeier, J., Pfeiffer, T., and Talamo, S. 2006. Santorini Eruption Radiocarbon Dated to 1627–1600 B.C. *Science* 312: 548.

Gervais, B.R., and MacDonald, M. 2001. Tree-ring and summer-temperature response to volcanic aerosol forcing at the Northern tree-line, Kola Peninsula, Russia. *The Holocene* 11: 499–505.

Grudd, H., Briffa, K.R., Gunnarson, B.E., and Linderholm, H.W. 2000. Swedish tree rings provide new evidence in support of a major, widespread environmental disruption in 1628 BC. *Geophysical Research Letters* 27: 2957–2960.

Grudd, H., Briffa, K.R., Karlen, W., Bartholin, T.S., Jones, P.D., and Kromer, B. 2002. A 7400-year tree-ring chronology in northern Swedish Lapland: natural climatic variability expressed on annual to millennial timescales. *The Holocene* 12: 657–665.

Guyette, R.P., and Cutter, B.E. 1991. Tree-ring analysis of fire history of a post oak savanna in the Missouri Ozarks. *Natural Areas Journal* 11: 93–99.

Guyette, R.P., Cutter, B.E., and Henderson, G.S. 1989. Longterm relationships between molybdenum and sulphur concentrations in red cedar tree rings. *Journal of Environmental Quality* 18: 385–389.

Guyette, R.P., Cutter, B.E., and Henderson, G.S. 1991: Longterm correlations between mining activity and levels of lead and cadmium in tree-rings of eastern red-cedar. *Journal of Environmental Quality* 20: 146–149.

Haas, G., and Muller, A. 1995. Radioecological investigations on tree rings of spruce. *The Science of the Total Environment* 173: 393–397.

Hagemeyer, J. 1993. Monitoring trace metal pollution with tree rings: a critical reassessment. In: Markert, B. (ed.), *Plants as biomonitors*: 541–563. New York: VCH Weinheim.

Hall, G.S., Yamaguchi, D.K., and Rettberg, T.M. 1990. Multielemental analysis of tree rings by inductively coupled plasma mass spectrometry. *Journal of Radioanalytical Nuclear Chemistry Letters* 146: 255–265.

Hammer, C.U., Clausen, H.B., and Dansgaard, W. 1980. Greenland ice sheet evidence of post-glacial volcanism and its climatic impact. *Nature* 288: 230–235.

Hammer, C.U., Clausen, H.B., Friedrich, W.L., and Tauber, H. 1987. The Minoan eruption of Santorini in Greece dated to 1645 BC? *Nature* 328: 517–519.

Hammer, C.U., Kurat, G., Hoppe, P., Grum, W., and Clausen, H.B. 2003. Thera eruption date 1645 BC confirmed by new ice core data? In M. Bietak (ed.), *The synchronisation of civilisations in the eastern mediterranean in the second millennium B.C. II. Proceedings of the SCIEM 2000—EuroConference, Haindorf, 2nd–7th of May 2001*: 87–93. Vienna: Österreichischen Akademie der Wissenschaften.

Hantemirov, R.M., and Shiyatov, S.G. 2002. A continuous multimillennial ring-width chronology in Yamal. *The Holocene* 12: 717.

Häsänen, E., and Huttunen, S. 1989. Acid deposition and the element composition of pine tree rings. *Chemosphere* 18: 1913–1920.

Hoffmann, E., Ludke, C., Scholze, H., and Stephanowitz, H. 1994. Analytical investigations of tree rings by laser ablation ICP-MS. *Fresenius Journal of Analytical Chemistry* 350: 253–259.

Hughes, M.K. 1988. Ice-layer dating of eruption at Santorini, *Nature* 335: 211–212.

Hutchinson, T.C., Watmough, S.A., Sager, E.P.S., and Karagatzides, J.D. 1998. Effects of excess nitrogen deposition and soil acidification on sugar maple (*Acer saccharum*) in Ontario, Canada: an experimental study. *Canadian Journal of Forest Research* 28: 299–310.

Huttunen, S., Karhu, M., and Torvela, H. 1983. Deposition of suphur compounds on forests in southern Finland. *Aquilo Ser Botanica* 19: 270–274.

Jonsson, A., Eklund, M., and Hakansson, K. 1997. Heavy metals of the 20th century recorded in oak tree rings. *Journal of Environmental Quality* 26: 1638–1643.

Keenan, D.J. 2003. Volcanic ash retrieved from the GRIP ice core is not from Thera. *Geochemistry, Geophysics, Geosystems* 4: 10.1029/2003GC000608.

Kuniholm, P.I., Newton, M.W., Griggs, C.B., and Sullivan, P.J. 2005. Dendrochronological Dating in Anatolia: The Second Millennium B.C. Ü. Yalçın (ed.), *Der Anschnitt, Anatolian Metal III*. Beiheft 18: 41–47. Bochum.

Kuniholm, P.I., Kromer, B., Manning, S.W., Newton, M., Latini, C.E., and Bruce, M.J. 1996. Anatolian tree rings and the absolute chronology of the Eastern Mediterranean 2220-2718BC. *Nature* 381: 780–782.

LaMarche, V.C., and Hirschboeck, K.K. 1984. Frost rings in trees as records of major volcanic eruptions, *Nature* 307: 121–126.

Latimer, S.D., Devall, M.S., Thomas, C., Ellgaard, E.G., Kumar, S.D., and Thien, L.B. 1996. Heavy metals in the environment-dendrochronology and heavy metal deposition in tree rings of bald cypress. *Journal of Environmental Quality* 25: 1411–1419.

Legge, A.H., Kaufman, H.C., and Winchester, J.W. 1984. Treering analysis by PIXE for a historical record of soil chemistry response to acidic air pollution. *Nuclear Instruments and Methods in Physics Research* 3: 507–510.

Legrand, M., Hammer, C., De Angelis, M., Savarino J., Delmas R., Clausen H., and Johnsen, S.J. 1997. Sulfur-containing species (methanesulfonate and SO_4) over the last climatic cycle in the Greenland Ice Core Project (central Greenland) ice core. *Journal of Geophysical Research* 102: 26663–26679.

Leuschner, H.H., Sass-Klaassen, U., Jansma, E., Baillie, M.G.L., and Spurk, M. 2002. Subfossil European bog oaks: population dynamics and long-term growth depressions as indicators of changes in the Holocene hydro-regime and climate. *The Holocene* 12: 695–706.

Lin, Z.Q., Barthakur, N.N., Schuepp, P.H., and Kennedy, G.G. 1995. Uptake and translocation of Mn and Zn applied on foliage and bark surfaces of balsam fir (*Abies balsamea* (L.) Mill.) seedlings. *Environmental and Experimental Botany* 35: 475–483.

Long, R.P., and Davis, D.D. 1989. Major and trace element concentrations in surface organic layers, mineral soil, and white oak xylem downwind from a coal-fired power plant. *Canadian Journal of Forest Research* 19: 1603–1615.

Lövestam, G., Johansson, E., Johansson, S., and Pallon, J. 1990. Elemental micro patterns in tree rings—a feasibility study using scanning proton microprobe analysis. *Ambio* 19: 87–93.

Maclauchlan, L.E., Borden, J.H., Cackette, M.R., and D'Auria, J.M. 1987. A rapid multisample technique for detection of trace elements in trees by energy-dispersive X-ray fluorescence spectroscopy. *Canadian Journal of Forest Research* 17: 1124–1130.

Manning, S.W., Bronk Ramsey, C., et al. 2006. Chronology for the Aegean Late Bronze Age. *Science* 312: 565–569.

Manning, S.W., Kromer, B., Kuniholm, P.I., and Newton, M.W. 2001. Anatolian tree rings and a new chronology for the east Mediterranean Bronze–Iron ages. *Science* 294: 2494–2495.

Markrt, B. 1993. *Plants as Biomonitors*. Cambridge: VCH.

Martin, R.R., Zanin, J.P., Bensette, M.J., Lee, M., and Furimsky, E. 1997. Metals in the annual rings of eastern white pine (*Pinus strobus*) in southwestern Ontario by secondary ion mass spectroscopy (sims). *Canadian Journal of Forest Research* 27: 76–79.

McClenahen, J.R., Vimmerstedt, J.P., and Scherzer, A.J. 1989. Elemental concentrations in tree rings by PIXE: statistical variability, mobility, and effects of altered soil chemistry. *Canadian Journal of Forest Research* 19: 880–888.

McCormick, M.P., Thomason, L.W., and Trepte, C.R. 1995. Atmospheric effects of the Mt Pinatubo eruption. *Nature* 373: 399–404.

Myre, R., and Camiré, C. 1994. Distribution of P, K, Ca, Mg, Mn and Zn in the stem of European larch and tamarack. *Annales des Sciences Forestieres* 51: 121–134.

Okada, N., Katayama, Y., Nobuchi, T., Ishimaru, Y., and Aoki, A. 1993. Trace elements in the stems of trees—comparisons of radial distributions among hardwood stems. *Mokuzai Gakkaishi* 39: 1119–1127.

Oliveira, H., Fernandes, E.A.N., and Ferraz, E. 1997. Determination of trace elements in tree rings of Pinus by neutron activation analysis. *Journal of Radioanalytical and Nuclear Chemistry* 217: 125–129.

Oppenheimer, C. 2003. Climate, environmental and human consequences of the largest known historic eruption: Tambora volcano (Indonesia) 1815. *Progress in Physical Geography* 27: 230–259.

Padilla, K.L., and Anderson, K.A. 2002. Trace element concentration in tree-rings biomonitoring centuries of environmental change. *Chemosphere* 49: 575–585.

Parker, D.E. 1985. Frost rings in trees and volcanic eruptions, *Nature* 313: 160–161.

Pearce, N.J.G., Westgate, J.A., Preece, S.J., Eastwood W.J., and Perkins W.T. 2004. Identification of Aniakchak (Alaska) tephra in Greenland ice core challenges the 1645 BC date for Minoan eruption of Santorini. *Geochemistry, Geophysics, Geosystems* 5: 10.1029/2003GC000672.

Pearson, C.L. 2006. Volcanic Eruptions, Tree Rings and Multielemental Chemistry: An Investigation of Dendrochemical Potential for the Absolute Dating of Past Volcanism. *BAR International Series* 10191.

Pearson, C.L., Manning S.W., Coleman, M.L., and Jarvis, K. 2005. Tree rings, volcanic eruptions and multi-elemental chemistry: Could Tree-Ring Chemistry Reveal Absolute Dates for Past Volcanic Eruptions? *Journal of Archaeological Science*, 32: 1265–1274.

Pearson, C.L., Manning S.W., Coleman, M.L., and Jarvis, K. 2006. A Dendrochemical study of *Pinus sylvestris* from Siljansfors Experimental Forest, Central Sweden. *Applied Geochemistry* 21: 1681–1691.

Penninckx, V., Glineur, S., Gruber, W., Herbauts, J., and Meerts, P. 2001. Radial variations in wood mineral element concentrations: a comparison of beech and pedunculate oak from the Belgian Ardennes. *Annals of Forest Science* 58: 253–260.

Percy, K.E., and Baker, E.A. 1988. Effects of simulated acid rain on leaf wettability, rain retention and uptake of some inorganic ions. *New Phytologist* 108: 75–82.

Prohaska, T., Stadlbauer, C., Wimmer, R., Stingeder, G., Latkoczy, C., Hoffmann, E., and Stephanowitz, H. 1998. Investigation of element variability in tree rings of young Norway spruce by laser ablation–ICPMS. *The Science of the Total Environment* 219: 29–39.

Pyle, D.M. 1989. Ice-core acidity peaks, retarded tree growth and putative eruptions. *Archaeometry* 31: 88–91.

Pyle, D.M., Beattie, P.D., and Bluth, G.J.S. 1996. Sulphur emissions to the stratosphere from explosive volcanic eruptions. *Bulletin of Volcanology* 57: 663–671.

Ragsdale, H.L., and Berish, C.W. 1988. The decline of lead in tree rings of *Carya* spp. in urban Atlanta, GA, USA. *Biogeochemistry* 6: 21–29.

Rampino, M.R., and Self, S. 1982. Historic eruptions of Tambora (1815), Krakatau (1883), and Agung (1963), their stratospheric aerosols and climatic impact. *Quaternary Research* 18: 127–143.

Robitaille, G. 1981. Heavy-metal accumulation in the annual rings of balsam fir *Abies balsamea* (L.) Mill. *Environmental Pollution* 2: 193–201.

Robock, A. 2002. The climatic aftermath, *Science* 295: 1242–1244.

Sadler, J.P., and Grattan, J.P. 1999. Volcanoes as agents of past environmental change. *Global and Planetary Change* 21: 181–196.

Salzer, M.W., and Hughes, M.K. 2007. Bristlecone pine tree rings and volcanic eruptions over the last 5000 yr. *Quaternary Research* 67(1): 57–68.

Saka, S., and Goring, D. 1983. The distribution of inorganic constituents in black spruce wood as determined by TEM-EDXA. *Mokuzai Gakkaishi* 29: 648–656.

Scuderi, L.A. 1990. Tree-ring evidence for climatically effective volcanic eruptions. *Quaternary Research* 34: 67–85.

Sear, C.B., Kelly, P.M., Jones, P.D., Goodess, C.M. 1987. Global surface-temperatures responses to major volcanic eruptions, *Nature* 330: 365–367.

Sheppard P.R., Ort. M.H., Anderson, K.C., Elson, M.D., Vazquez-Selem, L., Clemens, A.W., Little, N.C., and Speakman, R.J. (in press) Multiple dendrochronological signals indicate the eruption of Parícutin Volcano, Michoacán, Mexico. [Note added in proof: published 2008 in *Tree-Ring Research* 64(2): 97–108.]

Shortle, W.C., Smith, K.T., Minocha, R., Lawrence, G.B., and David, M.B. 1997. Acidic deposition, cation mobilization and biochemical indicators of stress in healthy red spruce. *Journal of Environmental Quality* 26: 871–876.

Smith, K.T., and Shortle, W.C. 1996. Tree biology and dendrochemistry. In J.S. Dean, D.M. Meko, T.W. Swetnam (eds.), *Tree Rings, Environment and Humanity*: 629–634. Radiocarbon, Department of Geosciences, Tucson, Arizona.

Southon, J. 2002. A first step to reconciling the GRIP and GISP2 ice-core chronologies, 0–14,500 yr B.P. *Quaternary Research* 57: 32–37.

Stewart, C., Norton, D.A., and Fergusson, J.E. 1991. Historical monitoring of heavy metals in kahikatea ring wood in Christchurch, New Zealand. *The Science of the Total Environment* 105: 171–190.

Stothers, R.B., and Rampino, M.R. 1983. Historic volcanism, European dry fogs and Greenland acid precipitation, 1500BC to AD1500. *Science* 222: 411–412.

Sunden, A., Brelid, H., Rindby, A., and Engstrom, P. 2000. Spatial distribution and modes of chemical attachment of metal ions in spruce wood. *Journal of Pulp and Paper Science* 26: 352–357.

Symeonides, C. 1979. Tree-ring analysis for tracing the history of pollution: application to a study in northern Sweden. *Journal of Environmental Quality* 8: 482–486.

Tendel, J., and Wolf, K. 1988. Distribution of nutrients and trace elements in annual rings of pine trees (*Pinus silvestris*) as an indicator of environmental changes. *Experientia* 44: 975–980.

Tomita, M., Katayama, Y., Nishimura, K., and Takada, J. 1990. Radial distribution of Sb and Cu in the stem of woody plants grown on the Sb-polluted soil as well as on the Cu-enriched soil near an ore dressing house of copper mine. *Radioisotopes* 39: 553–556.

Tommasini, S., Davies, G.R., and Elliott, T. 2000. Lead isotope composition of tree rings as bio-geochemical tracers of heavy metal pollution: a reconnaissance study from Firenze, Italy. *Applied Geochemistry* 15: 891–900.

Ünlü, K., Kuniholm, P.I., Chiment, J.J., and Hauck, D.K. 2005. Neutron activation analysis of absolutely-dated tree rings. *Journal of Radioanalytical and Nuclear Chemistry* 264(1): 21–27.

Vinther, B.M., Clausen, H.B., et al. 2006. A synchronized dating of three Greenland ice cores throughout the Holocene. *Journal of Geophysical Research*, 111: D13102, doi:10.1029/2005JD006921.

Vinther, B.M., Clausen, H.B., et al. (in prep). Reply to: Comment on Vinther et al. "A synchronized dating of three Greenland ice cores throughout the Holocene": The Minoan tephra is not present in the 1642 B.C. layer of the GRIP ice core. *Journal of Geophysical Research* (n.p.).

Ward, N.I., Brooks, R.R., and Reeves, R.D. 1974. Effect of lead from motorvehicle exhausts on trees along a major thoroughfare in Palmerston North, New Zealand. *Environmental Pollution* 6: 149–158.

Wardell, J.F., and Hart, J.H. 1973. Radial gradients of elements in white oakwood. *Wood Science* 5: 298–303.

Watmough, S.A. 1999. Monitoring historical changes in soil and atmospheric trace metal levels by dendrochemical analysis. *Environmental Pollution* 106: 391–403.

Watmough, S.A., and Hutchinson, T.C. 1999. Change in the dendrochemistry of sacred fir close to Mexico City over the past 100 years. *Environmental Pollution* 104: 79–88.

Watmough, S.A., and Hutchinson, T.C. 2002. Historical changes in lead concentrations in tree-rings of sycamore, oak and scots pine in north-west England. *The Science of the Total Environment* 293: 85–96.

Watmough, S.A., Hutchinson, T.C., and Evans, R.D. 1997. Application of laser ablation inductively coupled plasma mass spectrometry in dendrochemical analysis. *Environmental Science and Technology* 31: 114–118.

Watt, S.F.L., Pyle, D.M., Mather, T.A., Day, J.A., and Aiuppa, A. 2007. The use of tree-rings and foliage as an archive of volcanogenic cation deposition. *Environmental Pollution* 148: 48–61.

Yamaguchi D.K., and Grissino-Mayer, H.D. 1993. Comment on resolving volcanic activity of 20 Ma ago with relative accuracy of 1 yr from tree rings of petrified wood. *Geophysical Research Letters* 20: 2279–2280.

Zielinski, G.A., Mayewski, P.A., Meeker, L.D., Groenvold, K., Germani, M.S., Whirlow, S., Twickler, M.S. and Taylor, K. 1997. Volcanic aerosol records and tephrochronology of the Summit, Greenland ice cores. *Journal of Geophysical Research* 102: 26625–26640.

Zielinski, G.A., Mayewski, P.A., Meeker, L.D., Whitlow, S., Twickler, M.S., Morrison, M., Meese, D.A., Gow, A.J., and Alley, R.B. 1994. Record of Volcanism Since 7000 B.C. from the GISP2 Greenland Ice Core and Implications for the Volcano-Climate System. *Science* 264: 948–952.

[Note added in Proof: On the wood samples from Porsuk and the likely identification of a volcanic signal in the 17th century BC, see now: Pearson, C.L., Dale, D.S., Brewer, P.W., Kuniholm, P.I., Lipton, J., and Manning, S.W. 2009. Dendrochemical analysis of a tree-ring growth anomaly associated with the Late Bronze Age eruption of Thera. *Journal of Archaeological Science* 36: 1206–1214.]

Article submitted May 2007

Dendrochemistry of *Pinus sylvestris* Trees from a Turkish Forest

D. K. Hauck and K. Ünlü

Abstract: *The compositional analysis of tree-rings has the potential to provide a global, high resolution proxy data set for environmental change and archaeological dating markers such as volcanic eruptions. This paper considers the possible impact of the 1991 Pinatubo eruption and large Mediterranean dust storms on the chemistry of trees growing in western Turkey. For this purpose, Neutron Activation Analysis was used to determine the trace element concentrations of four Pinus sylvestris trees that were growing in the Çatacık forest in Turkey during the years 1980–2000. Element concentrations were compared to aerosol index data compiled by the Total Ozone Mapping Spectrometer (TOMS) Group at the Goddard Space Flight Center. Concentrations of gold, silver, and manganese appeared to provide the most promising chronologies for reconstructing large environmental events. The trees used in this study were obtained from the Malcolm and Carolyn Wiener Laboratory for Aegean and Near Eastern Dendrochronology at Cornell University.*

Introduction

Interest in the compositional analysis of tree rings began in the early 1970s with the need to document the nature and distribution of anthropogenic pollution (Baes and McLaughlin 1984). Since that time more sophisticated analysis methods and increased interest in large scale environmental phenomena have made dendrochemistry a promising tool for climate reconstruction (Watmough 1999). In addition, there would be many archaeological benefits to having an independently dated timeline of events such as volcanic eruptions. Some studies have already found evidence of volcanic signals in the chemistry of tree-rings (Pearson et al. 2005; Ünlü et al. 2005). However, the relationship between climate and element uptake must be understood, or at least documented, before dendrochemical chronologies can stand alone as an important source of proxy information.

There are two mechanisms currently being considered in the literature to explain why volcanic eruptions have an effect on element uptake by trees. The first is that eruptions inject large volumes of ash into the atmosphere that are generally deposited onto the terrain as a function of distance away from the eruption site. The amount of trace element enrichment in volcanic ash can be used as a chemical fingerprint to identify the volcanic origin of the ash (Steinhauser et al. 2006). Therefore, certain signature elements will have a higher availability to the tree and may be absorbed directly by the leaves or by the roots in greater amounts. A second effect of volcanic eruptions is to release large amounts of sulfur gases into the atmosphere that combine with water vapor to form acid rain (Graf et al. 1998). The subsequent increase in the acidity of ground water makes many trace metals more soluble and more likely to be absorbed passively along with the trees' water supply (Kabata-Pendias and Pendias 1992). While this effect temporarily creates a high dissolved metal content in the soil solution, it ultimately results in depletion of the soil as the soluble metals are washed out to streams and lakes. It is possible to imagine the two eruption effects (ash fallout and increased acidity) working in tandem to create an isolated chemical signal in tree rings which can be uniquely identified as volcanic (Pearson 2003).

There are some complicating factors in interpreting a dendrochemical chronology. Certain elements, such as aluminum for some species, are blocked by the tree when they surpass a threshold value in the soil (DeWalle et al. 1991). In addition, tree ring concentration is implicitly assumed to be a proxy for annual uptake by the tree. However, increasing and decreasing trends of elements in trees and spiking behaviors at the sapwood-heartwood boundary indicate that elements are being pushed around by the biological activities of the tree (Watmough 1997). Smearing of a signal over the sapwood rings or passive diffusion and active movement by ray cells may shift or mute an environmental chemical signal. For all these reasons it will be important to identify elements that work well for dendrochemistry and identify their behaviors in different tree species. Manganese (Mn) has already been identified by some researchers for its potential in environmental reconstruction (DeWalle et al. 1999; Pearson et al. 2006).

The compositional analysis technique utilized for this study is Neutron Activation Analysis (NAA) which provides quantitative values for major and trace elements in tree-rings. The sensitivity with which NAA can detect particular isotopes ranges between parts-per-million and sub-parts-per-billion, depending on the nuclear properties of the elements of interest and analysis parameters (Ünlü et al. 2005). NAA takes advantage of the fact that certain nuclei undergo a neutron capture reaction when bombarded by neutrons. The result of neutron capture is a new radioactive nucleus that usually decays by releasing a beta particle and possibly one or more gamma rays. The energy of the gamma ray(s) is characteristic of the radioactive nucleus and can be used to qualitatively identify the stable isotopes that were originally in the sample. Measurement of the number of gamma rays emitted by the sample along with knowledge of the detector efficiency, neutron flux, and probability of neutron capture allows quantitative analysis of the isotopes in the sample. NAA is considered to be a non-destructive technique because the number of individual nuclei that are converted through the neutron capture process is very small and the concentration of the original isotope remains virtually unchanged. Therefore, samples may be re-irradiated and counted at a later time for further analysis. NAA is highly suitable for dendrochemistry with the largest disadvantage being the necessity to break up tree samples into individual tree-rings. This introduces the possibility of contamination, and for slow-growing species, may make it difficult to acquire enough ring-wood for a single sample. The neutron source for performing NAA was the TRIGA Mark III Breazeale Nuclear Reactor at The Pennsylvania State University's Radiation Science and Engineering Center.

Experimental Procedure

Samples from four *Pinus sylvestris* trees that grew in the Turkish forest of Çatacık were chosen for analysis. These trees covered at least the years 1980–2000. This time span straddles the eruption of Pinatubo in 1991, which was the most climatically effective eruption during the span of instrumental record keeping in Turkey. The Total Ozone Mapping Spectrometer (TOMS) has been in place on various satellites since late 1978. TOMS provides daily information on ozone, sulfur, aerosols, and dust in the atmosphere across most of the globe (Guo et al. 2004). One problem exists with choosing recently grown tree samples which cover a short range of years. It has been pointed out that it may be difficult to judge a tree's normal historical behavior with respect to particular elements based on modern wood (S.W. Manning, pers. comm., 24 March 2006). This is especially true since regional sources of anthropogenic pollution have increased greatly in number and intensity in the past century. The tree samples from Çatacık were a convenient choice for the present analysis since the dendrochemistry group at Pennsylvania State University previously analyzed a 500+ year chronology from this forest (Ünlü et al. this volume). It was hoped that this would provide at least general information on typical element concentrations prior to large scale pollution.

The first three tree samples analyzed were CAT-28, CAT-16, and CAT-32. Each of these samples was divided into individual rings using clean cutting tools. Each ring sample was placed into a heat sealed polyethylene bag and irradiated in groups of forty for 4 hours at a thermal flux of $1.7 x 10^{13} n/cm^2/s$ at the core of the Breazeale Nuclear Reactor. After allowing the samples to decay for 1 day, they were removed from the polyethylene bags and placed into clean, un-irradiated polyethylene vials and counted for 1 hour each. This counting procedure allowed for the identification of sodium (Na), zinc (Zn), lanthanum (La), potassium (K), bromine (Br), gold (Au), and Mn. The tree sample CAT-16 was large enough to provide two samples for each ring. The two sub-samples CAT-16A and CAT-16B were irradiated and analyzed separately according to the above procedure.

While analyzing the results from the first three samples, it was found that the gold concentrations were very close to the minimum detectable concentrations and may have been subject to large statistical errors. For this reason it was decided to alter the counting procedure for the remaining tree sample, CAT-21. In addition, CAT-32 was re-irradiated and re-analyzed with the new counting procedure. Accordingly, the tree samples CAT-21 and CAT-32 were irradiated in smaller batches of 15 to 20 tree-rings and were counted for 2 hours each thus decreasing the detection limits for gold. As time allowed, each sample was counted for 15 minutes prior to the 2-hour count to obtain information about manganese content. Subsets of CAT-21 and CAT-32 were also counted for 1 day each to obtain information on scandium (Sc), chromium (Cr), iron (Fe), cobalt (Co), strontium (Sr), silver (Ag), antimony (Sb), cesium (Cs), and europium (Eu). The

Tree Sample	Years	Irradiation and Count Procedure
CAT-16	1972–2001	Divided into two samples, CAT-16A and CAT 16B. The rings in each sub-sample were irradiated and counted for 1 hour each.
CAT-28	1971–2001	Rings were irradiated and counted for 1 hour each.
CAT-21	1969–2001	Each ring was irradiated and counted for 15 minutes each, then 2 hours each, then 1 day each.
CAT-32	1940–2001	Rings were irradiated and counted for 1 hour each. Later they were re-irradiated and counted for 2 hours each, then 1 day each.

Table 1: Summary of the irradiation and counting procedure used for each tree sample.

Element	Isotope (abundance)	Average Concentration	Number of Rings
Na	Na-23 (100%)	3.6 ppm	148
K	K-41 (6.73%)	29.7 ppm	148
Sc	Sc-45 (100%)	2.1 ppb	69
Cr	Cr-50 (4.35%)	4.8 ppb	49
Mn	Mn-55 (100%)	43.2 ppm	147
Fe	Fe-58 (0.28%)	21.1 ppb	64
Co	Co-59 (100%)	57.1 ppb	69
Zn	Zn-68 (18.80%)	1.7 ppb	69
Br	Br-81 (49.31%)	25.9 ppb	147
Sr	Sr-84 (0.56%)	27.0 ppb	33
Ag	Ag-109 (48.16%)	5.7 ppb	64
Sb	Sb-123 (42.70%)	2.3 ppb	64
Cs	Cs-133 (100%)	1.9 ppb	60
La	La-139 (99.91%)	14.9 ppb	87
Eu	Eu-151 (47.80%)	0.5 ppb	30
Au	Au-197 (100%)	0.8 ppb	128

Table 2: Average element concentrations of four *Pinus sylvestris* trees

counting procedure for each tree sample is summarized in Table 1.

Data and Results

The elements that were identified in the tree-rings are listed in Table 2 along with their average concentration for all tree samples. The elemental concentration was calculated from the experimentally determined isotopic concentrations by assuming natural isotopic ratios. The isotopic ratio is given in parenthesis in the Isotope column. Due to variations in the counting procedure, not all isotopes were identified in all tree samples. Therefore, Table 2 also lists the number of rings that were used for calculating the average concentration value. The concentrations of elements in CAT-16 were assumed to be the average of the concentration values of CAT-16A and CAT-16B. The concentrations of Na, K, Br, Mn, and Au in tree sample CAT-32C were assumed to be the average of the concentration results from the first and second irradiation. The last tree ring of each tree sample which corresponds to the year 2001 was not used in the calculation since it typically has anomalously high concentrations of trace elements. This may be due to the ring's close proximity to the bark, which typically has high trace element content.

Estimates of the experimental error and tree-ring variation for some of the identified isotopes are summarized in Table 3. The experimental errors were estimated by comparing the results from the first and second irradiations of tree sample CAT-32. The error values represent the standard deviation of percent difference between the results from the first and second irradiations. The tree-ring variabilities were estimated by calculating the standard deviation of the percent difference between the results from tree samples CAT-16A and CAT-16B. The variability values contain contributions from both the experimental errors and actual variabilities within the tree ring. These values are meant to be estimates of the typical errors found in the data sets presented in this paper. The variability in Au concentration may be high due to poor statistics which was corrected in subsequent samples by using a longer count time. The largest components of the experimental error are differences in neutron flux, depending on the control rod positions, and changes in the sample size. The long-lived isotopes that did not have duplicate analyses and are not included in Table 3 are expected to have errors in the range of 10 to 20 percent. The detector system and analysis procedure were able to reproduce the concentrations of Na, Mn, K, and Br in a calibrated standard created at the National Institute of Standards and Technology with less than 20 percent error in any of the isotopes.

The manganese concentrations in each of the four tree samples are presented in Figure 1. The data from years 1940 to 1970 are included in Figure 1 to demonstrate the peak in manganese concentration which occurs in 1948 in tree sample CAT-32. There are also peaks in Zn, Na, and La concentration in 1949, although the Na peak occurs in a single ring and is relatively small. There is a peak in Bromine concentration which lasts from 1953 to 1960. Between the years 1941 and 1946 the K concentration increases from 10 ppm to a new continuum level of about 40 ppm. The rings between 1940 and 1960 may contain the heartwood-sapwood barrier. However, the position of the transition between living and non-living rings was not recorded when the samples were collected so this hypothesis cannot be verified. The concentration of Au shows no discernible special behavior in this time frame.

In the region of Figure 1 where the data series overlap, some key years have correlated Mn concentration peaks. The year 1989 stands out for having peaks in all four tree samples. The years 1978, 1985, and 1997 also show some degree of peak correlation between at least three tree samples. The other elements show a less convincing degree of correlation between the four tree samples. However, there are correlated Zn concentration peaks in 1978 and correlated K concentrations peaks in 1989 which occurred in three out of four of the tree samples.

The Sc and Cs concentrations for tree samples CAT-21 and CAT-32 are shown in Figure 2. The Sc and Cs peaks in 1980 also occur in the Na, Co, and Fe concentrations in tree sample CAT-21. The next notable year in Figure 2 is 1982 in which Cs and Sc in both trees are correlated. Sodium peaks were observed in 1982 in CAT-16 and CAT-21 and in 1983 in CAT-28. The peaks in CAT-32 Cs concentration and CAT-21 Sc concentration in 1992 are closely followed by peaks in Cs and Sc concentrations in CAT-21 and CAT-32, respectively, in 1993. The same Cs-Sc pattern occurs later in the years 1995 and 1996. The Fe concentration very closely follows that of Sc in both tree samples and could have been substituted for Sc in Figure 2.

The most anomalous behavior of all the trace elements that were measured was exhibited by Ag and Au in tree sample CAT-32. The Au concentration has peaks that occur in single rings and that are an order of magnitude above the gold continuum level. Contamination should be an obvious concern here. The silver concentration shows a similar pattern although the peaks are no more than five to six times higher than the continuum. None of the other elements in CAT-32 exhibit the strange behavior. The presence of the Ag and Au peaks was verified when tree sam-

Element	Estimated Error (%)	Estimated Variation (%)
Na	14	34
K	9	10
Mn	15	11
Zn	14	24
Br	11	26
La	16	30
Au	34	81

Table 3: Estimated error and tree-ring variation in element concentrations

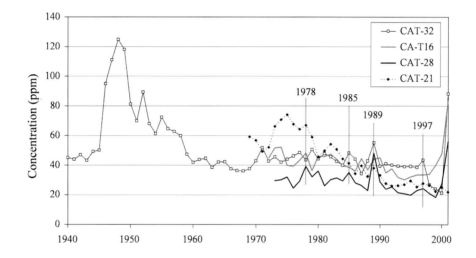

Figure 1: Manganese concentration in four tree samples from a Turkish forest. The peak observed in Mn concentration in 1948 is correlated with peaks in concentration in Zn, Na, and La in the same tree and may be due to the heartwood-sapwood barrier. The years 1978, 1985, 1989, and 1997 show some degree of Mn peak correlation between at least three out of four of the tree samples. There are also correlated Zn concentration peaks in 1978 and correlated K concentrations peaks in 1989.

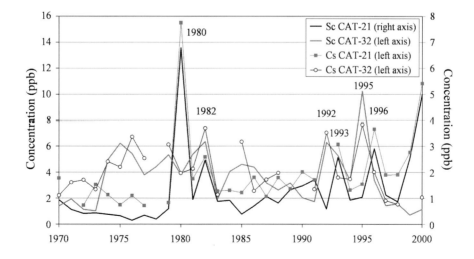

Figure 2: Concentrations of Sc and Cs in two tree samples from a Turkish forest. The Fe concentration closely follows that of Sc in both CAT-21 and CAT-32 and could have been used in place of Sc in this graph. The CAT-21 peaks in 1980 also appear in Na and Co concentrations. Sodium has correlated concentration peaks in 1982–1983 which may be related to the Sc and Cs peaks in 1982. The pattern of offset Sc and Cs peaks in 1992–1993 is repeated later in the years 1995 and 1996.

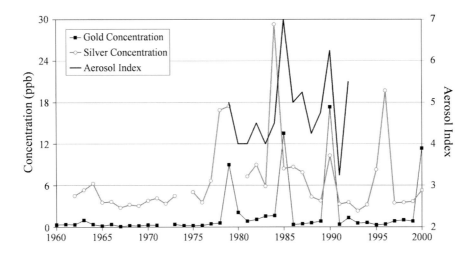

Figure 3: Gold and Silver concentration in a tree sample from a Turkish forest. The Au and Ag concentrations exhibited large peaks in single rings which was unique among all identified isotopes. The presence of the peaks was verified when tree sample CAT-32 was irradiated for a second time and re-analyzed. The average Aerosol Index for the months May, June, July, and August was calculated from the monthly average index values available from the TOMS website.

ple CAT-32 was irradiated for a second time and re-analyzed. Therefore, it is tempting to propose that Au and Ag have peaks in single years because they are relatively immobile in tree rings and do not suffer from the same migration effects as the other elements. The possibility of contamination will be studied further to eliminate this concern. Figure 3 shows the Ag and Au concentrations in CAT-32 along with the average Aerosol Index for the months May, June, July, and August over the Çatacık forest in Turkey. This data is available from the TOMS website (NASA/GSFC TOMS Team 1993).

Discussion

The elements found in tree rings in this study could be broken into categories based on the typical amount of variation from ring to ring. Some elements, including K, Br, and Zn vary by a small amount and have flat or gently sloping chronologies. Other elements have only moderate variations, including Co, Mn, and Na. Elements with a high degree of variability include Sc, Cr, Cs, Sb, and Fe which were measured with long (1 day) count trials. The noble metals Ag and Au demonstrate behavior unique from the other identified elements with large variations in one of the tree samples. The chemistry of Ag and Au prevents them from becoming mobile except under special conditions and this may explain the presence of peaks which occur in single tree-rings (Butt et. al. 1998).

The data obtained from this study resulted in the identification of key years in which there was some degree of correlation between elemental concentrations. The first interesting range of years is 1978 to 1980. The unique behavior during these years includes correlated Mn peaks in 1978, Ag and Au peaks in 1979, and peaks in the concentrations of several elements in CAT-21 in 1980. As shown in Figure 3, these peaks correspond roughly to the peak in Aerosol Index in 1980. A major contribution to the Aerosol Index over Turkey is African dust blowing across the Mediterranean Sea (Goudie and Middleton 2001). If this correlation is correct, one might expect that the heightened levels of Ag, Au, and Mn are due to direct fall-out of these elements as constituents of dust. For instance, it has been shown that an oak forest in northeastern Spain derives much of its nutrients from wet deposition of dust aerosols originating in Africa (Avila et. al. 1997). Two other ranges of years, 1984–1985 and 1989–1990, also have peaks in Au and Ag concentrations, correlation among Mn concentration peaks, and corresponding peaks in Aerosol Index. Volcanic eruptions may also contribute to the Aerosol Index although their presence is not obvious. Small peaks in the Aerosol Index in 1982 and 1992 may result from the eruptions of El Chichon and Pinatubo, respectively. As described below, these years also correspond to correlations between Sc, Cs, and Fe concentrations. The problem remains that the large Au and Ag peaks observed in CAT-32 were not observed in any of the other three tree samples.

The eruption of El Chichon in early April of 1982 corresponds to correlated peaks in Sc, Cs, and Fe concentrations in CAT-21 and CAT-32. Similarly, the eruption of Pinatubo in June 1991 and the eruption of Hudson, Cerro, in August of 1991 roughly correspond to peaks in Sc, Cs, and Fe in the same two tree samples but are slightly offset over the years 1992 and 1993. Finally, there is a correlation between Sc, Cs, and Fe in 1995–1996 which is correlated with a peak in Ag concentration in 1996. There were no large volcanic eruptions during this time and the gap in TOMS data prevents a possible correlation with the Aerosol Index. The fact that the Ag concentration increased in 1996 without a corresponding peak in gold concentration may provide an important clue to the type of event that occurred in this year.

Conclusions

The concentrations of several elements have been determined in portions of four *Pinus sylvestris* trees that grew in western Turkey. Some correlation in Mn, Na, K, and Zn concentration was apparent between the four tree samples especially in the years 1978, 1985, 1989, and 1997. The years 1978, 1985, and 1990 also correspond to peaks in Ag and Au concentration in one of the tree samples and the Aerosol Index over the sample site calculated by the TOMS group. However, a gap in TOMS data makes it impossible to investigate a possible correlation with the Ag and Mn concentration peaks in 1996 and 1997. Finally, correlated Sc, Cs, and Fe peaks appear in the year of the El Chichon eruption in 1982, in the year following the Pinatubo eruption of 1991, and in the year of 1995 during which there were no major volcanic eruptions. The peaks in 1995–1996 may be related to the phenomena that caused the Ag concentration peak in 1996.

While there are many indications of correlation between tree uptake and environmental events, there are also some troubling aspects. For instance, the peaks in Ag and Au concentration that correlate so nicely with the Aerosol Index appear in only one of the four tree samples tested. It must be determined whether some unique localized conditions of that tree made it an ideal recorder of environmental conditions.

Dendrochemistry seeks to create a superior proxy data set of past environmental events. However, it may be necessary to compare tree-ring elemental concentrations to appropriate proxy data sets in order to understand the chemical response of trees to environmental signals. For this purpose the TOMS data proved to be useful but of limited applicability because of the small range of years that it covers. This study indicates that it may be possible to identify signals from large volcanic eruptions and Mediterranean dust storms in tree-ring concentration data of Turkish trees. In particular, the concentrations of Mn, Ag, Au, Sc, and Cs appear to have potential for indicating environmental conditions. However, the interpretation of elemental concentration peaks in this study and in tree-rings in general is still uncertain. Further control studies will help pinpoint the conditions under which volcanic eruptions and other environmental events can be identified using tree-ring chemistry.

Acknowledgments

We would like to thank the Radiation Science and Engineering Center at Penn State for the reactor time necessary for completing the NAA measurements. We would also like to thank Peter I. Kuniholm for his supply of wood samples and collaboration on this project over the past eight years. We look forward to continued work with Peter and with Sturt Manning, whose dendrochemical experience has already provided helpful discussions and insights into the problem of tree-ring compositional analysis.

References

Avila, M., Alarcon, M., and Queralt, I. 1997. The chemical composition of dust transported in red rains—Contribution to the biogeochemical cycle of a holm oak forest in Catalonia (Spain). *Atmospheric Environment* 32: 179–191.

Baes, C.F., III, and McLaughlin, S.B. 1984. Trace Elements in Tree Rings: Evidence of Recent and Historical Air Pollution. *Science* 224: 494–497.

Butt, C.R.M., Gray, D.J., Lintern, M.J., Robertson, I.D.M., Taylor, G.F., and Scott, K.M. 1998. CRC LEME Open File Report 29: Gold and associated elements in the regolith—dispersion processes and implications for exploration. Wembley: CRC LEME, CSIRO.

DeWalle, D.R., Swistock, B.R., Sayre, R.G., and Sharpe, W.E. 1991. Spatial Variations of Sapwood Chemistry with Soil Acidity in Appalachian Forests. *Journal of Environmental Quality* 20: 486–491.

DeWalle, D.R., Tepp, J.S., Swistock, B.R., Sharpe, W.E., and Edwards, P.J. 1999. Tree-Ring Cation Response to Experimental Watershed Acidification in West Virginia and Maine. *Journal of Environmental Quality* 28: 299–309.

Goudie, A.S., and Middleton, N.J. 2001. Saharan dust storms: Nature and consequences. *Earth Science Reviews* 56: 179–204.

Graf, H.-F., Langmann, B., and Feichter, J. 1998. The contribution of Earth degassing to the atmospheric sulfur budget. *Chemical Geology* 147: 131–145.

Guo, S., Bluth, G.J.S., Rose, W.I., Watson, I.M., and Prata, A.J. 2004. Re-evaluation of SO2 release of the 15 June 1991 Pinatubo eruption using ultraviolet and infrared satellite sensors. *Geochemistry Geophysics Geosystems* 5: 4. [Online: http://www.agu.org/journals/gc/ (cited 25 March 2004)].

Kabata-Pendias, A., and Pendias, H. 1992. *Trace Elements in Soils and Plants*, 2nd edition. Boca Raton: CRC Press.

NASA/GSFC TOMS Team 1993. Aerosal Monthly Averages. Nimbus 7 TOMS Data and Images. [Online: http://toms.gsfc.nasa.gov/n7toms/nim7toms_v8.html (last cited 2 March 2007)].

Pearson, C.L. 2003. Volcanic Eruptions, Tree Rings and Multielemental Chemistry: An Investigation of Dendrochemical

Potential for the Absolute Dating of Past Volcanism. Ph.D. Dissertation, University of Reading.

Pearson, C.L., Manning, S.W., Coleman, M., and Jarvis, K. 2005. Can tree-ring chemistry reveal absolute dates for past volcanic eruptions? *Journal of Archeological Science* 32: 1265–1274.

Pearson, C.L., Manning, S.W., Coleman, M., and Jarvis, K. 2006. A dendrochemical study of *Pinus sylvestris* from Siljansfors Experimental Forest, Central Sweden. *Applied Geochemistry* 21: 1681–1691.

Steinhauser, G., Sterba, J.H., Bichler, M., and Huber, H. 2006. Neutron Activation Analysis of Mediterranean volcanic rocks—An analytical database for archaeological stratigraphy. *Applied Geochemistry* 21: 1362–1375.

Ünlü, K., Kuniholm, P.I., Chiment, J.J., and Hauck, D.K. 2005. Neutron Activation Analysis of Absolutely-Dated Tree Rings. *Journal of Radioanalytical and Nuclear Chemistry* 264: 21–27.

Watmough, S.A. 1999. Monitoring historical changes in soil and atmospheric trace metal levels by dendrochemical analysis. Environmental Pollution 106: 391–403.

Watmough, S.A. 1997. An evaluation of the use of dendrochemical analyses in environmental monitoring. *Environmental Reviews* 5: 181–201.

Article submitted April 2007

Neutron Activation Analysis of Dendrochronologically Dated Trees

K. Ünlü, P. I. Kuniholm, D. K. Hauck, N. Ö. Cetiner, and J. J. Chiment

Abstract: *Uptake of metal ions by plant roots is a function of the type and concentration of metal in the soil, the nutrient biochemistry of the plant, and the immediate environment of the root. Uptake of gold (Au) is known to be sensitive to soil pH for many species. Soil acidification due to acid precipitation following volcanic eruptions can dramatically increase Au uptake by trees. Identification of high Au content in tree rings in dendrochronologically-dated, overlapping sequences of trees allows the identification of temporally-conscribed, volcanically-influenced periods of environmental change. Ion uptake, specifically determination of trace amounts of gold, is being performed for dendrochronologically-dated tree samples utilizing Neutron Activation Analysis (NAA). We are trying to see whether the concentration of gold can be correlated with known environmental changes, e.g., volcanic activities, during historic periods. Several thousand wood samples have been scanned for gold concentration measurement. Samples containing elevated levels of gold will be analyzed again for short and long half-life elements, e.g., silver, copper, etc., to investigate other elemental signatures of environmental change using a Compton suppressed NAA system. The samples are activated at the 1 MW Breazeale Nuclear Reactor and counted at the NAA laboratory, both at Penn State. All samples used in this study have been independently dated by researchers at The Malcolm and Carolyn Wiener Laboratory for Aegean and Near Eastern Dendrochronology at Cornell University, where over 40,000 individually-dated wood samples with 4.5 million rings, spanning the period from 7000 BC to the present, are archived. Most of these samples are available for this project. By applying the NAA technique to this material, we hope to be able to define periods of global environmental stress during the past nine thousand years and deduce environmental and climatic history from the measured data.*

Introduction

Instrumental Neutron Activation Analysis (INAA) is used for the dendrochemistry project at the Breazeale Nuclear Reactor at the Pennsylvania State University to search for heightened signature element concentrations in tree rings (Baes and McLaughlin 1984; DeWalle et al. 1997; Guyette et al. 1991, 1992; Hagemayer 1993; Kagawa et al. 2002; Pearson 2003; Watmough 1997). Increased elemental concentrations may be an indication of environmental effects. The main goal of this study to try to identify the environmental effects on history that could possibly be obtained from dendrochemistry data. The initial focus element for identification was gold because of its suitability to INAA. Other elements are also investigated to find suitable signature elements for identification of volcanic eruptions and/or climate changes.

Uptake of metal ions by plant roots is a function of the type and concentration of metals in the soil, nutrient biochemistry of the plant, and immediate environment of the root. Uptake of gold (Au) is sensitive to soil pH for many species (USGS 1970; Stewart and Mckown 1995; Anderson et al. 1998, 1999). Soil acidification due to acid precipitation following volcanic eruptions may dramatically increase Au uptake. Volcanic aerosols from major eruptions stress plant communities in a variety of ways: darkness, acidity, increased or decreased rainfall. Plant stress invokes enzymatic antioxidant systems. Copper (Cu), an important trace nutrient in plants, is necessary for these systems. Plants do not discriminate in uptake among the Group 1b transition elements: Cu, silver (Ag), and Au. Once admitted to the plant Au is bound within the wood of the year and does not participate in biochemical reactions. Identification of high-Au tree rings in dendrochronologically-dated, overlapping sequences of trees may allow the identification—to the exact year—of temporally-conscribed, volcanically-produced periods of environmental change. We believe that a correlation with ion uptake in tree rings would allow us to extend the climatological record back thousands of years.

The Malcolm and Carolyn Wiener Laboratory for Aegean and Near Eastern Dendrochronology at Cornell University archives more than 40,000 individually-dated tree samples with 4.5 million rings from 109 forests in the eastern Mediterranean and former Soviet Union and some 600 archaeological sites. These samples span most of the period from 7000 BC to the present (Kuniholm 1990, 1996). All dendrochronogically dated samples at the Wiener Laboratory are available for this study.

NAA is the assay of choice for Au in organic material (Pillay 1976; Oliveira et al. 1997; Schaumloffel et al. 1996; Ünlü et al. 2005). Au has a large neutron cross-section and NAA can detect the element in the parts-per-billion (ppb) range. A dedicated NAA system was built at Cornell's Ward Center for Nuclear Sciences for this study with NSF and Cornell funds. Samples were activated at the core of Cornell's 500 kW TRIGA Mark II research reactor, and NAA measurements were made using a spectroscopy facility with an automatic sample changer system. After Cornell's decision to close the Ward Center for Nuclear Sciences, the prepared samples, gamma spectroscopy system, and sample changer were transferred to Pennsylvania State University, Radiation Science and Engineering Center (RSEC). Measurements have continued at Penn State using the 1 MW Breazeale Nuclear Reactor. A description of the Pennsylvania State University Neutron Activation Analysis Facility, various irradiation schemes, and analysis results of four samples each containing several hundreds of rings will be discussed.

Penn State NAA facility

The RSEC facilities include the Penn State Breazeale Reactor (PSBR), gamma irradiation facilities (In-pool Irradiator, Dry Irradiator, and Hot Cells), and various radiation detection and measurement laboratories. The PSBR, which first went critical in 1955, is the nation's longest continuously operating university research reactor. The PSBR is a 1 MW, TRIGA with movable core in a large pool with pulsing capabilities. The core is located in a 24 foot deep pool with $\sim 71,000$ gallons of demineralized water. A picture of the reactor core is shown in Figure 1. A variety of dry tubes and fixtures are available in or near core irradiations. A pneumatic transfer system is also available for irradiation of samples. When the reactor core is placed next to a D_2O tank and graphite reflector assembly near the beam port locations, thermal neutron beams become available for neutron transmission and neutron radiography measurement from two of the seven existing beam ports. In steady state operation at 1 MW, the thermal neutron flux is $1x10^{13} n/cm^2$ per second at the edge of the core and $3x10^{13} n/cm^2$ per second at the central thimble. The PSBR can also pulse with the peak flux for maximum pulse $\sim 6x10^{16} n/cm^2$ per second with pulse half width of ~ 10 per millisecond.

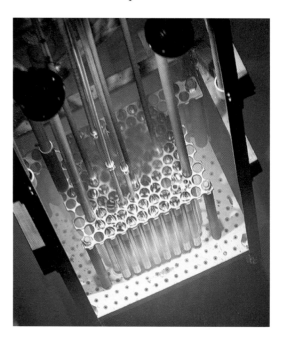

Figure 1: Penn State Breazeale nuclear reactor core.

The NAA samples were irradiated in Dry Irradiation Tubes (DITs) for long-lived isotope identification (Hauck 2005). For short-lived isotope identification a Pneumatic Transfer System (PTS) was used. Sample irradiation facilities at the Radiation Science and Engineering Center, and the NAA facility, are briefly described below. The PSU NAA facility consists of an Automatic Sample Handling System (ASHS), a High-Purity Germanium (HpGe) Detector, a Digital Spectrum Analyzer (DSA 2000) and Genie-2000 software.

Dry Irradiation Tubes

Two dry irradiation tubes were installed in the reactor core for irradiation of NAA samples. The DITs occupy approximately symmetric positions within the core. The bottom of the DITs was designed to be identical to the bottom fuel rod tips and fit into the bottom grid plate. The height of the DITs is approximately 21 feet and they bend to 6 inches off the centerline approximately one third of the way up. This allows for samples to be pulled out of the core while they decay but maintain a depth of 14 feet of water as a radiological shield. Flux measurements performed in the DITs at 10 W, 800 kW, and 1 MW show that the height of the flux peak is strongly dependent on the position of the control rods but fairly independent of core position within the reactor pool. Measurements were done both in the center of the reactor pool position and coupled core-D_2O tank position. The peak flux at 1 MW was approximately $1.5x10^{13} n/cm^2/s$ with DIT1 being consistently slightly higher than DIT2. The flux measured inside the polyethylene vials used for wood irradiation is approximately $1.2x10^{13} n/cm^2/s$, with the difference presumably from the position of the polyethylene vials and possibly from moderation by the polyethylene vial.

Pneumatic Transfer System

The Pneumatic Transfer System (PTS) allows samples to be sent from the NAA laboratory directly into the reactor core through a pneumatic tube loaded with carbon dioxide. A timer system provided by the reactor is set to the required irradiation time plus the 5 seconds travel time. At the end of the irradiation the sample is automatically returned to the hood in the NAA laboratory. The PTS is usually used to find isotopes with short half-lives and it is often preferred to transfer the sample from the hood to the detector as quickly as possible. For this reason it is easiest to count the sample in the vial in which it was irradiated. It is still necessary to remove the sample vial from the "rabbit," the larger vial designed specifically for use in the PTS. With practice, it is possible to begin counting the sample 40–50 seconds after the sample leaves the core. The time between irradiation and counting is measured by hand with a stopwatch.

Automatic Sample Handling System

The Automatic Sample Handling System (ASHS) brought from the Ward Center at Cornell was installed and tested at Penn State (Ünlü et al. 2005). Batch programming files have been written specifically for the dendrochemistry project and for general usage. These files allow the user to define the number of samples to be run in the ASHS in sequence and the amount of time that they are to be counted. Special presets can be used, for example real time vs. live time, and files can be automatically named and analyzed according to user specified values. The ASHS holds up to 90 samples at one time, allowing for a consistent throughput of samples, especially useful for such an extensive study as the dendrochemistry project.

High Purity Germanium Detector

A High Purity Germanium (HpGe) detector was used for gamma spectroscopy. The HpGe detector used for this study has a 36% relative efficiency and 1.8 keV resolution at the higher energy cobalt peak (1332 keV). This has allowed for more accurate isotope identification and area calculations. The detector is surrounded by 3000 lbs of pre-WWII lead bricks to block out environmental background radiation. Figure 2 shows the ASHS with HpGe detector with shielding and data acquisition and counting systems.

Data acquisition and analysis

A digital spectrum analyzer (DSA 2000) and state of the art gamma analysis software (Genie 2000) are used at the PSU NAA facility. The DSA 2000 works as the multi-channel analyzer (MCA), amplifier, and high voltage power source to the detector. The Genie 2000 Gamma Analysis Software package includes a set of advanced analysis algorithms for further processing of the gamma spectra acquired with the Genie 2000 Basic Spectroscopy Software. These algorithms provide a complete analysis of gamma ray spectra obtained from any type of gamma detector and also are accessible from the Genie 2000 interactive environment. Analysis execution utilizes data resident in the Configuration Access Method (CAM) data structures. Results from each algorithm are also stored in CAM files, making the resultant data file a complete record of the entire analysis.

Sample description and preparation

Tree samples were gathered and dated by the Malcolm and Carolyn Wiener Laboratory for Aegean and Near Eastern Dendrochronology at Cornell University. Members of the laboratory traveled to Turkey and Greece to obtain tree samples for dendrochronology and dendroanalysis. Tree samples chosen for further analysis through INAA were systematically cut into individual rings to produce samples which range from .1 to .25 grams in size. The rings were placed into polyethylene heat sealed bags, labeled with a permanent marker, and set aside for irradiation (Figure 3). Samples were irradiated for 4 Megawatt-hours at the DITs for long-lived isotope identification (e.g., ^{198}Au) and allowed to decay for 15 hours before counting. This allows for adequate activation of the gold without producing excessive amounts of radioactivity. Tree rings are irradiated in groups of 40. Therefore, counting samples for 1 hour each insures that they will all be counted before the gold has decayed by 1 half-life.

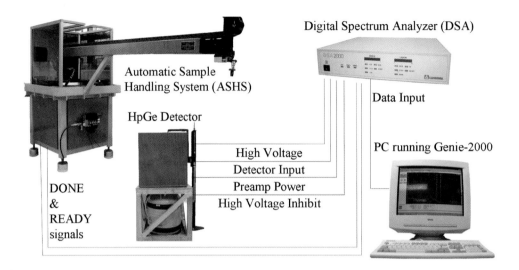

Figure 2: Automated sample handling system, HpGe detector with shielding and data acquisition, and processing equipment for PSU-NAA facility.

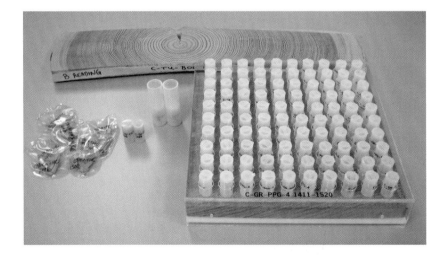

Figure 3: a) typical wood sample, b) samples cut from a single ring and heat-sealed in bags, c) samples after irradiation and weighing, but before counting, d) archived samples.

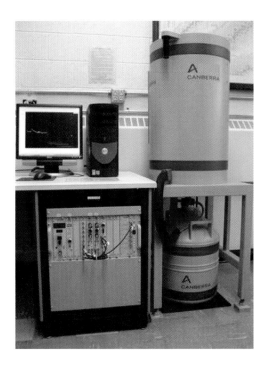

Figure 4: A picture of PSU Compton Suppression System showing data acquisition and analysis components, HpGe detector, and shield with NaI guard detectors.

PSU Compton Suppression System

A Compton Suppression System (CSS) was purchased and tested for usage in the dendrochemistry and other projects at RSEC (Çetiner 2008). Compton suppressors provide a tool to suppress the unwanted background. In order to reduce the contribution of scattered gamma-rays, the detector is surrounded by a guard detector. The two detectors are operated in anti-coincidence, which means that if an event occurs at the same time in both detectors, then the event is rejected. The guard detector catches the escaping photons, and the effect of those photons is subtracted from the background. The combination of a central HpGe detector and an annulus NaI(Tl) detector is called a Compton suppression spectrometer. The Compton Suppression System at PSU includes a HpGe detector, NaI guard detectors in a lead shield, NIM Bin/Power supply (Canberra Model 2100), PC desktop, and Genie 2000 software (Figure 4). A Canberra Model 3106D NIM high voltage power supply is used for operation with the HpGe detector.

Tree-ring analysis with NAA: irradiation schemes, results, and discussions

Three major irradiation and counting schemes have been employed for analyzing tree-rings with NAA (Hauck 2009). Each scheme was designed for identifying isotopes with a different range of half-lives. Certain isotopes or elements are observed in order to identify promising environmental markers. The results obtained using a variety of irradiation schedules will be discussed with an emphasis on the applicability of the identified isotopes to environmental monitoring. This discussion will also include the advantages and disadvantages of each irradiation schedule and the necessary steps in the experimental work itself.

Irradiation in the DITs

Much of the work completed for the dendrochemistry project focused on quantifying gold in tree-rings with irradiations suited to identify the ^{198}Au isotope which has a 2.7 day half-life. Following this goal, four tree samples each over 300 years in length were analyzed for the dendrochemistry project according to the following method. Tree samples were cut into individual rings and were irradiated in batches of 40 tree-rings for 4 Megawatt hours in the Breazeale Nuclear Reactor (BNR) core. The thermal and epithermal fluxes in the BNR core are approximately $1.7x10^{13} n/cm^2/s$ and $6x10^{11} n/cm^2/s$, respectively. Each batch was allowed to decay over night and then each tree-ring was counted for one hour each. This procedure allowed for the identification of ^{24}Na, ^{42}K, ^{56}Mn, ^{82}Br, ^{69}Zn, ^{140}La, and ^{198}Au. In one tree La was not detected consistently because it was present at or below the detection limits and Au was near the detection limits in all trees. Mn was detected but has a short enough half-life (2.6 hrs) that it had decayed away before the counting of each batch of samples was completed. All other elements (Na, K, Br, and Zn) were quantified well above the detection limits. The tree samples analyzed in this way and the average concentration results are provided in Table 1. The concentration values are in parts-per-million except where specified otherwise. The values in parenthesis are the number of rings that had been identified for each element and were used as part of the calculation of the average.

Gold behaved in a unique manner compared to the other elements listed in Table 1. Au showed no anomaly at the heartwood boundary (HWB) and has large peaks in single tree-rings. In most cases the other identified elements had either peaks or increases in the continuum level at the HWB. Au concentrations observed are shown in Figure 5 for four different samples analyzed.

The next step in the dendrochemistry project was the analysis of four small tree samples (30–60 years in length) from the Çatacık forest in Turkey. The period of analysis covered at least the years 1980–2002. The four Çatacık tree samples (C-TU-CAT-32, C-TU-

Tree Sample	C-GR-PPG-4A	C-TU-KLK-10B	C-TU-CAT-14C	P-SW-SEQ-2
Years Analyzed	1411–1979	1628–1999	1500–1979	1446–1948
Country	Greece	Turkey	Turkey	USA (CA)
Au*	8 ppb (467)	7 ppb (314)	4 ppb (445)	8 ppb (360)
Br	.08 (470)	.07 (355)	.06 (459)	.4 (368)
K	10 (548)	22 (369)	18 (459)	37 (370)
La	.003 (42)	.006 (203)	.02 (428)	.2 (369)
Mn**	7.4 (47)	11 (67)	21 (237)	14 (200)
Na	69 (557)	60 (369)	37 (459)	86 (374)
Zn	.67 (197)	1.6 (331)	1.5 (456)	1 (307)

Table 1: Average isotopic concentration in four tree samples irradiated in the DITs.
(*) Rows 4–10 contain the average isotopic concentrations in ppm units unless specified otherwise. The values in parentheses are the number of concentration values used for calculating the average.
(**) The Mn concentration was measured in a small percentage of the tree-ring samples due to its short half-life.

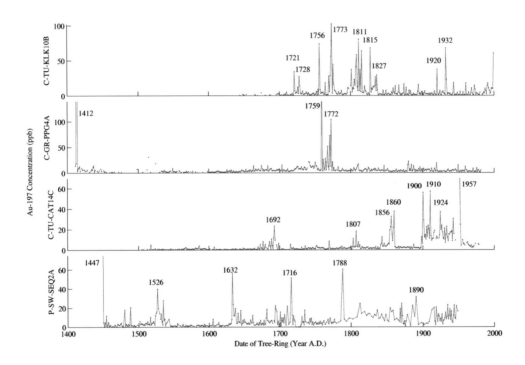

Figure 5: Gold concentrations in 4 different tree samples. Observed gold peaks at the date of tree rings are marked.

Tree Sample	C-TU-CAT-32	C-TU-CAT-21	C-TU-CAT-16	C-TU-CAT-28
Years Analyzed	1940–2002	1969–2002	1972–2002	1971–2002
Au	1.3 (61)	.19 ppb (30)	.42 ppb (30)	.14 ppb (16)
Br	27 ppb (62)	18 ppb (33)	18 ppb (30)	43 ppb (31)
K	41 (62)	20 (33)	19 (30)	30 (31)
La	27 ppb (62)	15 ppb (33)	19 ppb (30)	16 ppb (31)
Mn	51 (62)	44 (33)	42 (30)	29 (30)
Na	3.1 (62)	5.1 (33)	6.2 (30)	2.5 (31)
Zn	.78 (62)	.81 (33)	.36 (30)	.86 (31)

Table 2: Average isotopic concentrations obtained in four tree samples from the Çatacık Forest using irradiation in the DITs.

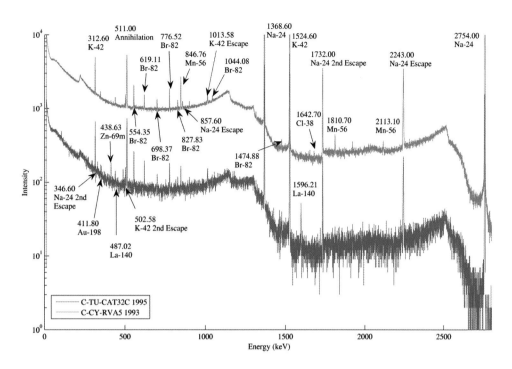

Figure 6: Typical gamma spectrum from tree-ring samples C-TU-CAT-32C (1995) and C-CY-RVA-5 (1993) for long irradiation and 1 hour count after 24 hours of decay.

CAT-21, C-TU-CAT-16, and C-TU-CAT-28) were analyzed according to the procedure above except for the following changes. It was found that the isotope with the most interesting chronologies, Au, was being detected at or near the minimum detectable concentration. Accordingly, the batch sizes were reduced to approximately 30 samples and each tree-ring was counted for 2 hours. This was sufficient for reliably recovering gold in most tree-rings and with better statistical accuracy. Doing so also decreased the La detection limits so that it could be more reliably identified. ^{56}Mn has a short enough half-life that it had decayed away before the samples were finished being counted. To make up for this, all samples were counted for 15 minutes each prior to starting the 2-hour counts.

The four tree samples that were analyzed from the Çatacık forest are listed in Table 2. A typical gamma spectrum for two samples analyzed is shown in Figure 6. The first tree sample analyzed was C-TU-CAT-32 without the new irradiation procedure standards and the initial results of this tree sample led to the implementation of the new procedure.

Large peaks in Au concentration were observed, and each ring was reanalyzed with the new procedure to check the reproducibility of the gold peaks. The remaining three samples were analyzed with the new procedure described above. Tree Sample C-TU-CAT-16 was divided in half lengthwise and each piece was cut into individual rings and analyzed separately. The results in Table 2 for C-TU-CAT-16 are the average of the results from the two sections, and the results for C-TU-CAT-32C represent the average of the results from the first and second irradiations.

At a later time, each of the tree-rings from C-TU-CAT-32 and C-TU-CAT-21 were recounted for one day each. Although this process was extremely time consuming (7 samples per week versus 40-60 per week with the 2-hour count time), a wealth of isotopic data was obtained. It resulted in the identification of 7 additional metals including ^{46}Sc, ^{59}Fe, ^{60}Co, ^{65}Zn, ^{110m}Ag ^{124}Sb, and ^{134}Cs. The concentrations of these long-lived isotopes in C-TU-CAT-32 and C-TU-CAT-21 are presented in Table 3. The Ag and Au concentrations in C-TU-CAT-32 were correlated with each other. Ag and Au are the only elements that had the sharp peaks above the level of their continua. It was hypothesized that the unique noble metal characteristics of these elements made them suitable for documenting events in the life of the tree. In particular, the Ag and Au compositions may have been documenting the severity of environmental effects such as volcanic activity and/or Mediterranean dust storms.

Ag has a short-lived isotope (^{108}Ag) which might be identified with short irradiations of tree rings. This study was pursued to determine if environmentally significant data, especially the concentration of Ag

	C-TU-CAT-32	**C-TU-CAT-21**
Years Analyzed	1962–2000	1969–2000
^{46}Sc*	1.6 ppb (39)	2.6 ppb (33)
^{59}Fe	16 ppb (33)	26 ppb (33)
^{60}Co	78 ppb (39)	32 ppb (33)
^{65}Zn	1.8 ppm (39)	1.7 ppm (33)
^{110m}Ag	6.7 ppb (37)	4.3 ppb (33)
^{124}Sb	2.8 ppb (38)	1.8 ppb (29)
^{134}Cs	1.9 ppb (34)	2.0 ppb (30)

Table 3: Average concentrations of long-lived isotopes in two tree samples.
(*) Rows 3–9 contain the average isotopic concentrations. The values in parentheses are the number of concentration values used for calculating the average.

in tree rings, could be determined in a more efficient manner through short irradiations.

Irradiation in the Pneumatic Transfer System (PTS)

A sample of *Pinus brutia* from the Roudhias Valley, Arodafna, Cyprus, identified as C-CY-RVA-5, was chosen for analysis in the PTS and obtained from the Malcolm and Carolyn Wiener Laboratory at Cornell (S.W. Manning 2007, pers. comm.). An activity prediction indicated that 1-minute irradiations at 900 kW (the maximum power allowed while using the PTS) may make it possible to quantify the amount of Ag in each tree-ring. The empty vials were labeled with a permanent marker and weighed. Individual tree-rings were then cut from the sample using a stainless steel knife chisel, cleaned with alcohol, and placed in the vials which were heat-sealed and re-weighed. Shortly before each run, sample vials were loaded into the rabbits for irradiation in groups of 25–35 at a time. Each sample was irradiated for 1 minute and counted for a live time of 200 seconds. The decay time was minimized as much as possible for each sample and was typically 40–60 seconds. After counting, each spectrum file was saved in the Genie-2000 for analysis at a later time.

Irradiation in the PTS allowed for the analysis of 14 isotopes including ^{18}O, ^{24}Na, ^{27}Mg, ^{28}Al, ^{37}S, ^{38}Cl, ^{42}K, ^{49}Ca, ^{52}V, ^{56}Mn, ^{64}Cu, ^{108}Ag, ^{128}I, and ^{41}Ar. The concentration of each isotope in the wood was calculated using Instrumental NAA and assuming a constant efficiency calibration performed for the bottom of the sample vial. The average concentration of each isotope and the number of rings used for the calculation are given in Table 4.

Ag was considered the most promising element that could be identified using short irradiations. ^{108}Ag has a convenient half-life (2.37 m) and, based on previous results, appears to be climatically significant. Ag was identified in tree sample C-CY-RVA-5 although there are many missing values (evidently due to decreasing concentration toward the center of the tree) at years earlier than about 1965.

Figure 7 shows the silver concentrations from tree samples C-TU-CAT-32 and C-CY-RVA-5 plotted together. It is evident that the chronologies are similar with large peaks within one year of each other in 1984–85, 1988–89, and 1996–97. It was hypothesized that Ag and Au maintain large peaks in single rings because they become soluble only under special chemical conditions and, once deposited in the tree, may be relatively immobile. A correlation of Ag concentration among different trees would support the hypothesis that Ag is an appropriate monitor of environmental conditions. Figure 7 loosely shows an apparent correlation between one tree in Cyprus and one tree in Turkey.

Most of the identified elements are relatively constant between 1865 and the bark, neglecting variability. The heaviest isotope identified was ^{128}I, which exhibited larger ring-to-ring variability than the other isotopes and is lacking the strong 1929 anomaly evident in many of the other elements.

It was also possible to measure the concentration of S which may be available to trees in greater amounts during acidic events (DeWalle et al. 1997; Wallner 1998). If the S concentration in tree-rings is found to record reliably the availability of S to the tree, it may be possible to differentiate between types of environmental events. The data would be similar to that obtained by from the S composition analysis of ice cores. Volcanic eruptions tend to release large amounts of S gasses and aerosols which are washed out of the atmosphere as acid rain. This may cause high solubility of metals and therefore greater availability to the tree (Bluth et al. 1993). Extended acidification may also cause leaching and depletion of these metals in the soil (Guyette et al. 1992, 1989; Johnson and Siccama 1983; Kogelmann and Sharpe 2006). The dry

Isotope	Concentration	Isotope	Concentration
^{17}O	632 (138)	^{41}K	12 (138)
^{23}Na	10 (138)	^{48}Ca	1.0 (138)
^{26}Mg	9.0 (138)	^{51}V	4 ppb (130)
^{27}Al	2.0 (138)	^{55}Mn	4.8 (138)
^{36}S	7 ppb (102)	^{65}Cu	.067 (119)
^{37}Cl	5.0 (138)	^{107}Ag	64 (56)
^{39}Ar	46 (138)	^{127}I	.068 (129)

Table 4: Average concentrations of short-lived isotopes in tree sample C-CY-RVA-5.

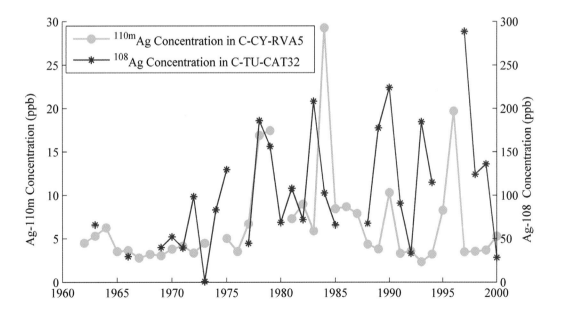

Figure 7: Comparison of Ag Concentration in C-TU-CAT-32 and C-CY-RVA-5.

deposition of dust may result in a greater availability of metals due to direct fallout with less of a S signal than for a volcanic eruption. Correlation studies may be important for comparing the S concentration or S buildup in the tree to concentrations of other elements.

In some tree-rings the amount of S in the wood was only slightly higher than the total amount of S in the polyethylene vial and the tree-ring sample. The variation of S in the blank vials contributed errors to the S concentration in the tree-rings. It may be possible to create better reliability in detecting S by reducing the vial size or extending the counting time.

Comparison of irradiation schemes

A comparison of each of the irradiation schemes used, their time requirements, and isotopic data obtained is provided in Table 5. Due to the smaller time constraints, flexible scheduling, and wealth of data obtained, short irradiations in the PTS were selected as the preferred irradiation scheme. It is not necessary to allow for time to count 20–30 samples between irradiations, making it easier to schedule reactor time. Another advantage of the short irradiations is the lower amount of radioactivity created. The longer irradiations in the DITs (discussed below) activated long-lived isotopes such as ^{60}Co in the wood. The activity of the samples was small but easily measurable after several months. However, the longest-lived isotopes activated in the PTS were K and Na (14 and 13 hours respectively). This made the samples suitable for non-controlled storage within a couple of days.

If all of the irradiation schemes are to be completed to achieve the full range of isotopic data available with NAA, it is advisable to begin analysis of samples with the short irradiations. Radio-chemical changes may be taking place during the long irradiations. For instance, ^{28}Al goes through more than 100 half-lives during a 4 hour irradiation. In addition, samples are less radioactive after the short irradiations and easier to prepare for subsequent irradiations.

The major problem seen with the short irradiations is the variability in the data. For most isotopes the concentration errors are expected to be about 25% at one standard deviation. This is slightly worse than the errors in the longer irradiations which were within 15-20% for most isotopes. It is believed that some of these errors come from changes in sample height and therefore changes in the gamma-ray detection efficiency. The height of each sample in its vial is dependent on the mass of the tree-ring sample and the size of the wood shavings.

If each wood sample had the same height in the vial, the concentrations would be off by a constant factor since an efficiency calibration for the bottom of the vial was used. However, variations in the height of the wood will introduce relative errors between samples. Error due to the efficiency variations could be eliminated by considering the ratio of elements in each tree ring. Oxygen concentration may serve as an appropriate comparison element for calculating ratios. Measuring the height of each sample and adjusting the efficiency calibration appropriately may eliminate some errors.

The minimum detectable concentration would be greatly reduced by decreasing the heights of the Compton shoulders in the background. The presence of the isotopes ^{49}Ca, ^{56}Mn, ^{28}Al, and ^{41}Ar all contribute significantly to the background under the smaller peaks such as Ag and I. Reducing the excess air in the vials would reduce the ^{41}Ar continuum. Ashing or digesting the samples so that they have a consistent geometry would make this easier. The presence of ^{56}Mn which is easily activated and has a long half-life compared to the other identified isotopes contributes a great deal of background. Any radiochemical separation that removed the Mn would be highly beneficial to detecting isotopes present in lesser amounts. Due to the presence of Mn it is very difficult to design a better scheme for short irradiations to identify Ag, making it a good candidate for use of a Compton Suppression System, also available at the RSEC.

Measurements with the Compton Suppression System

Elemental analysis of a wood sample from Porsuk, in the central Anatolian region of Turkey, with 819 Relative Gordion Year was performed by using the PSU Compston Supression System (CSS) in order to demonstrate the effects of the CSS system (Çetiner et al. 2008). The C-TU-POR-819 wood sample was activated at 1 MW in the PSU Breazeale Nuclear Reactor and then counted in the CSS. Figure 8 shows the comparison of suppressed and unsuppressed spectra with determined gamma peaks. As demonstrated in Figure 9, the 411.8 keV gold peak was buried in the Compton continuum without the suppression, whereas with the CSS, it was possible to resolve the gold peak.

Conclusions

Neutron Activation Analysis proved to be an important methodology for analyzing tree-ring samples for dendrochemistry. Our study showed that samples can be sensitively and nondestructively analyzed with NAA and the Compton Suppression System. Both systems are available at the Radiation Science and Engineering Center at Penn State, and analyses of

	Short Irradiations in the PTS*	Long Irradiations in the DITs* (short count time)	Long Irradiations in the DITs (long count time)
Irradiation time	1 minute	4 hours	4 hours
Count time	4 minutes	2 hours	1 day
Reactor time per sample	6 minutes	8 minutes	8 minutes
Samples per week	90	60	7
Elements Identified	O, Na, Mg, Al, S, Cl, K, Ca, V, Mn, Cu, Ag†, I	Na, K, Br, Zn, La, Au	Co, Zn, Zr, Cs, Sc, Fe, Ag

Table 5: A comparison of three different tree-ring irradiation procedures.
PTS = Pneumatic Transfer System, DIT = Dry Irradiation Tube
† Ag and La were only occasionally identified

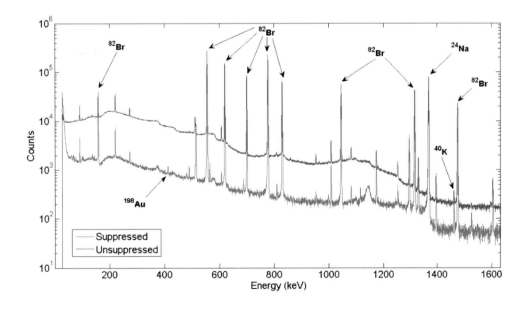

Figure 8: Comparison of C-TU-POR-3-819 wood sample gamma-ray spectrum with and without Compton suppression.

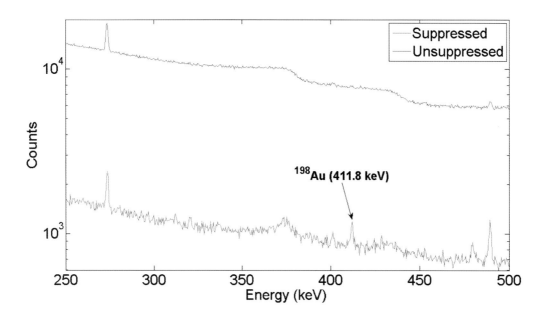

Figure 9: Expanded region of the C-TU-POR-3 spectrum showing the gold (Au) peak is visible under suppression.

various other tree-ring samples are continuing. For NAA of tree-rings, short irradiations in the Pneumatic Transfer System (PTS) were an efficient method of identifying a wide range of elements. It is recommended that analysis of any tree sample start by using the PTS, with further analysis of the tree sample then guided by the resulting isotopic information. Reducing the errors in the data would make it possible to identify small-scale true variability in the tree rings which may be relevant to environmental conditions.

Acknowledgments

This study is multidisciplinary in nature, involving contributions of many individuals. The authors thank the staff and students of the Malcolm and Carolyn Wiener Laboratory for Aegean and Near Eastern Dendrochronology at Cornell University, especially Meg Underwood and Pam Sullivan, for acquiring and preparing samples, and Paul Craven, Martin Moravek, Fred Decker, and Dr. Purushottam Dokhale of the Ward Center for Nuclear Sciences at Cornell University for helping to set up the automatic sample handling system, reactor operation, and initial data collection. We also thank the staff of the Radiation Science and Engineering Center at Penn State (Thierry Daubenspeck, Ron Eaken, Paul Rankin, and Mac Bryan) for helping to move equipment related to this project to Penn State, setting it up, testing, and performing reactor operations and numerous data collection. This study was supported by the National Science Foundation (NSF Award: SBR-9905389), Department of Energy (DE-FG07-05ID14702), the Ward Center for Nuclear Sciences at Cornell University, the Malcolm H. Wiener Foundation, and the Radiation Science and Engineering Center at Penn State University.

References

Anderson, C.W.N., Brooks, R.R., Stewart, R.B., and Simcock, R. 1998. Harvesting a crop of gold in plants. *Nature* 395: 553–554.

Anderson, C.W.N., Brooks, R.R., et al. 1999. Phytomining for nickel, thallium and gold. *J. Geochem. Explor.* 67: 407–415.

Baes, C.F., III, McLaughlin, S.B. 1984. Trace Elements in Tree Rings: Evidence of Recent and Historical Air Pollution. *Science* 224: 494–496.

Bluth, G.J.S., Schnetzler, C.C., Krueger, A.J., and Walter, L.S. 1993. The contribution of explosive volcanism to global sulfur dioxide concentrations. *Nature* 366: 327.

Cetiner, N.Ö. 2008. Specifications and Performance of the Compton Suppression Spectrometer at The Pennsylvania State University, MSc. Thesis, Pennsylvania State University.

Cetiner, N.Ö., Ünlü, K., and Brenizer, J.S. 2008. Compton Suppression System at Penn State Radiation Science And Engineering Center, *Journal of Radioanalytical and Nuclear Chemistry* 276(3): 615–621.

D'Arrigo, R., Wilson, R., Liepert, B., and Cherubini, P. 2008. On the 'Divergence Problem' in Northern Forests: A review of the tree-ring evidence and possible causes. *Global and Planetary Change* 60(3): 289–305.

DeWalle, D.R., Tepp, J.S., Pickens, C.J., Edwards, P.J., and Sharpe, W.E. 1995. Tree-Ring Chemistry Response in Black Cherry to Ammonium Sulfate Fertilization at two West Virginia Sites. In K.W. Gottschalk and S.L.C. Fosbroke (eds.) *Proceedings of the 10th Central Hardwood Forest Conference, 5-8 March 1995, Morgantown, WV*: 179–187. Gen. Tech. Rep. NE-197. Radnor, PA: U.S. Department of Agriculture, Forest Service, Northeastern Forest Experiment Station.

Fritts, H.C. 1976. *Tree Rings and Climate*. London: Academic Press.

Guyette, R.P., Cutter, B.E., and Henderson, G.S. 1992. Reconstructing soil pH from manganese concentrations in tree-rings. *Forest Science* 38: 727–737.

Guyette, R.P., Cutter, B.E., and Henderson, G.S. 1991. Long-Term Correlations between Mining Activity and Levels of Lead and Cadmium in Tree-Rings of Eastern Red-Cedar. *J. Environ. Qual.* 20: 146-150.

Guyette, R.P., Cutter, B.E., and Henderson, G.S. 1989. Long-Term Relationships between Molybdenum and Sulfur Concentrations in Redcedar Tree Rings. *J. Environ. Qual.* 18: 385–389.

Hagemeyer, J. 1993. Monitoring trace metal pollution with tree rings: a critical reassessment. In B. Markert (ed.) *Plants as biomonitors, Indicators for heavy metals in the terrestrial environment*: 541–563. Weinheim: VCH-Verlag.

Hauck, D.K. 2005. Neutron Activation Analysis of Dated Tree Rings. MSc Thesis, The Pennsylvania State University.

Hauck, D.K. 2009. Dendrochemistry: Seeing the Forest Through the Trees. PhD Dissertation, Pennsylvania State University.

Johnson, A.H., and Siccama, T.G. 1983. Acid deposition and forest decline. *Environ. Sci. Technol.* 17: 294A–305A.

Jonsson, A. Eklund, M., and Håkansson, K. 1997. Heavy Metals of the 20th Century Recorded in Oak Tree Rings. *J Environ. Qual.* 26: 1638–1643.

Kagawa, A., Aoki, T., Okada, N., Katayama, Y. 2002. Tree-Ring Strontium-90 and Cesium-137 as Potential Indicators of Radioactive Pollution. *J. Environ. Qual.* 31: 2001–2007.

Kogelmann, W.J., and Sharpe, W.E. 2006. Soil Acidity and Manganese in Declining and Nondeclining Sugar Maple Stands in Pennsylvania. *J. Environ. Qual.* 35: 433–441.

Jones, P.D., Briffa, K.R., Schweingruber, F.H. 1995. Tree-ring evidence of the widespread effects of explosive volcanic eruptions. *Geophysical Research Letters* 22: 1333–1336.

Kuniholm, P.I. 1990. Archaeological Evidence and Non-Evidence for Climatic Change. In S.K. Runcorn and J.-C. Pecker (eds.) *The Earth's Climate And Variability Of The Sun Over Recent Millennia. Phil. Trans. R. Soc. London, A*, 645–655.

Kuniholm, P.I. 1996. The Prehistoric Aegean: Dendrochronological Progress as of 1995, Absolute Chronology: Archaeological Europe 2500–500 B.C. *Acta Archaeologica* 67 Suppl. 1 327–335.

Lepp, N.W. 1975. The potential of tree-ring analysis for monitoring heavy metal pollution patterns. *Environmental Pollution* 9: 49–61.

Martinelli, N. 2004. Climate from dendrochronology: latest developments and results. *Global and Planetary Change* 40: 129–139.

Oliveira, H., Fernandes, E.A.N., and Ferraz, E. 1997. Determination of trace elements in tree rings of Pinus by neutron activation analysis. *Journal of Radioanalytical and Nuclear Chemistry* 217: 125–129.

Padilla, K.L., and Anderson, K.A. 2002. Trace element concentration in tree-rings biomonitoring centuries of environmental change. *Chemosphere* 49(6): 575–585.

Pearson, C.L. 2003. Volcanic Eruptions, Tree Rings and Multielemental Chemistry: An Investigation of Dendrochemical Potential for the Absolute Dating of Past Volcanism. Doctoral Thesis, University of Reading.

Pillay, K.K.S. 1976. Activation analysis and dendrochronology for estimating pollution histories. *Journal of Radioanalytical and Nuclear Chemistry* 32: 151–171.

Schaumloffel, J.C., Filby, R.H., and Moore, B.C. 1998. Ponderosa Pine Tree Rings as Historical Monitors of Zinc and Cadmium Pollution. *J. Environ. Qual.* 27: 851–859.

Schaumloffel, J.C., Filby, R.H., and Moore, B.C. 1996. Instrumental Neutron Activation Analysis of Tree Rings for Dendrochemical Studies. *J. of Radioanal. and Nucl. Chem.* 207(2): 425–435.

Stewart, K.C., Mckown, D.M. 1995. Sagebrush as a sampling medium for gold exploration in the Great Basin—evaluation from a greenhouse study. *J. Geochem. Explor.* 54: 19–26.

Symeonides, C. 1979. Tree-ring analysis for tracing the history of pollution: application to a study in northern Sweden. *J. Environ. Qual.* 8: 482–486.

Ünlü, K., Kuniholm, P.I., Chiment, J.J., and Hauck, D.K. 2005. Neutron Activation Analysis of Absolutely-Dated Tree Rings. *Journal of Radioanalytical and Nuclear Chemistry* 264: 21–27.

USGS 1970. Absorption of gold by plants. United States Geological Society Bulletin, B1314.

Wallner, G. 1998. Elements in tree rings of Norway spruce (Picca abies (L.) Karst.) as indicators for SO2 polluted sites at the East-Erzgebirge (Germany), *Journal of Radioanal. Nucl. Chem.* 238: 149–153.

Watmough, S.A. 1997. An evaluation of the use of dendrochemical analysis in environmental monitoring. *Environmental Reviews* 5: 181–201.

Article submitted October 2008

Third Millennium BC Aegean Chronology: Old and New Data from the Perspective of the Third Millennium AD

Ourania Kouka

Abstract: *The foundations for a relative chronology of the third millennium BC in the Aegean, namely for the Early Bronze Age (EB), were set already in the last century. However, the synchronization of cultural sequences in the various landscapes of the Aegean as well as their absolute dating was not successful until the late 1980s due to problematic stratification and to the limited evidence of radiocarbon data respectively. Sturt Manning's important contribution to the latter problem shed new light (Manning 1995). Archaeological evidence has furthermore demonstrated how regionalism and local traditions in material culture can create problems regarding the correlation even of neighboring landscapes within the Aegean. This paper discusses the necessity for archaeologists working in the third millennium AD to develop a new chronological frame beyond the tripartite system (EB I, II, III) and the definition of cultures (e.g. Keros-Syros). Based on the seminal works on Aegean chronology and particularly on data from recent, very well stratified excavations and their absolute datings, a new chronological code based on the centuries within the third millennium BC may now be defined.*

Introduction

The third millennium BC in the eastern Mediterranean and particularly in the Aegean is, in fact, synonymous with the period called by archaeologists the "Early Bronze Age." In current terminology, the EB in the Aegean starts toward the end of the fourth millennium BC. Its end may occur at the end of the third or at the beginning of the second millennium BC, depending on the geographic area (Warren and Hankey 1989: Table 2.1; Manning 1995: Figs. 1–2). Therefore, in this paper I will refer to select data and problems of the EB. I will also include some data of the preceding phase, namely the Final Neolithic (FN), that are important for establishing the beginning of the EB and the synchronisms within the Aegean in relative and absolute sense. Concerning the term Aegean let me underline in advance that in the code of Aegean prehistorians the Aegean includes the modern state of Greece, as well as the western coast of Asia Minor (modern Turkey). In this paper I will refer briefly to: I. The history of research on relative EB chronology until 2000; II. The research on absolute EB chronology before and after 2000; and III. The results of some new excavations in the central and the east Aegean that are important for the relative and absolute chronology in the fourth and third millennia BC; and IV. Closing remarks and suggestions for future work.

I. The Relative Chronology Until 2000

The foundations for a relative chronology of the Aegean Bronze Age were already set in the first two decades of the 20th century. More specifically, the relative chronological division of the Aegean Bronze Age was based on the tripartite chronological system introduced by Evans for Minoan Crete, which was also applied by Wace and Blegen for mainland Greece and the Cyclades in 1918 (Wace and Blegen 1916–1918, 186–189). The tripartite system was not used simply for dividing the Bronze Age of Crete, the Cyclades, and mainland Greece into Early, Middle, and Late periods, called respectively Early Minoan (EM), Early Cycladic (EC), Early Helladic (EH) etc. It was also used to distinguish three subphases in each of the above mentioned cultural periods, based exclusively on typological criteria for ceramics. Thus, the EH was divided into EH I, EH II, and EH III; the EC into EC I, II, and EC III; and the EM into EM I, EM II, and EM III. As archaeological research was expanding to Thessaly, Macedonia, the East Aegean islands, and Troy the chronological landscape of the EB was enriched particularly after the 1950s, when many of the previously excavated sites were published. Thus, the designation Early Thessalian I, II, and III (ETh) was introduced for Thessaly in 1959 by V. Milojčić, and the term EB I–III or EB 1–3 for northern Greece and the islands of the East Aegean. At some important sites, such as Troy, Poliochni, and Sitagroi, cultural layers with the names I, II, etc., were introduced to refine the chronological sequences even more. All these chronological systems were based almost exclusively on pottery typologies (Renfrew 1972: 116–134; Maran 1998: 37–53, Tafel 80–82; Alram-Stern 2004: 151–193).

In 1972 Renfrew broke away from this naming tradition. Taking into consideration all aspects of cultural life, he introduced a new system particularly for the EH, the EC, and the EB in the East Aegean focusing on the successive "cultural entities" with geographical and chronological meaning. Hence, he used the terms Grotta-Pelos for the EC I; Keros-Syros for the EC II; and Phylakopi I for the EC III (Renfrew 1972: 135-195, Table 9.II; 196–221, Tables 13.II, 13.III, 13.IV). This system was evolved furthermore by Doumas in 1977, who replaced the term "culture" with the term "group" and recognized eight successive groups in the EC (Doumas 1977: 25). Doumas renamed the Grotta-Pelos culture as the Pelos-Lakkoudes culture and proposed four chronological groups before Renfrew's Keros-Syros culture. He named each group after the site where the subordinate assemblage in question was best represented, namely Lakkoudes, Pelos, Plastiras, and Kampos. The last was considered to represent a transitional phase to the Keros-Syros culture of EC II. In 2000, the EC sequence was further modified by Rambach with new groups based on the study of graves, like the Panagia Complex, that precede the Kampos group and the Aplomata group that follows it (Rambach 2000b: 103–111: 186, 203–220, Abb. 22, Beil. 3; 265–268, 363, Abb.23, Beil. 4–5) (Table 1).

For mainland Greece Renfrew designated the EH I as the Eutresis culture and the EH II as the Korakou culture. For the EH III, he suggested two terms: Tiryns culture for the northwestern Peloponnesos (equal to Lerna IV) and Lefkandi I assemblage for the northern areas of mainland Greece (Aegina, Attica, Boeotia, Euboea, and Thessaly) (Renfrew 1972: 99–116, esp. 103–105). In 1979, Rutter underlined that the culture represented by the finds from Lefkandi I was contemporary with the last phases of the Korakou culture of EH II and not with the later Tiryns culture of the EH III period. He further specified that the Lefkandi I assemblage was contemporary with Renfrew's Kastri assemblage of the EC II Keros-Syros culture (Renfrew 1972: 180–183, Table 13.III; Rutter 1979: Table 3). These cultures belonged to a broader phenomenon of the presence of distinctive red- and black-burnished ceramics which had clear derivations from Anatolian prototypes (Rutter 1979).

From the 1980s until 2000 the critical study of old archaeological evidence as well as recently excavated material in mainland Greece and the Cyclades contributed enormously to the finer division of both Wace and Blegen's and Renfrew's chronological systems. Barber and MacGillivray (1980: 150–152 Table II, Ill. 2) subdivided the EC III into EC IIIA corresponding to Renfrew's Kastri assemblage, and EC IIIB, corresponding to Renfrew's Phylakopi I culture. In 1983 Rutter demonstrated that a major cultural hiatus separated Barber and MacGillivray's periods EC IIIA and EC IIIB (Rutter 1983, 6971, 75). In Rutter's view, this hiatus involved not simply a significant cultural discontinuity, but also a substantial gap in the EC period. And because the EC IIIA, namely the Kastri Group, was contemporary with the Lefkandi I assemblage on mainland Greece, he proposed that the Kastri group demonstrated an EC IIB phase (the EB IIA would then be the Keros-Syros culture), whereas the EC IIIA would represent an "EC III gap," and the EC IIIB of Barber and MacGillivray would be contemporary with Middle Helladic (MH) and Middle Minoan IA (MM). Therefore, he renamed the EC IIIB as MC I (Table 1).

The relative chronology of the EB in western Anatolia has been included in a tripartite system referring to the entire Anatolian peninsula in EB I, EB II, and EB III. Within this system two sites with rich stratigraphical sequences played the protagonistic roles: Troy in northwestern Anatolia with a much discussed stratigraphy (Korfmann and Kromer 1993: Abb. 1; Korfmann 2000: Abb. 6) and Tarsus in the southeast, located in Cilicia. Renfrew (1972: 127–132) synchronized the EB 1 in northwestern Anatolia with Troy I culture, the EB 2 with Troy II culture, and the EB 3 with Troy III-V. Korfmann included later Troy I-III in his "Maritime Troia-Kultur" (Korfmann 1996: 2, 22, Abb. 18). After the 1980s two versions of the tripartite system in Anatolia were developed; their main differences lie, though, in the subdivision of the three main subphases. Thus, the one proposed in 1988 by Efe based on the rich stratigraphy of Demircihüyük distinguishes the subphases as follows: EB 1, EB 2a, EB 2b, EB 3a, EB 3b (Efe 1988: Abb. 98). The other one, published in 1992 by Mellink, discerns the following subphases: EB IA, EB IB, EB II, EB IIIA, and EB IIIB (Mellink 1992: 213–219, Table 2–3). The latter subdivision was based on the comparative stratigraphy of major Anatolian

Settlement stratigraphy	Barber and MacGillivray 1980	Rutter 1983	Renfrew 1972	Renfrew 1972 Cultures	Doumas 1977 Cultures	Doumas 1977 Groups
Phylakopi A I	*EC I*	*EC I*	Kampos	Grotta-Pelos	Pelos-Lakkoudes	Lakkoudes Pelos Plastiras Kampos
Phylakopi I-I (A2) Agia Irini II	*EC II*	*EC II A* Keros-Syros		Keros-Syros	Keros-Syros	Syros
Agia Irini III Kastri	*EC IIIA*	*EC IIB* Kastri Group-Lefkandi I	Kastri assemblage			Kastri
Phylakopi I-ii I-iii	*EC IIIB*	*EC III MC I* (Chalandriani, Amorgos Phylakopi I)	Amorgos	Phylakopi I	Phylakopi I	Amorgos Phylakopi I
Phylakopi II	*MC*					

Table 1: Periodization of the EC period (Based on Barber and MacGillivray 1980: 143 Tab. 1; Rutter 1983: 75).

sites and in particular on the main cultural phases excavated at Tarsus and Troy. A first remark on these chronological schemes would be that they cannot be generalized for Anatolia in its entirety, which is a huge geographical area with a wide cultural diversity since the Pre-Pottery Neolithic (PPN). And for our study a question arising through these systems would be: Why should we base the chronology of coastal western Anatolia, which is located in the Aegean (Kouka 2002: 295–302, Tab. 1), on the sequences of the far away Tarsus?

The above remarks, as well as comments on the chronology of further geographical parts of the EB Aegean, were critically presented in 1995 together with the first published absolute datings of these sites in the seminal book of Manning, a reference book for archaeologists working on the Aegean EB (Manning 1995: 40–73, Fig. 1–2). In addition, Maran, having taken into consideration Manning's results on absolute chronology, undertook in 1998 the most detailed and critical presentation so far on the relative chronology of the Aegean FN/CH and EB sites (Maran 1998: 7–159, Taf. 80–81). Of importance for our further discussion are his special and extremely useful references on the FN and on the Lefkandi I–Kastri phases.

Regarding the Aegean FN two subphases are defined: the earlier phase (second half of the fifth millennium BC) is known in Thessaly as the Rachmani culture and in the middle and south Greek mainland and the Cyclades as the Attica-Kephala culture; the later one (fourth millennium BC) is known in Thessaly from Petromagoula, in the middle and southern Greek mainland from the North Slope of the Akropolis and Eutresis II, while in the Cyclades this later subphase can at least partially be synchronized with the Pelos-Lakkoudes phase (Maran 1998: 7–8, 25, 30–31, 152–153, Taf. 80–81) (Figure 1). Regarding the Lefkandi I assemblage or the Kastri group the studies of Manning and Maran clarified that the earlier defined cultural "assemblage" or "group" of Anatolian influence in the Cyclades and the East Greek Mainland littoral is in fact a long phase of about 300 years (26th–22nd centuries BC/2550–2200 BC), which started in the middle EH II (Lerna IIIC) and lasted until the beginning of the EH III (Lerna IV.1) (Maran 1998: 140–146, 153–159, Taf. 11–13, 80–81; Kouka 2002, 300–301, Tab. 1). This fact can also be understood if one takes into consideration the long presence of these "Anatolianizing" pottery types in their motherland, namely in western Anatolia.

In Anatolia there is also a discussion whether the horizon of this pottery is dating in the EB IIB or in the EB IIIA (Mellink 1986: pl. 16; 1992; Efe 2006). In fact, in all mentioned areas one can define an earlier and a later phase through the gradual appearance of the typical shapes of the Lefkandi I–Kastri phase: in the earlier phase the bell-shaped cups and the one-handled tankards were appearing, whereas in a later phase the assemblage has been enriched with the depas cups, the shallow bowls, the first wheelmade plates and the cut-away spouted jugs with globular bodies (Mellink 1986: pl. 16; Şahoğlu 2005b: 343–350, Table 1, 4–8). Maran called this phase "Wendezeit" (Time of change), meaning a cultural change (Maran 1998: 450–457). Thus, a while before 2000, besides the hitherto simply chronological and cultural labels, terms describing general cultural phenomena are making their appearance.

From the above comments, we recognize the confusion that such polyglot terminology has borne upon

Figure 1: Map with Aegean sites of the LN/CH and the EB.

the relative Aegean and Anatolian EB chronology at the end of the 20th century. The attempt to fit together all the old and the new subphases in the traditional tripartite system is obvious. As Oliver Dickinson noted in his book on the Aegean Bronze Age about the tripartite system: "The system has in fact become a bed of Procrustes, to which the material must be fitted willy-nilly" (Dickinson 1994: 9–22, Fig. 1.1–1.3) (Table 2).

II. The Absolute Chronology Before and After 2000

The synchronization of cultural sequences in the various landscapes of the Aegean as well as their absolute dating was problematic until the late 1980s due to the problematic stratification, as well as due to the limited amount of radiocarbon data respectively. According to the published calibrated dates from various landscapes of the Aegean before 2000, the EB should have started between 4000 and 3100, while the end should be put between 2100 and 1800 BC (Table 3)!

The problems of the beginning, of the end, and of the duration of each subphase of the EB were thoroughly discussed by Manning in 1995, shedding new light on old problems. After having discussed the old radiocarbon dates from Troy (published in 1981 by Quitta) the new ones published by Korfmann and Kromer in 1993, the very few dates published by Warren and Hankey in 1989, the nine thermoluminescence dates from Beşik-Yassı Tepe as well as the results of the Aegean Dendrochronology Project directed by Peter Kuniholm since 1973 (Manning 1997: 154–160, 165–166, 177–179), Manning concluded that the absolute dates for the Aegean Bronze Age were as follows: EB 1 = 3100–2650 BC; EB 2 = 2650–2200/2150 BC; and EB 3 = 2200/2150–2000 BC (Manning 1995: 141–153, Table 2) (Table 4).

Regarding the beginning of the EB in western Anatolia, this has been set up by Mellink (EB IA, EB IB, EB II, EB IIIA, EB IIIB) in 1992 at around 3400 BC and the end at about 2200 BC (Mellink 1992: Table 2–3) (Table 5).

The Troy Project set the beginning of Troy I at 2900 BC (Korfmann and Kromer 1993: Abb. 23), while in 2004 the TAY Project (the Turkish Archaeological Atlas) published calibrated dates from western Anatolia that put the beginning of Troy and Beşik-Yassı Tepe also before 3000 BC, at about 3100 BC (Erdoğu, Tanındı and Uygun 2003: Ek 2) (Table 6).

The most discussed calibrated dates during the 1990s were those of Troy published in 1993 by Korfmann and Kromer, since they were important for both the chronology of western Anatolia and the synchronization with the Cyclades and the Greek mainland. These were also discussed by Manning in 1997 together with the dates from Beşik-Yassı Tepe, Poliochni azzurro (2910–2672 BC), and Thermi I (3943–3195 BC, 2910–2780 BC). According to this data the EB I seems to have started on the East Aegean islands of Lemnos and Lesbos at around 3000 BC (Table 4). However, problems for the comparative chronology have been raised from the arguments of Korfmann and Kromer through the synchronization of mid-late Troy (Id-k) with the early Troy II (Manning 1997, 501–505, Table 2; Cf. Korfmann 2000, Abb. 6).

Apart from the comments of Manning on the radiocarbon and dendrodates, one should take into consideration the pottery evidence from Troy I-V that was unearthed in Troy after 1987 (*Studia Troica* 1991–1996), the study of which is currently in progress (Korfmann 2006). At this point we should not forget, on the one hand that there are differences indeed in the pottery of Troy Ia-k and Troy IIa-g as defined by Blegen et al. in 1950. Furthermore, these differences can stratigraphically be followed in western Anatolia in the rich EB strata at Liman Tepe/Klazomenae, in the Izmir Region (Şahoğlu 2002). Besides, six calibrated radiocarbon dates taken from an old profile at Tarsus have been published recently and have been compared with the ones from Troy (Aslı Özyar et al. 2005: Table 17–18). According to these dates Troy Ia-d can be dated in the time between 2824–2659 BC (Sample 1) while Troy Ie-l between 2625–2401 BC (Samples 2-6) (Aslı Özyar et al. 2005: 23, Fig. 21; 23–25, Fig. 22).

The absolute dates from the FN or CH and the EB Aegean and Anatolia have been enriched in the late 1980s and after 2000 through samples from new excavations from both the Aegean (Kastri on Thasos, Mikro Vouni on Samothrace, Palamari on Skyros, Grotta and Zas Cave on Naxos, Markiani on Amorgos, Skarkos on Ios, Proskynas in Lokris) and western Anatolia (Liman Tepe, Bakla Tepe, Çeşme-Bağlararası, Ulucak). Based upon the published dates we have a time span for the FN between 4700/4500–3300/3100 (Andreou, Fotiadis and Kotsakis 2001: 260, Table 1). Another one from Doliana in Epirus gives a dating of the site in the FN and the beginning of the EB: 3770–2925 (Alram-Stern 2004: 194) (Figure 1). Extremely important data for the absolute chronology of the FN and the EB in the Cyclades and therefore for the entire Aegean are coming from Daskaleio-Kavos, Markiani on Amorgos, Zas Cave on Naxos, and Akrotiri on Thera (Renfrew, Housley, and Manning 2006; Manning 2008). Their verification led Manning to the following results (Figure 1, Table 7).

Periods	Middle & South Mainland Greece	Cyclades	Northern Greece
Final Neolithic / Early Chalcolithic 2nd half 5th mill. BC	Attica – Kephala	Attica – Kephala Strophilas	Rachmani Palioskala Sitagroi III Dikili Tash II
Final Neolithic / Late Chalcolithic 4th mill. BC	Akropolis – North Slope Eutersis II	Pelos – Lakkoudes	Petromagoula Magoula Miktothivon
EH I 3100/3000-2700/2650 BC	Eutresis III-V Manika 1 Perachora-Vouliagmeni Talioti-Kephalari	Kampos Group (*bigger part*)	ETh I-Argissa I Servia 8 Kritsana I/II (part) Pentapolis I Sitagroi IV-Va (part) Dikili Tash IIIA
EH II early 2700/2650 –2550 BC	Lerna IIIA-B Tiryns FH II-früh Tsoungiza Lithares 6-7 Eutresis VI-VII Manika 2-3 Agios Kosmas A Tsepi	Kampos Group (*later part*) Keros – Syros Agia Irini II	Pefkakia 1-5 Kritsana I/II Sitagroi Va Pentapolis I
EH II middle 2550/2500 BC	Lerna IIIC Lefkandi I Leukas-R-Gräber	Agia Irini II Kastri	Pefkakia 6 Sitagroi Vb=EH II *ANATOLIAN EB2b*
EH II late *WENDEZEIT* 2500/2450 BC *NO SECURE DIVISION*	Lerna IIID Lefkandi I Thebes Group B Aghios Kosmas B Raphina House A Rouf Agios Dimitrios IIb	Agia Irini III Kastri	Pefkakia 7
REGIONALISM *WENDEZEIT*	Lerna hiatus Tiryns-Transitional Phase EH II/III Lefkandi I	Agia Irini III Kastri	Pefkakia 7
EH II/ EH III Transition Some EH III shapes *WENDEZEIT* 23rd cent. BC	Lerna IV.1 Lefkandi I Thebes Group B Aegina-Kolonna III	Kastri	Pefkakia 7 late
EH III 2200/2150 BC	Lerna IV Phase 1-3 Aegina-Kolonna IV-VI Olympia-Altis (Apsis houses) Olympia-New Museum (early EH III) Lefkandi 3	EC IIIB	Pefkakia MB Phase 2 Palamari House G

Table 2: The Aegean EB relative and absolute chronology (based on Maran 1998: Taf. 80-81, modified by the author).

REFERENCE	Beginning of EB Years BC	End of EB Years BC
Treuil 1983: Fig. 30	4000/3800	1900
Coleman 1992: Table 2, 4	3700/3500	2100
Warren & Hankey 1989: Table 2.1	3650/3600	1800
Pullen 1985: Table 3.5, 3.8	3200	2050
Manning 1995: 144 f., 168, Table 2	3200/3100	2000

Table 3: Calibrated datings for the beginning and the end of the Aegean EB until 1995.

EB periodization	Coleman 1992 Calibrated BC	Manning 1995 Calibrated BC	Manning 1995 EB periodization
EB I	3700-2900	3100/3000-2650	EB 1
EB II early (Lerna IIIA-B, Thebes Group A)	2900-2400 EC II: 3100-2400	2650-2450/2350	EB 2
EB II late (Lerna IIIC-D, Lefkandi I, Thebes Group B)	2400-2100	2450/2350-2200/2150	
EB III	2100-2000	2200/2150-2050/2000	EB 3
MB	2000-1900	2050/2000-1950/1900	MB

Table 4: Absolute datings of the Aegean EB (after Coleman 1992: Table 2, and Manning 1995: Fig. 2).

Mellink 1992 EB periodization	West and South Anatolia	Begin Calibrated BC
EB IA	Kumtepe Tarsus EB I	3400
EB IB	Troy I early Tarsus EBI	
EB II	Troy I Besiktepe Yortan Iasos Tarsus EB II	2700?
EB IIIA	Troy IIb-IIg Tarsus EB IIIA	2400?
EB IIIB	Troy III-V Tarsus EB IIIB	2200?

Table 5: Relative and absolute chronology of the Anatolian EB (Mellink 1992: Table 2-3).

Western Anatolia	Beginning of the period Calibrated BC
Kumtepe IB Beycesultan XVII	3500-2900
Demircihöyük C	3300-2900
Troy I	3100-2500
Troy II	2500-
Troy III	2300
Troy IV	2200
Troy V	2100

Table 6: Calibrated dates of Western Anatolia (based on Erdoğu, Tanındı and Uygun 2003: Ek 2).

Manning (2008) Cultural phase	Proposed time range Calibrated BC
EC I – Grotta Pelos-Lakkoudes phase	3100(+) to 2950
EC I/II – Kampos phase	c. 2950-2650
EC II – Keros-Syros phase	c. 2650-2500
EC III – Kastri phase	c. 2500-2250

Table 7: Absolute chronology of the EC (Manning 2008: 59).

III. New Results and New Aspects of the Third Millennium AD

Archaeological research in the Aegean during the last three decades as well as final publications of older excavations offer to archaeologists of the third millennium AD abundant data for relative chronology and comparative stratigraphy as well as an absolute chronology of the FN/CH and the subphases of the EB from sites whose stratifications vary in richness (Alram-Stern 2004).

Important for the FN or Late Neolithic II (LN) or CH are records from Makri on Thrace, Mikro Vouni on Samothrace, Myrina on Lemnos, Doliana in Epirus; Palioskala in the Karla Lake and Magoula at the Junction Mikrothivon in Thessaly; Zagani and Lambrika in Attica; the Euripides Cave on Salamis; Strophilas on Andros and the Zas Cave on Naxos; Chrysokamino and Kefala Petra in East Crete; Bakla Tepe and Liman Tepe in the Izmir Region (Alram-Stern 2004; Kouka 2008: 272–278, Fig. 27.1) (Figure 1). Regarding the EB the following sites offered good evidence for a comparative stratigraphy and absolute dates in the Aegean: Archondiko, Mandalo, and Xeropigado Koiladas in Macedonia; Proskynas in Lokris, Thebes, Tsepi at Marathon, Kolonna on Aegina, Petri and Tzoungiza in Nemea, Geraki in Laconia, Aigio-Helike, Olympia, and Nydri in Levkas in mainland Greece; Zas Cave and Grotta on Naxos, Skarkos on Ios and Akrotiri on Thera in the Cyclades; Chania, Poros Irakleiou, Kefala Petra, Petras in Crete; Mikro Vouni on Samothrace, Palamari on Skyros, Myrina and Koukonisi on Lemnos in the North and East Aegean (Kouka 2002; Alram-Stern 2004; Kouka 2008: 272–274, Figs. 27.2, 27.3, 27.4). Finally in western Asia Minor: Troy (*Studia Troica* 1991–1996; Korfmann 2006), Liman Tepe (Erkanal and Günel 1996; Erkanal, Artzy, and Kouka 2003; 2004), Bakla Tepe (Erkanal and Özkan 1999; Tuncel 2005), Çeşme-Bağlararası (Erkanal and Karaturgut 2004), Ulucak and Kulaksızlar in the Izmir region and Küllüoba (Efe 2006; 2007) in the Eskişehir Region (Kouka 2002; Harmankaya and Erdoğu 2002; ARKEOATLAS 2003; Schoop 2005) (Figure 1).

Keeping in mind the short discussion on relative and absolute chronology as presented above, let us focus on some new data that lead us to rethink some important moments of the current Aegean EB chronology.

Select evidence from the FN/CH and the EB

Firstly I will refer to the comparative chronology of the FN and the EB I on mainland Greece and the Cyclades. Regarding this I will focus on the pottery type of a lid with incised spiral decoration with white incrustation (Alram-Stern 2004: 752–753, Taf. 52). This pottery type is well known in the Baden culture of the southwest Balkans (Macedonia, Albania) as "Bratislava-type" and is dating in the period of Maliq IIIa, in absolute datings between 3600–3100 BC (Maran 1998: Taf. 1–4). Pots of this type were found so far in Doliana in Epirus, dating between 3770–2925 BC, in Petromagoula of Volos (Maran 1998: 344–346, Abb. 1, Taf. 1, 1–3, 73) and in Raxi at the lake of Xynias. The finds from Petromagoula led Maran in 1998 to designate a later stage of the FN in Thessaly, later than the Rachmani period. Recent finds from Palioskala at the Lake of Karla, as well as from Magoula at the Junction Mikrothivon, underline the existence of a phase later than Rachmani (Figure 1). This phase is in the Cyclades at least partially synchronous with the Pelos-Lakkoudes Phase (Table 1). Similar lids with spiral and star or sun motives were recently published by Pantelidou from the cemetery of Tsepi at Marathon (Pantelidou-Gofa 2005: Pl. 6, 18, 27, 29). This cemetery was in use particularly in the EB I and shows a Cycladic character. After Pantelidou, the majority of the material belongs to the Kampos phase. One of the most distinctive features of the Kampos phase, a cultural stage that Doumas set up as the transitional period from the EC I to the EC II, is the frying pan with straight sides and a Π-shaped handle (Doumas 1977: 25; Pantelidou-Gofa 2005: Pl. 9, 13, 16). The decoration of the Kampos frying pans with incised running spirals around a central star or sun resembles one of the lids known from Epirus and Thessaly. Tsepi is the only site where both types were found together and Pantelidou postulates that the frying pans almost coexisted with the simple lids (Pantelidou-Gofa 2005: 314–316). Therefore, one could assume, that the frying pan of the Kampos phase may represent the successor or the handleless lids of the final stage of the FN on mainland Greece. The existence of both types in Tsepi indicates that the Kampos phase should represent a much longer phase within the EC I, and that at least in Attica, followed the Akropolis North-Slope and Eutresis II period. In this respect one should not forget that a major interaction between the Cyclades, Attica, Boeotia, Euboea, Lokris (Proskynas) and the Argosaronic Gulf has been established since the Attica-Kephala cultural phase (Kouka 2008: 275–276). This was based on the exchange of metals and metal technologies and was expressed through more or less stronger affinities in the material culture. The Kampos phase dates according to Manning's latest absolute dates (2008: 58–59) between 3040–2630 or 2950–2650 BC (Table 7). The Kampos phase is followed by the Aplomata group, which precedes the Chalandriani group, namely the EC II Keros-Syros culture. Indicative features of this group are footed bowls, pyxides, and jugs with dark-on-light painted decoration as well as frying pans with incised and stamped running spirals or star/sun motives within frames with Kerbschnitt (a progressive type of the frying pan of the Kampos phase). Within this group the first sauceboats of EH inspiration have appeared (Rambach 2000b: 265–268, 363, Abb.23, Beil. 4–5).

The evidence at Liman Tepe

Let us move now to the Izmir region in order to investigate the archaeological records from the fourth and third millennia BC. The extensive survey and large excavations in Panaztepe, Liman Tepe, Bakla Tepe, Kocabaş Tepe and Çeşme-Bağlararası undertaken within the framework of the Izmir Region Excavations and Research Project (IRERP) since 1992, under the direction of Prof. Hayat Erkanal, Ankara University, opened new perspectives in Aegean prehistory. The fertile peninsula of Urla is located in the middle of the western Anatolian coastline. The area was rich in metal ores (copper, silver, lead, and gold) and was an ideal field for habitation since the Neolithic. Large-scale excavations in the IRERP region offer important data for an intra- and inter-site analysis of settlement history from the Neolithic (Araptepe, Barbaros) through the Bronze Age as well as for studying early urbanism in western Anatolia and its cultural interaction to the Aegean and the chronological sequence of western Anatolia (Figure 1). For the purposes of this paper we will focus on Liman Tepe. Liman Tepe (Klazomenae) (LMT) is located in the peninsula of Urla and has direct access to both the Anatolian plateau and the Aegean. Architectural remains revealed the following stratified levels (Şahoğlu 2005b, Figs. 1–2) (Tables 8–9).

Excavations at Liman Tepe revealed a flourishing urban harbor-settlement since EB I. Settlement planning, monumental fortification walls, massive house architecture, craft specialization, prestige objects, numerous imports from the Cyclades and the Greek mainland found in the rich levels of the EB I-late EB II settlements, all testify to the economic and political complexity and importance of Liman Tepe as an early urban center in this landscape (Figure 2). The site also contributed to the development of trade networks between Anatolia and the Aegean as well as to the establishment of the cultural *koine* in the North and East Aegean from the EB I through the EB II Periods. This koine reached its peak in the advanced EB II, when Liman Tepe became one of the biggest and richest urban cities in western Anatolia and the Aegean with a monumental fortification, a fortified harbor, a lower town, an administrative complex, and craft specialization (Kouka 2002: 6–7, 295–302).

Liman Tepe	Periodization
LMT VII (VII.4-VII.1)	Late Ch
LMT VI (1d-1c-1b-1a)	
VI 1d	EB I early-middle
VI 1c-VI 1b	EBI middle
VI 1b-VI 1a	EB I late
LMT V (3b-3a-2b-2a-1)	
V 3b-V 3a	EB II-early
V 2b-V 2a-V 1b	EB II-late
V 1a	EB II-final
LMT IV (2-1)	
IV 2	EB IIIa
IV 1	EB IIIb
LMT III (4, 3, 2/1)	MB
LMT II	LB

Table 8: The stratigraphical sequence at prehistoric Liman Tepe/Klazomenae.

DATE	ANATOLIA Kültepe	ANATOLIA Tarsus	ANATOLIA Beyce Sultan	CRETE	GREEK MAINLAND	CYCLADES	EASTERN AEGEAN ISLANDS	TROIA	LIMAN TEPE AREA - A	LIMAN TEPE AREA - B	PERIOD
2000 BC.			VIa		Early Helladic III	Middle Cycladic	Heraion V	V	?	LMT B IV-1	EBA IIIb
			VIII		Aegina V	Early Cycladic III (Phylakopi I)	Poliochni Brown Heraion IV	IV (?)			
	11b			Early Minoan IIB	Lerna IV Aegina IV		Heraion III	III		LMT B IV-2	EBA IIIa
	12	EBA III	XII c		Early Helladic II Late Lerna IIID	Early Cycladic IIb (Kastri Group) Ayia Irini III Palamari III Syros-Kastri Zas IV Mt. Kynthos	Poliochni Yellow Emborio I	Late	LMT A V-1	LMT B V-1a	End of EBA II
	13				Aegina III					LMT B V-1b	EBA II Late
					Lefkandi I / Manika Pevkakia VII			II			
2500 BC.	14		XIII a		Lerna IIIC Aegina II Pevkakia VI		Heraion II Emborio II		LMT A V-2a	LMT B V-2	
		EBA II		Early Minoan IIA Knossos Poros	Early Helladic II Early Lerna IIIB	Early Cycladic IIa Ayia Irini II Palamari II Chalandriani Mt. Kynthos	Poliochni Red Heraion I Thermi V	Early	LMT A V-2b		
	15						Emborio III Poliochni Green	Late	LMT A V-3a	LMT B V-3	EBA II Early
	16				Lerna IIIA						
	17		XVII				Thermi IV	I	LMT A V-3b		
	18				Aegina I	(Aplomata Group)	Polio. Blue evoluta Emborio IV Polio. Blue Archaic Thermi III	Middle	LMT A VI-1a	LMT B VI-1	EBA I Late
			XIX	Early Minoan I	Early Helladic I				LMT A VI-1b		EBA I Middle
		EBA I							LMT A VI-1c	?	EBA I Middle / Late
						Early Cycladic I (?)	Emborio V	Early (?) End of Kumtepe 1b (?)	LMT A VI-1d		
3000 BC.			XX				Thermi II Thermi I		LMT A VII		EBA I Early / Late Chal. (?)

Table 9: Comparative chronology of Liman Tepe, Anatolia and the Aegean (after Şahoğlu 2005b, Fig. 2).

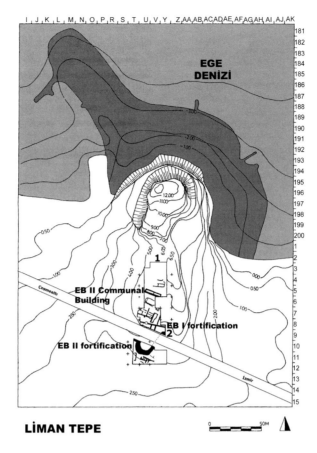

Figure 2: Liman Tepe. Topographical map.

The Late CH settlement—LMT VII

The Late CH settlement included four successive architectural levels (LMT VII. 4-VII.1), that were investigated recently in the northern part of the peninsula. The architecture of the earliest phase (LMT VII.4) revealed in particular round silos with narrow, white plastered mudbrick walls. The pottery included bowls with rims thickened on the interior and with pattern burnished decoration (Figure 4–missed Fig 3a; should renumber), cheese-pots as well as closed pots. Noteworthy among the small finds was the abundance of long blades and scrapers made with Melian obsidian. Similar obsidian blades, among them leaf-shaped obsidian arrow-heads, and fragments of a marble conical cup were found in the obsidian workshop of the site. The obsidian arrow-heads and the marble conical cups are typical for the Attica-Kephala culture in the central Aegean (Figure 5), but also known in the wider Izmir region in the marble workshop at Kulaksızlar (NE of Izmir) (Takaoğlu 2005). The finds of the third CH Phase (VII.3) included pattern burnished pottery, cheese pots, and closed pots with knobs at the upper part of their handles, a clay stamp seal and a clay figurine, but no important architecture.

The later phases (LMT VII.2-1) included a destroyed apsidal house with wattle-and daub walls, and rectangular constructions with pisé walls used for storage. Similar architecture is known from Emporio IX-VIII, Poliochni Black, Myrina, Bakla Tepe, and Kum Tepe. The most characteristic pottery of these phases included conical bowls with rolled rims and closed pots with white painted decoration (Figure 6), or fine incised decoration with white incrustation, bowls with carinated bodies and symmetric lugs, and a biconical rhyton. This pottery is typical for the Late CH in western Anatolia and the southeast Aegean and dates the latter two phases of LMT VII in a later stage than the Attica-Kephala culture, namely at around 3100 BC following Manning's (2008: 58–59) dates from the Cyclades (Tables 8–9. From Liman Tepe itself the analysis of radiocarbon samples is still in progress. Finally, in the latest phase (VII.1) an evolved agricultural economy and in situ metalworking of copper were also attested.

The late CH finds from Liman Tepe indicate the economic and social structures of the CH LMT, since they indicate the participation of western Anatolia in the common cultural and symbolic code within the Aegean as early as the fourth millennium BC. This

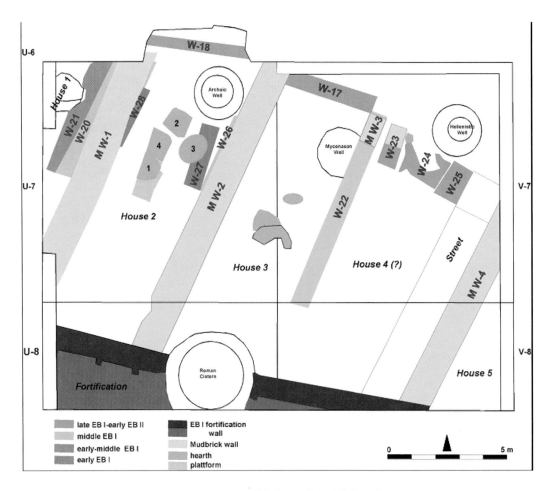

Figure 3: Liman Tepe VI. Plan of the EB I settlement.

code included metal jewellery and tools, leaf-shaped arrowheads of obsidian, jewellery made of *Spondylus gaederopus* (Kouka 2008: 58–59).

The EB I settlement—LMT VI

The Late CH settlement was destroyed by fire, and Liman Tepe was rebuilt in the EB I with totally new concepts in building materials, methods of construction, house types, and intra-site organization. Due to the geomorphology, the architectural organization of the EB I was not the same in the north and south parts of the settlement (Figures 2–3).

The southern part of the settlement included a freestanding, 3m high and 90cm wide and very well built defensive system strengthened with rectangular buttresses and a sloping supporting wall, a gate flanked by two trapezoidal bastions, and blocks of long-room rectangular houses in a radiating arrangement, typical for this period (Erkanal and Günel 1996, Çiz. 5, Res. 11–12). The fortification wall displays three main construction phases, dating from the early EB I (LMT VI d) to the very early EB II (LMT V3b).

The houses were in use longer, namely from the early EB I (VI d-VI a) until the late early EB II (LMT V3a) (Figure 3). The successive floors (3–6) examined so far in Houses 1–3 date to the early/middle (Troy Ib) and late EB I–early EB II (late Troy I, Ih) (Erkanal, Artzy, and Kouka 2003: 424–425, Res. 1; 2004: 165–168, Res. 1–3). The earliest levels of the EB I in these houses have not yet been investigated. These levels were fortunately reached in the north part of Liman Tepe.

The architectural concept in the northern part of the site was different. A stone terrace wall and high stone-built house walls belonging to rectangular houses indicate three architectural phases that succeeded the four LCH layers. They provide data for the earliest phases of the EB I that are lacking from the southern part. The construction of such a terrace wall must have acted as a protecting belt against erosion, surrounding the built area of an "upper settlement." This interpretation also suggests that the EB II topography of the site was being shaped much earlier. The city to the north of the buildings within the monumental fortifications was constructed on ter-

Figure 4: Liman Tepe VII.4. Pattern burnished pottery.

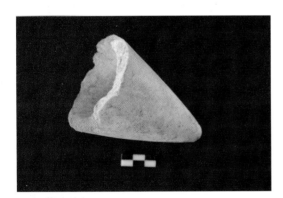

Figure 6: Liman Tepe VII. 2. Jugs with white painted decoration and rhyton.

Figure 5: Liman Tepe VII.3. Obsidian arrowheads, fragments of a marble conical cup and "cheese-pots."

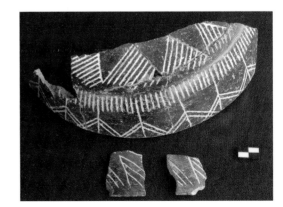

Figure 7: Liman Tepe VI. EC frying pans.

Figure 8: Liman Tepe VI. Characteristic pottery shapes of the EB I.

Figure 9: Liman Tepe VI. Imported EH and EC pottery.

races in keeping with the topography of the site. The earliest EB I architectural phase included rolled-rim bowls, and bowls with inverted rims as the most characteristic pottery shapes. Unlike examples from the later EB I periods, these bowls are thin-walled and quite delicate. The inverted part of the rim is also narrower than those of later periods. On other sites (e.g. Emporio), this shape has been mostly observed in Late CH or the transition between the Late CH/EB I periods. Rolled-rim or "S"-profile bowls are very rare, whereas the inverted rim bowls with string-hole lugs are quite abundant. Among this, pottery fragments of two different EC frying pans of the Kampos phase were discovered; the rims have incised, linear decoration of parallel lines, whereas the base is decorated with diagonally-hatched triangles (Figure 7). These finds suggest that the earliest EB I architectural phase at Liman Tepe is contemporary with the Kampos phase in the Cyclades.

On the house floors of the southern part of the settlement, particularly in Houses 2 and 3, the pottery included numerous pots typical for the EBI in the north and east Aegean (Figure 8) and many ceramic imports from the Cyclades and the Greek mainland (Erkanal, Artzy, and Kouka 2003: 424–425, Res. 1; 2004: 165–168, Res. 4), namely: EH sauceboats in Urfirnis ware, the earliest found to date in western Anatolia EC I/early EC II imports, such as dark-on-light painted pots similar to those known from the Aplomata Group, pyxides and small pithoi with incised, stamped and plastic decoration (Figure 9). These finds as well as the remarkable presence of cores and blades of Melian obsidian and of Naxian emery demonstrate the intensive trade contacts between Liman Tepe and the Central and South Aegean. Moreover, they are a reliable indicator of the economic activities and the social status of the inhabitants of Houses 2 and 3, houses associated with industrial activities, such as working of metal, bone, and flintstone (House 2) and textile production (House 3).

Social differentiation within EB I Liman Tepe is suggested by a clay stamp seal and a schematic stone figurine from the south part, as well as a clay figurine and the EC frying pans of the Kampos phase from the north part of the settlement and a golden band with slightly curved ends and incised decoration found in House 1 and dating to the middle EB I (Troy Ib-c), the earliest golden artefact so far in the eastern Aegean and in western Anatolia, much earlier than the "treasures" from Poliochni giallo and Troy IIg (EB II) (Erkanal, Artzy, and Kouka 2003: Res. 3). The existence of such a valuable artifact in one of the earliest phases of the EB I settlement is one of the most indicative features of the social stratification and the economic prosperity of Liman Tepe at the beginning of the EB. If we adopt the relative chronology—since there are so far no results from the radiocarbon samples from Liman Tepe—the EB I levels at Liman Tepe will be contemporary with the Kampos and with the Aplomata groups and at any case will be earlier than the Keros-Syros phase. Following Manning's new dates from the Cyclades, this would mean the EB I at Liman Tepe should be dated between 2950 until some time after 2650 BC. This would furthermore mean that the EB I at Liman Tepe corresponds very well with Poliochni azzurro and Thermi I as well as with Troy I early–middle (EB 1 of Efe 1988 and EB I B of Mellink 1992) (Tables 7–8).

The EB II settlement—LMT V

During the earlier phases of the EB II, of the period that Renfrew defined as the Keros-Syros culture and later as the period of an "international spirit" in the Aegean, the contacts of Liman Tepe with the central Aegean become more intensive (Kouka 2002: 299–301): Urfirnis sauceboats, fragments of marble vessels, marble figurines, a bronze pin with amphoriskos head, and Melian obsidian demonstrate these contacts (Şahoğlu 2005a: Fig. 3, 5, 8–11, 14). These earlier phases should be dated according to Manning (2008: 59) between 2650–2500 BC (Tables 7–8). The later EB

II at Liman Tepe, namely Rutter's Lefkandi I-Kastri Phase (EC IIB), called in Anatolian terms EB IIIA, could be studied very well in the deposits of the communal storage at Liman Tepe, the previously called "corridor house" (Erkanal and Günel 1996: 312–313, Res. 16–18) (Figure 2). This building belongs to the administrative complex of Liman Tepe during the EB II. The study of this material by Şahoğlu in 2002 demonstrated the existence of two phases: a) an earlier phase with the one-handled cups, the bell-shaped cups and the one-handled tankards; and b) a later phase, during which the late EB II service set has been enriched furthermore with depas cups, shallow bowls, the first wheelmade plates and the cut-away spouted jugs with globular bodies. In his recent study on trade networks between Anatolia and the Aegean Şahoğlu defines this late EB II period as "Anatolian Trade Network" phase (Şahoğlu 2005), which is contemporary with Maran's "Wendezeit." According to the new dates from the Cyclades this period should be dated between 2500–2250 BC (Tables 7–8).

The EB III period at Liman Tepe is also well-defined with gray and red burnished wares, the shapes of them indicating on the one hand the continuation from the preceding phase and on the other hand the first signs of the wheelmade red and gray wares of the MB (Şahoğlu 2002).

From the above presentation it becomes clear that the large-scale excavations at Liman Tepe offer a tight stratigraphical sequence and define a clear chronological frame for the EB in the western Anatolia littoral. The evidence here may also help clear the misunderstandings of the 1990s—in both relative and absolute sense—caused by the new evidence from Troy (the dates were recovered from the excavations of the "pinnacles" left unexcavated by Schliemann, Dörpfeld, and Blegen). Due to its location in the center of the East Aegean and its participation to limited or extended trade networks, the stratigraphy of Liman Tepe contributes significantly to the comparative chronology of western Anatolia, with the North and East Aegean islands, the Cyclades, and mainland Greece from the fourth through the second millennia BC.

Conclusions

Discussing chronological problems that cover over a millennium within 20 minutes is quite difficult, and certainly not always pleasant for an audience. Rather, it is a matter worthy of causing a headache, particularly since the study focuses on more than one chronological horizon, all possessing many different names, as does, for example, the late EB II. This horizon is known as the Lefkandi I assemblage, or the Kastri group, or the Lefkandi I-Kastri Phase, or the "Wendezeit," or the "Anatolian trade network" phase, that belongs either to the EB IIB or to the EB IIIA in Aegean terms, or to the EB IIIA in Anatolian terms. What all these conventional labels refer to is the same period: that is c. 2500–2250 BC. What then is the need for this plurality of terms for the same matter? This can cause only misinterpretations.

I think that the study of Aegean prehistory from its beginnings in 1870 to now has showed us that using the tripartite system and trying to fit cultures and groups in it causes more problems than it solves. Furthermore, archaeological evidence has demonstrated how regionalism and local traditions in material culture can create problems regarding the correlation of different landscapes (e.g Macedonia with Crete) within the Aegean, even in cases of close neighboring landscapes (e.g. Troy with Samos). I think that in the third millennium AD the abundance of archaeological evidence as outlined very briefly above allows us to develop a new, simpler chronological code, beyond labels based on numbers (EB I, II, III), cultures, and groups. Absolute dates from various sites of the Aegean will soon allow us to define a chronological frame based on centuries within the third millennium BC. Published data and forthcoming dates are indeed very promising. Thus, we will be able to speak of what happened for example in western Anatolia and Crete in the 26th or in the 23rd century BC and so on.

The establishment of a new and flexible chronological frame anchored both in the relative and the absolute chronology requires hard work and involves the following steps:

1. To study the internal/local stratigraphical sequences of well excavated and well published sites.

2. To undertake a comparative stratigraphy based on the most characteristic local artefact types as well as on imports.

3. And finally, to combine the comparative stratigraphy with the calibrated dates of the studied sites.

By adopting these steps we will be able to rethink the third millennium BC and the Aegean prehistory in general terms and go ahead with the writing of the prehistory of the Aegean as a part of the prehistory of the East Mediterranean, beyond the Babel of chronological labels, which is currently the situation for the Aegean, Anatolia, Cyprus, the Levant etc.

Reassessing the complicated division and the traditional chronological schemes is not meant to show lack of respect to the pioneers of Aegean archaeology. On the contrary. It is a sign that the science is alive and is not becoming fossilized! Collaboration with colleagues working in the Aegean and western Anatolia is going to be essential for this new direction in research. The upcoming international conference entitled "The Early Bronze Age in the Aegean: New Evidence" that will be held in Athens in 2008 may provide the suitable forum for these intellectual exchanges.

Acknowledgments

I am indebted to Prof. Dr. Hayat Erkanal, Ankara University, and Director of the Izmir Region Excavations and Research Project (IRERP) for inviting me to participate in the project since 2000 and for the permission to publish material from Liman Tepe. Besides, I would like to thank all the members of the IRERP team for collaboration in the field, and particularly Dr. V. Şahoğlu and Dr. R. Tuncel for discussions on the material from the CH and the EB phases of Liman Tepe. Finally, I am grateful to Prof. S. Manning for allowing me to use data of his (then) unpublished paper (Manning 2008).

References

Alram-Stern, E. 2004. Die Ägäische Frühzeit 2. Band–Teil 1 and 2. *Die Frühbronzezeit in Griechenland.* Wien: Österreichischen Akademie der Wissenschaften.

Andreou, S., Fotiadis, M., and Kotsakis, K. 2001. Review of Aegean Prehistory V: The Neolithic and Bronze Age of Northern Greece. In T. Cullen (ed.), *Aegaean prehistory: a review*: 259–327. Boston: Archaeological Institute of America.

2003. Son Kalkolitik ve İlk Tunç Çağı. *ARKEOATLAS* 2.

Barber, R.L.N., and MacGillivray, J.A. 1980. The Early Cycladic period: matters of definition and terminology. *American Journal of Archaeology* 84: 141–157.

Barber, R.L.N., and MacGillivray, J.A. 1984. The prehistoric Cyclades: a summary. In J.A. MacGillivray and R.L.N. Barber (eds.), *The prehistoric Cyclades: contributions to a workshop on Cycladic chronology (dedicated to J.L. Caskey)*: 296–302. Edinburgh: University of Edinburgh.

Blegen, C.W., Caskey, J.L., Rawson, M., and Sperling, J. 1950. *Troy. General Introduction. The First and Second Settlements.* Vol. I. Part 1: Text, I. Part 2: Plates. Princeton: Princeton University Press.

Broodbank, C. 2000. *An island archaeology of the early Cyclades.* Cambridge: Cambridge University Press.

Coleman, J.E. 1992. Greece, the Aegean and Cyprus. In R.W. Ehrich (ed.), *Chronologies in Old World Archaeology* Vol. 1: 247-288; Vol. 2: 203–229. Chicago: The University of Chicago Press.

Dickinson, O. 1994. *The Aegean Bronze Age.* Cambridge: Cambridge University Press.

Doumas, C.G., and La Rosa, V. (eds.) 1997. Η Πολιόχνη και η Πρώιμη Εποχή του Χαλκού στο Βόρειο Αιγαίο, Διεθνές Συνέδριο, Αθήνα 22-25 Απριλίου 1996. Athens: Scuola Archeologica Italiana di Atene –Πανεπιστήμιο Αθηνών, Τομέας Αρχαιολογίας και Ιστορίας της Τέχνης.

Efe, T. 1988. *Demircihüyük. Die Ergebnisse der Ausgrabungen 1975-1978. Bd. III, 2. Die Keramik 2 C. Die frühbronzezeitliche Keramik der jüngeren Phasen (ab Phase H).* Mainz: Philipp von Zabern.

Efe, T. 2006. Anatolische Wurzeln—Troia und die frühe Bronzezeit im Westen Kleinasiens. In M.O. Korfmann (ed.), *Troia—Archäologie eines Siedlungshügels und seiner Landschaft*: 15-28. Mainz: Philipp von Zabern.

Efe, T. 2007. The Küllüoba excavations and the cultural/political development of Western Anatolia before the second millennium B.C. In M. Doğan-Alparslan, M. Alparslan, and H. Peker (eds.), *VİTA. Festschrift in Honor of Belkis Dinçol and Ali Dinçol*: 251–267. Istanbul: Ege Yayınları.

Erdoğu, B., Tanındı, O., and Uygun, D. 2003. TAY: Türkiye Arkeolojik Yerleşmeleri 14C veri tabanı. Istanbul: Ege Yayınları.

Erkanal, H., and Günel S. 1996. 1994 Liman Tepe Kazısı, *XVII. Kazı Sonuçları Toplantısı* I: 305–328.

Erkanal, H., and Özkan, T. 1997. 1995 Bakla Tepe Kazıları, *XVIII. Kazı Sonuçları Toplantısı* I: 261-279.

Erkanal, H., and Özkan T. 1999. Excavations at Bakla Tepe. In. T. Özkan and H. Erkanal (eds.), *Tahtalı Baraji Kurtarma Kazısı Projesi (Tahtalı Dam Area Salvage Project)*: 108–138. Izmir: Simedya.

Erkanal, H., Artzy, M., and Kouka, O. 2003. 2001 Yılı Liman Tepe Kazıları, *XXIV. Kazı Sonuçları Toplantısı.* Vol. 1: 423–436.

Erkanal, H., Artzy, M., and Kouka, O. 2004. 2002 Yılı Liman Tepe Kazıları. *XXV. Kazı Sonuçları Toplantısı* 2: 165–178.

Erkanal, H., and Karaturgut, E. 2004. 2002 Yılı Çeşme Bağlararası Kazıları. *XXV. Kazı Sonuçları Toplantısı* 2: 153–164.

Harmankaya, S., and Erdoğu, B. 2002. *TAY: Türkiye Arkeolojik Yerleşmeleri–4a/4b: İlk Tunç Çağı.* Istanbul: TASK Vakfı Yayınları.

Korfmann, M., and Kromer, B. 1993. Demircihüyük, Beşiktepe, Troia—Eine Zwischenbilanz zur Chronologie dreier Orte in Westanatolien. *Studia Troica* 3: 135–171.

Korfmann, M. 1996. Troia—Ausgrabungen 1995. *Studia Troica* 6: 1–63.

Korfmann, M. 2000. Troia—Ausgrabungen 1999. *Studia Troica* 10: 1–34.

Korfmann, M.O. (ed.) 2006. *Troia—Archäologie eines Siedlungshügels und seiner Landschaft.* Mainz: Philipp von Zabern.

Kouka, O. 2002. *Siedlungsorganisation in der Nord- und Ostägäis während der Frühbronzezeit (3. Jt. v.Chr.). Internationale Archäologie* 58. Rahden-Westfalen.: VML Verlag.

Kouka, O. 2008. Diaspora, presence, or interaction? The Cyclades and the Greek Mainland from the Final Neolithic to Early Bronze Age II. In N.J. Brodie, J. Doole, G. Gavalas, and C. Renfrew (eds.), Ὁρίζων: *A colloquium on the prehistory of the Cyclades, Cambridge, 25th-28th March 2004*: 271–279. McDonald Institute Monograph Series. Cambridge: McDonald Institute for Archaeological Reasearch.

Manning, S.W. 1995. *The absolute chronology of the Aegean Early Bronze Age. Archaeology, radiocarbon and history.* Monographs in Mediterranean Archaeology 1. Sheffield: Sheffield Academic Press.

Manning, S.W. 1997. Troy, radiocarbon and the chronology of the northeast Aegean. In C.G. Doumas and V. La Rosa (eds.), Η Πολιόχνη και η Πρώιμη Εποχή του Χαλκού στο Βόρειο Αιγαίο, Διεθνές Συνέδριο, Αθήνα 22-25 Απριλίου 1996: 498–521, Athens: Scuola Archeologica Italiana di Atene –Πανεπιστήμιο Αθηνών, Τομέας Αρχαιολογίας και Ιστορίας της Τέχνης.

Manning, S. 2008. Some initial wobbly steps towards a Late Neolithic to Early Bronze III radiocarbon chronology for the Cyclades. In N.J. Brodie, J. Doole, G. Gavalas, and C. Renfrew (eds.), Ὁρίζων: *A colloquium on the prehistory of the Cyclades, Cambridge, 25th-28th March 2004*: 55–59. McDonald Institute Monograph Series. Cambridge: McDonald Institute for Archaeological Reasearch.

Maran, J. 1998. Kulturwandel auf dem griechischen Festland und auf den Kykladen im späten 3. Jahrtausend v.Chr., Teil

I-II. *Universitätsforschungen zur Prähistorischen Archäologie* Bd. 53. Bonn: Habelt.

Mellink, M. 1986. The Early Bronze Age in West Anatolia: Aegean and Asiatic Correlations. In G. Cadogan (ed.), *The End of the Early Bronze Age in the Aegean, Festschrift for J.L. Caskey*: 139–152. Cincinnati: Cincinnati Classical Studies N.S. 7.

Mellink, M. 1992. Anatolia. In R.W. Ehrich (ed.), *Chronologies in Old World Archaeology* Vol. 1: 207–220; Vol. 2: 171–184. Chicago: The University of Chicago Press.

Özyar, A., Danışman, G., and Özbal, H. 2005. Fieldseasons 2001-2003 Tarsus-Gözlükule interdisciplinary research project. In A. Özyar (ed.), *Fieldseasons 2001-2003 Tarsus-Gözlükule interdiciplinary research project*: 8–47. Istanbul: Ege Yayınları.

Pantelidou-Gofa, M. 2005. Τσέπι Μαραθώνος. Το Πρωτοελλαδικό νεκροταφείο, Βιβλιοθήκη της εν Αθήναις Αρχαιολογικής Εταιρείας Αρ. 235. Αθήναι: Η εν Αθήναις Αρχαιολογική Εταιρεία.

Pullen, D.J. 1985. Social organization in the Early Bronze Age Greece: a multidimensional approach. Ph.D. Dissertation, Indiana University, University Microfilms International. Michigan: Ann Arbor.

Rambach, J. 2000a. Kykladen I. Die Frühe Bronzezeit—Grab- und Siedlungsbefunde. *Beiträge zur ur- und frühgeschichtlichen Archäologie des Mittelmeer-Kulturraumes (BAM)* 33. Bonn: Habelt.

Rambach, J. 2000b. Kykladen II. Die Frühe Bronzezeit—Frühbronzezeitliche Beigabensittenkreise auf den Kykladen: Relative Chronologie und Verbreitung, *Beiträge zur ur- und frühgeschichtlichen Archäologie des Mittelmeer-Kulturraumes (BAM)* 34. Bonn: Habelt.

Renfrew, C. 1972. *The Emergence of Civilisation. The Cyclades and the Aegean in the Third Millennium B.C.* London: Methuen and Co Ltd.

Renfrew, C., Housley, R., and Manning, S. 2006. The absolute dating of the settlement. In L. Marangou, C. Renfrew, C. Doumas, and G. Gavalas (eds.), *Markiani, Amorgos. An Early Bronze Age fortified settlement. Overview of the 1985–1991 investigations*, The British School at Athens Supplementary Volume NO. 40: 71–76. Athens: The British School at Athens.

Rutter, J.B. 1979. Ceramic change in the Aegean Early Bronze Age. The Kastri Group, Lefkandi I and Lerna IV: a theory concerning the origin of EH III ceramics. *UCLA Institute of Archeology Occasional Papers* 5. Los Angeles: UCLA Institute of Archeology.

Rutter, J.B. 1983. Some observations on the Cyclades in the later third and early second millennia. *AJA* 87: 69–75.

Şahoğlu, V. 2002. Liman Tepe Erken Tunç Çağı seramiğinin Ege arkeolojisindeki yeri ve önemi. Ph.D. Dissertation, Ankara University.

Şahoğlu, V. 2005a. Interregional contacts around the Aegean during the Early Bronze Age: new evidence from the Izmir region. *Anadolu/Anatolia* 27: 97–120.

Şahoğlu, V. 2005b. The anatolian trade network and the Izmir region during the Early Bronze Age. *American Journal of Archaeology* 24: 339–360.

Schoop, U.-D. 2005. *Das anatolische Chalkolithikum. Eine chronologische Untersuchung zur vorbronzezeitlichen Kultursequenz im nördlichen Zentralanatolien und den angrenzenden Gebieten*. Remshalden: Verlag Albert Greiner.

Studia Troica 1–16, 1991–2006. Mainze am Rhein: P. von Zabern.

Takaoğlu, T. 2005. *A Chalcolithic marble workshop at Kulaksızlar in Western Anatolia: an analysis of production and craft specialization*. BAR International Series, 1358. Oxford: Archeopress.

Treuil, R. 1983. *Le Néolithique et le Bronze Ancien égéens. Les problèms stratigraphiques et chronologiques, les techniques, les hommes*. Athens: Bibliothéque des Écoles françaises d'Athènes et de Rome, 248.

Tuncel, R. 2005. Bakla Tepe Geç Kalkolitik Cağ seramiği. Ph.D. Dissertation, Hacettepe University, Ankara.

Wace, A.J.B., and Blegen, C.W. 1916-1918. The pre-Mycenaean pottery of the mainland. *BSA* 22: 175-189.

Warren, P., and Hankey, V. 1989. *Aegean Bronze Age chronology*. Bristol: Bristol Classical Press.

Article submitted July 2007

Middle Helladic Lerna: Relative and Absolute Chronologies

Sofia Voutsaki, Albert J. Nijboer, and Carol Zerner

Abstract: *This paper will present the first results of the radiocarbon analysis of human bones from Middle Helladic sites in the Argolid. The main aim of the analysis is to provide a coherent set of radiocarbon data from several sites and from all sub-phases of the MH period. In this paper we concentrate on the results from Lerna, and we discuss the problems arising when integrating relative and absolute dates. It is the first time that a systematic large-scale program of ^{14}C analyses from MH sites has been undertaken; in fact, the mainland has been largely ignored in the recent chronological debate in Aegean prehistory. We therefore hope that our analyses will not only refine the MH sequence, but will also add a new dimension to the debates surrounding the chronology of the Aegean Bronze Age.*

1. Introduction

This paper presents the results of radiocarbon analysis from human skeletal material from Middle Helladic (hereafter MH) Lerna, and discusses the problems arising when comparing and integrating sequences of absolute and relative dates. The radiocarbon analysis presented here is part of a wider interdisciplinary project: the MH Argolid Project, financed by the Netherlands Organization for Scientific Research and the University of Groningen, the Netherlands. The aim of the wider project is to reconstruct the social organization of MH communities and to interpret the important social, political and cultural changes that took place in the southern Greek mainland during the MH period and the transition to the LH (see Voutsaki 2005; and http://www.MHArgolid.nl). This is being pursued by means of an integrated analysis of funerary, skeletal, and settlement data from the MH Argolid. The radiocarbon analysis is part of the examination of the funerary data, as the samples analyzed have been taken from human remains.

2. The Analysis

2.1 Introduction

Before presenting the aims and methods of the analysis, we would like to start with some introductory remarks on the position of radiocarbon analysis in Aegean prehistory. Despite the fact that significant progress has been made in this field, there are still serious problems that hamper the integration of relative and absolute dates in the Aegean. It has been pointed out already several times (i.e. Manning 1996: 29) that collaboration between archaeologists and scientists rarely occurs. Radiocarbon analyses are often placed in an Appendix at the end of the final publication, but are not really integrated into the research design of archaeological projects (although notable exceptions exist). Second, the coverage of the different periods is very uneven: while there is a substantial number of measurements for the Early Bronze Age (Manning 1995), there are very few dates for the earlier part of the Middle Bronze Age. Again, this was stressed already in the 1970s (Cadogan 1978: 20), but three decades later the situation has changed only marginally (see also Manning 1996: 30). The situation is better in MB III and especially in LB I (Höflmayer 2005) because of the debate surrounding the eruption of the Thera volcano; not surprisingly, there are fewer dates from the later part of the LBA (Wiener 1998; Manning and Weninger 1992). In addition, the coverage of different regions is clearly uneven: there are many dates from the Cyclades, but by far the majority comes from Akrotiri. There are considerably fewer measurements from Crete, while the mainland has remained virtually absent from the chronological debates in Aegean archaeology (Manning 2005: 113). Moreover, there are many sporadic measurements, rarely more than one or two from sites that have been

Sample	Lerna Grave No.	Blackburn Grave No.	Skel No.	Age BP	$^{13}\delta$(‰)	%C
GrA-28046	BD 27	12	Lerna 103	3830 ±35	-19.21	42.1
GrA-28051	BD 14	32	Lerna 91	3730 ±35	-19.08	43.1
GrA-28213	DE 71 and 72	25	Lerna 239 (double burial; 239 buried first)	3640 ±45	-19.70	43.8
GrA-28054	D 22	47	Lerna 52	3595 ±35	-19.28	41.4
GrA-28053	BD 9	123	Lerna 87	3595 ±35	-19.38	41.0
GrA-28045	H 1	136	Lerna 31	3585 ±35	-20.30	41.1
GrA-28050	BD 19	80	Lerna 95	3560 ±35	-19.00	43.3
GrA-28039	A 1	152	Lerna 2	3545 ±35	-19.15	42.2
GrA-28048	BD 21	79	Lerna 97	3535 ±35	-19.67	42.8
GrA-28041	D 20	52	Lerna 50	3530 ±35	-19.74	41.8
GrA-28044	B 13	58	Lerna 44	3520 ±35	19.56	42.2
GrA-28040	D 9	182	Lerna 28	3510 ±35	-19.50	43.0
GrA-28159	DE 55	100	Lerna 198	3510 ±40	-19.44	40.7
GrA-28211	J 5	63	Lerna 218	3510 ±50	-19.96	41.9
GrA-28043	DE 64	103	Lerna 204	3495 ±40	-20.05	43.9
GrA-28157	BE 30	22	Lerna 137 (5 skeletons buried at once)	3475 ±40	-19.50	39.9
GrA-28160	J4 A	84	Lerna 216 (double burial; 216 upper grave; later)	3440 ±40	-19.77	40.8

Table 1: The quality of the ^{14}C results. Please note: the Laboratory Sample Number is followed by the Lerna Grave Number to allow easy identification of the archaeological context.

extensively excavated over longer periods, but very few series of measurements and even fewer complete sequences. Again, this situation was lamented in the 1970s (Betancourt and Weinstein 1976: 330; Betancourt et al. 1978: 202) but has not really been rectified (Manning 1996: 29). Many of the old measurements are not always reliable, as they were taken before the significant improvements in radiocarbon dating procedures (Manning 1998: 301). Finally, very few radiocarbon dates come with good and extensive contextual information. While there is a lot of discussion about short-lived (e.g. seeds) versus long-lived samples (e.g. wood, charcoal), less attention is paid to the fact that a sample may come from inside a floor, from the floor deposit, or from the destruction layer above the floor—but these are significant differences that need to be taken into account when integrating absolute and relative dates. Once more, this problem has been raised repeatedly (i.e. Betancourt and Weinstein: 331; see also pertinent remarks by Whitelaw 1996: 233).

We see therefore that these problems have been recognized since the 1970s, but there are still few systematic and problem-oriented programs of ^{14}C analysis in Aegean prehistory. Our project hopes to redress this situation: We are undertaking an extensive program of analyses, taking a series of samples from different MH sites in the Argolid. We sample six burials from each sub-phase of the MH period (i.e. 6 from MH I, 6 from MH II, 6 from MH III), and we try to include burials that can be placed in the earlier or later part of each sub-phase (though this is not always possible). In this way, we can set up a compendium of dates for all the sites we are studying and reconstruct whole sequences for the entire MH period. While here we present only the results from Lerna, we are sampling all sites with a substantial number of burials. So far we have taken 19 samples from Lerna, 7 from Argos-Aspis, 12 from Asine-East cemetery, and 9 from Asine-Barbouna (the cemeteries in Aspis and Asine are not in use throughout the MH period, hence the smaller number of samples). We also plan to analyze skeletons from the Argos "tumuli," and possibly the prehistoric cemeteries at Mycenae and Midea.

We analyze human bones, and thereby avoid the problems encountered when analyzing long-lived materials such as wood or charcoal. By sampling human skeletons (more often than not from single burials)

Sample	Lerna Grave No.	Blackburn Grave No.	Skel No.	Age BP	$^{13}\delta(‰)$	%C
GrA-28261	B 21A	88	Lerna 67	3700 ±45	-23.50	0.8

Table 2: Result with deficient quality parameters.

rather than wood, charcoal, or seeds found in settlement layers, we date a specific depositional episode rather than settlement deposits which may have accumulated over a period of unknown duration (Nijboer and van der Plicht 2008). We use the AMS (Accelerator Mass Spectrometry) rather than the conventional ^{14}C method; the latter may be more accurate, but it requires much larger samples (250g versus 5g necessary when using the AMS method). We decided to use AMS because of the need to preserve the Lerna skeletal assemblage for future generations of researchers. The analyses have been carried out at the Centre of Isotope Research, University of Groningen (for requirements see Nijboer and van der Plicht 2008).

We take samples exclusively from well excavated and extensively documented cemeteries, and notably from well preserved tombs which can be dated with reasonable accuracy. In the case of Lerna, in particular, we sample only tombs with clear stratigraphic contexts whose relative date has been carefully controlled (and sometimes revised) by Carol Zerner. Of course, we should not underestimate the complexity of the MH sequence at Lerna; we are dealing with a large intramural cemetery with a complex history of use (Blackburn 1970; Zerner 1978), and with several areas used interchangeably for burial and for habitation (Milka, in press). This situation has certain advantages, as many graves have clear stratigraphic associations with earlier or later houses. However, these stratigraphic associations may sometimes provide only a *terminus ante* or *post quem*. In effect, when integrating absolute and relative dates, we compare two ranges of possible dates: just as ^{14}C dates come with a certain margin of error, the relative date of a grave may also sometimes span more than one ceramic sub-phase (e.g. a grave may be MH I or MH II early). Interpreting radiocarbon dates in an archaeological context involves precisely this: trying to reconcile two ranges of possible dates. Therefore, slight modifications of the relative date in the light of radiocarbon results are permissible—as long as we stay within the range dictated by both the stratigraphy of the site and the ^{14}C measurements.

2.2 The Results

The quality of the radiocarbon data, as can be seen in Table 1, is good (for a definition of quality parameters, see Nijboer and van der Plicht 2008).

Only one sample from Lerna (Table 2) did not contain any collagen and the measurement was based on the residue. This sample has not been included in the analysis.

It should also be added that we have carried out an extensive program of stable isotopes analysis (we have sampled 48 burials from MH Lerna alone) in order to establish the diet of the MH population in Lerna (see Voutsaki et al., in press; Triantaphyllou et al. 2008). The results, in particular the rather low δ15 N values, allow us to establish that the inhabitants of Lerna did not rely on marine resources during the MH period. In this way, we can rule out the "reservoir effect" (Lanting and van der Plicht 1998) and increase confidence in the accuracy of our results (Figure 2).

2.3. Discussion of the results

The aims of the analysis are to increase the chronological resolution of the analysis of funerary and settlement data, and to refine the MH chronological sequence. What we want to achieve is a more accurate definition of both the boundaries of the MH period and its internal sub-divisions. Therefore, the discussion here will proceed by examining one by one the chronological divisions of the MH period. Figure 2, which presents once more the results of the analysis, as well as an indication of their relative date, will be the basis of the discussion.

(i) The EH III/MH I boundary The accepted date for the beginning of the MBA has been placed in 2100/2000 BC, i.e. around or just before the beginning of the 2nd millennium (Cadogan 1978: 213; Warren and Hankey 1989: 124; Manning 1995). We have taken three measurements from tombs which had a MH I relative date: the results from DE 71 & 72 and BD 14 fall clearly within the accepted range, but suggest that the beginning of the period should be placed around 2100 BC rather than at 2000 BC. In contrast, the result from BD 27 (Figure 3), the burial of a child of 2-3 years old accompanied by a one-handled cup

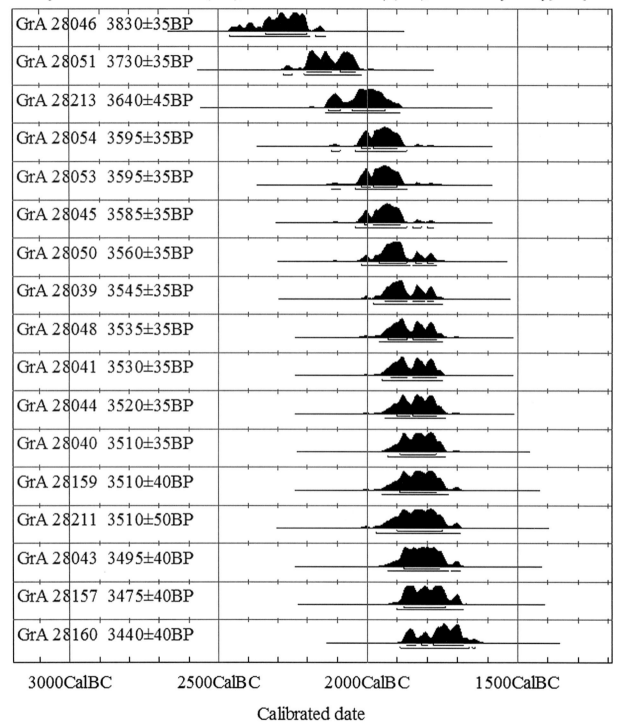

Figure 1: The results of the analysis.

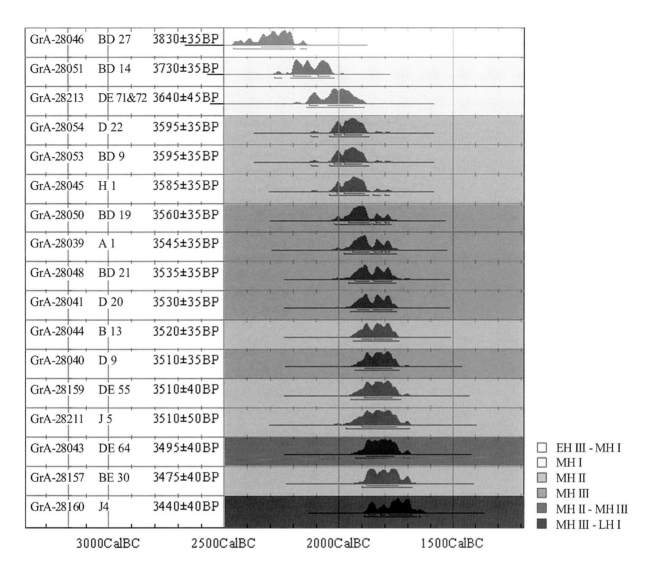

Figure 2: The results of the analysis: relative and absolute dates.

(Figure 4), has a range of 2460-2140 BC at 2σ probability level (95.4%) (Figure 5). While the grave had at first been dated to MH I, possibly MH I early, its absolute date suggests that it may have to be placed earlier, i.e. in the late EH III period, or in the transition between EH III/MH I. Indeed the cup could perfectly well belong to the EH III period. A renewed examination of the stratigraphic context by Carol Zerner concluded that a date in late EH III or in the EH III/MH I transition is equally acceptable.

We see therefore that the absolute dates may sometimes be used to reconsider the relative date, as long as one has confidence in the sample and its context—otherwise, we may enter a circular argument, as has often been done in the past. As we stressed above, the relative dating often covers a certain range, especially when (as is so often the case in the MH period) the grave contains either no diagnostic offerings, or no offerings whatsoever. As we stressed above, the integration of relative and "absolute" dates is often an exercise in comparing and attempting to reconcile two ranges of possible dates.

(ii) The MH I/MH II division The beginning of the MH II period is placed probably around 1900 BC (Dietz 1991: 317; Rutter 2001: 106). The Lerna results confirm this: all three MH I measurements lie before 1900 BC. We could therefore tentatively conclude that the MH I period lasts approximately from 2100 to 1900 BC. Some caution is of course necessary, as we have only very few measurements from the MH I period. In order to corroborate this date, we ought to take more samples from MH I burials. Unfortunately, outside Lerna, EH III–MH I burials have been found only during the old excavations in the Lower Town of

Figure 3: Child burial in grave BD 27.

Figure 4: One-handled cup found in BD 27.

Asine (Frödin and Persson 1938; Nordquist 1987), but the skeletons from those excavations have gone missing. While we could sample more EH III burials from Lerna, they are all neonates and may not be suitable for ^{14}C analysis. Needless to say, we also need to compare the Lerna analysis with results recently obtained from contemporary sites, e.g. Kolonna in Aegina (for the results of the recent investigations see Gauss and Smetana, in press).

(iii) The MH II/MH III division is dated 1750–1720 BC by Manning (1995), while Dietz places the transition around 1800 BC (Dietz 1991: 317). Unfortunately, the Lerna results cannot help us decide on this issue: the results fall too close together and we therefore cannot distinguish between MH II and MH III. Two examples can illustrate the problem: BD 19 (Figure 6), a cist grave containing an adult man buried with a Cycladic jug (Figure 7) and a strainer jug, is firmly dated in MH III because of the shape of the jug; in addition, the fact that the grave is opened above a MH II house confirms this date. The ^{14}C result (Figure 8) has a range of 2020–1770 BC at 2σ probability level (95.4%) which is compatible with a date in MH III, but cannot exclude a date in MH II.

In contrast, BE 30, a pit grave containing five burials which seem to have been buried simultaneously, is quite firmly dated in MH II: it was opened into the ruins of a MH I–II house, while in MH III another house was built on top. However, the result of 1900–1680 BC at 2σ probability level (Figure 9) is compatible with a date in MH II, but cannot exclude a MH III date.

The inability to distinguish the MH II and MH III periods in the absolute sequence and the rather broad range of ^{14}C dates has to do with the shape of the calibration curve in this period and the relatively short duration of these sub-phases. Here we evidently reach the limits of precision of the radiocarbon method (see Warren 1996: 283–284; admitted also by Manning 1996: 30) which may be overcome only with the use of Bayesian statistics.

Therefore, while there are important differences between MH II and MH III in terms of ceramic changes and historical developments, the MH II–III boundary cannot be accurately defined in our analysis. It should be emphasized that we encounter the same problem in Asine–East Cemetery which is also in use in MH II and MH III (Voutsaki et al., in press). We may be able to move forward by concentrating on the burials whose relative dates are most secure and by combining measurements from different graves belonging to the same phase. However, there are problems: despite widespread practice, this procedure is not really statistically valid. It should also be stressed that defining what constitutes a most secure context is not straightforward. There are cases where stratigraphic associations are quite secure, as in the graves in Lerna that are in use after a house is abandoned and before a new house is built in the same location. The situation in extramural cemeteries is very different. While there are fewer stratigraphic restrictions, extramural graves sometimes (but not always) contain more (and more diagnostic) offerings, as they date mostly to late MH II and MH III. Therefore, separating the most securely dated graves is not an easy task; we have to accept that there are degrees of certainty and accuracy in relative dating. This is an important point, which should be taken into account when combining measurements or applying different statistical methods on groups of measurements.

(iv) The MH III/LHI boundary. The duration of the MH III period and the transition to the LH I period are heavily debated. Establishing synchronisms between the mainland and Minoan Crete is not straightforward (Girella, in press); in fact, the Cretan sequence itself in MMIII–LMI is subject to debate (see

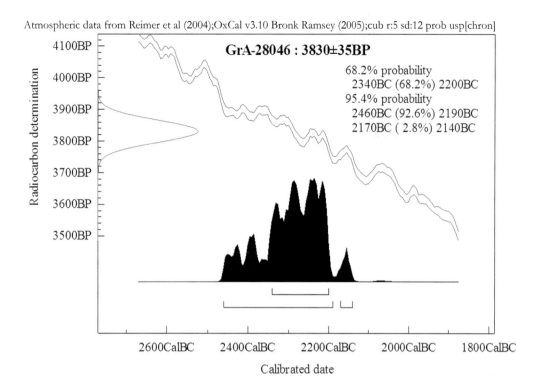

Figure 5: ^{14}C result from grave BD 27.

Figure 6: Adult man buried in grave BD 19.

Figure 7: Cycladic jug from BD 19.

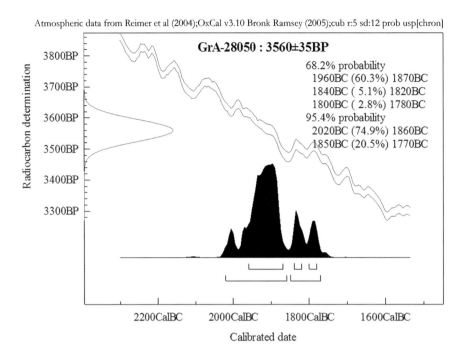

Figure 8: ^{14}C result from grave BD 19.

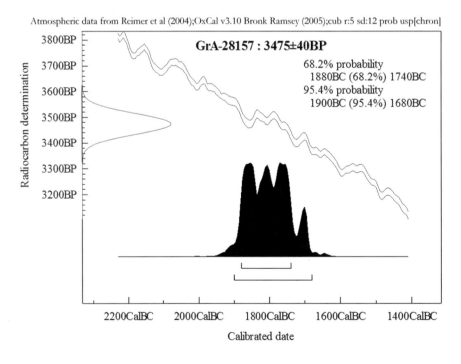

Figure 9: ^{14}C result from grave BE 30.

papers in Felten et al. 2007). On the other hand, there are few secure synchronisms with the Near East (Manning 1996: 17). According to the "Low Chronology" (Warren and Hankey 1989; Warren 1996; Bietak 2003; Wiener 2003) the transition takes place around 1600 BC, while the "High Chronology" (Betancourt 1987; most recent results, Manning et al. 2006) places it around 1700 BC.

All Lerna samples from graves with a relative date in MH III (5 samples) produce results which are earlier than 1700 BC. The only grave to cross the 1700 BC boundary (Figure 12) is J 4A, a semi-cist grave with an adult man (Figure 10) accompanied by a rather undiagnostic bowl (Figure 11). The grave is dated to the "Shaft Grave era," i.e. MH III–LH I, because of its stratigraphic associations (the grave is opened partly on top of MH II grave J 4B).

Figure 11: Bowl from J 4A.

Figure 10: Adult man buried in grave J 4A.

These results may render support to the "High Chronology" (Manning et al. 2006). But again some caution is necessary: we have only five measurements from Lerna, and there are few other comparable measurements from the mainland. The results from Tsoungiza are rather inconsistent (Bronk Ramsey et al. 2004), while many graves in Asine-East Cemetery span precisely this transition MH III–LH I. The Asine results cannot be discussed here in detail, but it should be pointed out that in Asine the relative date of graves is less secure: we are dealing with extramural graves which have fewer stratigraphic associations, while at the same time they rarely contain diagnostic offerings (see Voutsaki et al., in press).

3. Summary and Conclusions

We have suggested that the MH I period lasted from 2100 to 1900 BC. If we accept that the MH III period finished around 1700 BC, then the date proposed by Dietz for MH II (1900–1800 BC) seems preferable. We need to be aware, of course, that this suggestion is based more on common sense than on actual results. If we want to summarize our results, we come up with the sub-divisions presented in Table 3. It cannot be emphasized enough that these sub-divisions are only tentative, as the analysis is still in progress.

MH I	2100?–1900 B.C.
MH II	1900–1800? B.C.
MH III	1800?–1700 B.C.

Table 3: Tentative sub-divisions of the MH period

To conclude: the radiocarbon analysis of MH burials from Lerna has produced a tight and coherent sequence with a good correspondence between absolute and relative dates. Therefore, despite the considerable difficulties encountered when integrating relative and absolute chronologies, progress can be made—even if, in the particular case, we have not been able to distinguish the MH II from the MH III burials. The case we are dealing with, the Middle Bronze Age mainland, is particularly difficult because there are problems with both the relative and the absolute sequences. On the one hand, the stratigraphic problems in both intramural and extramural cemeteries, the rarity of offerings in the graves, the uneven (and imperfectly understood) rate of change in MH ceramics, the differences in the ceramic assemblages between regions and even between sites in the same region, and the difficulties in establishing synchronisms with the Minoan or Cycladic sequences mean that relative dating is rather insecure. On the other hand, the shape of the calibration curve in this period, the relatively short duration

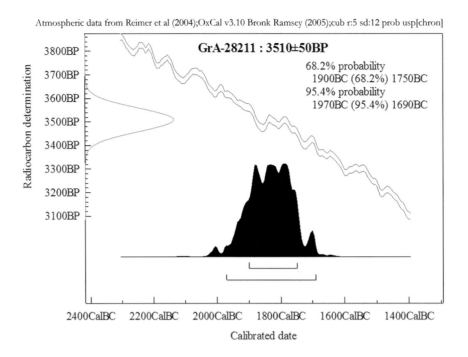

Figure 12: ^{14}C result from grave J 4A.

of the sub-phases of the MH period, but also the rarity of comparative data and the virtual absence of systematic programs of radiocarbon analysis in the MH mainland add to the inherent problems of applying the radiocarbon method (discussed by Warren 1998: 324; Wiener 2003). We therefore have to adopt a balanced approach, and try to avoid the polarization which has characterized in recent years the debate surrounding the chronology of the Aegean Bronze Age. Both relative and absolute dates come with a certain range; neither method is infallible, and neither method is fully precise. However, progress can be made and the two sets of data can be integrated, if we undertake systematic, extensive, and problem-oriented programs of analyses, and if we consider the nature of the samples—and their archeological contexts—very closely.

Acknowledgments

We would like to thank Sturt Manning for inviting us to participate in the Conference in honour of Peter Kuniholm, which proved a very stimulating and (largely thanks to Peter Kuniholm's contagious enthusiasm!) deeply enjoyable event. Further thanks to the entire team at Cornell, and particularly to Mary Jaye Bruce, for their hospitality.

We are grateful to the Netherlands Organization of Scientific Research and the University of Groningen, the Netherlands, for their generous funding of the MH Argolid Project. We would like to express our thanks to the successive Ephors of the 4th Ephorate of Classical and Prehistoric Antiquities, Mrs. Zoi Aslamatzidou and Mrs. Anna Banaka, as well as the Department of Conservation, Greek Ministry of Culture, for granting us a permit to re-examine and sample the MH burials from Lerna. We thank the American School of Classical Studies, as well as Dr. M. Wiencke, Dr. C. Zerner, and Dr. E. Banks for granting us permission to study and sample the Lerna skeletons. We would also like to acknowledge the assistance of the staff at the 4th Ephorate, particularly Dr. Alkistis Papadimitriou. The personnel in the Museum of Argos have been extremely helpful; we thank them all. We are grateful to Todd Whitelaw for his insightful corrections on the first draft. Eleni Milka has assisted with the tabulation of the contextual data, Tomek Hertig, the project assistant, has produced the graphs, and Siebe Boersma has helped with the illustrations. Finally, we would like to thank the editors for their patience.

References

Betancourt, P.P. 1987. Dating the Aegean Late Bronze Age with radiocarbon. *Archaeometry* 29(1): 45–49.

Betancourt, P.P., and Weinstein, G. 1976. Carbon-14 and the beginning of the LBA in the Aegean. *American Journal of Archaeology* 80: 329–348.

Betancourt, P.P., Michael, H.N., and Weinstein, G.A. 1978. Calibration and the radiocarbon chronology of Late Minoan 1B. *Archaeometry* 20(2): 200–203.

Bietak, M. 2003. Science versus archaeology: Problems and consequences of High Aegean Chronology. In M. Bietak (ed.), *The synchronization of civilizations in the eastern Mediterranean in the second millennium B.C. Proceedings of the SCIEM 2000—EuroConference, Haindorf, 2–7 May 2001*: 23–33. Vienna: Österreichischen Akademie der Wissenschaften.

Blackburn, E.T. 1970. Middle Helladic graves and burial customs with special reference to Lerna in the Argolid. Ph.D. Dissertation, University of Cincinnati.

Bronk Ramsey, C., Manning, S.W., and Galimberti, M. 2004. Towards high-precision AMS: progress and limitation. *Radiocarbon* 46(1): 325–344.

Cadogan, G. 1978. Dating the Aegean Bronze Age without radiocarbon. *Archaeometry* 20(2): 209–214.

Dietz, S. 1991. *The Argolid at the transition to the Mycenaean Age: Studies in the chronology and cultural development in the Shaft Grave period.* Copenhagen: The National Museum of Denmark.

Felten, F., Gauss, W., and Smetana, R. (eds.) 2007. Middle Helladic Pottery and Synchronisms. *Proceedings of the International Workshop SCIEM 2000, Salzburg, 31 October–2 November 2004.* Vienna: Österreichischen Akademie der Wissenschaften.

Frödin, O., and Persson, A.W. 1938. *Asine. The Results of the Swedish Excavations, 1922–1930.* Stockholm: Generalstabens Litografiska Anstalts Förlag.

Gauss, W., and Smetana, R. In press. Aegina Kolonna in the Middle Bronze Age. In G. Touchais, A. Philippa-Touchais, S. Voutsaki, and J. Wright (eds.), *MESOHELLADIKA: The Greek mainland in the Middle Bronze Age. Proceedings of the International Conference, Athens, 8–12 March 2006.* Supplément, Bulletin de Correspondance Hellénique.

Girella, L. In press. MH III and MM III: Ceramic synchronisms in the transition to the Late Bronze Age. In G. Touchais, A. Philippa-Touchais, S. Voutsaki, and J. Wright (eds.), *MESOHELLADIKA: The Greek mainland in the Middle Bronze Age. Proceedings of the International Conference, Athens, 8–12 March 2006.* Supplément, Bulletin de Correspondance Hellénique.

Höflmayer, F. 2005. Die absolute Datierung der späten Mittelbronzezeit und der frühen Spätbronzezeit in der Ägäis. Magister Dissertation, University of Vienna.

Lanting, J.N., and van der Plicht, J. 1998. Reservoir effect and apparent ^{14}C ages. *The Journal of Irish Archaeology* IX: 151–165.

Manning, S.W. 1995. *The absolute chronology of the Aegean Early Bronze Age: Archaeology, radiocarbon, and history.* Sheffield: Sheffield University Press.

Manning, S.W. 1996. Dating the Aegean Bronze Age: without, with and beyond, radiocarbon. In K. Randsborg (ed.), *Absolute chronology: Archaeological Europe 2500-500 BC*: 15–37. Acta Archaeologica 67. Copenhagen: Munksgaard.

Manning, S.W. 1998. Aegean and Sardinian Chronology: radiocarbon, calibration and Thera. In M.S. Balmuth and R.H. Tykot. (eds.), *Sardinian and Aegean Chronology. Towards the resolution of relative and absolute dating in the Mediterranean. Proceedings of the International Colloquium, Medford, Massachusetts, 17–19 March 1995*: 297–307. Studies in Sardinian Archaeology 5. Oxford: Oxbow Books.

Manning, S.W. 2005. Simulation and the date of the Theran eruption: outlining what we do and do not know from radiocarbon. In A. Dakouri-Hild and S. Sherratt (eds.), *AUTOCHTHON: Papers presented to O.T.P.K. Dickinson on the occasion of his retirement*: 97–114. BAR International Series 1432. Oxford: Archaeopress.

Manning, S.W., and Weninger, B. 1992. A light in the dark: archaeological wiggle matching and the absolute chronology of the close of the Aegean Bronze Age. *Antiquity* 66: 636–663.

Manning, S.W., Bronk Ramsey, C., Kutschera, W., Higham, T., Kromer, B., Steier, P., and Wild, E.M. 2006. Chronology for the Aegean late Bronze Age 1700-1400 BC. *Science* 312(5573): 565–569.

Milka, E. In press. Burials upon the ruins of abandoned houses in the MH Argolid. In G. Touchais, A. Philippa-Touchais, S. Voutsaki, and J. Wright (eds.), *MESOHELLADIKA: The Greek mainland in the Middle Bronze Age. Proceedings of the International Conference, Athens, 8-12 March 2006.* Supplément, Bulletin de Correspondance Hellénique.

Nijboer, A.J., and van der Plicht, H. 2008. The Iron Age of the Mediterranean: recent radiocarbon research at the University of Groningen. In D. Brandhern and M. Trachsel (eds.), *A new dawn for the dark Age? Shifting paradigms in Mediterranean Iron Age chronology. Proceedings of the XV Congress of the International Union for Prehistoric and Protohistoric Sciences, Lisbon, 4-9 September 2006*: 103–118. BAR International Series 1871. Oxford: Archaeopress.

Nordquist, G.C. 1987. A Middle Helladic village: Asine in the Argolid. *Boreas, Uppsala Studies in Ancient Mediterranean and Near Eastern Civilization* 16. Uppsala.

Rutter, J.B. 2001. Review of Aegean Prehistory II: The prepalatial Bronze Age of the southern and central Greek mainland. In T. Cullen (ed.), *Aegean Prehistory: a review. American Journal of Archaeology Supplement 1.* Boston: Archaeological Institute of America.

Triantaphyllou, S., Richards, M., Zerner, C., and Voutsaki, S. 2008. Isotopic dietary reconstruction of humans from Middle Bronze Age Lerna, Argolid, Greece. *Journal of Archaeological Science* 35: 3028–3034.

Voutsaki, S., Dietz, S., and Nijboer, A.J. In press. Radiocarbon analysis and the history of the East Cemetery, Asine. *Opuscula*.

Warren, P. 1996. The Aegean and the limits of radiocarbon dating. In K. Randsborg (ed.), *Absolute chronology: Archaeological Europe 2500-500 BC*: 283–290. Acta Archaeologica 67. Copenhagen: Munksgaard.

Warren, P., and Hankey, V. 1989. *The absolute chronology of the Aegean Bronze Age.* Bristol: Bristol Classical Press.

Whitelaw, T. 1996. Review of Manning, *The absolute chronology of the Aegean Early Bronze Age: Archaeology, radiocarbon, and history. Antiquity* 70: 232–234.

Wiener, M.H. 1998. The absolute chronology of Late Helladic IIIA2. In M.S. Balmuth and R.H. Tykot. (eds.), *Sardinian and Aegean Chronology. Towards the resolution of relative and absolute dating in the Mediterranean. Proceedings of the International Colloquium, Medford, Massachusetts, 17-19 March 1995*: 309–319. Studies in Sardinian Archaeology 5. Oxford: Oxbow Books.

Wiener, M.H. 2003. Time out: The current impasse in Bronze Age archaeological dating. In K. Polinger Foster and R. Laffineur (eds.), *METRON. Measuring the Aegean Bronze Age. Proceedings of the 9th International Aegean Conference, New Haven, 18-21 April 2002*: 363–399. Aegaeum 24. Liège and Austin.

Zerner, C.W. 1978. The beginning of the Middle Helladic Period at Lerna. Ph.D. Dissertation, University of Cincinnati.

Article submitted July 2007

Absolute Age of the Uluburun Shipwreck: A Key Late Bronze Age Time-Capsule for the East Mediterranean

*Sturt W. Manning, Cemal Pulak, Bernd Kromer, Sahra Talamo,
Christopher Bronk Ramsey, and Michael Dee*

Abstract: *By integrating radiocarbon and dendrochronological investigations, we can provide a high-resolution date in the later 14th century* BC *for the time of the last voyage of the extraordinary Late Bronze Age sailing vessel found wrecked at Uluburun near Kaş off the southern coast of Turkey: approximately 1320±15* BC*. This shipwreck was in a remarkable state of preservation because it lay on a steep underwater slope at a considerable depth (42–52m, with some artefacts scattered to 61m). The ship's cargo forms one of the largest and wealthiest assemblages known from the period, including a key link to the Amarna-period Egyptian Queen, Nefertiti. Our precise absolute dating provides an important chronological marker for the Amarna period in Egypt and across the Ancient Near East, resolving a number of areas of debate or contention in the scholarly literature.*

1. Introduction

An important (and at present unique) ancient shipwreck was excavated between 1984 and 1994 in deep water at Uluburun, near Kaş, off the southern coast of Turkey, by the Institute of Nautical Archaeology (Bass 1986; 1987; Bass et al.1989; Pulak 1988; 1998; 2001; 2005a; 2005b; 2008). The original vessel was around 15 m in length and would have carried some 20 tons of cargo (for a reconstruction, see Pulak 2005: 60 fig. 11; 2008: 293 fig. 94). Underwater excavation revealed an extraordinary assemblage of over 15,000 catalogued artefacts.

The famous shipwreck at Uluburun has been a noteworthy and exciting, albeit vexing, topic for the Malcolm and Carolyn Wiener Laboratory for Aegean and Near Eastern Dendrochronology for many years. Early on, Cemal Pulak submitted wood samples to Peter Kuniholm for analysis. However, the establishment of a direct tree-ring date as proposed in the 1990s (e.g. Pulak 1996; 1998: 213–214) has proved incorrect with further work (or at least unsupported and over-optimistic, and in need of more evidence: Wiener 2003: 244–246). As a result, work on an integrated radiocarbon and dendrochronological dating approach was undertaken. Initial work on dating some of the ship's timbers by such radiocarbon wiggle-matching was reported by Newton et al. (2005) and Newton and Kuniholm (2005). In this paper in honor of Peter Ian Kuniholm—friend, teacher, colleague—we are pleased to present the outcome of a robust, integrated dating program combining radiocarbon analysis and dendrochronology to best date both the ship's timbers

Raw materials recovered include about 10 tons of copper ingots, at least a ton of tin ingots, more than half a ton of terebinth resin in approximately two-thirds of the 150 Canaanite jars aboard, 175 ingots of glass, ebony logs, ostrich eggshells, elephant tusks, hippopotamus teeth, logs of African blackwood, and various other food, craft, or medicinal items. In addition to these raw materials, manufactured goods found on the wreck include a range of ceramics (Syro-Palestinian, Cypriot, and Mycenaean Greek), faience cups, copper alloy vessels, objects in ivory and gold, jewellery, and the earliest known examples of wooden writing boards (diptychs). The origin and destination of the ship have been actively sought within the world of the east Mediterranean, in the central Levantine coast and the Aegean, respectively (Pulak 1998; 2005c; 2008). The scale of wealth present suggests that this was perhaps an elite or royal shipment of cargo and that the ship was engaged in high-level exchange (e.g. Pulak 2005b; 2008, and, to some extent, Bachhuber 2006), along the lines of those to be inferred from the 14th-century BC Amarna letters recording royal diplomatic contact in the Ancient

---

as well as the final cargo (and hence to make an estimate of the date when the ship sank). This study replaces previous statements on the dating of the ship by dendrochronology and/or radiocarbon.

Near East between Egypt and other states and rulers (Moran 1992). Indeed, the wreck yielded a unique gold scarab bearing the cartouche of the Amarna-period Egyptian queen Nefertiti (Weinstein 1989), linking it directly to this general time period (and providing a *terminus post quem*—or date after which—for the shipwreck from during or after her reign).

The Uluburun ship represents an incredible time-capsule and has become a key source of evidence for study of numerous aspects of Bronze Age Mediterranean history, trade, interrelations at all levels, and especially for maritime interaction and technology (e.g. Bass 1986; 1987; 1991; 1998; Bass et al. 1989; Pulak 1998; 1999a; 2001; 2005; 2008; Wachsmann 1998: 303–307; Cleary and Meister 1999; Yalçın et al. 2005; Cucchi 2008; Welter-Schultes 2008). The high-resolution absolute dating of this shipwreck, and especially of its last cargo and voyage, would provide a key chronological marker-point for the synthesis of the history, archaeology, and art of the wider East Mediterranean region. In particular, given the rich international cargo, a precise date for the last voyage would have important implications for the dating of material culture across the region from Egypt to Greece, and it would provide a key test for the validity both of the long established conventional proto-historical and archaeological chronologies estimated for Egypt, Cyprus, and the Aegean, as well as various claims for radical alternatives made in recent decades.

2. Integrated tree-ring and radiocarbon dating of the Uluburun ship

This report presents a comprehensive, high-precision dating program to establish directly the approximate calendar age of the Uluburun ship, especially that of its last voyage. Previous suggestions (Kuniholm et al. 1996; Wiener 1998; Manning 1999: 344–345) of a possible direct dendrochronological date for some timbers aboard the ship have proved, with further examination and additional data and development of regional tree-ring sequences, to be without good dendrochronological support; these are hereby withdrawn (cf. Manning et al. 2001: 2535 n.38; Wiener 2003a: 244–245). A previous report of some initial radiocarbon wiggle-match work on timbers from the vessel (Newton et al. 2005; Newton and Kuniholm 2005) is much expanded here, and dates on a range of the short-lived sample material from the vessel's last voyage are incorporated into a comprehensive dating model. For this project we developed an integrated research design to date the ship combining:

(i) Radiocarbon wiggle-match dating (Bronk Ramsey et al. 2001; Galimberti et al. 2004) of several short tree-ring sequences from long-lived wood either comprising the ship's timbers (specifically, its keel; for the ship's hull construction, see Pulak 1999a; 1999b; 2003) or from aboard the ship (dunnage, or in one case perhaps an element of the ship), which set *terminus post quem* ranges for the final voyage of the ship; with

(ii) Radiocarbon dating of short- or shorter-lived materials or elements on board the ship when it sank. These materials include fittings or other functional components (wicker-work, a rope fragment made of grass from the Gramineae family) or actual cargo such as olive seeds, leaves, terebinth resin, and thorny burnet (a dense, spiny shrub native to central and eastern Mediterranean used as dunnage or bedding material between the hull and the cargo of copper ingots). These elements should set a very close *terminus post quem* for, or even in several cases theoretically date the year of, the last voyage of the ship.

The sets of radiocarbon evidence are assessed within a comprehensive Bayesian analytical model (using the approach and software of OxCal: Bronk Ramsey 1995; 2001; 2008; 2009) in order to combine the known relative time-order of the sample materials with the radiocarbon ages obtained, and to yield the best dating estimates from the simultaneous resolution of the linked multiple dating probabilities. An interesting additional issue is that we may test the validity of conventional Egyptian chronology (the date range for Queen Nefertiti) against the combined tree-ring and radiocarbon evidence from the Uluburun ship (and vice versa).

3. The samples of short-lived materials

We obtained eight radiocarbon measurements on samples of short-lived cargo materials from the final use of the ship: Figure 1 (lower), Table 1 (Hd-23129, 23132, 23162, OxA-15022, 15024, 15026, 15025, 15065). These sample material types and dates vary a little, but should all date the final use period of the ship either to the year or within a few years at most (see further discussion in section 5 (ii) below). All ages obtained are broadly similar, with quality control provided both by: (i) the comparable findings of two different laboratories (Heidelberg and Oxford); and (ii) new measurements of known age German Oak run around the same time at Heidelberg and Oxford which show generally good agreement with each other, although a little older on average compared to the IntCal98/04 average values and indicating somewhat more curve amplitude (Figures 2, 3).

The data from these eight short-lived material samples can be combined together, consistent with

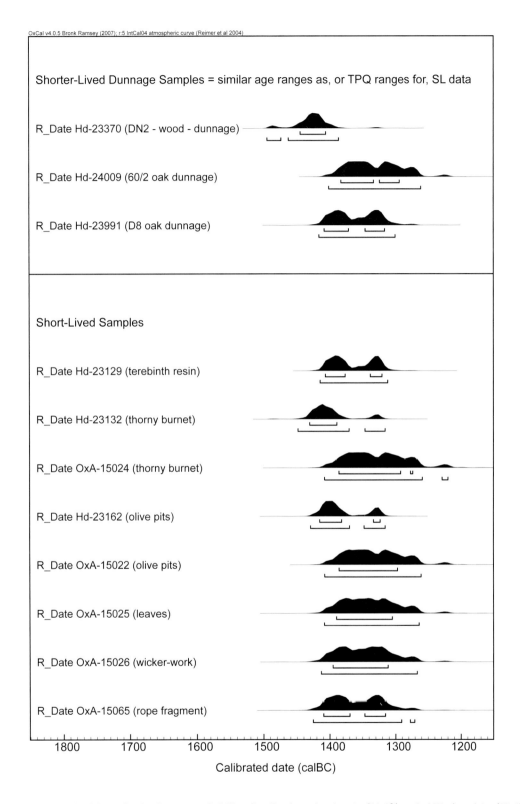

Figure 1: The individual calibrated calendar age probability distributions showing 1σ (68.2% probability) and 2σ (95.4% probability) ranges (upper and lower lines under each histogram respectively) for (i) (upper 3 histograms) 3 measurements on shorter-lived or short tree-ring sequence wood dunnage samples from the Uluburun ship, which should either be approximately the same age as, or set *terminus post quem* ranges for (perhaps very close in some cases), the ship's last voyage and the shipwreck, and (ii) (lower 8 histograms) 8 measurements on short-lived samples from the Uluburun shipwreck—these samples should closely date the final voyage time interval (the same year or next year for samples like the leaves and olive seeds, and somewhere from the same year to the next couple or few years for the other samples). Calibration to calendar years employs the IntCal04 radiocarbon calibration dataset (Reimer et al. 2004) and the OxCal software (Bronk Ramsey et al. 1995; 2001; 2008) version 4.0.5.

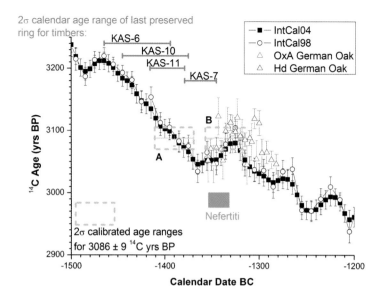

Figure 2: IntCal04 (Reimer et al. 2004), black squares, and IntCal98 (Stuiver et al. 1998), black hollow circles, radiocarbon calibration curves for the period 1500 to 1200 BC. 1σ (68.2% confidence) error bars shown. Note the inversion, or wiggle (to older radiocarbon ages), centred around 1325 BC (see further in Figure 3). The 2σ (95.4% confidence) calibrated calendar age ranges from IntCal04 for the weighted average ^{14}C age of the 8 short-lived samples found as contents from the Uluburun ship's last voyage (for the individual dates, see Figure 1, lower; for the weighted average, see Figure 4) are shown by the two cyan boxes (1411–1369 BC and 1358–1315 BC; the 1σ ranges are 1403–1377 BC and 1337–1321 BC). The ^{14}C wiggle-matched calibrated calendar age ranges at 2σ confidence for the last preserved ring of the four dendrochronological samples from the Uluburun ship's structure or from the wood materials carried on the ship (dunnage, etc.) (see Figures 8 and 9) are shown in orange (these dendro samples from the Uluburun ship have the laboratory identification code of KAS); these set *termini post quos* for the final voyage of the ship (see text for discussion). The standard reign period of Amenhotep IV (Akhenaten) and so of Nefertiti as queen of Egypt is also indicated (Hornung et al. 2006: 492, 206–208, 477–478). A gold scarab of Nefertiti was found among the contents of the Uluburun ship (see text below); hence some part of her reign acts as a *terminus post quem* for the last voyage of the ship. See text below for discussion. (Note: a modified date range for Amenhotep IV and Nefertiti some 11 years later may become necessary given recent work—see footnote 2 and text below.)

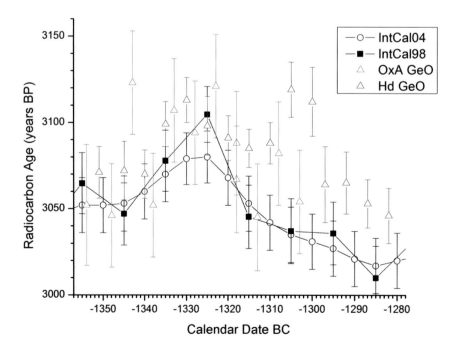

Figure 3: Detail from the plot in Figure 2 showing the region of the radiocarbon age inversion (or wiggle) in the region around 1325 BC. The new Heidelberg (part of a larger set) and Oxford measurements on German Oak use known age samples from Augsfeld supplied by Michael Friedrich. 1σ error bars shown.

Lab ID	Sample Name	Sample Material	Start Ring	End Ring	$\delta^{13}C‰$	^{14}C Age (yrs BP)	SD
Hd-22632	C-TU-KAS-3&11	Wood – *Cedrus libani*	1030	1040	-24.57	3252	25
Hd-22633	C-TUKAS-3&11	Wood – *Cedrus libani*	1050	1060	-24.38	3210	22
Hd-22642	C-TU-KAS-3&11	Wood – *Cedrus libani*	1070	1080	-24.36	3157	22
Hd-22559	C-TU-KAS 1&10	Wood – *Cedrus libani*	1009	1016	-24.4	3244	20
Hd-22591	C-TU-KAS 1&10	Wood – *Cedrus libani*	1054	1063	-24.73	3187	18
Hd-22592	C-TU-KAS-1&10	Wood – *Cedrus libani*	1064	1073	-25.13	3173	23
Hd-22593	C-TU-KAS 1&10	Wood – *Cedrus libani*	1074	1094	-25.23	3148	21
Hd-22604	C-TU-KAS-6A&C	Wood – *Cedrus libani*	1024	1034	-24.96	3276	23
Hd-22588	C-TU-KAS-6A&C	Wood – *Cedrus libani*	1035	1044	-25.04	3243	19
Hd-22580	C-TU-KAS-6A&C	Wood – *Cedrus libani*	1045	1054	-25.42	3235	25
Hd-22816	C-TU-KAS-7	Wood – *Cedrus libani*	1012	1022	-24.03	3122	23
Hd-24113	C-TU-KAS-7	Wood – *Cedrus libani*	1022	1032	-24.4	3092	25
Hd-24114	C-TU-KAS-7	Wood – *Cedrus libani*	1032	1042	-24.18	3087	20
Hd-23345	C-TU-KAS-7	Wood – *Cedrus libani*	1042	1049	-23.96	3076	22
Hd-22815	C-TU-KAS-7	Wood – *Cedrus libani*	1049	1059	-23.9	3078	15
Hd-23370	DN 2 (dunnage)	Wood – *Quercus coccifera*			-27.86	3145	23
Hd-24009	DN 60/2 (dunnage)	Wood – *Quercus coccifera*			-27.68	3050	23
Hd-23991	D 8 (dunnage)	Wood – *Quercus coccifera*			-26.79	3083	21
Hd-23129		Terebinth resin			-25.77	3087	16
Hd-23132	# 3932.01	Thorny burnet - *Sarcopoterium spinosum*			-26.51	3122	23
OxA-15024	#11268/11314/11426	Thorny burnet - *Sarcopoterium spinosum*			-25.31	3054	27
Hd-23162		Olive pits			-25.22	3104	17
OxA-15022		Olive pits			-24.37	3051	29
OxA-15026	#9387 (wicker work)	Wood - *Nerium oleander*			-26.84	3071	29
OxA-15025	# 3932.02	Leaf - *Quercus* sp.			-27.72	3061	28
OxA-15065	#11369	Rope fragment - Gramineae			-13.90	3085	28

Table 1: Samples and radiocarbon (^{14}C) data employed in this study. Source laboratories: Oxford Radiocarbon Accelerator Unit – OxA, and Heidelberg Radiocarbon Laboratory – Hd. Each timber sample is recorded in terms of an arbitrary relative sequence beginning with ring 1001.

the hypothesis that they could represent the same radiocarbon age at the 95% confidence level, to offer a more precise weighted average radiocarbon age estimate of 3086 ± 9 BP for the final cargo or last voyage (Ward and Wilson 1978) (see Figures 4 and 5). Without any other constraints, this weighted average radiocarbon age indicates a calendar date range with the current IntCal04 radiocarbon calibration curve (Reimer et al. 2004) and the OxCal calibration software v.4.0.5 (Bronk Ramsey 1995; 2001; 2008) of either about 1411–1369 Cal BC or about 1357–1315 Cal BC at 2σ (95.4% confidence): see Figure 4. The date range employing the previous IntCal98 (Stuiver et al. 1998) radiocarbon dataset is shown for comparison in Figure 5; IntCal98 employs similar underlying data for this period, but with a less sophisticated and less smoothed modelling. The bi-modal possible ranges reflect the shape of the radiocarbon calibration curve at this period (the record of past natural atmospheric radiocarbon derived for this time period from known-age tree-ring archives), in particular the pronounced short-term radiocarbon age inversion (a "wiggle") in the region around 1325 BC (see Figure 2). The wiggle is even more apparent in the previous, less-smoothed IntCal98 calibration dataset (Figures 2, 5) and is also reported in contemporary Aegean tree-rings (Manning et al. 2003; 2005). This wiggle is furthermore even more apparent in recent measurements of absolutely dated German Oak (from Augsfeld, kindly provided by Michael Friedrich) made at Heidelberg and Oxford

(Figure 3), both of which may indicate a slightly larger and somewhat longer inversion period (or plural wiggles) in the late 14th to early 13th centuries BC.

4. Wiggle-matching the tree-ring sequences and setting the *terminus post quem* for the construction of the Uluburun ship

To test and refine this age, and to resolve the bi-modal dating ambiguity noted above independently in radiocarbon terms, we analysed several long-lived *Cedrus libani* timber samples from the ship,[1] with the expectation that the last extant tree-rings on these samples would set *terminus post quem* ranges for the construction of the ship and, in turn, indicate the possible calendar ages of the short-lived samples from the ship's last voyage. Uluburun hull wood and their species identification are in Liphschitz and Pulak (2007/2008: 75).

These samples are:

- KAS-6A&C (hereafter KAS-6, with 105 years of growth represented);

- KAS-1&10 = Uluburun Lot number 6010 (hereafter KAS-10, with 108 years of growth represented);

- KAS-3 & 11 = Uluburun Lots number 6574 and 6594 (hereafter KAS-11, with 118 years of growth represented)

These three samples comprise dunnage (or chocks for wedging the cargo) from the wreck, and, in one case, KAS-6, perhaps a ship element, a frame-timber.

The final sample comes from the ship's keel:

- KAS-7, with 66 years of growth represented (the sample is also of *Cedrus libani*)

Dendrochronological sequences were measured for each timber (Figure 6). The cross-matches are not decisive or strong in either statistical or visual terms between any of these timbers (and their often erratic growth), nor between any of these timbers and other conifer chronologies from the region. It should be noted that the samples are far from perfect for dendrochronology given extensive damage by shipworm (*Teredo navalis*), which makes reading the tree-ring record challenging (Figure 7). Hence, to be conservative, we have chosen to treat each timber as independent in this study. No sapwood or bark or other features indicating outermost tree-rings are preserved. The samples' outer surfaces were worn and abraded, and thus an unknown number of tree-rings have been lost.

The critical sample is the keel (KAS-7). This is the one sample integral to the ship; it is the very foundation of its structure, given the ship's shell-based construction where the planks are joined with mortise-and-tenon to the keel (or spine of the vessel) and to each other (Pulak 1999a; 1999b; 2003). Examination of the hull remains shows that the keel is an original piece; there is no evidence for any kind of repair. For the keel to have been replaced, the mortise-and-tenon joints would have had to be cut and replaced, a radical overhaul which would leave unmistakable traces. In shell-based hull construction, the framing is secondary. Thus, even if KAS-6 is a frame-timber, it is not as integral to the ship as is the keel. The various 'dunnage' samples could be any age. In principle, the outermost tree-ring could be from the year of the last voyage of the ship (and therefore later than the date of the ship's construction); alternatively, this material could be recycled or re-used wood from years or even many decades earlier. We do not know the answer (i.e. age) as *a priori* information, and only a scientific dating can inform us.

Thus, for the wood-dendro samples, the key information is the dating of the last preserved tree-ring on the keel (KAS-7). This will set a *terminus post quem* for the construction of the ship, since there is no bark or sapwood.

Several fixed sequences of approximately 10-year increments of wood were extracted from each of these timbers for radiocarbon dating (for details, see Table 1), and the known sequences of radiocarbon dates obtained were then matched against the IntCal04 radiocarbon calibration curve (Reimer et al. 2004) to offer best age estimates for each timber (Figure 8). The last preserved tree-rings for each timber date (see Figures 8 and 9) within 2σ ranges as follows:

- KAS-6: 1465-1394 Cal BC;

- KAS-10: 1416-1379 Cal BC;

- KAS-11: 1445-1375 Cal BC; and

- KAS-7: 1379-1345 Cal BC.

The ship's final voyage must have occurred at some point after the latest of these age ranges—that is, after somewhere between 1379–1345 BC. Perhaps surprisingly, it is the ship's keel (KAS-7) which provides the most recent *terminus post quem*. This indicates either that the other wood samples (dunnage, or possibly a frame-timber in one case) were either old or re-used material, and/or have lost a number of outer tree-rings (whether pre- or post-deposition). The keel sets the

---

[1] This designation covers the overall *Cedrus libani* grouping, including *Cedrus libani* var. *libani* from Lebanon and western Syria, and also *Cedrus libani* var. *stenocoma* from southern Turkey, and *Cedrus libani* var. *brevifolia* from Cyprus.

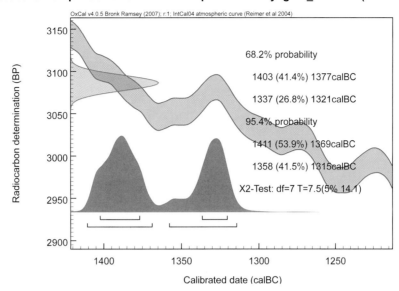

Figure 4: The calibrated calendar age ranges for the weighted average of the 8 short-lived samples in Figure 1 as an estimate of the date of the last voyage of the Uluburun ship (the last voyage was reasonably soon after this date range) (see cyan boxes in Figure 2 also). The lines under the histogram show the 1σ and 2σ ranges respectively. We see a bi-modal possibility due to the radiocarbon age inversion (wiggle) centred around 1325 BC (which is more pronounced in IntCal98). Calibration employs the IntCal04 radiocarbon calibration dataset (Reimer et al. 2004) and the OxCal software (Bronk Ramsey 1995; 2001; 2008) version 4.0.5 with curve resolution set at 1. The same radiocarbon data are shown calibrated with the previous IntCal98 (Stuiver et al. 1998) radiocarbon calibration dataset in Figure 5. For discussion of the inversion, see the caption to Figure 5.

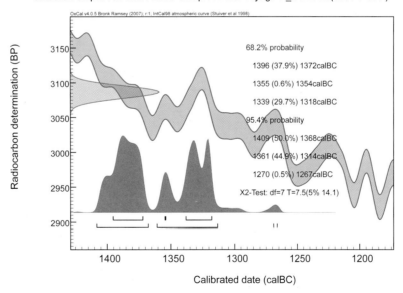

Figure 5: As Figure 4, but employing the IntCal98 (Stuiver et al. 1998) radiocarbon calibration dataset. Note especially that the age inversion (wiggle) centered around 1325 BC is more pronounced in the IntCal98 dataset (which lacks the smoothing of the IntCal04 data set). This seems noteworthy. For example, given that the final cargo/voyage of the Uluburun ship must at least post-date the accession of Nefertiti as queen (since her scarab was on the boat), and thus be after about 1353 BC (Hornung et al. 2006) or 1342 BC (see footnotes 2 and 3 and text below), the later of the calibrated dating sub-ranges (1339–1318 BC at 1σ and 1361–1314 BC at 2σ), and thus the age inversion centered around 1325 BC, would seem to be the likely date range where the short-lived samples from the last voyage of the Uluburun ship should belong. This age inversion is prominent in the Aegean record (Manning et al. 2005) and in the recent Heidelberg and Oxford data on German oak (Figure 3). This issue, and the question of how the Nefertiti date correlates with the radiocarbon information, is discussed further in the text below (section 7).

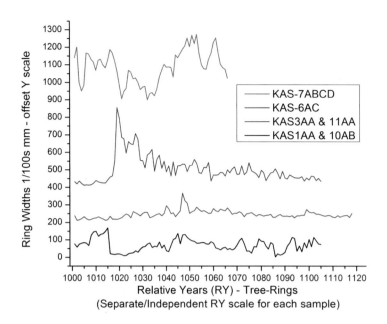

Figure 6: Dendrochronological sequences for the tree-ring samples (all *Cedrus libani*) employed in this study. Examples of erratic growth are common and samples are difficult to read or measure due to damage from shipworm (*Teredo navalis*) (see Figure 7). Each is treated as an independent sequence in this study since there are no convincing cross-dates among the samples. Details of the tree-rings radiocarbon dated from the samples for the dendro-radiocarbon wiggle-matches are given in Table 1.

Figure 7: Section of KAS-7, from the keel of the ship, illustrating the extensive damage due to holes bored by shipworm (*Teredo navalis*), which make reading the tree-ring record a challenge (Newton et al. 2005: Abb.1).

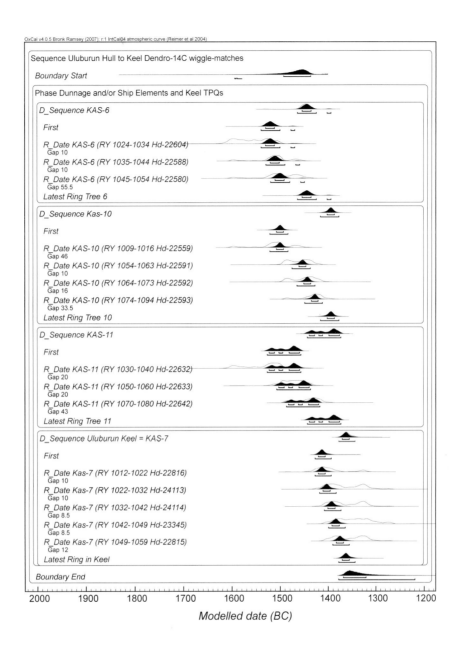

Figure 8: Bayesian fixed-Sequence (dendro-wiggle-match) analysis for the tree-ring sample sequences extracted from timbers KAS-6 (frame-timber from the Uluburun ship or dunnage), KAS-10 (dunnage), and KAS-11 (dunnage) and KAS-7 (the Uluburun ship's keel), calculating the calibrated calendar age range for the last (most recent) preserved tree-ring in each sample. The hollow (outline) distributions show the calibrated ages for each individual sample on its own; the solid black distributions within these show the calculated ranges applying the Bayesian model based on the known tree-ring (calendar year) intervals between the samples. The horizontal lines under each distribution indicate the 1σ and 2σ confidence, calibrated calendar age ranges using IntCal04 (Reimer et al. 2004) and OxCal (Bronk Ramsey 1995; 2001; 2008) version 4.0.5 with curve resolution set at 1. Each run of such an analysis produces very slightly different results; therefore, a typical outcome is shown here. To test for problems and outliers we used the OxCal agreement index. This is a calculation of the overlap of the simple calibrated distribution versus the distribution after Bayesian modelling. If the overlap falls below 60%, it is approximately equivalent to a combination of normal distributions failing a χ^2 test at the 95% confidence level. The OxCal agreement index values are indicated in parentheses, and all surpass an approximate minimum 95% confidence threshold. For the specific sequences: KAS-6 n = 3 yields 156.9 v. minimum threshold value at $\approx$ 95% confidence level of 40.8, KAS-10 n = 4 yields = 153.1 v. minimum threshold value at $\approx$95% confidence level of 35.4, KAS-11 n = 3 yields = 89.9 v. minimum threshold value at $\approx$95% confidence level of 40.8, and KAS-7 n = 5 yields = 148.2 v. minimum threshold value at $\approx$95% confidence level of 31.6. The agreement index value for each individual sample is shown. The approximate 95% confidence threshold value is $\geq$ 60. The placement of each of the samples on the IntCal04 radiocarbon calibration curve is shown in Figure 9.

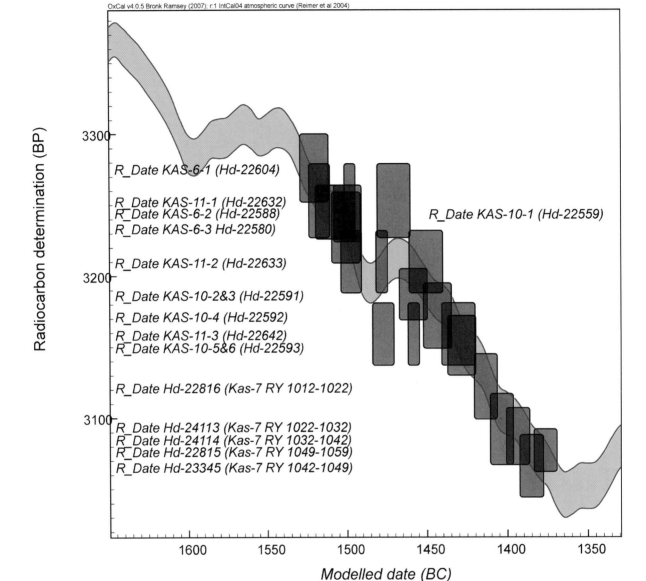

Figure 9: The modelled 1σ (68.2%) most likely calendar placements of each of the Uluburun samples in Figure 8 is shown against the IntCal04 radiocarbon calibration curve (Reimer et al. 2004). Note that the most recent tree-rings lie variously from the late 15th century through to the mid-14th century BC. In particular, the last preserved ring from the keel (KAS-7) lies at 1372–1357 Cal BC at 1σ (68.2%) confidence (1379–1345 Cal BC at 2σ 95.4% confidence). These samples thus occupy the slope of the radiocarbon calibration curve at this period (anchored here by the earlier tree-rings of each sample which must lie variously in the late 16th through later 15th centuries BC; see also Figure 8). This in turn means that the ambiguity in date for the short-lived samples (Figures 1, 2, 4, 5) is resolved: they must date later than the last preserved tree-ring of the ship itself (and thus after the last ring from the keel at 1372–1358 Cal BC at 1σ or 1379–1345 Cal BC at 2σ). Therefore, the otherwise possible date range of about 1411–1369 Cal BC (2σ) for the short-lived samples from the final voyage must be dismissed in favour of the alternative later possible date range of 1357–1315 Cal BC (2σ). Hence, the ambiguity created by the shape of the radiocarbon calibration curve and especially the late 14th-century BC age inversion (wiggle) is resolved.

relevant *terminus post quem* range for the construction of the ship: 1372–1357 Cal BC at 1σ or 1379–1345 Cal BC at 2σ. (If IntCal98, Stuiver et al. 1998, is employed, the dates for the last tree-ring of KAS-7 from typical runs of the analysis model in Figure 8 are very similar: e.g. 1370–1353 Cal BC at 1σ and 1377–1339 Cal BC at 2σ). As noted above, there was no bark or sapwood or other indication of outermost rings on this sample (and the process of its shaping likely removed outer rings); thus the *terminus post quem* provided here dates the last extant ring, and the actual date of the felling of the tree used to make the keel would be several years more recent than this.

5. Temporal relationships from the context of the last tree-ring in the keel to the dunnage samples and to the short-lived samples

(i) Dunnage samples and frame-timber(?) sample. The keel sample (KAS-7) is the only one integral to the ship and indisputably part of its primary construction (Pulak 1999a; 1999b; 2003). The possible frame-timber (KAS-6) might be as well, but this is less certain. The other "dunnage" samples, both the ones wiggle-matched above (KAS-10, KAS-11) and also the three other (shorter-lived/shorter-sequence) samples of wood dunnage material from the wreck—Hd-23370 (DN2 wood dunnage), Hd-24009 (60/2 oak dunnage), and Hd-23991 (D8 oak dunnage) (see Figure 1, upper)—have no context-based relationship with the date of the ship's construction and thus no relationship to the *terminus post-quem* date offered by the keel. They could be (1) older material that has been re-used, (2) roughly contemporary material (and when they yield older ages it could be argued that they have lost outer tree-rings), or (3) material dating as late as the year of the last voyage of the ship. We do not know from the context.

The outermost preserved tree-rings of these dunnage samples can at best be stated to provide *terminus post quem* ranges for the last voyage of the ship, with the length of the 'post' dependent on which of the three previously-noted scenarios in fact really applies. This is the assumption/model employed for the dunnage samples in Model A below.

(ii) Short-lived sample matter from the ship (from its last voyage). The short-lived sample material (Figure 1, lower; Table 1) comprises a variety of items. Some, like the olive pits, might be provisions, or could relate to cargo. In either case, they are likely to relate to the final voyage of the ship, or at most to the last few trips, and to no more than a time window of a few years. The terebinth resin was part of the ship's cargo, carried in Canaanite jars; again, it is likely to date from the final voyage, with a plausible dating window of no more than a few years. The thorny burnet dunnage again likely relates to the most recent voyage or voyages. It is hard to imagine this material surviving in usable form for many voyages, no more than a few years at most. The leaves come from young tree branches also used as dunnage, but again likely relate to the current or recent voyage and packing of cargo; they are likewise unlikely to have lain around on the ship for more than a few years in reasonable condition. The possible exceptions to this scenario of all such items belonging to the latest voyage, or at most last few voyages, or a horizon of a few years at most, are the wicker-work and the grass rope fragment samples. The growth period involved in each case is probably annual to no more than a few years (given the materials employed to make ropes at the time: Wachsmann 1998: 254, and the likely materials involved in the wicker-work/matting from the fence along the side of the boat: Pulak 1992: 11). These items were presumably used as long as they lasted in fit condition. Ropes of the available technology in the Late Bronze Age did not last long, as at least some of the finds of large numbers of Late Bronze Age stone anchors lost on the sea-bed indicate, and a lifetime of more than a few years seems unlikely. The wicker-work is more difficult to judge. Again, it seems unlikely in the taxing conditions at sea that it would have lasted for more than a few years, but perhaps this could be a time horizon of up to a decade or so, rather than of one to a few years. It is impossible to know.

A reasonable and fairly conservative assumption is that all the short-lived sample materials lie in a time horizon of no more than about 10 years (i.e. none was more than ten years old by the time of the last voyage of the ship). Thus, collectively, they should define a relatively short time-period immediately prior to the date the ship sank.

All this material should also represent constituent ages (total periods of growth) of only one to a few years at most. Thus the ages should, within one to a few years, define the last voyage.

Even if the Uluburun ship was on its maiden voyage, it is probable that the last preserved tree-ring in the keel (a *terminus post quem*, adding missing rings, sapwood and bark to the actual cutting date of the tree, and then ship construction) is older than the short-lived samples (even if only by a year or a few years on the maiden voyage scenario). The wicker-work was presumably only made of short (or shorter)-lived materials cut in the year (or year before) construction; the food, cargo and packing likewise would relate to no earlier than the year or year or two before construction (even on a maiden voyage scenario), and the ropes were likely new for the vessel and regardless could not have been very old as they typically did not last long. In all likelihood, this was not the ship's maiden voyage. This being the case, it is all the more certain that the last tree-ring preserved on the keel sets a *terminus post quem* for all the short-lived sample materials on the ship. We assume this temporal sequence where the date for the last ring of the keel (KAS-7), and the dates for the last rings or ages of the other wood samples from the ship (whether ship elements or dunnage), act as a *terminus post quem* for the short-lived samples in Model B. We expect this model to yield the most realistic estimate of the dating of the last voyage of the ship.

Examination of Figures 1 (upper), 2, 8, and 9 reveals that the latest samples from the ship's keel (KAS-7), and several of the dunnage samples, date in the same possible calendar time range as the earlier of the two possible dating ranges found for the short-lived cargo and dunnage samples shown in Figures 2, 4, and 5 (the older possible range labelled as "A" in Figure 2). In other words, there is overlap in the later 15th through mid-14th centuries BC. However, as argued, since we may reasonably assume that the short-lived items on board the ship as contents or fittings during the last voyage of the ship were at least later than the last preserved decade of tree-rings in the keel of the ship, this situation indicates that it is the later of the two calendar date ranges possible for the short-lived cargo/packing/fittings samples which must apply. Thus the ambiguity shown in Figures 2, 4, and 5 can be approximately resolved (in favour of range "B" in Figure 2). In turn, we can best estimate the last voyage of the Uluburun ship as either effectively the same date as, or (better) immediately following, this group (a phase) of short-lived samples from the final cargo or dunnage on board. However, the majority of these short-lived contents samples should most likely be understood as closely defining this final horizon, that is, the year of, or a couple of years before and including, the last voyage, rather than being uniformly distributed throughout the preceding period.

6. Bayesian analysis of all the radiocarbon evidence to best define the date of the Last Voyage (LV) of the Uluburun ship

We can quantify the observations made in the previous section through Bayesian analyses combining all the Uluburun data sets discussed. We consider two models based on the discussions in the previous section:

Model A. This employs all the data and (i) lets the dates of the last preserved tree-rings or the ages of the various dendro/wood samples (keel, frame-timber(?), dunnage) set a minimum (oldest possible age) estimate for the Last Voyage, and (ii) lets the short-lived samples do the same, but does not assume that the short-lived samples must necessarily post-date the latest tree-ring/wood dates. This model allows for the maiden voyage scenario (as one extreme) and realistically sets minimum (old as possible) parameters for the discussion.

Model B. This employs all the data in Model A for the hull timbers of the ship, or for wood found on the ship, to provide a *terminus post quem* boundary ("Ship Fitted Out") for the phase "Contents Ship Last Voyage" of short-lived sample matter from the last voyage. As the most recent key element, the wiggle-match date for the last extant tree-ring of the keel (KAS-7) in effect sets this *terminus post quem* for the contents phase. The Last Voyage of the ship is dated as the boundary (LV) immediately after the "Contents Ship Last Voyage" phase of short-lived samples. This model likely calculates a realistic estimate of the date of the Uluburun ship's last voyage.

Model A makes two further assumptions. First, rather than treat the short-lived samples as forming a uniform distribution within a last voyage phase, we more realistically assume that it is likely that the majority of these samples (everything except perhaps the wicker-work sample?) will tend to date towards the end of this phase—that is within the same year as, or one or so years before, the last voyage. Hence the date for the Last Voyage (LV) may be best estimated using the Tau_Boundary model in OxCal (Bronk Ramsey 2009), where a group of dated samples are assumed to be exponentially distributed rising to a maximum event probability at the end event—which we define as the Last Voyage, or LV, of the Uluburun ship.

Second, we have to consider the relevant time constant, that is, the average age of the samples dated within the phase. Above (section 5.ii), we proposed a time period of 10 years as a conservative range to cover all the ages of the short-lived materials. We use this 10 years model as our best estimate. With the Tau_Boundary, this means that one or two samples can be older, even substantially older, but the rest should be within 0–10 years of age—this is the nature of the exponential distribution on the Tau end which runs to infinity. However, to be cautious and to see how a longer time period estimate offsets the calculated outcome, we also consider time periods of 25 years and 50 years.

Model A, and the outcome with a 10-year time constant for the short-lived sample material, is shown employing IntCal04 (Reimer et al. 2004) in Figure 10, with the modelled age range for the LV shown in detail in Figure 11: 1333–1319 Cal BC (1σ) and 1381–1364 (10.0%), and 1341–1312 (85.4%) Cal BC (2σ). The same model is shown employing IntCal98 (Stuiver et

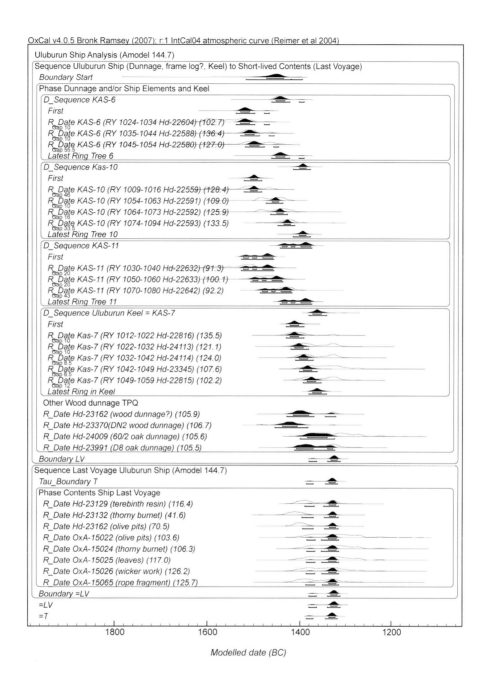

Figure 10: Model A with time constant of 10 years for the short-lived samples from the last voyage of the ship (see text) and IntCal04 (Reimer et al. 2004). Bayesian Sequence analysis using OxCal 4.0.5 and curve resolution = 1 (Bronk Ramsey 1995; 2001; 2008) for all the Uluburun data in order to best estimate the date range for the Last Voyage (LV) of the Uluburun ship (Figure 11). Model A does not assume that the short-lived samples from the ship's last voyage necessarily have to be more recent than the last tree-rings in the dunnage or ship elements (frame-timber and keel). This is in contrast to Model B, and covers, for example, the case that the ship was brand new when it sank. The time constant, that is, the assumed average (growth) age of the short-lived samples from the ship, is the main variable. Common sense would indicate that this should be a fairly short time, and the text proposes 0–10 years as plausible. The figure shows the analysis with a 10-year time period as the constraint. Outcomes when the constraint is 25 years, or 50 years, are shown in Table 2 for comparison. A Tau_Boundary model (see text, and Bronk Ramsey 2009) is applied, whereby it is assumed that the majority of the short-lived samples lie towards the date of the last voyage. The hollow (outline) distributions show the calibrated ages for each individual sample on its own; the solid black distributions within these show the calculated ranges applying the Bayesian model indicated. The horizontal lines under each distribution indicate the 1σ and 2σ confidence calibrated calendar age ranges. Note: every run of a sequence analysis achieves very slightly different results, the above being a typical example. The agreement index value for each individual sample is shown also (in parentheses). The approximate 95% confidence threshold value is $\geq 60\%$. Only sample Hd-23132 is slightly inconsistent, and would prefer a slightly older age; however, given the *terminus post quem* from the keel timber especially, we may instead suspect that this short-lived sample represents a near miss for the marked ^{14}C age inversion ("wiggle") region centred around 1325 BC (see Figures 2, 3); the same argument probably also informs the measured ^{14}C age for Hd-23162. The identical radiocarbon data and analysis are shown employing the previous IntCal98 radiocarbon calibration dataset (Stuiver et al. 1998) in Figure 12.

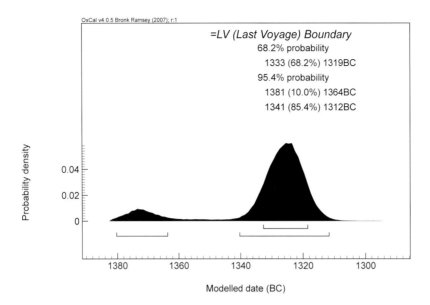

Figure 11: The modelled age estimate for the Last Voyage (LV) of the Uluburun ship from Figure 10.

al. 1998) in Figures 12 and 13, with the LV calculated as 1335–1325 (42.7%) and 1323–1316 (25.5%) Cal BC (1σ), and 1357–1348 (9.6%) and 1339–1312 (85.8%) Cal BC (2σ). Table 2 shows the modelled age ranges for the Last Voyage from Model A, given allowances of 10 year, 25 year, and 50 year time periods for the short-lived sample material from the final voyage of the ship.

Model B estimates the Last Voyage as a boundary immediately after the (uniform, and not Tau_Boundary model) phase of the short-lived material, with these placed as after the wood elements of the ship and the other wood material on board (and principally after the last ring of the keel). The overall model outcome employing IntCal04 (Reimer et al. 2004) is shown in Figure 14, with the modelled age range for the LV shown in detail in Figure 15: 1332–1311 Cal BC (1σ) and 1340–1289 Cal BC (2σ). The same model is shown employing IntCal98 (Stuiver et al. 1998) in Figures 16 and 17, with the LV calculated as 1333–1307 Cal BC (1σ) and 1343–1279 Cal BC (2σ).

With regard to Figures 10, 12, 14, and 16, we see that with IntCal04 one sample (Hd-23132) does not offer a satisfactory agreement index value (it would prefer to be a little older), whereas, with IntCal98, all the samples have satisfactory agreement index values. This may indicate that Hd-23132 is a near-miss for the marked radiocarbon age inversion (or wiggle) region around 1325 BC (see Figures 2 and 3). This wiggle is relatively smoothed away in the IntCal04 radiocarbon dataset, whereas it is more pronounced in the non-smoothed IntCal98 dataset (and in the new Oxford and Heidelberg data in Figure 3).

The striking observation is the similarity of the likely date ranges calculated for the LV across both Models A and B (Figures 11, 13, 15, 17). In Model A, if the length of the time interval for the short-lived samples is increased, then a small probability occurs in the earlier 14th century BC, but the most likely range nonetheless remains firmly in the later 14th century BC. This is clear when the time interval allowed is shorter (e.g. the 10 years of Figures 11, 13) and the most likely (1σ) ranges are 1335–1333 to 1319–1316 Cal BC. In Model B, which more likely captures the dating reality with the Last Voyage somewhat after the construction of the ship, all the analyses clearly find a later 14th-century BC range, with the most likely range about 1333–1332 to 1311–1307 Cal BC (a high-precision dating with a total range of 22–25 years) at 1σ or 1343–1340 to 1289–1274 Cal BC (with a 54–66 year range) at 2σ.

These ranges provide a likely good, close, and highly resolved dating for the LV of the Uluburun ship.

This late 14th-century BC date for the last voyage of the Uluburun ship is independently obtained solely from the radiocarbon, tree-ring, and context-based knowledge of the necessary sequence of the samples. Conveniently, and offering a strong independent reinforcing of its likely approximate validity, this late 14th century BC date range is also very compatible with the artefact/archaeological date assessments of the final voyage based on the material culture items recovered from the wreck (e.g. Wiener 2003a: 245–246; Bass et al. 1989; Pulak 2008).

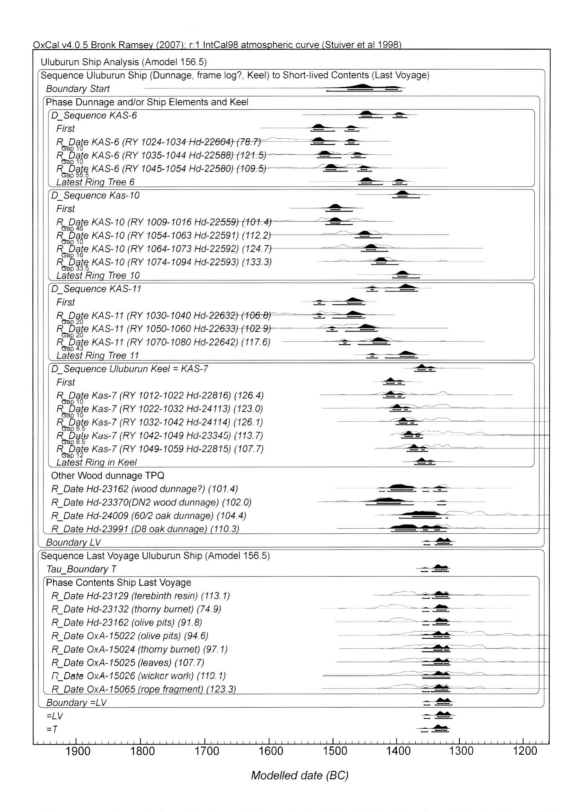

Figure 12: As Figure 10, but with the calibration analysis employing the IntCal98 radiocarbon calibration dataset (Stuiver et al. 1998). A very similar set of outcomes results. For other details, see the caption to Figure 10.

Tau Boundary Interval	IntCal04 1σ Cal BC	IntCal04 2σ Cal BC	IntCal98 1σ Cal BC	IntCal98 2σ Cal BC
10 years	1333-1319	1381-1364 (10.0%) 1341-1312 (85.4%)	1335-1325 (42.7%) 1323-1316 (25.5%)	1357-1348 (9.6%) 1339-1312 (85.8%)
25 years	1371-1365 (6.1%) 1335-1315 (62.1%)	1377-1347 (25.8%) 1341-1310 (69.6%)	1373-1366 (11.0%) 1354-1352 (2.5%) 1335-1314 (54.8%)	1378-1344 (35.2%) 1339-1309 (60.2%)
50 years	1371-1359 (16.6%) 1335-1314 (51.6%)	1376-1309	1372-1365 (9.1%) 1357-1349 (10.4%) 1335-1313 (48.6%)	1376-1308

Table 2: Modelled calendar age ranges for the Last Voyage (LV) of the Uluburun ship from Model A (see text and Figures 10–13). Where there is more than one sub-range, if there is a clearly much more likely sub-range, then it is underlined. Each run of such analyses produces very slightly different outcomes; typical results are shown. Model A does not assume that the short-lived samples from the ship's last voyage necessarily have to be more recent than the last tree-rings in the dunnage or ship elements (frame-timber and keel*); this is in contrast to Model B and covers, for example, the case that the ship was newly built when it sank. The time constant allocated to the last-voyage samples, that is, the average age of the short-lived samples from the ship, is the main variable. These materials are all likely less than 10-years old and are in fact likely to range from a single year's growth to a maximum of a few years' growth in total. Thus, a time constant of 10 years seems plausible and realistic. However, to be conservative, we have also considered longer intervals of 25 and 50 years. This 10-year time constant is different than the issue of time duration (i.e, cutting/use/production) of the various short-lived samples; that is, how long is the time window of the Last Voyage, and where within this time window (the Phase) might the majority of the samples belong? Common sense would indicate that this time period should be a fairly short time. Short-lived materials and cargo/use items would likely represent the current voyage, or at most be leftovers or re-use from the previous few voyages. The wicker-work and the rope fragment might be a little longer duration in use, but given maritime conditions of the time (see text), it seems likely that these would have had to be renewed or replaced within a few years. The thorny burnet, on board as packing, again likely reflects the last voyage, but could be re-used; it is nevertheless unlikely to have survived for such re-use for more than a few trips or a few years in total. A total time window for the short-lived material on the ship's last voyage of less than about 10 calendar years is probably quite realistic and even generous, and most of the samples likely fall into an even tighter time window of just the last one or few years. It thus appears realistic to consider a Tau_Boundary model where it is assumed that the majority of the samples lie close to the end of the Phase (i.e. the time of the Last Voyage). This is the suggested likely scenario as shown in Figures 10–13. The most likely dating range or sub-range is consistent across all the models (1335/33 to 1325/19/16/15/14/13 Cal BC at 1σ), but as the time constant is increased, the modelled range widens a little. * Note: the keel (KAS-7) does require at least a very short *terminus post quem* factor in reality (ignored in Model A). Even if the ship was newly built, the keel is missing bark or sapwood, and thus at least a few (or more) years must lie between the last extant tree-ring in the keel and when the tree was actually cut (bark). There would then be additional time before the wood was used for the keel and before the ship sailed (even if this was its maiden voyage). This short *terminus post quem* is likely longer than the time period in which the short-lived sample matter was lying around or stored, but clearly could be almost the same time period (as Model A allows for; contrast this with Model B).

7. Uluburun dating, Nefertiti, and Egyptian and Near Eastern chronology

Although small, one of the most notable finds from the Uluburun ship is of a unique gold scarab of Nefertiti, wife of Amenhotep IV (Akhenaten) (Weinstein 1989; 2008; Bass 1987: 731–732). This object can now offer an important test for the relationship between the Uluburun last voyage radiocarbon-based dating (above) and the standard historical chronology for Egypt. The earliest date for the production of this scarab is from early in the reign of her husband Amenhotep IV. Weinstein (1989: 27) argues for an earliest date (or *terminus post quem* for the item) from about years 2 or 3 to certainly year 5 of Amenhotep IV's reign. Amenhotep IV's accession is conventionally dated to about 1353 BC (Hornung et al. 2006: 492, 477), with the lowest recent "mainstream" published scholarly date at 1340 BC (Helck 1987). The last date for its production is around the time Amarna was abandoned (and when the worship of the Aten was abandoned), which is about three years after the death of Amenhotep IV. (There are wine vintages attested at Amarna for 13 years under Amenhotep IV/Akhenaten, equating to years 5 through 17 of his reign, and then for a further three vintages: Hornung et al. 2006: 207). A date around 1332–1331 BC results, likely more or less when the child Tutankhamun becomes ruler; there is an absence of explicit evidence for the date of his accession (for a summary of the evidence, see Hornung et al. 2006: 208, 477). By this time Nefertiti was either dead, or, with the abandonment of the worship of the Aten and move of the capital, a scarab naming Nefertiti and the Aten would no longer have been produced (Weinstein 1989: 27). New work may, however, change the "conventional" position, and instead suggests an accession date for Amenhotep IV at 1342 BC and for Tutankhamun at 1321–1320 BC.[2] The scarab could not have existed

---

[2] Recent work on evidence of Horemheb's last attested year appears to indicate that the length of his reign may have to be re-assessed downwards from the usual 28 years of reign to perhaps a reign of no more than about 14 or 15 years (David Warburton and Rolf Krauss, pers. comms. 2008, 2009; contrast this with comments in Hornung et al 2006: 476–477). Taking

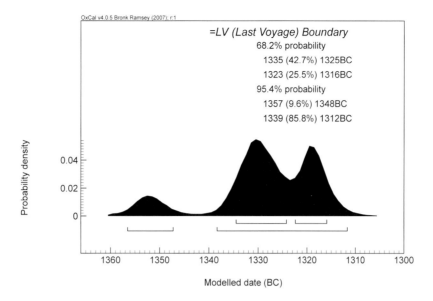

Figure 13: The modelled age estimate for the Last Voyage (LV) of the Uluburun ship from Figure 12.

before this Amarna period, so, according to the latest work, not before about 1340 BC, previously not before about 1350 BC. However, the scarab could, of course, have been in circulation at any time after her death. Weinstein (1989: 27–29) further speculates that the form of the writing and the title used may indicate a more specific date late in the reign of Amenhotep IV, or in the year or two immediately after his death (if and/or when Nefertiti perhaps became co-regent in Amenhotep IV/Akhenaten years 15–16 and/or perhaps was ruler/king subsequently, although these are all contentious points).[3] In this case, a date around the mid-1330s BC (or mid-later 1320s BC on the new chronological position noted in footnote 2) would be called for. The scarab was found quite worn in places, especially on the back, but still retained traces of lines representing wing ribbing (so the extent and use-time of the "wear" is open to interpretation). This being the case, the scarab may have been around for some years or decades before this last voyage and have been "bric-a-brac" by this time (Weinstein 1989: 23). Equally, however, this worn state could be explained in other ways, and the scarab might have only been a few years, to a few decades, old at the time of the shipwreck. Weinstein argued that the scarab probably belonged to an Egyptian official or a member of his family, and was disposed of after the end of the Amarna period, after which it ended up with a merchant, and on the ship, some years or even a couple of decades later in the post-Amarna period. This is clearly a plausible scenario, and the combination of tree-ring and radiocarbon evidence would suggest that the scarab reached the seabed with the ship either during the later Amarna period or in one of the next few decades following it, with either the conventional Egyptian dates or the revised dates (see footnote 2).

The accession of Amenhotep IV thus sets a clear *terminus post quem* for the shipwreck. We can therefore test the compatibility of the conventional proto historical date estimates for Nefertiti against the above radiocarbon-based chronology from the Uluburun ship. To be potentially valid, the proto-historical dates must be older, and not more recent, than the age range found above for the shipwreck. We find exactly this situation: the conventional dates, or recently modified conventional dates (e.g. Weinstein 1989: 17–29; Hornung et al. 2006; Helck 1987; Kitchen 1996; 2007; von Beckerath 1997; Krauss and Warburton, personal communications—see footnote 2), are either a little older than, or contemporary with, the age range determined by radiocarbon. For example:

---

the other evidence for the New Kingdom (e.g. lunar information) into account, this would mean reducing the "conventional" date for Amenhotep IV to about 1342 BC (i.e., to more or less where Helck 1987 placed it). A similar reduction would apply for Tutankhamun (to 1321–1320 BC). A study on this new chronology ("The basis for the Egyptian dates") by Krauss and Warburton will appear in a forthcoming volume edited by D. Warburton entitled *Time's Up! Dating the Minoan Eruption of Santorini* (Monographs of the Danish Institute at Athens 10.)

[3] Various as yet unproven claims have been made, for example, for Nefertiti to be identified with the kings named Ankhetkheprure, Nefernefruaten or Smenkhkare, and especially the woman ruler Ankhetkheprure Nefernefruaten (Hornung et al. 2006: 207). But she may well have died (as queen—that is king's wife), and the woman ruler during a short period between Amenhotep IV/Akhenaten and Tutankhamun could instead be one of Akhenaten's other wives or daughters: e.g. Kiya or Merytaten. For a good discussion of the poorly understood period of the few years around and following the death of Akhenaten, see Allen 2009. An online version of this volume is available at: http://history.memphis.edu/murnane/.

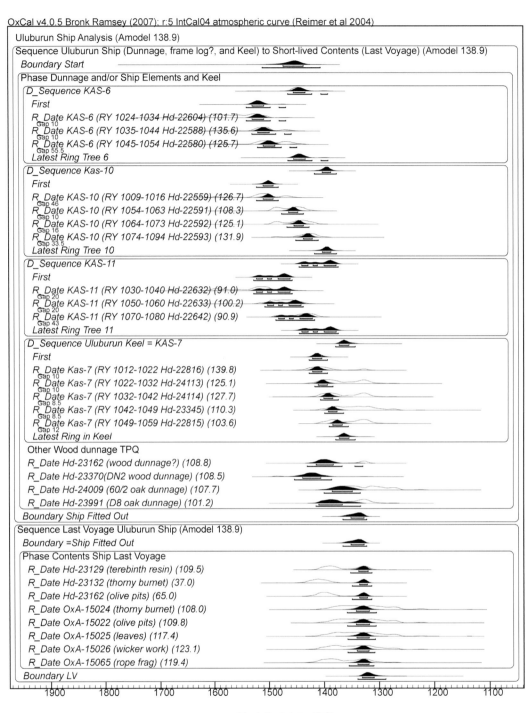

Figure 14: Model B and IntCal04 (Reimer et al. 2004). Bayesian Sequence analysis using OxCal 4.0.5 and curve resolution = 1 (Bronk Ramsey 1995; 2001; 2008) for a sequence where the wood elements of the ship (the keel: KAS-7) and other wood items on the ship act to define when the ship was built/fitted out, and set a *terminus post quem* for the short-lived items from the ship from its last voyage (Phase Contents Ship Last Voyage). The Last Voyage (LV) is calculated as a boundary immediately subsequent to this phase (see Figure 15). For other details on how to read the figure, see the caption to Figure 10.

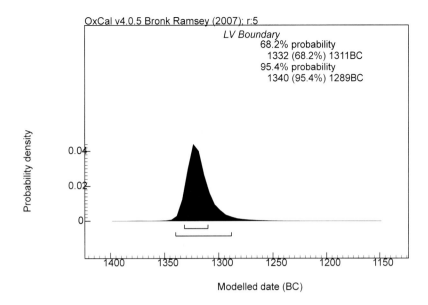

Figure 15: The modelled age estimate for the Last Voyage (LV) of the Uluburun ship from Figure 14

(i) Conventional TPQ/production range 1353–1332/1 BC, earlier than or equal to the most likely radiocarbon 1σ ranges (see previous section) of Model A: 1335/33–1319/16 Cal BC, or Model B: 1333/32–1311/07 Cal BC (or, referring to the 2σ or most likely sub-range of the overall 2σ ranges, Model A: 1341/39–1312 Cal BC, and Model B: 1343/40–1289/74 Cal BC); or

(ii) Revised conventional TPQ/production range 1342–1321/20 BC, earlier than or equal to the most likely radiocarbon 1σ range of Model A: 1335/33–1319/16 Cal BC, or Model B: 1333/32–1311/07 Cal BC (or, referring to the 2σ or most likely sub-range of the overall 2σ ranges, Model A: 1341/39–1312 Cal BC, and Model B: 1343/40–1289/74 Cal BC).

In reverse, proposed chronologies for Egypt which posit dates for the accession of Amenhotep IV substantially later than the conventional dates, and, in particular, later than about 1311–1307 BC (the most likely 1σ range from Model B), or, at the extreme dates of variously about 1309 BC, 1308 BC, 1289 BC or 1274 BC (the latest date in any of the 2σ ranges from either models A or B), appear incompatible with the robust radiocarbon-based chronology summarised in Figures 10–17 and Table 2, and so may be rejected. (These include, for example, chronologies which are some 70 years later than the conventional dates, as in Hagens 2006, and certainly those radical ultra-low chronologies which suggest dates a couple of centuries later again: James et al. 1991; Rohl 1995.) In turn, because of the inter-linked correspondence recorded in the Amarna archive between the Egyptian kings and contemporary rulers of Babylonia (especially) and Assyria, the Hittites, the Mitanni, Alashiya (usually regarded as Cyprus), and various other Levantine entities (Moran 1992), the above finding, which requires at least the conventional range of dates for Nefertiti and Amenhotep IV (Akhenaten), also provides a similar requirement for the chronologies of Babylonia and Assyria (summaries in von Beckerath 1994: 23–24; Klinger 2006: 313–319). Radically later (more recent) dates for the civilizations of second to first millennium BC Egypt, and the linked ancient Near East in general, are thus incompatible with the substantive and independent integrated dendro-radiocarbon wiggle-match evidence presented here for the Uluburun ship.

8. Discussion and Conclusions

The high-resolution integrated dendro-radiocarbon methods produce a date between about 1335–1332 to 1319–1307 BC (1σ) or 1343–1339 to 1312–1274 BC (2σ) (overall range of options from the analyses above, with the most likely range 1335/1333–1319/1316 BC on Model A and 1333/1332–1311/1307 BC on Model B at 1σ for the last voyage of the Uluburun ship. With its rich cargo, the ship now provides a key independent chronological marker for the east Mediterranean region. It independently confirms the approximate absolute dating of the well-documented Amarna period in Egypt and its contemporaries in the ancient Near East, in the mid-later 14th century BC. In the Aegean, it dates the Late Helladic IIIA2

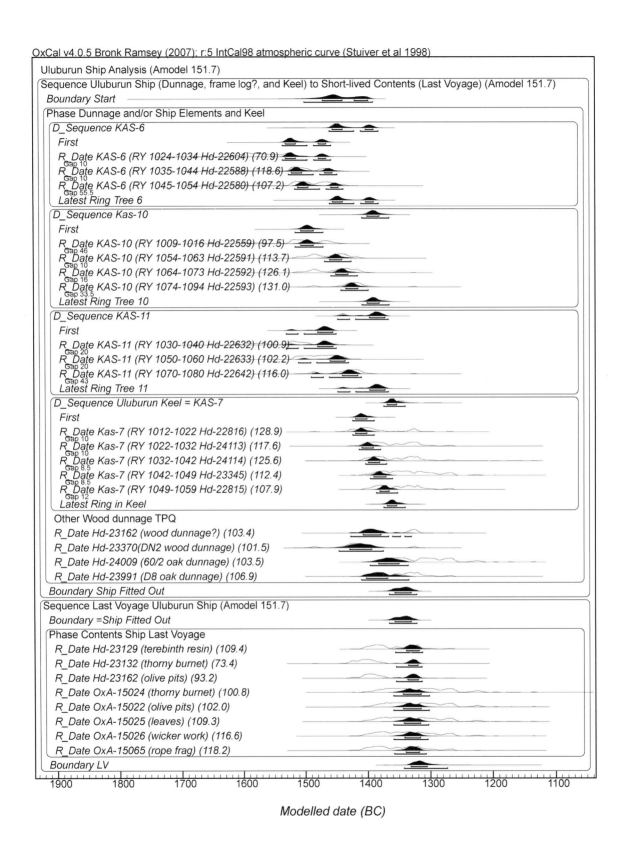

Figure 16: As Figure 14, but with the calibration analysis employing the IntCal98 radiocarbon calibration dataset (Stuiver et al. 1998). A very similar set of outcomes results. For other details of how to read the figure, see the caption to Figure 10.

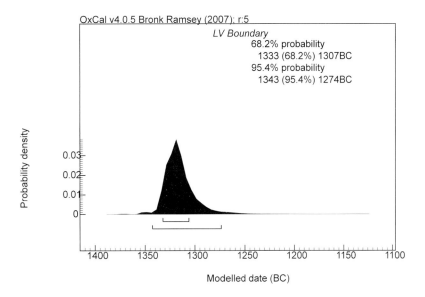

Figure 17: The modelled age estimate for the Last Voyage (LV) of the Uluburun ship from Figure 16.

period (Bass et al. 1989; Warren and Hankey 1989: 148–154; Wiener 1998; 2003a).

The ability of radiocarbon to resolve this Amarna-period time-capsule precisely, and in good agreement with standard recent assessments of the historical information regarding the ancient chronologies of Egypt and the Ancient Near East, is important. The Amarna period is the best documented (e.g. Moran 1992; Murnane 1995) and most securely cross-dated short time horizon in the Near East from the whole of the Bronze Age, where, critically, documents at Amarna combined with those known from Mesopotamia attest that Amenhotep IV and his father Amenhotep III were contemporaries of Burnaburiash of Babylonia and (thence) also of Ashshur-Uballit I of Assyria (Klinger 2006: 313–319; von Beckerath 1994: 23–24). This nexus of links therefore ties the Amarna period dating (above) to the Babylonian and Assyrian chronological traditions. The Amarna letters also permit direct links to other rulers such as the Hittite king Shuppiluliuma I; hence the dendro-radiocarbon dating of the Uluburun ship confirms his mid later 14th century BC date, and those for others linked to the Amarna time-frame. There is also a very secure tie to Aegean chronology through the large number of Aegean ceramics found at Amarna itself, or in contexts closely associated to this short time-period (Hankey 1997; Wiener 1998; 2003a). The fact that an integrated radiocarbon and dendrochronological analysis can resolve a precise date in accord with the archaeology and history when there is good, extensive, and replicated archaeological and historical information, suggests that the former is a good guide to dating the latter (and radiocarbon dates from Amarna itself are also compatible with the historical dating: Manning 2006: 335–338). Where we then have archaeological or historical evidence that is much less secure, unclear, sparse, or non-replicated, we might even venture to suspect that large-scale integrated dendrochronological and radiocarbon analyses might offer the best guide to absolute dates.

It is of course noticeable that the good agreement of the integrated dendrochronological and radiocarbon analysis for the Uluburun ship with the dating of the Amarna period in Egypt and the Ancient Near East stands in stark contrast with the situation in the Aegean and east Mediterranean in the 17th to early 15th centuries BC. For the 17th to early 15th centuries BC, scholarship has for some years noted an apparent disagreement between the radiocarbon-based chronology versus conventional archaeological-historical assessments and estimates for dates in the Aegean and on Cyprus (Betancourt 1987; Manning 1988; 1999; Friedrich et al. 2006 and this volume; Manning et al. 2006 and this volume; Bietak 2003; Bietak and Höflmayer 2007; Wiener 2003b and this volume).

What is going on? There is no apparent difference on the radiocarbon side—and indeed the same two (of three) radiocarbon laboratories and their quality controls (see Manning et al. 2006: Supporting Online Material) have provided the data in both the Uluburun case (above) and the most recent major studies of the 17th to 15th century BC case (Manning et al. 2006; Friedrich et al. 2006). Some have suggested or asserted that perhaps volcanic CO_2 or another mechanism somehow affected the dates on samples from Santorini, but the data seem in fact to indicate the contrary, and no actual positive evidence has been produced to support these claims (see the contrasting discussions of Friedrich et al. 2006, and this volume, Manning et al. 2006, and this volume, and Wiener, this volume). Even a supposed Santorini-specific problem cannot explain away the evidence from elsewhere in the Aegean. As Manning (et al. 2006, see also this volume) observed, even excluding all the evidence from Santorini, the other Aegean radiocarbon data indicate a similar chronology (and one at odds with the conventional archaeological-historical dates for the 17th to earlier 15th centuries BC). Keenan (2002) claimed that upwelling of stagnant radiocarbon-depleted deep water in the Mediterranean caused radiocarbon dates to be too old before about year 0 in the Mediterranean. The case lacks any evidential support (Manning et al. 2002) and is clearly disproved back to the 14th century BC by the Uluburun data (above); it is also noticeable that east Mediterranean radiocarbon dates from earlier in the second millennium BC (e.g. Marcus 2003; or Voutsaki et al., this volume), or from the third millennium BC (e.g. Manning 1995; 2008), do not seem too old in any systematic or significant way. The "problem" seems to lie in the 17th to early 15th centuries BC.

If we look to the archaeology, the notable difference is the quality and quantity of data at issue for the interpretation of the material culture correlations between the Aegean and Cyprus, and Egypt and its historical chronology in the 17th and 16th centuries BC. (By the time we reach the 15th century BC, things come together, and can even agree well: Manning n.d.) For the Amarna period there is a multi-strand, secure, Egyptian historical chronology linking with Babylonia and Assyria (and their independent chronological traditions). There is furthermore a vast web of material culture linkages (including a large cache of direct Aegean-Egyptian linkages from Tell el-Amarna itself, and other associated linkages: Hankey 1981; 1997; Wiener 1998; 2003a). In contrast, for the Late Bronze I to II periods, there are far fewer, or no (for the Late Minoan IA period) clear and direct linkages, and thus there is much more flexibility or ambiguity in the conventional interpretations of cultural associations, especially for the earlier LBI stage (Kemp and Merrillees 1980; Betancourt 1987; Manning 1988; 1999; 2007; Manning et al. 2006). Since the present study demonstrates that sophisticated integrated dendrochronological and radiocarbon analysis offers a precise chronology compatible with secure and well-based archaeological-historical data and assessment for the east Mediterranean region, we may have to consider the possibility that a similar integrated analysis on appropriate samples (e.g. Manning et al. 2006; Friedrich et al. 2006) could offer better guidance for other periods in the Aegean-east Mediterranean where the archaeological-historical linkages are less secure.

In the meantime, we approach a near-fixed point for the archaeology of the east Mediterranean, agreed by science and archaeology: a date for the last voyage of the Uluburun ship, and for its extraordinary contents, in about 1335–1332 to 1319–1307 BC (1σ) or about 1343–1339 to 1312–1274 BC (2σ) (overall range of options from the analyses above, with the most likely range 1335/1333–1319/1316 BC on Model A and 1333/1332–1311/1307 on Model B at 1σ). If we then generalise these various date estimates and slightly different approaches, and the use of the two calibration datasets, we can offer an approximate round numbers estimate for the last voyage of the Uluburun ship of about 1320 ±15 BC (covering all the 1σ ranges).

Acknowledgments

We thank the NSF, NEH, NSERC, NERC, the Institute of Nautical Archaeology (INA) and Texas A & M University, the Institute for Aegean Prehistory, the Malcolm H. Wiener Foundation, and other patrons of the Aegean Dendrochronology Project for their funding and support of this work. We thank and acknowledge especially the honorand of this volume, Peter Ian Kuniholm, and also Maryanne Newton, for their fundamental work on the Uluburun samples over many years (e.g. Newton et al. 2005), and we thank Laura Steele for her substantial laboratory work on the Uluburun samples. We thank N. Liphschitz for plant and tree species identification. We thank Michael Friedrich for supplying samples of known-age German Oak. We thank Peter Kuniholm for comments on the text.

References

Allen, J.P. 2009. The Amarna Succession. In P. Brand and L. Cooper (eds.), *Causing His Name to Live: Studies in Egyptian Epigraphy and History in Memory of William J. Murnane. Culture and History of the Ancient Near East 37*. In press. Leiden: Brill.

Aruz, J., Benzel, K., and Evans, J.M. (eds.) 2008. *Beyond Babylon: Art, Trade, and Diplomacy in the Second Millennium B.C.* New York/New Haven: The Metropolitan Museum of Art/Yale University Press.

Bachhuber, C. 2006. Aegean Interest on the Uluburun Ship. *American Journal of Archaeology* 110: 345–363.

Bass, G.F. 1986. A Bronze Age Shipwreck at Ulu Burun (Kaş): 1984 Campaign. *American Journal of Archaeology* 90: 269–296.

Bass, G.F. 1987. Oldest Known Shipwreck Reveals Splendors of the Bronze Age. *National Geographic* 172: 693–733.

Bass, G.F. 1991. Evidence of Trade from Bronze Age Shipwrecks. In N.H. Gale (ed.), *Bronze Age Trade in the Mediterranean*: 69–82. SIMA 90. Göteborg: Paul Åströms Förlag.

Bass, G.F. 1998. Sailing between the Aegean and the Orient in the Second Millennium BC. In E.H. Cline and D. Harris-Cline (eds.), *The Aegean and the Orient in the Second Millennium: Proceedings of the 50th Anniversary Symposium, Cincinnati, 18-20 April 1997. Aegaeum 18*: 183-191. Liège: Université de Liège, Histoire de l'art et archéologie de la Grèce antique; University of Texas at Austin, Program in Aegean Scripts and Prehistory.

Bass, G., Pulak, C., Collon, D., and Weinstein, J. 1989. The Bronze Age Shipwreck at Ulu Burun. *American Journal of Archaeology* 93: 1–29.

Beckerath, J. von. 1994. Chronologie des ägyptischen Neuen Reiches. *Hildesheimer ägyptologische Beiträge 39*. Hildesheim: Gerstenberg Verlag.

Beckerath, J. von. 1997. *Chronologie des pharaonischen Ägypten. Die Zeitbestimmung der ägyptischen Geschichte von der Vorzeit bis 332 v. Chr.* Mainz: Philipp von Zabern.

Betancourt, P.P. 1987. Dating the Aegean Late Bronze Age with Radiocarbon. *Archaeometry* 29: 45–49.

Bietak, M. 2003. Science Versus Archaeology: Problems and Consequences of High Aegean Chronology. In M. Bietak (ed.), *The Synchronisation of Civilisations in the Eastern Mediterranean in the Second Millennium B.C. II. Proceedings of the SCIEM 2000—EuroConference Haindorf, 2nd of May-7th of May 2001*: 23–33. Contributions to the Chronology of the Eastern Mediterranean II, Vienna: Verlag der Österreichischen Akademie der Wissenschaften.

Bietak, M. and Höflmayer, F. 2007. Introduction: High and Low Chronology. In M. Bietak and E. Czerny (eds.), *The Synchronisation of Civilisations in the Eastern Mediterranean in the Second Millennium B.C. III. Proceedings of the SCIEM 2000—2nd EuroConference, Vienna 28th of May–1st of June 2003*: 13–23. Vienna: Verlag der Österreichischen Akademie der Wissenschaften.

Bronk Ramsey, C. 1995. Radiocarbon Calibration and Analysis of Stratigraphy: The OxCal Program. *Radiocarbon* 37: 425–430.

Bronk Ramsey, C. 2001. Development of the Radiocarbon Calibration Program OxCal. *Radiocarbon* 43: 355–363.

Bronk Ramsey, C. 2008. Deposition Models for Chronological Records. *Quaternary Science Reviews* 27: 42–60.

Bronk Ramsey, C. 2009. Bayesian Analysis of Radiocarbon Dates. *Radiocarbon* 51: 337–360.

Bronk Ramsey C., van der Plicht, J., and Weninger, B. 2001. 'Wiggle Matching' Radiocarbon Dates. *Radiocarbon* 43: 381–389.

Cleary, M., and Meister, M.J. (eds.) 1999. *Cargoes from Three Continents: Ancient Mediterranean Trade in Modern Archaeology.* Boston: Archaeological Institute of America.

Cucchi, T. 2008. Uluburun Shipwreck Stowaway House Mouse: Molar Shape Analysis and Indirect Clues about the Vessel's Last Journey. *Journal of Archaeological Science* 35: 2953–2959.

Friedrich, W.L., Kromer, B., Friedrich, M., Heinemeier, J., Pfeiffer, T., and Talamo, S. 2006. Santorini Eruption Radiocarbon Dated to 1627–1600 B.C. *Science* 312: 548.

Galimberti, M., Bronk Ramsey, C., and Manning S.W. 2004. Wiggle-match Dating of Tree Ring Sequences. *Radiocarbon* 46: 917–924.

Hagens, G. 2006. Testing the Limits: Radiocarbon Dating and the End of the Late Bronze Age. *Radiocarbon* 48: 83–100.

Hankey, V. 1981. The Aegean Interest in El Amarna. *Journal of Mediterranean Anthropology and Archaeology* 1: 38–49.

Hankey, V. 1997. Aegean Pottery at El-Amarna: Shapes and Decorative Motifs. In J. Phillips (ed.), *Ancient Egypt, the Aegean, and the Near East: Studies in Honour of Martha Rhoads Bell*: 193–218. San Antonio: Van Siclen Books.

Helck, W. 1987. 'Was kann die Ägyptologie wirklich zum Problem der absoluten Chronologie in der Bronzezeit beitragen?' Chronologische Annäherungswerte in der 18. Dynastie. In P. Åström (ed.), *High, Middle or Low? Acts of an International Colloquium on Absolute Chronology Held at the University of Gothenburg 20th-22nd August 1987, Part 1*: 18–26. *Studies in Mediterranean Archaeology and Literature Pocket-book 56*. Göteborg: Paul Åströms Förlag.

Hornung, E., Krauss, R., and Warburton, D.A. (eds.) 2006. *Ancient Egyptian Chronology.* Leiden: Brill.

James, P., Thorpe, I.J., Kokkinos, N., Morkot, R., and Frankish, J. 1991. *Centuries of Darkness.* London: Jonathan Cape.

Keenan, D.J. 2002. Why Early-historical Radiocarbon Dates Downwind from the Mediterranean are Too Early. *Radiocarbon* 44: 225–237.

Kemp, B.J., and Merrillees, R.S. 1980. *Minoan Pottery in Second Millennium Egypt.* Mainz am Rhein: Philipp von Zabern.

Kitchen, K.A. 1996. The Historical Chronology of Ancient Egypt, a Current Assessment. *Acta Archaeologica* 67: 1–13.

Kitchen, K.A. 2007. Egyptian and Related Chronologies— Look, No Sciences, No Pots! In Bietak, M. & Czerny, E. (eds.), *The Synchronisation of Civilisations in the Eastern Mediterranean in the Second Millennium B.C. III. Proceedings of the SCIEM 2000—2nd EuroConference, Vienna 28th of May–1st of June 2003*: 163–171. Vienna: Verlag der Österreichischen Akademie der Wissenschaften.

Klinger, J. 2006. Chronological Links between the Cuneiform World of the Ancient Near East and Ancient Egypt. In E. Hornung, R. Krauss and D.A. Warburton (eds.), *Ancient Egyptian Chronology*: 304–324. Leiden: Brill.

Kuniholm, P.I., Kromer, B., Manning, S.W., Newton, M., Latini, C.E., and Bruce, M.J. 1996. Anatolian Tree Rings and the Absolute Chronology of the Eastern Mediterranean, 2220–718 BC. *Nature* 381: 780–783.

Liphschitz, N., and Pulak, C. 2007/2008. Wood Species Used in Ancient Shipbuilding in Turkey: Evidence from Dendroarchaeological Studies. *Skyllis—Zeitschrift für Unterwasserarchäologie* 8 (1-2): 74–83.

Manning, S.W. 1995. *The Absolute Chronology of the Aegean Early Bronze Age: Archaeology, History and Radiocarbon. Monographs in Mediterranean Archaeology 1.* Sheffield: Sheffield Academic Press.

Manning, S.W. 1999. *A Test of Time: the Volcano of Thera and the Chronology and History of the Aegean and East Mediterranean in the Mid Second Millennium BC.* Oxford: Oxbow Books.

Manning, S.W. 2006. Radiocarbon Dating and Egyptian Chronology. In E. Hornung, R. Krauss, and D.A. Warburton (eds.), *Ancient Egyptian Chronology*: 327–355. Leiden: Brill.

Manning, S.W. 2007. Clarifying the 'High' v. 'Low' Aegean/Cypriot Chronology for the Mid Second Millennium BC: Assessing the Evidence, Interpretive Frameworks, and Current State of the Debate. In Bietak, M. & Czerny, E. (eds.), *The Synchronisation of Civilisations in the Eastern Mediterranean in the Second Millennium B.C. III. Proceedings of the SCIEM 2000—2nd EuroConference, Vienna 28th of May–1st of June 2003*: 101–137. Vienna: Verlag der Österreichischen Akademie der Wissenschaften.

Manning, S.W. 2008. Some Initial Wobbly Steps towards a Late Neolithic to Early Bronze III Radiocarbon Chronology for the Cyclades. In N.J. Brodie, J. Doole, G. Gavalas, and C. Renfrew (eds.), *Ὁρίζων: A Colloquium on the Prehistory of the Cyclades*: 55–59. Cambridge: McDonald Institute for Archaeological Research.

Manning, S.W. n.d. Beyond the Santorini Eruption: Some Notes on Dating the Late Minoan IB Period on Crete, and Implications for Cretan-Egyptian Relations in the 15th Century BC (and Especially LMII). In D. Warburton (ed.), *Time's Up! Dating the Minoan Eruption of Santorini*. In press. Monographs of the Danish Institute at Athens, Vol.10.

Manning, S.W., Barbetti, M., Kromer, B., Kuniholm, P.I., Levin, I., Newton, M.W., and Reimer, P.J. 2002. No Systematic Early Bias to Mediterranean ^{14}C Ages: Radiocarbon Measurements from Tree-ring and Air Samples Provide Tight Limits to Age Offsets. *Radiocarbon* 44: 739–754.

Manning, S.W., Bronk Ramsey, C., Kutschera, W., Higham, T., Kromer, B., Steier, P., and Wild, E. 2006. Chronology for the Aegean Late Bronze Age. *Science* 312: 565–569.

Manning, S.W., Kromer, B., Kuniholm, P.I., and Newton, M.W. 2003. Confirmation of Near-absolute Dating of East Mediterranean Bronze-Iron Dendrochronology. *Antiquity* 77 (295): [http://antiquity.ac.uk/ProjGall/Manning/manning.html].

Manning, S.W., Kromer, B., Kuniholm, P.I., and Newton, M.W. 2001. Anatolian Tree-Rings and a New Chronology for the East Mediterranean Bronze-Iron Ages. *Science* 294: 2532–2535.

Manning, S.W., Kromer, B., Talamo, S., Friedrich, M., Kuniholm, P.I., and Newton, M.W. 2005. Radiocarbon Calibration in the East Mediterranean Region. In T.E. Levy and T. Higham (eds.), *The Bible and Radiocarbon Dating Archaeology, Text and Science*: 95–103. London: Equinox.

Marcus, E. 2003. Dating the Early Middle Bronze Age in the Southern Levant: A Preliminary Comparison of Radiocarbon and Archaeo-historic Synchronisms. In Bietak, M. (ed.), *The Synchronisation of Civilisations in the Eastern Mediterranean in the Second Millennium BC (II). Proceedings of the SCIEM2000 EuroConference Haindorf, May 2001*: 95–110. Vienna: Verlag der Österreichischen Akademie der Wissenschaften.

Moran, W.L. 1992. *The Amarna Letters*. Baltimore: Johns Hopkins Press.

Murnane, W.J. 1995. *Texts from the Amarna Period in Egypt*. Atlanta: Scholars Press.

Newton, M.W., and Kuniholm, P.I. 2005. The Wood from the Kaş, Shipwreck at Uluburun: Dendrochronological and Radiocarbon Analyses through 2004. Paper Read at the 2005 Annual Meeting, Archaeological Institute of America. Abstract: [http://www.archaeological.org/webinfo.php?page=10248\&searchtype=abstract\&ytable=2005\&sessionid=2A\&paperid=84].

Newton, M.W., Talamo, S., Pulak, C., Kromer, B., and Kuniholm, P. 2005. Die Datierung des Schiffswracks von Uluburun. In Ü. Yalçın, C. Pulak, & R. Slotta (eds.), *Das Schiff von Uluburun. Welthandel vor 3000 Jahren*: 115–116. Bochum: Deutsches Bergbau-Museum.

Pulak, C. 1988. The Bronze Age Shipwreck at Ulu Burun, Turkey: 1985 Campaign. *American Journal of Archaeology* 92: 1–37.

Pulak, C. 1992. The Shipwreck at Uluburun Turkey: 1992 Excavation Campaign. *The INA Quarterly* 19/4: 4–11, 21.

Pulak, C. 1996. Dendrochronological Dating of the Uluburun Ship. *The INA Quarterly* 23/1: 12–13.

Pulak, C. 1998. The Uluburun Shipwreck: An Overview. *International Journal of Nautical Archaeology* 27: 188–224.

Pulak, C. 1999a. The Late Bronze Age Shipwreck at Uluburun: Aspects of Hull Construction. In W. Phelps, Y. Lolos, and Y. Vichos (eds.), *The Point Iria Wreck: Interconnections in the Mediterranean ca. 1200 BC. Proceedings of the International Conference, Island of Spetses, 19 September 1998*: 209–238. Athens: Hellenic Institute of Marine Archaeology.

Pulak, C. 1999b. Hull Construction of the Late Bronze Age Shipwreck at Uluburun. *The INA Quarterly* 26/4: 16–21.

Pulak, C. 2001. The Cargo of the Uluburun Ship and Evidence for Trade with the Aegean and Beyond. In L. Bonfante and V. Karageorghis (eds.), *Italy and Cyprus in Antiquity: 1500–450 BC*: 13–60. Nicosia: The Costakis and Leto Severis Foundation.

Pulak, C. 2003. The Uluburun Hull Remains. In H.E. Tzalas (ed.), *Tropis VII. Proceedings of the 7th International Symposium on Ship Construction in Antiquity (27 August–31 August, 2000, Pylos)*: 615–636. Athens: Hellenic Institute for the Preservation of Nautical Tradition.

Pulak, C. 2005a. Das Schiffswrack von Uluburun und seine Ladung. In Ü Yalçın, C. Pulak, and R. Slotta (eds.), *Das Schiff von Uluburun—Welthandel vor 3000 Jahren*: 53–102. Bochum: Deutsches Bergbau-Museum Nr.138.

Pulak, C. 2005b. Discovering a Royal Ship from the Age of King Tut: Uluburun, Turkey. In G. Bass (ed.), *Beneath the Seven Seas: Adventures with the Institute of Nautical Archaeology*: 34–47. London: Thames and Hudson.

Pulak, C. 2005c. Who Were the Mycenaeans aboard the Uluburun Ship? In R. Laffineur and E. Greco (eds.), *EMPORIA. Aegeans in the Central and Eastern Mediterranean. Proceedings of the 10th International Aegean Conference, Athens, Italian School of Archaeology, 14-18 April 2004, Aegaeum 25*: 295–312. Liège: Université de Liège, Histoire de l'art et archéologie de la Grèce antique; University of Texas at Austin, Program in Aegean Scripts and Prehistory.

Pulak, C. 2008. The Uluburun Shipwreck and Late Bronze Age Trade. In J. Aruz, K. Benzel, and J.M. Evans (eds.), *Beyond Babylon: Art, Trade, and Diplomacy in the Second Millennium B.C.*: 288–305. New York/New Haven: The Metropolitan Museum of Art/Yale University Press.

Reimer, P.J., Baillie, M.G.L., Bard, E., Bayliss, A., Beck, J.W., Bertrand, C.J.H., Blackwell, P.G., Buck, C.E., Burr, G.S., Cutler, K.B., Damon, P.E., Edwards, R.L., Fairbanks, R.G., Friedrich, M., Guilderson, T.P., Hogg, A.G., Hughen, K.A., Kromer, B., McCormac, G., Manning, S., Bronk Ramsey, C., Reimer, R.W., Remmele, S., Southon, J.R., Stuiver, M., Talamo, S., Taylor, F.W., van der Plicht, J., Weyhenmeyer, C.E. 2004. IntCal04 Terrestrial Radiocarbon Age Calibration, 0-26 Cal Kyr BP. *Radiocarbon* 46: 1029–1058.

Rohl, D.M. 1995. *A Test of Time. Volume One: The Bible—from Myth to History*. London: Century.

Stuiver, M., Reimer, P.J., Bard, E., Beck, J.W., Burr, G.S., Hughen, K.A., Kromer, B., McCormac, G., Plicht, J. van der, and Spurk, M. 1998. INTCAL98 Radiocarbon Age Calibration, 24,000-0 Cal BP. *Radiocarbon* 40: 1041–1083.

Wachsmann, S. 1998. *Seagoing Ships & Seamanship in the Bronze Age Levant*. College Station: Texas A & M University Press.

Ward, G.K., and Wilson, S.R. 1978. Procedures for Comparing and Combining Radiocarbon Age Determinations: A Critique. *Archaeometry* 20: 19–31.

Warren, P., and Hankey, V. 1989. *Aegean Bronze Age Chronology*. Bristol: Bristol Classical Press.

Weinstein, J. 1989. III. The Gold Scarab of Nefertiti from Ulu Burun: Its Implications for Egyptian History and Egyptian-Aegean Relations. In G.F. Bass, C. Pulak, D. Collon, and J. Weinstein (eds.), The Bronze Age Shipwreck at Ulu Burun: 1986 Campaign: 17–29. *American Journal of Archaeology* 93.

Weinstein, J. 2008. Nefertiti Scarab. In J. Aruz, K. Benzel and J.M. Evans (eds.), *Beyond Babylon: Art, Trade, and Diplomacy in the Second Millennium B.C.*: 358. New York/New Haven: The Metropolitan Museum of Art/Yale University Press.

Welter-Schultes, F.W. 2008. Bronze Age Shipwreck Snails from Turkey: First Direct Evidence for Oversea Carriage of Land Snails in Antiquity. *Journal of Molluscan Studies* 74: 79–87.

Wiener, M.H. 1998. The Absolute Chronology of Late Helladic IIIA2. In M.S. Balmuth and R.H. Tykot (eds.), *Sardinian *and Aegean Chronology: Towards the Resolution of Relative and Absolute Dating in the Mediterranean*: 309–319. *Studies in Sardinian Archaeology V*. Oxford: Oxbow Books.

Wiener, M.H. 2003a. The Absolute Chronology of Late Helladic III A2 Revisted. *Annual of the British School at Athens 98*: 239–250.

Wiener, M.H. 2003b. Time Out: The Current Impasse in Bronze Age Archaeological Dating. In K.P. Foster and R. Laffineur (eds.), *Metron: Measuring the Aegean Bronze Age*: 363-399. *Aegaeum 24*. Liège and Austin.

Yalçın, Ü., Pulak, C., and Slotta, R. (eds.). 2005. *Das Schiff von Uluburun—Welthandel vor 3000 Jahren*. Bochum: Deutsches Bergbau-Museum Nr. 138.

[Note added in Proof: This paper has employed the current standard IntCal04 radiocarbon calibration curve (and considered the previous IntCal98 curve). However, Figure 3 shows that some recent Heidelberg measurements, especially, on known-age German oak, provide radiocarbon ages somewhat older than IntCal04 in the period 1360–1260 BC (see further in Kromer, B., Manning, S., Friedrich, M., Talamo, S., and Trano, N. n.d. ^{14}C calibration in the 2nd and 1st millennia BC—Eastern Mediterranean Radiocarbon Comparison Project (EMRCP). *Radiocarbon*, in press). When these data are included in a future IntCal modelling exercise (with the other available data), they may slightly raise the radiocarbon ages for IntCal in this interval. If so, this may allow the date for the Last Voyage to become slightly more recent. Hence, if there is room for movement on the date of 1320 ± 15 BC stated above, it is toward a slightly more recent date (i.e., toward c. 1300 BC).]

Article submitted April 2009

How About the Pace of Change for a Change of Pace?

Jeremy B. Rutter

Abstract: *A procedure for measuring variability in the pace of artifactual change during the Aegean Bronze Age is outlined, using ceramics as an examplar of a specific artifact category. Thanks to ever more precise estimates for the durations of culture-historical phases, and to more detailed understanding of total ceramic variability during those phases, the metrics proposed for assessing the pace of change in Aegean prehistoric material culture in a comparative fashion should become increasingly more reliable with time. A few ways in which a better appreciation for differences in its pace might enhance our overall understanding of cultural change are briefly explored.*

One unfortunate side effect of the intense focus over the past fifteen years on the question of the absolute date of the Theran volcano's eruption at some point early in the Aegean Late Bronze Age has been a comparative lack of recent interest in exploring the highly variable rate of cultural change in this region of the eastern Mediterranean throughout the third and second millennia BCE. For no matter when the Theran eruption is ultimately to be dated in calendar years, most authorities would now agree that no immediate, fundamental, or obvious changes in cultural trajectories can be attributed to the Theran eruption aside from the abandonment of the island of Thera itself as a place to live.[1]

What, in fact, are the moments at which pronounced cultural changes occurred during the Aegean Bronze Age? How can absolute dates, especially of the precision that dendrochronology may eventually be able to furnish us with, help us to investigate such periods of momentous change? If one of our goals as archaeologists is to assess and explain change in material culture, and as part of this endeavor to differentiate between periods during which significant changes took place rapidly and periods when change was comparatively slight and slow, what kind of analytical approaches to the data recovered from fieldwork as well as experimental work in the laboratory should we adopt and why?

Peter Kuniholm, whose pioneering work we are honoring in this volume, has through the Aegean Dendrochronological Project devoted thirty years to providing us with ever sharper measures of the duration of bygone archaeological periods by determining the dates of more and more structural as well as natural timbers. The Malcolm and Carolyn Wiener Laboratory for Aegean and Near Eastern Dendrochronology at Cornell by its very name indicates the region of the world where this refinement of our ability to date closely and accurately has been concentrated, in addition to identifying the generous donors who have made and continue to make much of this progress possible. As a specialist in Aegean Bronze Age ceramics, I would like in this contribution to focus on how pottery might be exploited as arguably the most sensible as well as sensitive artifactual category for the purposes of establishing a semi-quantitative measure of cultural change during the roughly two millennia (ca. 3000–1000 BC) encompassed by the Bronze Age in this part of the eastern Mediterranean world.

So, what are the moments of pronounced cultural change during the Aegean Bronze Age? One answer to this is that it depends very much on one's specific location, since the Aegean is not at all culturally homogeneous, even if intercultural contacts were frequently very strong and the bearers of the various cultural traditions represented in the region knew of each other from a very early stage of the Early Bronze Age onwards. For example, the relatively sudden collapse of the so-called Korakou culture of the Early Helladic II

---

[1] For some recent commentaries on this issue, see Driessen 2002; Manning and Sewell 2002; Warren 2006; Manning et al., this volume; Wiener, this volume.

Greek mainland doesn't appear to have had much of an impact on Minoan Crete. Nor can the destruction of Troy VI or VIIa, despite the continuing power of the Greek oral tradition, be shown to have had any noteworthy impact on Mycenaean Greece. The ups and downs of Minoan palatial civilization, to be sure, had certain pan-Aegean repercussions, as did the collapse of Mycenaean palatial civilization at the end of the 13th century. But as often as not, particular regions and cultures within the Aegean followed independent paths of advancement and decline.

More importantly, what do I mean by a phrase such as "pronounced cultural change"? This choice of language makes it sound as though some quantity of cultural change is at issue, along with how quickly such changes may have taken place. For such a phrase to have real meaning, then, I ought to have some way of quantifying archaeological change—that is, change in material culture—not only at particular moments (as, for example, on either side of a short-lived event) but also over some lengthier period (that is, a rate of change per some unit of time).

Why is it that we archaeologists seek to identify and then focus upon such "moments of pronounced change"? Since it is one of the universally acknowledged goals of our discipline to explore and explain cultural change, intervals of time when such change is comparatively intense are of inherent interest to us. But the very fact that neither the nature nor the extent of such change is a constant over time introduces the need for some metric that will allow us to evaluate different kinds and paces of material cultural change. So can we identify a workable unit of time and one or more categories of material culture in terms of which change can be usefully quantified for the purposes of comparative analysis?

Let us begin with the problem of the artifactual category. Pottery—that is, a corpus of ceramic containers—is an extremely useful vehicle for assessing change in material culture because of its peculiar combination of utility, fragility, and value. Pots are useful when whole, but not very expensive to produce in most environments in comparison to containers produced from other materials. They are relatively fragile, however, especially when small, and thus are routinely subject to breakage from frequent handling or exposure to sudden changes in temperature, and when fragmentary or broken, they are ordinarily valueless. Many varieties of pottery thus need to be replaced frequently, resulting in the production of large quantities of these containers. Competition between the producers of the regularly required replacements results in constant appeals being made for the attention of potential consumers in terms of aesthetics, functionality, or basic affordability—and so the decorative elaboration, physical form, and technological sophistication of these artifacts can and often does vary enormously, not only at one and the same time but also through time.

Due to the considerable variability among ceramic containers, it would not make much sense simply to count up the total number of changes observable in the ceramic repertory from phase to phase, however such phases are to be defined and irrespective of the precise kinds of change that might be involved. To do so would be to multiply the inappropriate comparison of apples and oranges hundreds of times over. Instead, it is important to isolate those variables in a complete ceramic corpus that can be validly compared diachronically within a given cultural tradition or else cross-culturally at a given moment in time.

Let us consider a way of tabulating the changes between two or more ceramic corpora that might yield useful results. I suggest beginning with the identification of a series of major functional classes into which any collection of containers may be subdivided (Table 1). The specific identities of the particular set of functional classes I have listed are not as important as the need to devise a system whereby the selected artifactual category at any particular moment in time is being compared with that of the preceding phase in as informative a way as possible insofar as assessing the degree and the nature of the change from one phase to the next is concerned.[2] The functional classes used here tend to clump those vessel forms produced in similar fabrics and subjected to similar breakage hazards. For example, such a separation into functional classes will avoid the possibility of pithoi being compared to thin-walled drinking cups, or of wide-mouthed jugs used for cooking being compared to narrow-necked jugs used for dispensing wine or water.

Under the heading of each functional class are presented some notional numbers of shapes that can be identified, respectively, as common, occasional, or rare in any given phase. These numbers could be supplemented by total numbers of discrete sizes, as the heading over the relevant columns suggests, were shape volumes to prove of potential significance. A second breakdown of the entire ceramic corpus could also be undertaken in terms of the pottery's decorative elaboration, if this should prove desirable for the purposes of highlighting changes in ornamentation as opposed

---

[2]The differentiation between open and closed shapes under the heading of tablewares is intended to distinguish preliminary pouring activities (from closed vessels such as jugs, amphoras, and jars) from those associated more directly with the consumption of food and drink (from open vessels such as cups, bowls, plates, etc.). This distinction has often been a feature of ceramic classification systems devised for Aegean Bronze Age ceramic corpora (Furumark 1941; Blegen et al. 1950, 1951, 1953, 1958; Rutter 1995; Wiencke 2000).

Ceramic Phase: LH <yyy> Region (or Site, or Culture): Estimated Duration of Phase: 50 years				Preceding Ceramic Phase: LH <xxx> Estimated Duration of Phase: 100 years				Changes in Nos. [Totals/ New Shapes]	Changes as % of Later Phase Nos. [Totals/ New Shapes]	Rate of Change per Unit Time [Decade?]
FUNCTIONAL CLASS	No. of Shapes [& Sizes?]			FUNCTIONAL CLASS	No. of Shapes [& Sizes?]					
	C (N)	O (N)	R (N)		C	O	R			
Tablewares, open	6 (4)	6 (2)	4 (4)	Tablewares, open	8	4	2	-2/4(C), +2/2(O), +2/4(R)	-33/67(C), +33/33(O), +50/100(R)	TBCAA
Tablewares, closed	2 (1)	4 (1)	2 (2)	Tablewares, closed	4	4	5	-2/1(C), 0/1(O), -3/2(R)	-100/50(C), 0/25(O), -150/100(R)	TBCAA
Storage				Storage						
[Transport]				[Transport]						
[Serving]				[Serving]						
Cooking	2 (0)	2 (0)	0 (0)	Cooking	2	3	0	0/0(C), -1/0(O), 0/0(R)	0/0(C), -50/0(O), 0/0(R)	TBCAA
[Miscellaneous]				[Miscellaneous]						

KEY: C = common; N = conceptually new form; O = occasional; R = rare; TBCAA = to be calculated as appropriate.

Table 1: A sample quantification of ceramic change from one phase to the next on the basis of shape changes within individual functional classes.

to form (Table 2). A third breakdown according to fabric and mode of manufacture as a way of providing insights into significant technological changes might also be considered worthwhile. Whatever the precise identity of the variable chosen for investigation, the corresponding numbers for the immediately preceding ceramic phase need to be provided so that the extent and nature of the changes from one phase to the next can be quantified, whether as an increase or a decrease. The changes may then be expressed as either a total figure per functional class (or decorative mode, or technologically oriented category) or as a percentage of the total shapes (or patterns, or wares) of the later of the two ceramic phases being compared. However the changes are to be quantified, one would also need to quantify novelties alongside changes of other kinds, hence the double sets of notional figures on the handout. Insofar as shape is concerned, what I mean by a novelty is a conceptually altogether new form.[3]

The sort of procedure I am outlining need not be restricted to pottery, but could in theory be applied to any artifactual category. In each and every case, the relevant question would be how many levels of the category would need to be explored in order for any genuinely informative measure of total change and rate of change to be generated, as well as whether that category is fully enough represented and well enough studied to yield adequate data. Do stone tools, jewelry, figurines, metal tools and weapons, or hearths and ovens lend themselves to this kind of quantitative analysis? Do they exhibit sufficient evidence of change from phase to phase to amplify or correct in any significant way the picture already derived from ceramics?

Enough is now understood about the Bronze Age pottery of at least some sub-regions of the central and southern Aegean for information along the lines I have sketched out in Tables 1–2 to be compiled for numerous different phases. Indeed, in some cases the record of ceramic development is essentially continuous throughout the third and second millennia BC, uninterrupted by any serious gaps in our knowledge. Ceramic phases vary in length, however, depending on the sub-region in question, and they tend to be increasingly longer, as well as more fuzzily defined, the further one proceeds back in time. Thus our ability to count up total numbers of shapes or decorative patterns per phase, just like our estimates of the durations of certain phases, are highly variable in terms of their precision. In the second half of the Late Bronze Age, thanks to encyclopedic works by scholars such as Furumark (1941) and Mountjoy (1986, 1999), the numbers of new shapes and patterns manifested per phase can be gauged quite closely, as can the duration of most phases to within several decades. A nominal "rate of change" can thus be calculated for each phase, say, during the Late Helladic III era in terms of changes per decade, using the duration of the later of the two periods in the comparison for this purpose. Such a rate can be calculated for each of the various ceramic categories being differentiated (that is, each functional class, and for common, occasional, and rare shapes within each class, for example) or al-

---
[3] For a recent, exemplary investigation of such a ceramic innovation, see Rotroff 2006, a study that draws upon the innovation theory popularized by Spratt (1982, 1989).

Ceramic Phase: LH <yyy> Region (or Site, or Culture): Estimated Duration of Phase: 50 years				Preceding Ceramic Phase: LH <xxx> Estimated Duration of Phase: 100 years				Changes in Nos.	Changes as % of Later Phase Nos.	Rate of Change per Unit Time
DECORATIVE MODE	No. of Patterns [& Motifs? Syntaxes?]			DECORATIVE MODE	No. of Patterns [& Motifs? Syntaxes?]					
	C (N)	O (N)	R (N)		C	O	R			
Painted: dark-on-light				Painted: dark-on-light						
Painted: light-on-dark				Painted: light-on-dark						
Painted: polychrome				Painted: polychrome						
Incised				Incised						
Impressed				Impressed						
Plastic: banding				Plastic: banding						
Plastic: pellets				Plastic: pellets						
Plastic: barnacle				Plastic: barnacle						
[Particular combinations of the above]				[Particular combinations of the above]						

Table 2: A sample quantification of ceramic change from one phase to the next on the basis of pattern changes within individual decorative modes.

ternatively for combinations of categories that may for whatever reason be considered meaningful in aggregate. In practice, such rates would best be expressed as ranges, since both the numbers of changes and the durations of the phases in decades are themselves ranges rather than precisely determinable quantities. Given the present state of our knowledge, we can already produce figures for some sub-regions, and even some individual sites, that may be considered quite reliable as well as narrow in their ranges. Of course, at the other end of the spectrum we will either not be able to provide a figure at all, or else its range will be so absurdly wide or unreliable as to be of little or no value.

Having gone to all this trouble to come up with a number of total changes, as well as a rate of change based upon the length of the period we are focusing on, what do we have that we weren't already using and how will it be useful for the purposes of explaining change in material culture? The roughly quantitative measures of cultural change I am proposing, though specific to a particular artifactual category (pottery, in the case of the Aegean Bronze Age), are sensitive to several different activities (including consumption of food and drink, food preparation, storage, and bulk transport of commodities).[4] The range of ceramic variables included in these calculations is flexible, so that both the extent and rate of change can be targeted at as many or as few of such aspects as vessel shape, size, decoration, or mode of production as a researcher cares to investigate. Interpretation of the variability in the particular metrics chosen may involve very different explanatory frameworks. For example, substantial changes in the decorative motifs decorating open tableware shapes would presumably involve factors that might not come into play at all in a discussion of a significant shift in cooking shapes or an abrupt change in the capacities, but not the basic forms, of pithoi. But by insisting on the inclusion of effectively all functional classes of ceramic containers, we will necessarily be adopting a holistic approach to identifying, quantifying, and seeking to explain the nature, degree, and pace of change across the full spectrum of a culture's ceramics within a given period of time. If, by contrast, we were to focus on just one or two functional classes at the expense of the rest, the range of issues we would be able to address would obviously be restricted in such a way as to impoverish and render less informative our investigation of ceramic change on virtually every level.

As greater progress in the establishment of absolute dates for specific ceramic phases is made, the ranges of the rates of change we are calculating should

---

[4] The American excavators of Troy devised a system of ceramic classification that allowed them to comment on the numbers of continuing as opposed to new shapes in four functionally differentiated categories from throughout the stratified sequence at that site, although the duration of the settlement represented by each major stratum was not taken into account in the assessment of ceramic change from one to the next: Blegen, Caskey, and Rawson 1950: 56–76 and figs. 129–132, esp. 57 table 6 and fig. 223 [Troy I]; 224–241, esp. 225 table 12, fig. 370 [Troy II]; Blegen, Caskey, and Rawson 1951: 22–34, 23 table 7, fig. 59 [Troy III]; 122–136, 123 table 14, fig. 154 [Troy IV]; 237–249, 238 table 21, fig. 238 [Troy V]; Blegen, Caskey, and Rawson 1953: 39–76, figs. 292–295 [Troy VI]; Blegen, Boulter, Caskey, and Rawson 1958: 25–44, 20 table 9, figs. 214–217 [Troy VIIa]; 159–176, 156–157 table 18, fig. 218 [Troy VIIb].

become narrower. And as ceramic specialists proceed to define more and more finely tuned ceramic phases, the durations of which will necessarily become progressively shorter than are the lengths of those ceramic phases currently recognized, the extents of ceramic change as well as their rates will become more precise. Episodes of more intense ceramic change will be more and more starkly highlighted and chronologically pinpointed, while eras of minor ceramic change will also become more sharply defined. Our metrics of artifactual change should therefore improve in quality, and so in utility as tools for the explanation of cultural change, as refinements in both absolute and relative chronologies continue to take place.

Fair enough, one might respond—but what do these metrics really mean? That is, how can they be used to confront questions of cultural change in any significant, much less novel way? Well, first of all, these metrics—both the percentage of change under any discrete heading as well as the pace of change under that heading—will permit ceramic phases of whatever duration to be compared in terms of a common measure of change. Thus the various stages of development within a given culture can be assigned a relative degree of and rate of change. The variability of the measures will surely call for comment and interpretation. Moreover, the degree and rapidity of change at any particular period among all the distinct cultural traditions conventionally recognized within the Aegean will presumably differ, inviting further comment and interpretation. And of course variety in the long-term pattern through time of such variability among the Aegean's constituent cultures might provoke additional investigation and explanation.

The spatial scale at which artifactual change could be evaluated might also be shifted around, with some potentially interesting results. For example, within the most thoroughly explored and published regions of the Bronze Age Aegean, one might imagine comparing degrees and rates of change at individual sites against those determined to exist for the culture as a whole, with the aim of determining if more diverse or changeable ceramic repertoires tend to be focused in a particular region or at a particular level of the site hierarchy manifested by the culture. In all of these applications of the metrics being calculated the absolute values of the metrics in question will have no meaning in themselves, but will derive any significance they may have purely from the difference or similarity they exhibit with comparably calculated figures obtained for other periods, locations, site types, or cultures. That is, these metrics would be used exclusively in a comparative fashion.

The approach to assessing ceramic change outlined here doesn't seem all that radical, I hope you will agree, so why haven't we been doing something like this routinely for a long time? One answer to this, of course, is that we have been doing similar things with checklists of changes in specific ceramic features from phase to phase for a long time.[5] But such checklists typically take only very limited portions of the entire ceramic repertoire of a phase into consideration and tend to be unsystematic in their approach in a variety of other ways as well. More comprehensive approaches to the full ceramic corpora or chronologically successive phases have not been possible until quite recently because we simply have not known these corpora in sufficient detail to be able to generate reliable numbers for more than a handful of sites, much less whole cultures. Even now, there are plenty of embarrassingly large gaps in our spatial as well as chronological coverage. In addition, our control of absolute chronology has also been only marginal—and as the still unresolved debate over the date of the Theran eruption shows, we still have a long way to go on that front as well! But we can now, in my view, begin to think of how to quantify change within an individual artifact category in an objective fashion. And I would argue that when we do this, we should do our best to include the full range of variability in that category. So in the case of pottery, we should devise an approach to quantification that targets all ceramic containers and not just those we find most often, consider most beautiful, or recognize most easily.

In closing, I may as well be candid about where and when in the Aegean Bronze Age I personally would choose to apply this attempt to quantify ceramic change. Naturally, I would want to explore a period of rapid and wide-ranging change—the transition from Early Helladic II to III on the Greek Mainland, or from Late Minoan IB to II on Crete, or the Late Bronze IIIC era throughout the Aegean. But that would leave me with a series of numbers describing great change but with no numbers describing instances of minimal change—and I'm going to need both (as well as everything in between) to get anything of real value out of this process. Even more important to recognize, of course, is that just about everything I've had to say here involves pottery first and foremost, and yet a culture's ceramic containers are only a small dimension of its material self-expression. But I will have to leave it to others more knowledgeable than myself to decide whether this notion of metrics quantifying change is genuinely applicable to other artifactual categories or aspects of material culture. Applying this idea just to ceramics is likely to be quite challenging enough!

---

[5] A recently published example of this sort of checklist approach to ceramic change applied to the LH IIIB pottery of the Greek mainland is Vitale 2006: 198, Table 1.

References

Blegen, C.W., Caskey, J.L., and Rawson, M. 1950. *Troy I: The First and Second Settlements*. Princeton University Press.

Blegen, C.W., Caskey, J.L., and Rawson, M. 1951. *Troy II: The Third, Fourth, and Fifth Settlements*. Princeton University Press.

Blegen, C.W., Caskey, J.L., and Rawson, M. 1953. *Troy III: The Sixth Settlement*. Princeton University Press.

Blegen, C.W., Boulter, C.G., Caskey, J.L., and Rawson, M. 1958. *Troy IV: The Seventh Settlement*. Princeton University Press.

Driessen, J. 2002. Towards an Archaeology of Crisis: Defining the Long-term Impact of the Bronze Age Santorini Eruption. In R. Torrence and J. Grattan (eds.), *Natural Disasters and Cultural Change*: 250–263. London/New York: Routledge.

Furumark, A. 1941. *The Mycenaean Pottery: Analysis and Classification*. Stockholm: K. Vitterhets Historie och Antikvitets Akademien.

Manning, S.W., and Sewell, D.A. 2002. Volcanoes and History: A Significant Relationship? The Case of Santorini. In In R. Torrence and J. Grattan (eds.), *Natural Disasters and Cultural Change*: 264–291. London/New York: Routledge.

Mountjoy, P.A. 1986. *Mycenaean Decorated Pottery*. A Guide to Identification. Göteborg: Paul Åströms Forlag.

Mountjoy, P.A. 1999. *Regional Mycenaean Decorated Pottery*. Rahden/Westfalen: Verlag Marie Leidorf.

Rotroff, S.I. 2006. The Introduction of the Moldmade Bowl Revisited: Tracing a Hellenistic Innovation. *Hesperia* 75: 357–378.

Rutter, J.B. 1995. *Lerna. A Preclassical Site in the Argolid III: The Pottery of Lerna IV*. Princeton University Press.

Spratt, D.A. 1982. The Analysis of Innovation Processes. *JAS* 9: 79–84.

Spratt, D.A. 1989. Innovation Theory Made Plain. In S.E. van der Leeuw and R. Torrence (eds.), *What's New? A Closer Look at the Process of Innovation*: 245–257. London: Routledge.

Torrence, R., and Grattan, J. (eds.) 2002. *Natural Disasters and Cultural Change*. London/New York: Routledge.

van der Leeuw S.E., and Torrence, R. (eds.) 1989. *What's New? A Closer Look at the Process of Innovation*. London: Routledge.

Vitale, S. 2006. The LH IIIB – LH IIIC Transition on the Mycenaean Mainland: Ceramic Phases and Terminology. *Hesperia* 75: 177–204.

Warren, P.M. 2006. The Date of the Thera Eruption in Relation to Aegean-Egyptian Interconnections and the Egyptian Historical Chronology. In E. Czerny, I. Hein, H. Hunger, D. Melman, and A. Schwab (eds.), *Timelines. Studies in Honour of Manfred Bietak*: 305–321. Louvain: Peeters.

Wiencke, M.H. 2000. *Lerna. A Preclassical Site in the Argolid IV: The Architecture, Stratification, and Pottery of Lerna III*. Princeton University Press.

Article submitted March 2007

Archaeologists and Scientists: Bridging the Credibility Gap

Elizabeth French and Kim Shelton

Abstract: *The excavation history of the site of Mycenae illustrates well the interaction over time between archaeology and archaeological science, the uneven nature of the relationship and the role played by chance and luck of discovery. Though little has been achieved so far in the fields in which Peter Kuniholm and his Cornell dendrochronology laboratory specialize, ceramic analysis has been very successful and the current excavation with its unique archaeological importance seems to offer more hope for the 21st Century.*

Introduction

It was in Ankara in the 1970s that the French family at the BIAA (British Institute of Archaeology at Ankara) became friends with the Kuniholm family at ARIT (American Research Institute in Turkey). We overlapped in many fields, not least our archaeological interests, as the BIAA was at that time working in conjunction with the British Academy Major Research Project on Early Agriculture. This included tree coring and our inauspicious expeditions to the Beynam woods (Figure 1) of which Peter reminded us at his feast. It is a pleasure to offer in honour of Peter a short paper using the site of Mycenae to which we have devoted much of our lives as an example of the interplay between archaeology and science.

The Past

Mycenae has been very variously served in this regard over the years. Schliemann used every modern method known to him, particularly photography, but including analysis of metal samples by Dr. Percy of the Royal School of Mines whose report is published in full (Schliemann 1878, 367-378). Tsountas' work has been discussed elsewhere (Shelton 2006); Wace suffered from a strictly classical education (French 2006) though he did have his skeletal material examined (Wace 1932). We do not know, however, what happened to the material after it was returned from Sweden to Greece, which is unfortunate as it is now in great demand. Only in the second half of the 20th century did scientific studies proliferate. But then we suf-

Figure 1: Members of the BIAA coring trees in the Beynam Woods.

fered from one of the major problems in this field. The archaeologist cannot foretell what he will find. During the 19 seasons that I was excavating at Mycenae we found no wood good for anything but species identification and only five samples of carbonized wood which were submitted to Pennsylvania for dating.

Unfortunately they have proved useless as evidence. First, the Lab never got back to us before publication to ascertain if there had been any change in the very preliminary context notes submitted with the samples, and thus when quoted, they are given all the wrong settings. Moreover, two of the samples are from the same context; a duplicate was taken as there was possible contamination of the original. This is carefully noted in the trench supervisor's notebook and recorded with a separate excavation sample number but the Lab gives only their own numbering and there is no cross reference to ours. (Similarly the samples taken from the Plaque of Amenhotep III have also become muddled. See Phillips 2007.) Now after years of post-excavation work I (French) can tell a great deal about exactly where each of these five samples came from and would love to have them accurately dated—but as singletons this is not possible.

We did find Linear B texts to my father's great delight and, of course, plenty of pottery. It is in the study of this that Mycenae has made its major contribution. I must confess to profound skepticism when the early OES analyses of Mycenaean Transport Stirrup Jars suggested that those full of olive oil found in the House of the Oil Merchant had been imported from Crete. Coals to Newcastle indeed! I felt that the data from the sherd collection of the Ashmolean Museum in Oxford were not really a sound basis of comparison. Luckily I was able to set up with the Department of Chemistry at the University of Manchester a program of NAA to establish a database of Mycenaean fabrics (French, et al., 1984). This was to be accompanied by parallel work in petrography (by John Riley of Southampton University) but that ran out of funding. The Manchester work under Professor Bill Newton with Vin Robinson and Alex Hoffmann was carried out partly as a series of third year and PhD projects. It has resulted in a large and widely accessible database, which now forms part of that in use by Professor Momsen at Bonn, though the Manchester program itself came to a very final end with the decommissioning of the university's research reactor. And just to re-assure scientists that archaeologists can accept unpalatable results: I have finally acknowledged that the Mycenae jars with their caps and sealings must originate in Crete though I feel that we have not yet paid enough attention to the commercial implications of such a trade.

The Present

As the field and publication director of our part of the site in recent years, I have a different problem: scientists with good ideas for potential programs who ask the impossible. Not only can an archaeologist not guarantee what he finds but also the only context he can assign to any individual piece (except in the most exceptional circumstances) is a date of deposition, not of production. Moreover, there are some things which are not acceptable to museum authorities: items of display quality may not be allowed for sampling, and any sampling that takes place must not destroy either the appearance or the evidence of the item. Thus, though it was extremely interesting, a program that required "well dated and well preserved samples of Mycenaean glass" was not viable. We do have a lovely mould (Figure 2) assumed to be for the production of gold and glass decorative plaques.

Figure 2: Mould for gold and glass jewelry from the Cult Centre at Mycenae.

Indeed some glass residue appeared to be preserved on it. This has been sampled twice, on each occasion accompanied by a report of excellent scientific value but which, by venturing into unsubstantiated archaeological interpretation, becomes risible. The key to resolving much of this is communication, what government in the UK calls "joined-up" thinking. If we discuss programs well in advance, agree to procedures (and stick to them), consider potential clashes of priorities, use correct referencing and terminologies, and, above all, look both ways before presentation or publication, then there is a future.

Figure 3: Wood *in situ* in Petsas House, Room Tau, south wall.

The Future

Indeed all may change in the 21st century. There is new material coming out of Mycenae that could potentially be of good scientific and chronological value, and the hope is that communication and cooperation will be both free-flowing and much improved. As discussed above, one area of scientific investigation at Mycenae that has been fruitful is pottery analysis. Even in this area, however, the chronological value has not been as important as the sourcing of clays, etc. Our ceramics are still stylistically dated in their own right and can only be tied down using direct correlates to scientifically derived dates. In the Petsas House excavation (*PAE*, Iakovidis 2000, 63–66; 2001, 49–55; 2002, 18–19; 2003, 23–24, and 2004, 24–26; *Ergon* 2005, 35–38; 2006, 28–31) we have plenty of pottery and a superb chance to coordinate and communicate with scientists and scientific studies. Even at the time of initial discovery (Papadimitriou and Petsas 1950, 1951), the potential for dating was recognized from the huge corpus of ceramics from late in the Late Helladic IIIA 2 period. The remarkable quantity and range of finds does indeed bode well for its use, plus we do seem to know in part what we will find. We have pots, lots of pots together with substantial evidence of international contact and exchange, and there is wood (Figure 3). The carbonized wood remains have been recovered from the timber-frame structure of the walls, from ceiling beams and most recently (2006) from an upper storey floor. There will certainly be more and our research design can and will address this recovery agenda.

There remain, though, a number of issues that will cause bumps in the cooperative road ahead. The excavation and sampling of the wood continues to be difficult for most field archaeologists and specialist presence is desired, especially when one can predict a significant amount of material. Another bump, more likely a hurdle, is the gap or gulf in dating represented by the nature of the evidence. This wood, if it can be dated closely through dendrochronology and/or ^{14}C dating, will tell us the construction date of Petsas House, while the pottery was deposited during the destruction of the building. These two events, I can already tell you from stylistic and stratigraphic sherd evidence, occurred as much as 100 years apart.

One of the most exciting aspects of Petsas House and its pots is the correlation of the material stylistically and chronologically with the Mycenaean material from Tell el Amarna (Hankey 1997) and that from the Uluburun shipwreck (Rutter pers. com.), especially the stirrup jars (Figure 4). The three contexts make up a chronological horizon that provides evidence for contemporary contact and trade. Both Amarna and Uluburun have been dated scientifically,

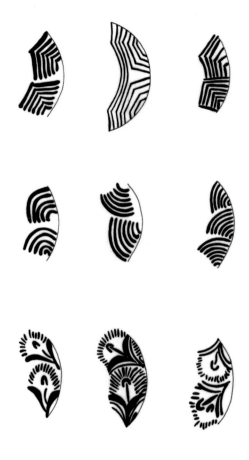

Figure 4: The patterns in the shoulder zone of Stirrup Jars of LH IIIA2: Tell el Amarna, Ulu Burun, Petsas House.

and as presented in this conference the dates fit very well into the more "traditional dating" of LH IIIA 2 in the later 14th century BCE. [For Uluburun dating, see Manning, Pulak et al., this volume. The Amarna data will be published elsewhere—*eds.*]

Ultimately though, we can be confident in our ceramic horizon, regardless of where exactly the scientists tie it down, as long as all three are tied down together. It may be here that the fundamental difference between archaeologist and physical scientist lies: it is for the linkages that we will go to the stake. If they are to be questioned, the implications must be fully considered and understood by all.

References

French, E.B., Newton, G.W.A, Robinson, V.J., and Scourtelli, A. 1984. Provenance of LH IIIB Pottery from the Argolid. *Ancient Greek and Related Pottery*, Allard Pierson Series, Vol. 5: 12–15. Amsterdam: Allard Pierson Museum.

French, E.B. 2006. Changing Aims and Methods in Archaeology during the last 100 years: A Family Viewpoint. In

P. Darque, M. Fotiadis, and O. Polychronopoulou (eds.), *Mythos: La Préhistoire égéenne du XIXe au XXIe siècle après J.C*: 259–265. BCH Supplément 46, Paris.

Hankey, V. 1997. Aegean Pottery at El-Amarna: Shapes and Decorative Motifs. In J. S. Phillips, L. Bell, and B. B. Williams (eds.), *Studies In Honor of Martha Rhoads Bell*: 193–217. San Antonio TX: Van Siclen Press.

Iakovidis, S. 2000–2004. Anaskaphai en Mykenais. *Praktika tes en Athenais Archaiologikes Etaireias*.

Papadimitriou, I., and Petsas, P. 1950. Anaskaphai en Mykenais. *PAE* 203–233.

Papadimitriou, I., and Petsas, P. 1951. Anaskaphai en Mykenais, *PAE* 192–196 and pl. III.

Phillips, J.S. 2007. The Amenhotep III 'Plaques' from Mycenae: comparison, contrast and a question of chronology. In M. Bietak and E. Czerny (eds.), *The Synchronisation of Civilisations in the Eastern Mediterranean in the Second Millennium B.C. III, Proceedings of the SCIEM 2000–2nd EuroConference, Vienna, 28th of May–1st of June 2003*: 479–493. Vienna: Verlag der Österreichischen Akademie der Wissenschaften.

Schliemann, H. 1878. *Mycenae; A narrative of researches and discoveries at Mycenae and Tiryns*. London: J. Murray.

Shelton, K.S. 2006. The Long Lasting Effect of Tsountas on the Study of Mycenae. In P. Darque, M. Fotiadis, and O. Polychronopoulou (eds.), *Mythos: La Préhistoire égéenne du XIXe au XXIe siècle après J.C*: 159–164. BCH Supplément 46, Paris.

Wace, A.J.B. 1932. Chamber Tombs at Mycenae. *Archaeologia* 82 Oxford.

Article submitted January 2007

Central Lydia Archaeological Survey: Documenting the Prehistoric through Iron Age periods

Christina Luke and Christopher H. Roosevelt

Abstract: *The Central Lydia Archaeological Survey (CLAS) is a regional survey project focused on a ca. 350 square kilometer area surrounding the Gygaean Lake (modern Marmara Gölü) in the Hermos (modern Gediz) River valley of central western Turkey. Following an overview of the paleoenvironmental and archaeological approaches of the project, this paper presents some preliminary archaeological results relating to early prehistoric through Iron Age times. Stone tools of the Paleolithic period represent the earliest remains of human activity in the region. Although Neolithic sites are known in immediately neighboring areas, the earliest sites of permanent occupation in central Lydia date to the Late Chalcolithic and Early Bronze Age, when links to east and west are reflected in material cultural assemblages. During the Middle and Late Bronze Age, rapidly increasing socio-political complexity is marked by the development of a network of fortified sites, several of which are associated with unfortified lower settlements and bear remains indicative of broad Aegean (Mycenaean) and Anatolian (Hittite) interaction, and one of which must have been a regional capital, owing to its large size and complexity. While the fortified sites were abandoned at the end of the Bronze Age, perhaps after a fiery conflagration at a few, some of the unfortified settlements continued to be occupied into the Iron Age. Along with the continued use of the area for occupation and subsistence activities in this time, central Lydian landscapes came to be dominated by the monumental tumuli of Lydian kings, and later of other elites, based at Sardis, the Iron Age capital of the region.*

Introduction

The Central Lydia Archaeological Survey (or CLAS) began in 2005 and is scheduled to continue at least through a fifth season in 2009. The ultimate goal of CLAS is to record accurately the locations of sites of cultural activity in the area of central Lydia in western Turkey surrounding Marmara Gölü, the ancient Gygaean Lake (Figure 1), the immediate hinterland of Sardis, which was capital of the kingdom and later empire of Lydia in the seventh and sixth centuries BCE, and later still the capital of an Achaemenid Persian province. Within this territory of lacustrine and alluvial flood plains, rolling hills, and rugged, upland foothills the project aims to understand better the multi-purpose use of the landscape for settlement, burial, and resource exploitation from prehistoric to modern times and shifts in such use in response to environmental, socio-political, economic, and/or other notable changes.

Our approach is multipronged, incorporating programs of both paleoenvironmental and archaeological research of varying strategies along with consultation of what textual sources survive. In particular, we are interested in assessing the possible interplay between cultural and environmental dynamics through time and their integral relationship to the rise and fall of settlements. Also, our broad project framework falls within the developing field of landscape archaeology, in which both tangible and intangible landscapes are explored and in which relationships between the land and how humans interact with and on it are seen as tightly integrated. We see the tangible landscape as providing the stage on which human decision-making plays out, based sometimes on environmental factors and sometimes on pre-existing palimpsests of meaning associated with particular places and particular people. Thus while we include a paleoenvironmental research focus, we understand cultural dynamics

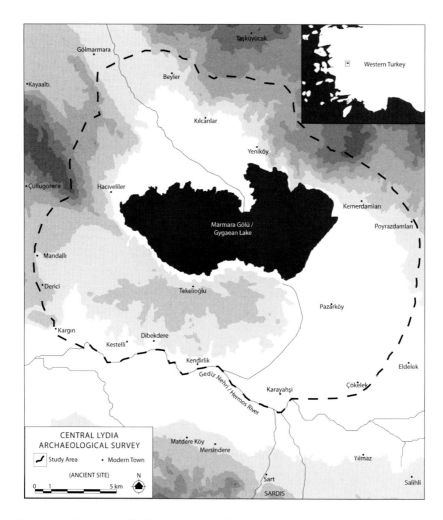

Figure 1: Map of the CLAS study area, with inset showing location in western Turkey.

within landscape archaeological frameworks that move beyond environmental determinism.

The purpose of this article is to introduce the project after its first few field seasons and to present in summary form its history, its approaches to studying the central Lydian past, and some preliminary results and interpretations. As is clear from the title, our chronological focus for the purposes of this article spans early prehistory through the Iron Age, and our interpretations are based primarily on data collected during the 2005–2007 seasons.

The History of CLAS

The Central Lydia Archaeological Survey was designed subsequent to a research project conducted in 2000–2001 on greater Lydia, an inland region of western Anatolia that has almost always served as a crossroads of corridors of communication and trade that connected the Aegean with the Anatolian plateau (Roosevelt 2006b; Roosevelt and Luke 2006a). For that work we conducted a targeted survey of tumuli, or burial mounds, of the Lydian and Achaemenid periods and worked with provenienced museum collections and archives to identify locations of settlement in greater Lydia, suggesting also what types of human–environment interaction were perhaps most important in these areas with respect to resource procurement, communications, and regional control. This was a very extensive study that begged for a follow-up focusing more intensively on a smaller subset of Lydia.

In 2004 we chose the area of central Lydia surrounding the Gygaean Lake for this more intensive study owing partly to its location within the immediate hinterland of Sardis and partly to its inclusion of Bin Tepe, Turkish for "The Thousand Mounds," a vast tumulus cemetery commonly thought to be the royal cemetery of Sardian kings and located between the Gygaean Lake and the Gediz, or ancient Hermos, River (Figure 1). In addition, local archaeological sequences stretching back to the Late Bronze Age, around 1400 BCE, were well known from Sardis, which had been the subject of continuous research and excavation for nearly 50 years (Hanfmann 1983; Greenewalt and Rautman 2000). Paradoxically little archaeological work had focused on periods preceding Sardis's rise to imperial status; and equally little work had focused on the area around the Gygaean Lake despite some 40 years of knowledge of its early prehistoric importance in the region derived from short-term projects associated with the Sardis Expedition and the Manisa Museum (e.g., Hanfmann 1967, 1968; Mitten and Yüğrüm 1968, 1971, 1974). Such work had demonstrated the presence of Early Bronze Age through Roman activities in the area in addition to the more prominent Lydian- and Achaemenid-period tumuli.

Central Lydia piqued our interest also because of its rich geological and historical record of environmental change: in addition to the formation of the Gygaean Lake itself some time within human memory—between 6000 and 3000 BCE according to a recent estimate (Hakyemez, Erkal, and Göktaş 1999)—several culture-changing droughts are documented in historical sources. These changes probably had profound and lasting effects on local populations. Thus over a contemporary time period, the study area offers the potential to illuminate a complex history of environmental dynamics alongside a rich and long archaeological record of human activities reflective of waxing and waning socio-political complexity and economic strategies ranging from the adoption of sedentary and agricultural ways of life, to the rise of urbanism, the emergence of empire, and intermittent periods of societal collapse, each marked by varyingly differentiated sites and settlement patterns and changes in artifact assemblages.

Paleoenvironmental Approaches and Preliminary Results

Beginning with our paleoenvironmental research program, the very definition of the study area makes clear our emphasis on and interest in the Gygaean Lake. The approximately 350 sq km study area is defined by a cost-distance from the lakeshore derived using Geographic Information System (GIS) tools that establish an equal contour of energy expenditure for one moving away from the lake—roughly equivalent to a 3–4 hour walk in any one direction from the lakeshore. Our underlying assumption in establishing the study area in this way is that the lake and its surrounding landscapes must have provided abundant resources in antiquity just as they do today: aquatic life including carp, pikeperch, crayfish, and eel exploited for regional consumption; abundant reeds employed in local production of various forms of woven matting; arable soils cultivated along seasonally receding lake margins; muds and clays for architectural and ceramic use; the water itself used for various purposes; and probable sacred values associated with this large inland water body (Curtius 1853; Robert 1982; Munn 2006). The Hermos River, too, would have provided a number of advantages to local populations: a partially navigable river system through the large Hermos River plain, fish, fowl, and reeds, and seasonal flooding, again, that replenished resources for agricultural purposes. The immediate alluvial plain between Sardis and Bin Tepe and the shores of the Gygaean Lake were uninhabitable marshland most of the year, prior to the construction of dams on the Gygaean Lake (completed in 1944), on the Hermos River (completed in 1960), and on two tributaries of the Alaşehir (or ancient Kogamos) River (completed in 1967 and 1977). Thus, local inhabitants of the early modern periods and earlier into prehistory would necessarily have been focused on both the benefits and dangers associated with these water systems, including severe flooding and drought resulting from climatic extremes, and pestilence associated with perennially marshy wetlands.

Local myths and pseudo-historical records indicate the occurrence of at least three culture-changing droughts in the history of central Lydia, one causing the migratory exodus of half of the local population (Herodotus 1.94; Strabo 1.3.4; Nicolas of Damascus (Jacoby 1923: 90 F 49)). Local environmental data is needed to corroborate these historical sources and to determine any effects on landscapes around the Gygaean Lake. Modern reports of seasonal lake-level fluctuations on the order of 3–5m prior to the installation of the Gygaean Lake dam in 1944 attest the volatility of lake levels, and the lake has been completely desiccated by regional droughts as recently as 1986 and again in 1994. The effects of a six-month drought were readily apparent between just May and September of 2007 (Figure 2).

Documentation of long-term fluctuations in lake levels associated with drought might clarify the extent and timing of drought and famine spread across the eastern Mediterranean posited to have occurred at the ends of the Early and Late Bronze Ages (Nüzhet Dalfes, Kukla, and Weiss 1997; Weiss 1982; Neumann and Parpola 1987; Shrimpton 1987). The proposed severity of such droughts disrupted long-distance exchange and contact, and is said to have contributed to societal collapse resulting from, among other things, resource depletion and its associated effects. Evaluation of environmental data from central Lydia can contribute to these larger questions in the eastern Mediterranean and elsewhere about the relationship between social and environmental trajectories, and it

Figure 2: Comparison of Gygaean Lake levels from May to September 2007 (September view courtesy K.C. Cooney).

will also eventually provide information about the stability of local subsistence resources.

The Gygaean Lake is the largest and the historically best-known lake in Lydia. It is underlain and bounded by formations of the Menderes Massif that consist of gneisses, schists, slates, quartzites, marbles, and limestones that underwent varying degrees of metamorphism some time in Precambrian or late Paleozoic times (Schuiling 1962; Brinkmann 1971: 172–73; Brinkman 1976). Extensive Neogene faulting (beginning sometime in the Miocene or Pliocene epoch) produced the fault-dropped half-graben valley in which the lake is situated today, bounded by residually high ridges and foothills (Brinkmann 1971: 171–90; Sullivan 1989: 28–41; Yusufoğlu 1996: 12–13). Alluvial deposition covered the half-graben as it subsided into the Pleistocene epoch of the Quaternary period, and recent research suggests that continued southward rotation of the Gediz half-graben in the Middle to Late Holocene epoch resulted in massive sedimentation of the Hermos River. These events caused a lake to form in a previously alluvial valley north of the modern ridge of Bin Tepe, achieving the present form of the Gygaean Lake sometime before the historical period, between 6000 and 3000 BCE (Hakyemez, Erkal, and Göktaş 1999: 550). The transformation of the central Lydian landscape from an alluvial valley to a lake basin must have had profound effects on local populations, providing opportunities for new economic strategies at the household level, including the exploitation of the vast resources of the lake, and may have played a significant role in a broader issue of prehistoric human development: the spread of sedentary and agricultural lifestyles from the Near East to Europe. Western Anatolia is commonly viewed as the land bridge over which agricultural peoples passed to Europe (Lichter 2005a, 2005b), yet more research in western Anatolia is needed to support this idea. The Gygaean Lake is the largest lake between the Lakes District, to the southeast, and the Aegean and Marmara Seas, to the west and north, respectively—all areas of known Neolithic activity (Özdoğan and Başgelen 1999). If our current understandings of local settlement history (see below) and the time frame of lake formation are correct, then lake formation and the first long-term occupation were roughly contemporary. The precise dating of this environmental transformation is thus of great significance, and it may be related also to wider patterns of environmental and landscape dynamics in Anatolia and the Near East.

In 2006 questions relating to the formation and long-term stability or volatility of the lake were addressed with programs of lake-sediment coring and bathymetric (or underwater topography) survey. Both activities focused on the western part of the lake and produced a good amount of preliminary data to be supplemented with that from ongoing geomorphological work. In addition to a rough bathymetric map, the transect of cores revealed a clear sequence of lake-sediment stratigraphy that provides tantalizing new evidence for a climatic correlation to cultural developments during specific periods. Of special interest was a long enduring desiccation event that began after the first half of the fifth century CE, according to two independent AMS radiocarbon analyses (Besonen and Roosevelt, 2008).

Already in 1987 Marcus Rautman cautiously attributed to climate change shifts in subsistence practices at Sardis beginning in the fifth century CE. In particular, he suggested that decreases in water supply, evidenced by shifts in water-procurement systems at Sardis from pressure-fed aqueducts to localized cisterns and household tanks, and declines in long-term cash crops, evidenced by a nearby pollen core, may

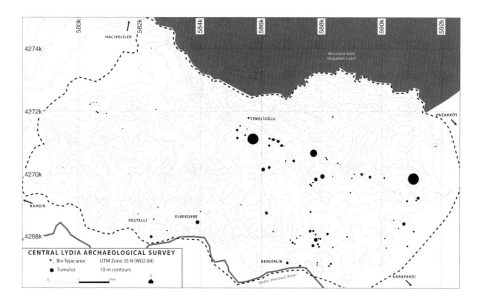

Figure 3: Map of tumuli in Bin Tepe.

have been the results of climate change (Rautman 1987). Our extensive survey efforts in central Lydia from 2005–2007 (see below) have revealed many small-scale sites of Late Roman or Early Byzantine date probably reflective of agricultural activities, but the exact nature of those activities and their intensity relative to earlier and later periods remains poorly understood.

For the meantime, then, it seems we have tantalizing new evidence for a climatic factor in cultural developments at Sardis and beyond in the Late Antique period, yet further correlation, of course, must await continuing paleoenvironmental research. Additionally, we must await further results for refining our interpretations about the formation and earlier history of the lake, which remain admittedly speculative at this point.

Archaeological Approaches

Moving on to our archaeological research program, we present below our approaches in order of our varying research strategies: work related to tumuli in Bin Tepe; extensive field survey and aerial reconnaissance; and intensive site survey.

Tumulus-Related Research

As already stated, the area located between the Gygaean Lake and the Hermos River bears the local name of Bin Tepe because of the many tumuli that punctuate its otherwise gently rolling landscape. The largest of the tumuli are associated with Lydian kings of the sixth century BCE based on both ancient testimony and their monumentality: as a frame of reference, of the famous three Egyptian pyramids on the Giza Plateau, only the Great Pyramid of Khufu is larger by volume than the tumulus of Alyattes.[1] Yet, in addition to the few large tumuli, many more, smaller tumuli dating to the Achaemenid period populate the landscape. Just how many tumuli there were in Bin Tepe, however, was unclear until recently. Previous non-systematic exploration in Bin Tepe in the eighteenth through twentieth centuries resulted in the excavation and study of a few tumuli, but the total tumulus count ranged from between 60 or so, up to around 130 (see von Olfers 1858; Choisy 1876; Sayce 1880; Butler 1922).

In 2005 CLAS conducted a systematic and targeted survey of these monuments using hand-held GPS receivers and 1:10,000 scale field sheets (see Roosevelt and Luke 2006b). Using standardized forms, digital cameras, and traditional measuring instruments, we documented the location, diameter, height, and condition of each tumulus, and noted anything of special interest, including locations and dimensions of looters' tunnels, intervisibility, and the accessibility and composition of chamber tomb complexes, among other details. As of 2005, a total of 116 tumuli were still standing in the area, and at that time we knew of 3 more tumuli that had been completely destroyed since their earlier publication (Figure 3). In addition

---

[1] If the volume of a cone equals one third base area times height (or $1/3(\pi r^2)(H)$), the volume of the tumulus of Alyattes (D = c. 361m; H = c. 70m) is roughly 2,388,257 m^3. If the volume of a square pyramid equals one third base area times height (or $1/3(L^2)(H)$), the volume of the Pyramid of Khufu (L = c. 230m; H = c. 148m) is roughly 2,609,733 m^3 and the volume of the Pyramid of Khafre (L = c. 215m; H = c. 144m) is roughly 2,218,800 m^3.

to exterior features, the interior tomb complexes of several tumuli were newly documented in 2005 and 2006 as well (Figure 4).

Owing to the severe degree of tumulus looting in the area, in 2006 we initiated a program of monitoring or reassessing the conditions of previously documented tumuli as part of a broader program working with local communities and law enforcement to help develop programs of cultural heritage protection. Each year we have revisited between 17 and 20% of the tumuli in Bin Tepe (Kersel, Luke, Roosevelt 2008). Each year roughly half of these tumuli show evidence of recent looting and destruction since the previous year, showing the current and ongoing nature of the problem (Roosevelt and Luke 2008a and forthcoming). In addition to on-the-ground monitoring, we have begun to analyze aerial photographs from 1949, 1994, and 1995, and a QuickBird satellite image from 2006 to attempt to understand better the current and historical conditions of tumulus landscapes in the area (Özgüner and Roosevelt 2008). Our preliminary work on this front suggests that the pace of tumulus destruction has quickened only recently (Figure 5).

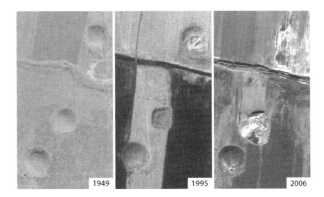

Figure 5: Aerial photographic sequence showing the conditions of tumuli in Bin Tepe in 1949, 1995, and 2006.

Figure 4: View of chamber tomb complex in the BT05.58 tumulus, with architect C.A. Wait.

Extensive Field Survey and Aerial Reconnaissance

Before CLAS began in 2005 our understanding of diachronic settlement patterns was limited to 11 multicomponent sites with dates ranging from the Early Bronze Age through Roman times. Evidence for these sites derived from previous research and, primarily, from the work of the Sardis Expedition, which conducted excavations and non-systematic survey in the area in the 1960s. We now know of over 200 areas of ancient cultural activity including sites of occupational, funerary, and other uses, as well as areas of "off-site" activities (Roosevelt 2007; Roosevelt and Luke 2008a). Such activities range in date from the Paleolithic period to Ottoman or early modern times, with notable peaks of settlement and prosperity in the Middle–Late Bronze Age, the Iron Age, and then again in the Medieval period and later. Our strategies for recovering this range of evidence point to our evolving, reflexive approach that has incorporated multiple survey methods dictated by varying topography, ground cover, and land use and that has intended to maximize coverage in as systematic a fashion as possible.

In the rolling hills of Bin Tepe and in the lacustrine flood plain that are both of intensive agricultural use today, we adopted a method we call "single-track survey" in which survey units, or SUs, were defined by modern fields and were somewhat randomly selected according to which fields had been recently plowed and/or provided good surface visibility. The name derives from the method of one person walking a "single track" through each survey unit, recording his or her path with a hand-held GPS unit, observing an approximately 2m wide swath on either side of the path, counting all encountered cultural artifacts, and collecting all lithics, feature sherds, and other diagnostics, including those presenting new ceramic fabrics (Roosevelt 2007). We do not claim anything approaching 100% coverage or even a stratified probabilistic sample, yet we have been pleased with the preliminary results of the method.

We adopted a second method of walking tracks primarily along ridges for the more diffuse and rugged terrain of the western and northern foothills of the study area. Survey units here were most usually dictated by a combination of topography and in places dense ground cover, and this was certainly less systematic, with obvious gaps in coverage owing to these factors in addition to restricted access to land (Roo-

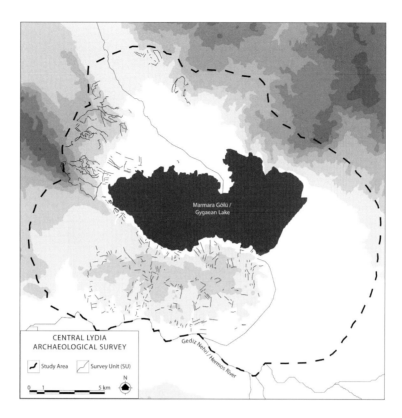

Figure 6: Map of "single-track" survey units (SUs) walked by CLAS in 2005 and 2006.

sevelt and Luke 2008a). In total, we have surveyed 450 survey units to date (Figure 6).

Subsequent to the documentation of survey units, we identified also what we call Points of Interest, or POIs, which usually correspond to some more identifiable type of archaeological site: settlement areas, cemeteries, quarries, and wells, for example. Settlement areas were identified as POIs during field survey and also upon later site revisitation and/or upon analysis of associated finds. In total, we have recorded 73 POIs to date (Roosevelt 2007; Roosevelt and Luke 2008a). An additional method of identifying POIs is aerial reconnaissance using a 2006 QuickBird satellite image with subsequent ground-truthing or targeted investigation during the survey season. This method has recovered both ancient sites, identified by clear evidence of structures or broader landscape transformations, as well as many abandoned early modern villages to be the subject of future studies.

Intensive Site Survey

The final survey approach to be mentioned here is intensive site survey. Two sites were subjected to intensive survey efforts in 2007, and others will receive similar treatment in later seasons. As with our other modes of survey, our intensive site survey methods are multiple in order to maximize the recovery of information using non-invasive techniques. We have conducted what we call "microtopography survey" to record archaeological remains expressed on the surface by slight and often almost imperceptible changes in elevation using a highly precise Real-Time Kinematic (RTK) GPS system. We have also employed geophysical survey methods to identify subsurface remains revealed by slight variations in their geophysical properties. In addition, we have conducted more traditional survey methods including systematic find-density counts and collections within grids laid out over the surface of sites.

Preliminary Archaeological Results

Data collected from 2005–2007 illustrate the rich history of central Lydia. While CLAS has a holistic and diachronic focus, the initial phases of analysis presented here concentrate on understanding the reasons why people originally settled in the region and the processes leading to the rise of social complexity. The imperial rise of Sardis speaks to the known political strength of the region in the Iron Age. The imperial realm of Lydia stretched far beyond Sardis, while the early center of power probably included only the immediate rural hinterland of the city. Our study area in central Lydia lies on the cusp of the overlapping territorial spheres associated with the Lydian Kingdom and later Empire, and the use of its landscapes by Iron Age Lydians is now well attested. Aquatic life and other resources associated with the Gygaean Lake were most likely of critical importance to Sardians, and the revered Lydian cavalry may even have pastured horses in the open and rolling hills of Bin Tepe. The monumental tumuli of high-status and royal Lydians in Bin Tepe most conspicuously reflect the Iron Age importance of the region that was likely tied up in its sacred significance. Yet, why and how this landscape became important for Iron Age populations needs careful consideration. The results from CLAS show that Bin Tepe and central Lydia in general were not restricted to funerary functions, with a number of contemporary and later settlements seemingly nestled in the shadows of funerary monuments.

The data for Neolithic occupation in western Anatolia is proving to be very rich. Recent and ongoing work at Limantepe (Erkanal and Erkanal 1983), Ulucak Höyük (Çilingiroğlu et al. 2004), and Çukuriçi Höyük (Horejs 2008), for example, show a vibrant and bustling population, interacting with other areas in Anatolia. Furthermore, discoveries in the Akhisar plain, in the eastern end of the Hermos River valley, and in the Alaşehir River valley indicate the presence of a number of settlements (e.g., French 1969; Meriç 1993). Many of these same sites continue to be occupied into the Bronze Age, alongside new settlements, with those along the immediate Aegean littoral eventually becoming part of the greater Minoan and Mycenaean spheres in the second millennium BCE. Furthermore, evidence shows connections with interior Anatolia, particularly political relationships with the Hittite imperial regime. The contemporary sites of Miletus, Bademgediği Tepe (perhaps Hittite Puranda), Troy, and Beycesultan all show varying degrees of Hittite and Mycenaean contacts, if not much more direct relationships (e.g., Niemeier 1998; Meriç and Mountjoy 2001 and 2002; Meriç 2003). How the data from central Lydia integrates with the Early through Late Bronze Age developments so well known from other areas of western and central Anatolia, and how these developments might shed light on why and how Sardis became an imperial power in the Iron Age are primary areas of continuing research.

Early Prehistory: Paleolithic through Early Bronze Age

Preliminary analyses of lithics collected from three areas suggest that a human or early human presence in the environs of the Gygaean Lake dates at least as early as Middle Paleolithic times (c. 100,000–40,000 years ago), with scattered finds also suggesting Mesolithic activities (c. 12,000-9,000 years ago) (Figure 7). Remains of both these periods are known from Bozyer (POI05.43), which appears to be the earliest and longest enduring site of activity in the study area, with abundant remains of the later periods, as well. A few stray finds indicate also Late Neolithic activities in the area, yet still no sedentary sites have been identified (Cooney 2008). If Neolithic villages were situated near seasonally flooded, marshy areas along the edges of the incipient Gygaean Lake, as we would expect based on Neolithic settlement locations elsewhere in Anatolia, they would lie concealed beneath the waters of the lake today.

Widespread occupation and rural activities in central Lydia are witnessed first in the Early Bronze Age (late fourth or early third millennium BCE; see Mitten and Yüğrüm 1968, 1971, 1974), with both tell and flat sites situated around the shores of the Gygaean Lake, suggesting its formation by this time, as well as along the northern margin of the Hermos River valley. The preliminary survey work of CLAS and ceramic analyses conducted by Daniel Pullen help identify at least nine Early Bronze Age settlements, with seven located in the low, rolling hills of Bin Tepe and two to the north and west of the lake; one of these sites, Bozyer (POI05.43) may even bear evidence of Late Chalcolithic activities. From these extensive survey data, it appears that Early Bronze Age settlements were located in agriculturally viable areas and along the lakeshore itself. These sites probably represent towns or very small centers, with intensifying land-use indicated by modest numbers of off-site scatters. In this regard they resemble contemporary Early Bronze Age sites in the Aegean (Davis 2001; Rutter 2001; Watrous 2001) and other areas of central and western Anatolia (Yakar 1999; Efe 2003a, 2003b), yet there is no evidence of the pronounced urbanization of other sites marked by public buildings and monumental fortifications (e.g., Warner 1994; Lochner 2001; Tartaron, Pullen, and Noller 2006). The number of EBA sites identified by CLAS is still too small to establish size hierarchies with any certainty, yet preliminary results from the Instrumental Neutron Activation Analysis (INAA) of Early Bronze Age ceramics from a number of sites indicate that production was decentralized (Boulanger et al. 2008). Craft production tends to be a good indicator of site hierarchy, and the decentralized modes of production suggested by INAA should indicate that we are dealing with small hamlets, rather than with major centers that would have controlled centralized production (compare Efe 2003b).

Connections between distant regions in Anatolia began to flourish in this time, as evidenced by what some refer to as the "Anatolian Trade Network" in the EB II period (Şahoğlu 2005) and the "Great Caravan

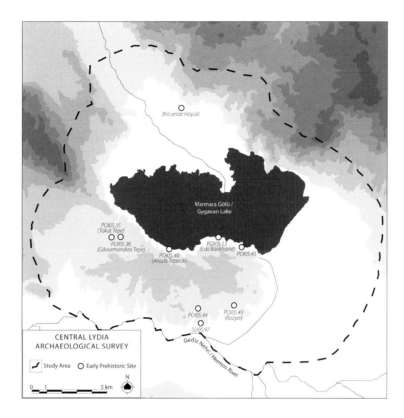

Figure 7: Map of early prehistoric (Paleolithic through Early Bronze Age) sites in central Lydia.

Route" in the EB III period (Efe 2007)—exchange routes that connected sites on the Aegean littoral, such as Limantepe and Baklatepe, to those on the interior plateau, such as Beycesultan and Küllüoba, and to other sites further east. Furthermore, the increasing centralization of western Anatolian EBA sites such as at Limantepe, Aphrodisias, and Beycesultan (Çevik 2007) suggest the increasing complexity of social and political spheres of interaction located just east and west of central Lydia. From the geographical perspective alone Early Bronze Age settlement in central Lydia can be expected: the area sits in between these two expanding cultural spheres. From the local perspective, the region offered nice places in which to live. If the Gygaean Lake formed sometime between 6,000 and 3,000 BCE, as discussed above, then the period of dramatic landscape transformation appears to have ended just prior to the initial rise in population. The specific phases of the Early Bronze Age to which the sites in central Lydia date, however, remains unclear. The lack of characteristic EB II wares coupled with the evidence for decentralized production and hamlet-sized settlements may suggest that these sites date to the EB I phase; yet chronological clarity awaits further analysis. The location of Early Bronze Age settlements along the lakeshore, at any rate—specifically at Ahlatlı Tepecik and Eski Balıkhane—appears to corroborate the general date of lake formation and the general model of early populations choosing to settle in resource-rich areas.

Middle to Late Bronze Age

At the end of the Early Bronze Age, when many sites in Greece were abandoned (Rutter 2001) and when societal collapse characterized much of the eastern Mediterranean (Manning 1997; Nüzhet Dalfes, Kukla, and Weiss 1997), most settlements in central Lydia were also abandoned. While social and environmental factors affecting the transition from this period to the Middle and Late Bronze Ages are still unclear, preliminary survey results indicate that subsequent patterns of settlement are strikingly different from their predecessors. At some point in the second millennium BCE at least 14 sites were newly settled along the shore of the Gygaean Lake, in its foothills to the north and west, and overlooking the Hermos River valley to its south (Figure 8). All of these sites are located in places with strategic visibility or proximity to communication routes, and at least four occupy upland areas of natural defensibility. These four sites also bear architectural remains of fortifications and large buildings set within enclosure walls. While such developments are more typical of both Early and Late Bronze Age sites in mainland and Aegean Greece (Davis 2001; Karageorghis and Morris 2001;

Site	Area inside fortification (hectares)
Kızbacı Tepesi	0.8
Gedevre Tepesi	1.2
Tınaztepe	2.6
Asartepe	4
Kaymakçı	8.6

Table 1: Areas of fortified second millennium BCE sites in central Lydia (in hectares).

	Kaymakçı	Gedevre Tepesi	Asartepe	Tınaztepe	Kızbacı Tepesi
Kaymakçı		4.5	9.5	11.3	14
Gedevre Tepesi	4.5		5.5	11.3	13.2
Asartepe	9.5	5.5		8	10
Tınaztepe	11.3	11.3	8		3
Kızbacı Tepesi	14	13.2	10	3	

Table 2: Distances between second millennium BCE sites in central Lydia (in kilometers).

Rutter 2001; Rehak and Younger 2001; Shelmerdine 2001; Simpson and Hagel 2006), they would not be out of place in Middle Bronze Age western Anatolia, judging from contemporary levels at Beycesultan and Troy (Lloyd and Mellaart 1965) and other sites further east (see Bier 1973). These sites appear to represent increasing socio-political complexity and territorial competition associated with state formation in the neighboring Hittite and Minoan and Mycenaean spheres that occurs in the last phases of the Middle Bronze Age and through Late Bronze Age. Consistent with this interpretation, off-site distributions of artifacts in central Lydia indicate continued intensification of agricultural practices; the sizes of fortified sites (ranging from 0.8 to 8.6 hectares in area) suggest a clear hierarchy; and the larger sites show interior structural differentiation reflective of increasing social differentiation. One site in particular in central Lydia is substantially larger and more complex than the others: Kaymakçı (POI06.01).

Kaymakçı covers the entire lower spur of Gür Dağ, due west of the Gygaean Lake. From this vantage point, one can see a wide expanse of the Hermos River plain to the south, entry points to the plains of Gölmarmara and Büyükbelen, to the north and west, respectively, and all other fortified second millennium BCE sites in the study area. From the peak above Kaymakçı, one can even see another strongly fortified site in northern Lydia: Şahankaya, or "Falcon Rock," probably the strategic stronghold referred to in Hittite sources as "Eagle Rock," with notable early features including a monumental, even "cyclopean," fortification wall and a somewhat later Achaemenid-period fire-altar (Roosevelt 2005; Roosevelt and Luke 2008b).

Microtopographic and geophysical survey of Kaymakçı conducted in 2007 revealed a citadel of 8.6 hectares enclosed within strong fortification walls (Figure 9). The boundaries of the unfortified areas of the site have yet to be pinpointed, but a lower settlement appears to extend down the slopes and possibly as far as the lakeshore to the east-northeast, and settlement and cemeteries extend down slope to the west-southwest, as well (Figure 10). The fortifications and interior building complexity, so apparent in the QuickBird satellite image, suggest that it was planned and also the ability to mobilize the large labor forces that must have been necessary to modify the landscape in so dramatic a fashion. Inside the walls numerous building foundations are readily apparent, and most likely include domestic structures, storage facilities, roads, and plazas, as well as interior enclosures and gates, and possible megaron-like structures. At 8.6 hectares the fortified area is larger than that of contemporary Troy, Mycenae, Gla, and Beycesultan, yet certainly not as large as Hattuşa nor other major sites to the east (Figure 11). At present, then, Kaymakçı represents the largest known Middle to Late Bronze Age site in western Anatolia.

From Kaymakçı one can see three or four other contemporary ridge- and hilltop sites: Gedevre Tepesi (POI07.01), Asartepe (POI06.24), Kızbacı Tepesi (POI07.04), and possibly Tınaztepe (see Figure 8). While these are much smaller in size than Kaymakçı, each ranging between one and four hectares, all are fortified and thus reveal the presence of a network of fortified sites ringing the Gygaean Lake (see Table 1). Site location appears to have been driven by several factors. All of these sites are situated on prominent ridges, all can be seen from Kaymakçı, and all are located at intervals of three to eight kilometers (see Table 2). While no materials have yet been collected from Tınaztepe to prove its Bronze Age history (as of 2007), its defensible location on a ridge

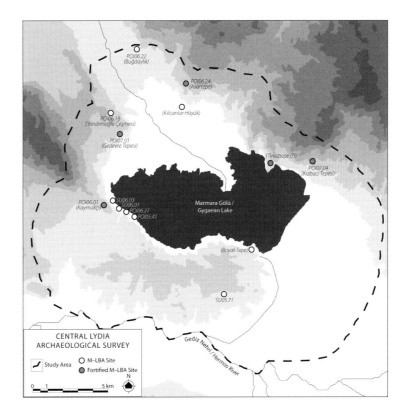

Figure 8: Map of Middle to Late Bronze Age sites in central Lydia.

extending from the foothills into the Gygaean Lake, its overt landscape transformations suggestive of fortifications, and its placement providing a visual link between Asartepe, to the west, and Kızbacı Tepesi, to the east, all suggest its participation in this second millennium network.

At this time the dating of the second millennium sites in central Lydia is imprecise owing to the lack of previously developed ceramic typologies for the region. The best parallels for our material are found in publications of material from Beycesultan to the east (Lloyd and Mellaart 1965; Mellaart and Murray 1995), Panaztepe to the west (Günel 1999), and from various other sites in western Anatolia and along the Aegean littoral (e.g., Troy, Larisa, and Bayraklı; Bayne 2000). The material suggests that our second millennium citadels should all date to the later phases of the Middle Bronze Age, down into the fourteenth or thirteenth century of the Late Bronze Age. Such dates are established from broad stylistic similarities with ceramics, including bowls with ring-foot bases and basket handles that bear deep incisions at their attachment points ("slit handles"), present in both plain red, "gold wash," and gray ware varieties common to western Anatolian ceramic traditions; and painted wares in limited numbers that suggest ties to Aegean, and specifically Mycenaean, spheres (see Mountjoy 1998). We cannot yet, unfortunately, suggest a chronological seriation of the fortified sites, yet we can say with a good degree of certainty that occupation at all of these sites lasted into the Late Bronze Age, and then ceased by its end. Fragments of burnt and vitrified mudbrick found at three of the fortified sites (Kaymakçı, Gedevre Tepesi, and Kızbacı Tepesi) suggest that their abandonment came following site conflagrations, perhaps at the hands of hostile aggressors.

Iron Age

Contrary to the general pattern of societal collapse throughout the eastern Mediterranean at the end of the Late Bronze Age (see Fischer et al. 2003), several of the larger second millennium BCE sites in central Lydia, along with Sardis itself, continue to be occupied through the transition to the Iron Age in the early first millennium BCE (Figure 12). While in the second millennium BCE the focus of regional power appears to have resided in the fortified sites surrounding the Gygaean Lake, by the eighth century BCE the clear focal point of the entire region had become Sardis, located c. 10km to the south, on the southern edge of the Hermos River valley. During this period a Lydian Kingdom, later to become the Lydian Empire, crystallized at Sardis alone out of a handful of potentially contending towns. What made Sardis preeminent is

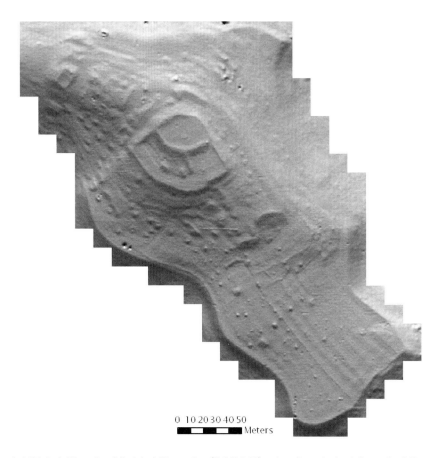

Figure 9: Hillshaded Digital Elevation Model of Kaymakçı (POI06.01) using data derived from the Microtopography survey.

unclear, but its extremely defensible position, overlooking a wide corridor of communication and controlling a rich source of electrum (the natural alloy of gold and silver), may have been a significant factor.

Sardis's subsequent rise to imperial status by the mid sixth century BCE was marked by monumental terraces for elite buildings, centralized workshops, and sprawling residential areas within a strongly fortified settlement area that had ballooned to over 100 hectares, far larger than any earlier settlement in the region. In addition Sardian products were exported outside of Lydia, and imported goods—especially luxury items—reached Sardis from foreign states.

These developments marked the first time in central Lydia in which a primary center was not located around the Gygaean Lake, yet the dramatic nucleation and urbanization of Sardis did not completely empty its hinterland of inhabitants. Around the Gygaean Lake at least 13 small and dispersed sites representing villages or hamlets probably controlled by elite land-owners continued to be occupied, and at least nine off-site scatters of ceramic material probably indicate continued intensive agriculture, now perhaps practiced both for local subsistence and to supply Sardis with surplus goods. These sites define an Iron Age settlement pattern spread over both upland and lowland areas, with a few of the lowland second millennium sites still inhabited. There was a marked shift away from the earlier fortified citadels, however: no fortified sites of the Iron Age have been documented around the Gygaean Lake, most likely because of the shift in political power to Sardis.

As in earlier periods, settlements and sites of all types could be found on all sides of the Gygaean Lake, and a number of household-based economies presumably interacted together within the growing state and imperial networks centered on Sardis. Preliminary evidence provides a working lens into socio-political differentials in the area. Ceramic wares imported from outside Lydia or of extremely fine types are rare in the hinterland while they are more common at Sardis, yet fine Lydian pottery is found in the hinterland, suggesting that long-distance exchange and contact was funneled through the capital, while local pottery may have been in part locally produced. For example, Lydian "streaky glaze" wares are ubiquitous in Lydia and may have been produced in several different locations (e.g., Gürtekin-Demir 2002); the Lydian variety of "Ephesian" ware, on the other hand, is rarer, and was perhaps produced at Sardis alone (Greenewalt

Figure 10: QuickBird satellite image of settlement at Kaymakçı (POI06.01) outside the fortification walls.

1970). Thus, by the Iron Age patterns in ceramic production and consumption are highly differentiated and are indicative of changes in socio-political complexity.

Beyond a new settlement pattern, the Iron Age is marked also by a new tradition of burial that is also clearly reflective of status differentiation. Whereas second millennium cemeteries appear to have been inconspicuous areas of cist and pithos burial associated with nearby settlements (e.g., Kaymakçı and Boyalı Tepe), the Iron Age witnessed the monumental transformation of the central Lydian landscape, especially Bin Tepe, into a vast tumulus cemetery. All indications are that the earliest tumuli of Bin Tepe were among the largest ever built, the tumuli of the legendary Lydian king Alyattes and his kin. These constructions appear to correspond in date to the consolidation of a territorial empire by the late seventh or early sixth century BCE that spanned most of western Anatolia and the Aegean littoral, and they reflect the growing power of that empire with impressive monumentality (compare Sinopoli 1994; Ashmore and Knapp 1999; Knapp 1986).

As the Lydian Empire was absorbed into the larger Achaemenid Persian Empire in the mid sixth century BCE, Bin Tepe was transformed further into a cemetery of tumuli not just for royalty, but also for other elite social groups. In this way Bin Tepe came to be dominated by numerous tumuli that survived mostly intact into the early twentieth century, but that are becoming increasingly endangered by looting and agricultural activities that have left standing only 116 examples today (Roosevelt and Luke 2006a and 2006b).

Thus, it appears that this landscape was at least partially considered sacred burial ground, even though it still must have been utilized by the living as well, as shown by the discovery of numerous sites within it. Of particular interest are the questions of why Bin Tepe was selected to be the royal cemetery of Sardis, and why it continued to be used for high-status burial during the Achaemenid period.

The Iron Age inhabitants of central Lydia must have been cognizant of the prior status of the area around the Gygaean Lake, and the royal Sardians appear to have ideologically co-opted such ancestral status with their construction of monumental tumuli that effectively laid claim to it through dominating the surrounding landscapes (Figure 13). The ancestral power of the lake itself is indicated by the probable derivation of its name from a word meaning something like "grandfather" or "ancestor" in Luwian (the probable second millennium BCE language of the area), and by Homer's late eighth century claim that the lake was the mother of the ancestral Lydians (Iliad 2.864–86). That Lydian elites of the Achaemenid period still felt the need to make similar claims to ancestral ties can be inferred from the locations of their tumuli near each of the second millennium fortified citadels—what would have seemed to them to be the most conspicuous remains of the glorious central Lydian past (compare Renfrew 1973: 542; Wheatley 1995: 174–76; Bradley 2000). One such tumulus was even built atop a second-millennium terrace at Kaymakçı, incorporating second-millennium material into its earthen fill. The selection of other areas for tumulus burial,

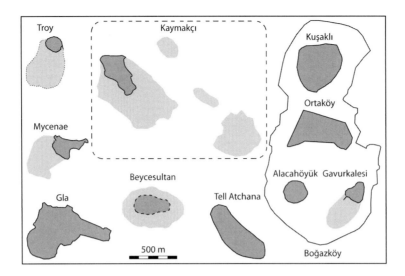

Figure 11: Chart showing comparison of the size of Kaymakçı (POI06.01) with that of contemporary sites.

including in Bin Tepe, was probably related to the presence of the earlier tumuli of Lydian kings, which by the Achaemenid period may already have been apt subjects of emulation.

In addition to highlighting the endurance of local Lydian traditions into the Achaemenid period, the tumuli of Bin Tepe and elsewhere in central Lydia reflect increased social differentiation in the countryside, probably resulting from the need of the Achaemenid administration at Sardis to cultivate and maintain rural elite support. We should envision the occupants of such tumuli to have been wealthy or noble estate-owners who controlled burgeoning local economies that contributed to Sardis's and Persia's largess as well as maintained local subsistence needs. Such rural elites maintained their importance in part owing to their embodiment of long-term traditions descending from previous generations of central Lydian inhabitants. Viewed in this light, the central Lydian hinterland of Sardis became a dynamic mosaic of differing social groups—both living and dead—that was integral to the power of Sardis in both the Lydian and Achaemenid periods.

Discussion

The Central Lydian Archaeological Survey has attempted to incorporate a range of strategies to understanding long-term cultural and environmental dynamics in the area: paleoenvironmental research, single-track and upland survey, high-platform remote sensing and ground-truthing, intensive site survey of varying forms, materials analyses and textual study, among others. First and foremost these approaches have allowed us to sketch out the beginnings of a regional culture history, and the preliminary results of our paleoenvironmental and archaeological programs presented above suggest the promise of continued research around the Gygaean Lake. Yet, pulling all our various strands of data together is going to take some time, and here we offer some further preliminary interpretations by way of conclusion.

Only one site, Bozyer (POI05.43), has rich data for pre-Bronze Age eras, and the postulated formation of the Gygaean Lake by the beginning of the third millennium corresponds well to the initial period of increased population. By the Early Bronze Age I period a number of sites are located in the rolling hills of Bin Tepe and along the shores of the Gygaean Lake and/or in close proximity to the river plain and marshlands. These small hamlets must have functioned along models of household economies with production of resources focused at the site level, rather than controlled by large political institutions; yet to be published ceramic analyses corroborate these suggestions (Boulanger et al. 2008).

The Middle to Late Bronze Age marks a dramatic shift in settlement patterns with one large site, Kaymakçı, acting as the major node among a network of smaller strongholds. Again, yet to be published ceramic analyses support the likely centrality of Kaymakçı with respect to production (Boulanger et al. 2008). Its size and complexity alone would suggest its role as a social and political hub, and contemporary archives and monuments of the Hittite world may be brought to bear on the subject.

Recent work on Hittite texts, seals, and the monuments at Karabel in western Lydia has provided strong support for the identification of the Hermos River valley with the Bronze Age Seha River Land, a kingdom of Arzawa and later Hittite vassal state, the location of which had been open to interpretation

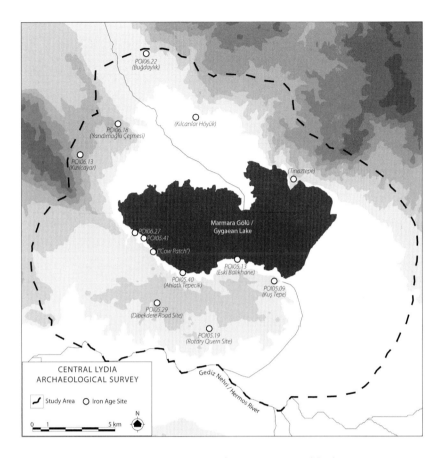

Figure 12: Map of Iron Age sites in central Lydia.

for some time (Hawkins 1998). We suggest here that if the equation of the Hermos River valley with the Seha River Land is correct, then Kaymakçı was probably the regional capital of the Seha River Land. As the Hittite texts describe, the region functioned as a key diplomatic node in the overall political infrastructure of western Anatolia (e.g., Beckman 1999). The Seha River Land served as one node of interaction (political, economic, and social) between Mira, to the south, and Wilusa, the probable seat of Troy, to the north, during a period of continual change in which Ahhiyawan (or Mycenaean) and Hittite power and influences were continually being negotiated. Future work will allow us to explore the rise of this kingdom and later vassal state, particularly with respect to the role of the Hittites, with the material record suggesting degrees and directions of interaction. At the very least, Hittite archives document a royal marriage alliance with a king of the Seha River Land. What we do not know, of course, but hope to illuminate in coming years, is how local populations viewed such a relationship with the Hittites. Hittite archives give us only one perspective.

The landscapes of Lydia during the second millennium BCE supported a thriving population. The rise of the political system and social networks of the Seha River Land outlined by Hittite texts corresponds to the rise of urbanism in central Lydia. In light of this data, the rise of Sardis, once believed to be the nascent political stronghold of the region, has to be reconsidered. Yet, we also have to consider why people chose to move from the rich environs of the Gygaean Lake to Sardis in the first millennium BCE. Perhaps an extended period of drought rendered former farmlands less fertile than the foothills surrounding Sardis, watered by perennial streams flowing north from the Tmolos or Bozdağ Mountain range. Perhaps the discovery of native electrum in the immediate region of Sardis offered a key resource desired by many; perhaps routes of communication began to shift, making Sardis a more practical and central location. Perhaps the threat from enemies demanded a location that could support a larger population within a walled-enclosure, as could the acropolis and foothills of Sardis. Perhaps it was a combination of all of these things.

Yet, while Sardis grew, why was Bin Tepe chosen for the location of royal and high-status burials? A long history of occupation and the significance of the second millennium strongholds may have created an enduring local lore that identified the area as the an-

Figure 13: View of the Gygaean Lake and the tumulus of Alyattes (Kocamutaf Tepe) to the south, with the Acropolis (left) and Necropolis (right) hills of Sardis in the background against the silhouette of the Tmolos (Bozdağ) Mountain range.

cestral homeland. As Homer tells us, the ancestral Lydians were born of the Gygaean Lake, and we are only now beginning to see the truth in his words.

Acknowledgments

This article is dedicated to Peter Ian Kuniholm, who, in addition to his role as dendrochronological pioneer, has always instilled in his students an eager interest in the archaeology of Anatolia. We write with great thanks to Peter for his initial and ongoing encouragement and support of our project and for his continued friendship.

Research presented in this article was supported by the National Science Foundation under Grant No. 0649981. We would like to acknowledge and thank also other generous sponsors, Boston University, the American Research Institute in Turkey, the Marion and Jasper Whiting Foundation, and anonymous donors. We are especially grateful to the General Directorate of Cultural Heritage and Museums of the Ministry of Culture and Tourism, the Republic of Turkey, for permissions and logistical support, and to project supporters, participants, and consultants. Included in the latter groups are the following: Manisa Museum Director M. Tosunbaş, and her staff; the Archaeological Exploration of Sardis, especially its former and current directors, C. H. Greenewalt, jr., and N. D. Cahill, respectively, E. Gombosi, K. Kiefer, and T. Yalçınkaya; Ministry of Culture Representatives N. İşçi (2005), E. Yılmaz (2006), and E. Özçelik (2007); field participants H. Alkan, M. R. Besonen, K. Cooney, N. P. Özgüner, N. Y. Rifkind, B. R. Vining, C. Wait, and N. Wolff; and consultants and collaborators M. T. Boulanger, M. Glascock, A. Ramage, M. L. Rautman, and C. Runnels.

References

Ashmore, W., and Knapp, A.B., eds. 1999. *Archaeologies of Landscape: Contemporary Perspectives.* Malden, MA: Blackwell.

Bayne, N. 2000. *The Grey Wares of North-West Anatolia in the Middle and Late Bronze Age and the Early Iron Age and Their Relation to the Early Greek Settlements.* Bonn: Dr. R. Habelt.

Beckman, G. 1999. *Hittite Diplomatic Texts.* 2nd Edition. Atlanta: Scholars Press.

Besonen, M.R. 1997. The Middle and Late Holocene Geology and Landscape Evolution of the Lower Acheron River Valley, Epirus, Greece. M.S. Thesis, University of Minnesota.

Besonen, M.R., Rapp, G., and Jing, Z. 2003. The Lower Acheron River Valley: Ancient Accounts and the Changing Landscape. In J. Wiseman and K. Zachos (eds.) *Landscape Archaeology in Southern Epirus, Greece* I, 199–263. Hesperia Supplement 32. Princeton: American School of Classical Studies at Athens.

Besonen, M.R., and Roosevelt, C.H. 2008. Living in Central Lydia: Changing Environmental Conditions. Paper presented in The Landscapes of Lydia, Western Turkey: Preliminary Results of the Central Lydia Archaeological Survey, 2005–2007, Society for American Archaeology, Annual Meeting, Vancouver, BC; March 2008.

Bier, C.M. 1973. The Excavations at Korucutepe, Turkey, 1968–70: Preliminary Report. Part II: The Fortification Wall. *Journal of Near Eastern Studies* 32(4): 424–34.

Bradley, R. 2000. *An Archaeology of Natural Places.* London and New York: Routledge.

Brinkmann, R. 1971. The Geology of Western Anatolia. In A.S. Campbell (ed.) *Geology and History of Turkey*, 171–190. Tripoli: Petroleum Exploration Society of Libya.

Brinkmann, R. 1976. *Geology of Turkey*. Trans. by I. Woodall. Stuttgart: Enke.

Boulanger, M.T., Glascock, M.D., Luke, C., Pullen, D.J., and Roosevelt, C.H. 2008. Ceramic Production in Central Lydia from the Early Bronze Age through the Late Lydian Period: Preliminary Results from Instrumental Neutron Activation Analysis (INAA). Paper presented in The Landscapes of Lydia, Western Turkey: Preliminary Results of the Central Lydia Archaeological Survey, 2005–2007, Society for American Archaeology, Annual Meeting, Vancouver, BC; March 2008.

Butler, H.C. 1919. Protection for the Historic Monuments and Objects of Art in Nearer Asia. *The Art Bulletin* 2(1): 46–58.

Butler, H.C. 1922. *Sardis I. The Excavations, Part 1: 1910–1914*. Leiden: E. J. Brill, Ltd.

Choisy, A. 1876. Note sur les tombeaux Lydiens de Sardes. *Revue archéologique* 32: 73–81.

Cooney, K.C. 2008. Central Lydian Lithic Technologies from the Upper Paleolithic through the Bronze Age. Paper presented in The Landscapes of Lydia, Western Turkey: Preliminary Results of the Central Lydia Archaeological Survey, 2005–2007, Society for American Archaeology, Annual Meeting, Vancouver, BC; March 2008.

Curtius, E. 1853. Artemis Gygaia und die lydischen Fürstengräber. *Archäologische Zeitung* 11: 148–161.

Çevik, Ö. 2007. The Emergence of Different Social Systems in Early Bronze Age Anatolia: Urbanisation versus Centralisation. *Anatolian Studies* 57: 131–140.

Çilingiroğlu, A., Derin, Z., Abay, E., Sağlamtimur, H., and Kayan, İ. 2004. Ulucak Höyük: Excavations Conducted Between 1995 and 2002. *Ancient Near Eastern Studies* Supplement 15. Paris: Peeters.

Davis, J. 2001. The Islands of the Aegean. In T. Cullen (ed.) Aegean Prehistory: A Review, 19–76. *American Journal of Archaeology* Supplement 1. Boston: Archaeological Institute of America.

Efe, T. 2002. The Interaction Between Cultural/Political Entities and Metalworking in Western Anatolia during the Chalcolithic and Early Bronze Ages. In Ü. Yalçın (ed.) *Anatolian Metal* II, 49–65. *Der Anschnitt*, Beiheft 15. Veröffentlichungen aus dem Deutschen Bergbau-Museum Bochum, no. 109. Bochum: Deutsches Bergbau-Museum.

Efe, T. 2003a. Küllüoba and the Initial Stages of Urbanism in Western Anatolia. In M. Özdoğan, H. Hauptmann, and N. Başgelen (eds.) *From Villages to Towns. Studies Presented to Ufuk Esin*, 265–282. İstanbul: Arkeoloji ve Sanat Yayınları.

Efe, T. 2003b. Pottery Distribution within the Early Bronze Age of Western Anatolia and its Implications upon Cultural, Political (and Ethnic?) Entities. In M. Özbaşaran, O. Tanındı, and A. Boratav (eds.) *Archaeological Essays in Honour of Homo amatus: Güven Arsebük*, 87–103. İstanbul: Ege Yayınları.

Efe, T. 2007. The Theories of the 'Great Caravan Route' between Cilicia and Troy: the Early Bronze Age III Period in Inland Western Anatolia. *Anatolian Studies* 57: 47–64.

Erkanal, A., and Erkanal, H. 1983. Vorbericht über die Grabungen 1979 im prähistorischen Klazomenai/Limantepe. *Hacettepe Üniversitesi Edebiyat Fakültesi Dergisi* 1983: 163–178.

Fischer, B., Genz, H., Jean, É., and Köroğlu, K. (eds.) 2003. *Identifying Changes: The Transition from Bronze to Iron Ages in Anatolia and Neighboring Regions. Proceedings of the International Workshop, Istanbul, November 8–9, 2002*. Istanbul: Türk Eskiçağ Bilimleri Enstitüsü.

French, D.H. 1969. Prehistoric Sites in Northwest Anatolia II. The Balıkesir and Akhisar/Manisa Areas. *Anatolian Studies* 19: 41–98.

Greenewalt, C.H., jr. 1970. Orientalizing Pottery from Sardis: the Wild Goat Style. *California Studies in Classical Antiquity* 3: 55–89.

Greenewalt, C.H., jr., and Rautman, M.L. 2000. The Sardis Campaigns of 1996, 1997, and 1998. *American Journal of Archaeology* 104: 643–681.

Günel, S. 1999. *Panaztepe II: M.Ö. 2. bine tarihlendirilen Panaztepe seramiğinin Batı Anadolu ve Ege arkeolojisindeki yeri ve önemi / Die Keramik von Panaztepe und ihre Bedeutung für Westkleinasien und die Ägäis im 2. Jahrtausend*. Ankara: Türk Tarih Kurumu Basımevi.

Gürtekin-Demir, R.G. 2002. Lydian Painted Pottery at Daskyleion. *Anatolian Studies* 52: 111–143.

Hakyemez, H.Y., Erkal, T., and Göktaş, F. 1999. Late Quaternary Evolution of the Gediz and Büyük Menderes Grabens, Western Anatolia, Turkey. *Quaternary Science Reviews* 18: 549–554.

Hanfmann, G.M.A. 1967. The Ninth Campaign at Sardis (1966). *Bulletin of the American Schools for Oriental Research* 186: 17–52.

Hanfmann, G.M.A. 1968. The Tenth Campaign at Sardis (1967). *Bulletin of the American Schools for Oriental Research* 191: 2–41.

Hanfmann, G.M.A. (ed.) 1983. *Sardis from Prehistoric to Roman Times. Results of the Archaeological Exploration of Sardis, 1958–1975*. Cambridge, MA: Harvard University Press.

Horejs, B. 2008. Çukuriçi Höyük. A New Excavation Project in the Eastern Aegean. In *Aegeo-Balkan Prehistory*. 2 July 2008. ⟨http://www.aegeobalkanprehistory.net/article.php?id_art=9⟩

Jacoby, F. 1923. *Fragmente der griechischen Historiker*. Berlin: Weidman.

Karageorghis, V., and Morris, C.E. 2001. *Defensive Settlements of the Aegean and the Eastern Mediterranean After c. 1200 B.C. Proceedings of an International Workshop held at Trinity College Dublin, 7–9 May, 1999*. Nicosia/Trinity College Dublin: The Anastasios G. Leventis Foundation.

Kersel, M., Luke, C., and Roosevelt, C.H. 2008. Valuing the Past: Perceptions of Archaeological Practice in Lydia and the Levant. *Journal of Social Archaeology* 8: 298–319.

Knapp, A.B. 1986. *Copper Production and Divine Protection: Archaeology, Ideology, and Social Complexity on Bronze Age Cyprus*. Göteborg: P. Åströms Förlag.

Latacz, J. 2001. *Troy and Homer*. Oxford: Oxford University Press.

Lichter, C. (ed.) 2005a. *How Did Farming Reach Europe? Anatolian–European Relations from the Second Half of the 7th through the First Half of the 6th Millennium CAL BC. Proceedings of the International Workshop, Istanbul, 20–22 May 2004*. Byzas 2. Istanbul: Deutsches Archäologisches Institüt.

Lichter, C. 2005b. Western Anatolia in the Late Neolithic and Early Chalcolithic: the Actual State of Research. In C. Lichter (ed.) *How Did Farming Reach Europe? Anatolian–European Relations from the Second Half of the 7th through the First Half of the 6th Millennium CAL BC. Proceedings of the International Workshop, Istanbul, 20–22 May 2004*, 59–74. Byzas 2. Istanbul: Deutsches Archäologisches Institüt.

Lloyd, S., and Mellaart, J. 1962. *Beycesultan* I. London: British Institute of Archaeology at Ankara.

Lloyd, S., and Mellaart, J. 1965. *Beycesultan* II. London: British Institute of Archaeology at Ankara.

Lochner, I. 2001. Die Frühbronzezeitlichen Siedlungsbefunde in Aizanoi. *Archäologischer Anzeiger* 2001: 270–294.

Manning, S. 1997. Cultural Change in the Aegean c. 2200 B.C. In H. Nüzhet Dalfes, G. Kukla, and H. Weiss (eds.) *Third Millennium B.C. Climate Change and Old World Collapse*, 149–171. Berlin/New York: Springer.

Mellaart, J., and Murray, A. 1995. *Beycesultan* III. London: British Institute of Archaeology at Ankara.

Meriç, R. 1993. Pre-Bronze Age Settlements of West-Central Anatolia (An Extended Abstract). *Anatolica* 19: 143–150.

Meriç, R. 2003. Excavations at Bademgediği Tepe (Puranda) 1999–2002: A Preliminary Report. *Istanbuler Mitteilungen* 53: 79–98.

Meriç, R., and Mountjoy, P.A. 2001. Three Mycenaean Vases from Ionia. *Istanbuler Mitteilungen* 51: 137–141.

Meriç, R., and Mountjoy, P.A. 2002. Mycenaean Pottery from Bademgediği Tepe (Puranda) in Ionia: A Preliminary Report. *Istanbuler Mitteilungen* 52: 79–98.

Miller Rosen, A. 1997. What has Drought Wrought? *Sciences* 37: 47.

Mitten, D.G., and Yüğrüm, G. 1968. Excavations at Ahlatlı Tepecik on the Gygean Lake, 1968. *Dergi* 17(1): 125–127.

Mitten, D.G., and Yüğrüm, G. 1971. The Gygean Lake, 1969: Eski Balıkhane, Preliminary Report. *Harvard Studies in Classical Philology* 75: 191–195.

Mitten, D.G., and Yüğrüm, G. 1974. Ahlatlı Tepecik beside the Gygean Lake. *Archaeology* 27: 22–29.

Mountjoy, P.A. 1998. The East Aegean–West Anatolian Interface in the Late Bronze Age: Mycenaeans and the Kingdom of Ahhiyawa. *Anatolian Studies* 48: 33–67.

Munn, M. 2006. *The Mother of the Gods, Athens, and the Tyranny of Asia. A Study of Sovereignty in Ancient Religion.* Berkeley: University of California Press.

Neumann, J., and Parpola, S. 1987. Climatic Change and the Eleventh–Tenth-Century Eclipse of Assyria and Babylonia. *Journal of Near Eastern Studies* 46: 161–182.

Niemeier, W.D. 1998. The Mycenaeans in Western Anatolia and the Problem of the Origins of the Sea Peoples. In S. Gitin, A. Mazar, and E. Stern (eds.) *Mediterranean Peoples in Transition: Thirteenth to Early Tenth Centuries BCE*, 17–65. Jerusalem: Israel Exploration Society.

Nüzhet Dalfes, H., Kukla, G., and Weiss, H. eds. 1997. *Third Millennium B.C. Climate Change and Old World Collapse.* Berlin/New York: Springer.

Özdoğan, M. 1998. Ideology and Archaeology in Turkey, in L. Meskell (ed.) *Archaeology Under Fire. Nationalism, Politics and Heritage in the Eastern Mediterranean and Middle East*, 111–121. London: Routledge.

Özdoğan, M., Hauptmann, H., and Başgelen, N. (eds.) 2003. *From Village to Cities: Early Villages in the Near East.* İstanbul: Arkeoloji ve Sanat Yayınları.

Özdoğan, M., and Başgelen, N., eds. 1999. *Neolithic in Turkey, the Cradle of Civilization: New Discoveries.* İstanbul: Arkeoloji ve Sanat Yayınları.

Özgen, İ., and Öztürk, J. 1996. *The Lydian Treasure: Heritage Recovered.* İstanbul: Uğur Okman for [the] Republic of Turkey, Ministry of Culture, General Directorate of Monuments and Museums.

Özgüner, N.P., and Roosevelt, C.H. 2008. Aerial Photography and High-Resolution Satellite Imagery: Assessing Landscape and Land-Use Changes in Central Lydia. Paper presented in The Landscapes of Lydia, Western Turkey: Preliminary Results of the Central Lydia Archaeological Survey, 2005–2007, Society for American Archaeology, Annual Meeting, Vancouver, BC; March 2008.

Rautman, M. 1987. Problems of Land Use and Water Supply in Late Antique Lydia. Paper presented at the Thirteenth Annual Byzantine Studies Conference, The Ohio State University, Columbus, OH; 5–8 November 1987.

Rehak, P., and Younger, J.G. 2001. Neopalatial, Final Palatial, and Postpalatial Crete. In T. Cullen (ed.) *Aegean Prehistory: A Review*, 383–465. *American Journal of Archaeology* Supplement 1. Boston: Archaeological Institute of America.

Renfrew, C. 1973. Monuments, Mobilization and Social Organization in Neolithic Wessex. In C. Renfrew (ed.) *The Explanation of Culture Change: Models in Prehistory. Proceedings of a Meeting of the Research Seminar in Archaeology and Related Subjects held at the University of Sheffield*, 539–558. London: Duckworth.

Robert, L. 1982. Documents d'Asie Mineure. XXI: Au Nord de Sardes. *Bulletin de correspondance hellénique* 106: 334–378.

Roosevelt, C.H. 2002. Lydian and Persian Period Site Distribution in Lydia. Paper Presented at the Annual Meeting of the Archaeological Institute of America, Philadelphia, PA; January 2002.

Roosevelt, C.H. 2003a. Tumulus Tomb Complexes, Distribution, and Significance in Lydian and Persian Period Lydia. Paper Presented at the Annual Meeting of the Archaeological Institute of America, New Orleans, LA; 5 January 2003.

Roosevelt, C.H. 2003b. Lydian and Persian Period Settlement in Lydia. PhD Dissertation, Cornell University. Ann Arbor, MI: UMI Dissertation Services.

Roosevelt, C.H. 2005. Şahankaya in Northern Lydia, Turkey. Paper Presented at the Annual Meeting of the Society for American Archaeology, Salt Lake City, UT; 1 April 2005.

Roosevelt, C.H. 2006a. Symbolic Door Stelae and Graveside Monuments in Western Anatolia. *American Journal of Archaeology* 110(1): 65–91.

Rooesvelt, C.H. 2006b. Tumulus Survey and Museum Research in Lydia, Western Turkey: Determining Lydian- and Persian-Period Settlement Patterns. *Journal of Field Archaeology* 31(1): 61–76.

Roosevelt, C.H. 2007. Central Lydia Archaeological Survey: 2005 Results. *Araştırma Sonuçları Toplantısı* 24(2): 135–154.

Roosevelt, C.H., and Luke, C. 2006a. Looting Lydia: the Destruction of an Archaeological Landscape in Western Turkey. In N. Brodie, M. Kersel, C. Luke, and K.W. Tubbs (eds.) *Archaeology, Cultural Heritage, and the Antiquities Trade*, 173–187. Gainesville, FL: University Press of Florida.

Roosevelt, C.H., and Luke, C. 2006b. Mysterious Shepherds and Hidden Treasures: The Culture of Looting in Lydia. *Journal of Field Archaeology* 31(2): 185–198.

Roosevelt, C.H., and Luke, C. 2008a. Central Lydia Archaeological Survey: 2006 Results. *Araştırma Sonuçları Toplantısı* 25(3): 305–326.

Roosevelt, C.H., and Luke, C. 2008b. New Light on the Archaeology of Central Western Anatolia in the Middle and Bronze Age. Paper Presented at the Annual Meeting of the American Schools of Oriental Research, Boston, MA; November 2008.

Roosevelt, C.H., and Luke, C. Forthcoming. Central Lydia Archaeological Survey: 2007 Results. *Araştırma Sonuçları Toplantısı* 26 (expected May 2009).

Rotroff, S.I. 1978. Hellenistic Athenian Pottery: 'Megarian Bowls'. *Current Anthropology* 19(2): 387–388.

Rutter, J. B. 2001. The Prepalatial Bronze Age of the Southern and Central Greek Mainland. In T. Cullen (ed.) *Aegean Prehistory: A Review*, 95–147. *American Journal of Archaeology* Supplement 1. Boston: Archaeological Institute of America.

Sayce, A.H. 1880. Notes from Journeys in the Troad and Lydia. *Journal of Hellenic Studies* I: 75–93.

Schuiling, R.D. 1962. On Petrology, Age, and Structure of the Menderes Migmatite Complex (SW-Turkey). *Maden, Tetkik, ve Arama Bulletin* 58: 71–84.

Shelmerdine, C.W. 2001. The Palatial Bronze Age of the Southern and Central Greek Mainland. In T. Cullen (ed.) *Aegean Prehistory: A Review*, 329–377. *American Journal of Ar-*

chaeology Supplement 1. Boston: Archaeological Institute of America.

Shrimpton, G. 1987. Regional Drought and the Economic Decline of Mycenae. *Echos du Monde Classique/Classical Views* 31(6): 137–177.

Simpson, R.H., and Hagel, D.K. 2006. *Mycenaean Fortifications, Highways, Dams and Canals. Studies in Mediterranean Archaeology* Vol. 133. Sävedalen. Paul Åströms Förlag.

Sinopoli, C.M. 1994. The Archaeology of Empires. *Annual Review of Anthropology* 23: 159–180.

Sullivan, D.G. 1989. Human-Induced Vegetation Change in Western Turkey: Pollen Evidence from Central Lydia. PhD Dissertation, University of California, Berkeley. Ann Arbor, MI: UMI Dissertations.

Şahoğlu, V. 2005. The Anatolian Trade Network and the Izmir Region during the Early Bronze Age. *Oxford Journal of Archaeology* 24(4): 339–361.

Tartaron, T.F., Pullen, D., and Noller, J.S. 2006. Rillenkarren at Vayla: Geomorphology and a New Class of Early Bronze Age Fortified Settlement in Southern Greece. *Antiquity* 80(307): 145–160.

Vining, B.R., and Roosevelt, C.H. 2008. Geophysical and Microtopographical GPS Survey at Kaymakçı, a Regional Capital of the Second Millennium BCE. Paper presented in The Landscapes of Lydia, Western Turkey: Preliminary Results of the Central Lydia Archaeological Survey, 2005–2007, Society for American Archaeology, Annual Meeting, Vancouver, BC; March 2008.

von Olfers, J.F.M. 1858. Über die Lydischen Königsgraber bei Sardes und den Grabhügel des Alyattes, nach dem Bericht des Kaiserlichen General-Consuls Spiegelthal zu Smyrna. *Abhandlungen der Königlichen Akademie der Wissenschaften zu Berlin* 1858: 539–556.

Warner, J.L. 1994. *Elmalı Karataş* II. *The Early Bronze Age Village of Karataş*. Bryn Mawr, PA: Bryn Mawr College.

Weiss, B. 1982. The Decline of Late Bronze Age Civilization as a Possible Response to Climatic Change. *Climatic Change* 4: 173–198.

Weiss, H. 2000. Beyond the Younger Dryas. In G. Bawden and M. Reycraft (eds.) *Environmental Disaster and the Archaeology of Human Response*, 75–98. Albuquerque, NM: Maxwell Museum of Anthropology.

Weiss, H., Courty, M.-A., Wetterstrom, W., Guichard, F., Senior, L., Meadow, R., and Curnow, A. 1993. The Genesis and Collapse of Third Millennium North Mesopotamian Civilization. *Science* 261: 999–1004.

Wheatley, D. 1995. Cumulative Wiewshed Analysis: a GIS-Based Method for Investigating Intervisibility, and its Archaeological Application. In G. Lock and Z. Stančič (eds.) *Archaeology and Geographic Information Systems: A European Perspective*, 171–185. London: Taylor and Francis.

Yakar, J. 1999. The Socio-Economic Significance of Regional Settlement Pattern in Early Bronze Age Anatolia—An Archaeological Assessment. In *International Symposium on Settlement and Housing in Anatolia through the Ages, Istanbul, Turkey, June 5–7, 1996*, 505–511. Istanbul: Ege Yayınları.

Yusufoğlu, H. 1996. Northern Margin of the Gediz Graben: Age and Evolution, West Turkey. *Turkish Journal of Earth Sciences* 5: 11–23.

Article submitted May 2008

The Chronology of Phrygian Gordion

Mary M. Voigt

Abstract: *For most archaeological sites that do not have texts and calendars, chronology is based primarily on relative dating: stratigraphic sequences, artifact comparisons with other sites, and sometimes informed guesses that link specific layers to well-dated historical events documented in adjacent regions. Other methods of absolute dating such as radiocarbon and dendrochronology are highly desirable, but may not be available if appropriate samples of charcoal and/or tree segments are not preserved. Gordion, home of the historical king Midas, provides a good example of the ways that archaeologists construct historical narratives in the absence of texts, and the ways in which an absolute date obtained from a rich suite of radiocarbon samples can drastically revise ideas about the relationship between archaeological materials and historical events.*

The identification of the modern flat mound called Yassıhöyük as Gordion, capital of the Phrygian kingdom during the first millennium BC, is secure despite the absence of textual evidence from the site. The mound and surrounding settled areas match descriptions of Gordion's location in ancient texts and this geographical link is supported by archaeological finds: the recovery of elite buildings and rich tombs dated to the 8th century BC, the time when the Phrygian king Midas ruled according to Assyrian records (Körte 1897; Körte and Körte 1904; Sams 2005: 10; see also DeVries 1980: 34 n.7; DeVries in Rose and Darbyshire forthcoming).

Modern excavation at Gordion began in 1950 under the direction of Rodney S. Young, sponsored by the University of Pennsylvania Museum. His initial research goals were stated as follows:

> Much of the early history of Phrygia is conjectural and the purpose of the University Museum in digging at Gordion is to throw some light on the obscure period when the Phrygian Kingdom was at its height, and to determine what was the relation of the Phrygians to the Hittite Empire which came before, and what their influence [was] on early Greek culture, in its formative state at the time of their greatest expansion (1953c: 20; see also Young 1951: 4).

Young carried out 17 seasons of excavation before his accidental death in 1974 (DeVries 1980; 1981: xxxv-xxxvi). He cleared more than 2.5 hectares on Yassıhöyük (also referred to as the Citadel Mound, see Figure 1), investigated the walled Lower Town to the south of the Citadel, and opened 31 burial mounds or tumuli.

As with most archaeological projects, the results of several seasons of excavation led to a reformulation of goals. Young discovered an occupation level that had been destroyed by fire and was therefore unusually well-preserved; he linked this "Destruction Level" to King Midas and to a raid by Kimmerian nomads in the early seventh century BC that was inferred from ancient texts. Deposits immediately above the Destruction Level were very badly disturbed and poorly preserved with the exception of a Hellenistic level attributed to Galatian "barbarians" who held little interest for Young. Deposits related to the Late Bronze Age–Iron Age (or Hittite–Phrygian) transition were of considerable interest but were deeply buried and lay beneath well-preserved Destruction Level buildings that were unique in Anatolia. Young therefore re-focused his work on the Destruction Level, with minimal attention to anything above it and only a very small sounding to investigate Hittite–Phrygian relationships.

Early in 1974 Young appointed Keith DeVries as his successor and Gordion Project Director. After Young's death in October, DeVries decided that the highest priority for the project should be to analyze and publish the now massive corpus of data from the site, work which he supervised until 1987 when the project again changed direction. In 1988 G. Kenneth Sams was appointed project director with responsi-

Figure 1: Aerial photograph of Gordion showing the citadel of Yassıhöyük and the adjacent Lower Town to the south. On the Citadel, the large excavated area with standing buildings was cleared by Rodney S. Young between 1950 and 1973. The cone-shaped light-colored areas surrounding the mound are dumps of soil.

bility for conservation and continued publication of the Young material. I was appointed as director for new field research which began in 1988 (Voigt 2005: 22–28). One of the primary goals set for renewed excavation on Yassıhöyük was to construct an improved stratigraphic and chronological sequence. The need for better control of the archaeological record was suggested by DeVries, who was responding to difficulties encountered by scholars working with poorly dated or inadequately documented contexts, especially within levels of the site above the Destruction Level (personal communication 1987).

We did not begin work in 1988 with the idea that there was anything significantly wrong with the absolute chronology for Gordion, which was based on links that Young had made between specific deposits and historical events; the latter carried calendrical dates that were generally accepted. We were instead concerned with building a more precise internal/relative chronology with short time-stratigraphic units characterized by relatively common diagnostic artifact types (e.g., pottery) (Figure 2; see also Voigt 2007: Figs. 6–

7). These diagnostics could be used to date materials and contexts from the Young years as well as surface collections to be obtained through survey.

Framing the Question

There are two major Iron Age contexts that are critical for an understanding of Phrygian Gordion. The first is the Destruction Level, which terminates the Early Phrygian period or Phase 6A in the recently defined Yassıhöyük Stratigraphic Sequence (YHSS: Table 1 and Figure 3; see also Voigt 1994). The Young team assigned a date of ca. 700 BC to the destruction, a nice round number based on one of several ancient texts (see below, and DeVries in Rose and Darbyshire forthcoming for an analysis of the texts and the dates they supply). The second context is the much larger and architecturally more elaborate city built above the 6A/Destruction Level ruins (Figure 4). The date of the beginning of this phase, now known archaeologically as the Middle Phrygian period or YHSS 5, has shifted several times (see below) but in 1988 most

Figure 2: The Yassıhöyük Stratigraphic Sounding or YHSS carried out in 1988–89 consisted of two sets of trenches. In the foreground is the Lower Trench Sounding, located within the Outer Court of the Early Phrygian/YHSS 6A palace complex (Fig. 3); the LTS obtained a sample of occupation dating from the Middle Bronze Age (YHSS 10) to the beginning of the Early Phrygian period (YHSS 6B). In the background is the Upper Trench Sounding (UTS) which sampled occupations from the Early Phrygian/YHSS 6A Destruction Level to the modern surface of the mound.

YHSS Phase	Period Name	Approximate Dates
0	Modern	1920s
1	Medieval	10th–15th centuries AD
2	Roman	1st century BCE–4th century AD
3A	Later Hellenistic	ca. 260–100 BC
3B	Early Hellenistic	ca. 330–ca. 260 BC
4	Late Phrygian	ca. 540–ca 330 BC
5	Middle Phrygian	after 800–ca. 540 BC
6A-B	Early Phrygian	900–800 BC
7	Early Iron Age	1100–900 BC
9–8	Late Bronze Age	1400–1200 BC
10	Middle Bronze Age	1600–1400 BC

Table 1: The Yassıhöyük Stratigraphic Sequence (YHSS) 2008

scholars placed the construction of Middle Phrygian Gordion in the late seventh century BC.

The absolute date of both the destruction and reconstruction has recently been revised (DeVries et al. 2003) to the surprise of many, and the consternation of a few (Muscarella 2003; Keenan 2004; see also Çilingiroğlu and Sagona 2007). Peter Kuniholm played a critical role in providing information and support that led to the Gordion re-dating, and this article is dedicated to him with many thanks, and admiration for the vision, energy, and tenacity that resulted in a dendrochronological sequence for the Mediterranean area.

The new absolute chronology for Gordion is ultimately based on radiocarbon determinations and dendrochronological dates, but the impetus to run the radiocarbon dates came from an accumulation of evidence gathered in traditional ways including fine-grained stratigraphic observations, artifact seriation, and comparisons of material culture with that at other sites (Voigt 2007; DeVries 2005; DeVries 2007, 2008; Sams 2007). The same suite of archaeological methods was employed by Young and his colleagues. The University of Pennsylvania radiocarbon lab was founded in 1951, and Young soon submitted samples to be dated. The earliest from the Destruction Level were charcoal, and, calculated with a half life of 5800 and with large standard deviations, produced dates that were consistent with the 700 BC date that was expected (Kohler and Ralph 1961: 361). In 1961, short-lived samples (charred seeds and textile fragments from Megaron 3) were submitted; calculated with the standard 5730 half-life, the dates were too early, and so were filed away and forgotten for 40 years (Rose

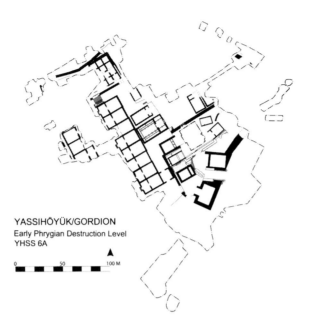

Figure 3: Plan of the Early Phrygian/YHSS 6A Palace Quarter at the time of a catastrophic fire. The individual megarons that make up the Terrace Building to the west were numbered from south to north, so that Terrace Building 2 is the second building in the row. At the time of the fire, the Phrygians were engaged in a construction project around the main gate referred to as the YHSS 6A Unfinished Project.

and Darbyshire forthcoming; see also Stuckenrath et al. 1966: 352). Young also recognized the potential of dendrochronology as a dating method as soon as he opened Tumulus MM (1958: 148, note 17), and he later sent timbers from the Early Phrygian settlement to the University of Arizona (1968: 240). While the many dendrochronological samples taken from Tumulus MM were consistent with each other, they were "floating" and thus could not provide an absolute date, even in 1981 when the tomb excavations were published (Young et al. 1981: 95-96; see also Kuniholm and Newton in Rose and Darbyshire forthcoming).

Thus the shift in the absolute dates of the Destruction Level and subsequent rebuilding resulted not from a difference in methodology, but from a difference in the relative weighting of the historical and archaeological records, and in the kind of samples that the archaeologists had to work with—itself a reflection of scholarly interests and values. In order to show how the Gordion chronology was built and rebuilt, this paper seeks to: 1) describe the ways in which the initial relative and absolute dates for the Early Phrygian destruction and the subsequent Middle Phrygian reconstruction were established; 2) highlight some of the problems that these initial dates posed for the understanding of excavated materials as analysis and excavation proceeded; and 3) describe the way in which the old dates came to be questioned by current members of the Gordion Project and then rejected in favor of a new chronology.

The "Old" Chronology

There is ample documentation to show how the initial chronological sequence for Gordion was constructed by Rodney Young. Although at the time of his death Young had not completed a single final report on his work, he published comprehensive preliminary reports for each full-scale season up to 1967 which describe changing goals and new finds as well as his interpretation of these finds. The following section quotes heavily from his work in order to show not only his conclusions, but the assumptions and arguments on which those conclusions were based in an effort to show why the old chronology was so weakly supported and thus so easily discarded once the 700 BC date for the Destruction Level was seriously challenged.

The archaeology of historic periods requires a consideration of information from both text and dirt, and the balance between these shifts depends on factors that include the adequacy of historical records and the inclinations and assumptions of the archaeologist. For Young, trained as a Classical archaeologist in the

Figure 4: Plan of the Middle Phrygian/YHSS 5 palace quarter. The rebuilt citadel used very different construction techniques from those of the Early Phrygian period (Fig. 3), but the overall plan of the two periods is virtually identical.

1930s (DeVries 1980: xv), the historical record was primary. This point of view is expressed over and over again in his publications: the role of archaeology was to provide a substantive material context for the dates and events recorded by ancient historians, or to document events that were not attested in the historical record, but which fit within the framework provided by texts (e.g. Young 1953a). This privileging of historical data had a profound impact on his construction of the Gordion chronology, since Young frequently assigned dates to construction levels based on a perceived correspondence between archaeological remains and the kind of settlement he expected to find based on texts (see below). Once these historical links between construction levels and historical events were made, they were viewed as sufficiently strong to overcome another date suggested by the archaeological evidence. The following example, although related to a chronological period beyond the scope of this paper, illustrates the kind of logic that was used to discard archaeological evidence when considering the date of the Destruction Level.

In 1953, Young found a small structure (the "Hearth Building" or Building E) that was part of the Middle Phrygian reconstruction, which Young at that time attributed to the Persians (Young 1955: 6; see also below). The building had been rebuilt several times and had a central hearth; this plus an absence of living quarters led to its interpretation as a shrine, and since Persian in date, to its identification as a structure related to the Persian fire cult. The Hearth Building/Building E was eventually abandoned, and a house was built above it; within this house was a group of pots including a lydion. Young's dating the abandonment of the Hearth Building/Building E has to take into account the date provided by the lydion which he tells us is of a type found elsewhere in the fifth century and found at Gordion in fourth century contexts. Nevertheless he concludes that:

> While we might hesitate to date this decadent lydion as late as after 333 B.C., the logical time for the destruction of the sanctuary and the abandonment of the fire cult which it sheltered would seem to be the coming of Alexander....(Young 1955: 6).

This example is informative for three reasons. First, it shows the degree to which Young relied on the historical record to interpret his finds, especially during the early years of his project. Second, it illustrates the way in which he clearly separated data and interpretation, and presented data that could lead to alternative interpretations; he presents arguments, not assertions. As a result we not only know what he thought but why he thought it, and it is this that gives his reports great value, even when the conclusions he drew were wrong. Third, it provides a reminder that the published corpus from Young's excavations are preliminary reports, and that as the excavated sample grew, he discarded old ideas in favor of new ones that better fit the current data set. Thus (as is the case with any excavation) anyone working with preliminary reports must start with the most recent, and use these to evaluate and reinterpret earlier publications. It is also true that with every season of excavation, the accumulated archaeological sample and repetitive patterns within this sample led Young to an increasing reliance on the archaeological record—an example of the "unpretentious pragmatism" which characterized his work at Gordion (DeVries 1980: xvii). To continue with our chronological example, by the late 1960s Young realized that the Middle Phrygian rebuilding might well predate the Achaemenids (see below) and that the late Iydions and similar pottery provided a legitimate date in the fifth or fourth century for what was indeed an Achaemenid period structure above the Hearth Building. It is this process of rebuilding the Gordion chronology that I examine in the next section of this paper.

An Initial Chronology for the Citadel Mound

A sounding located roughly in the center of the eastern half of the Citadel Mound was dug 16 meters down from the surface to the water table by Machteld Mellink in 1950. This limited sample was used to construct an archaeological sequence made up of six "levels" or periods extending from the Early Bronze Age to the Hellenistic period (Young 1950; 1951). Young also found Roman and substantial Medieval remains on the western half of Yassıhöyük (1951: 4, 7) and even in the Main Excavation area (1956: 250), but these periods were never included in his numbered sequence. In Mellink's trench, the most recent or first level was assigned to the Hellenistic period, identified by coins and a relatively well-known ceramic assemblage. Beneath was a second level that was soon attributed to the Persians (Young 1950: 198; 1953c: 20; see also below) and below that, Young's first sighting of the Phrygians:

> Unlike the buildings of the archaic second level, which had been badly plundered in later times by people in search of building material, this Phrygian building of the third level gives promise of being recovered in good condition. The debris over its floor, largely of broken bricks and rubble, contained also ash and cinders, perhaps suggesting a destruction by fire which could conceivably be brought into connection with the invasion of Asia Minor by the Cimmerians at the beginning of the seventh century, bringing an end to the hegemony of the Phrygian kingdom (Young 1951: 12).

Young was consistent in his view that the most impressive finds archaeologically must represent the period when the site was at its political and artistic height based on texts, primarily texts and legends from the west. Thus when he had excavated a substantial part of the gate building in 1953 he laid out the following argument for its date:

> Thus far the evidence proves only that this building was considerably earlier than the time of the Persian Empire. Since, however, it is on a scale even larger than that of the Persian Gate...and since it was constructed entirely of dressed stone without crude brick, it obviously must date from a period of great prosperity and power. On historical grounds it should be assigned to the time of the greatest expansion of the Phrygian Kingdom....The Phrygian Kingdom was...at the apex of its power toward the end of the eighth century, when it apparently extended as far to the southeast as the Taurus and was in contact with Assyria. This period of power was undoubtedly the time of the adornment and fortification of its capital city. It would thus seem safe on historical grounds to assume a date in the eighth century BC (or earlier) for the building of the Phrygian city wall and gate (Young 1953b: 16).

Young did not find any evidence of fire within the gate building in 1953, an observation that leads him to speculate on the relationship between the end of his "Phrygian" level and the nomads he thought he saw in 1950:

> It will be interesting to see [in future seasons] whether any evidence appears of a sack or destruction by the Cimmerian invaders, whose destructive activities have been much heralded in literature, but of which almost no traces have been found in excavation (Young 1953b: 18).

Support for his proposed link between the Phrygian Level and Kimmerians was found in the 1955 season, when Young began clearing an area well inside the gateway where he finally found abundant evidence for burning in the buildings beneath the clay (Young 1956a: 262–263). A sounding into one building produced a large series of pots, a "closed deposit" which provided archaeological evidence that could be used to establish an absolute date. Young compared this pottery to that from Tumulus III excavated by the Körte brothers, which had a "commonly accepted" date of 700 BC (1956a: 263, n. 24). The origin of this commonly accepted date illustrates another problem: Young cites Ekrem Akurgal for the dating of the Tumulus K–III material, offering no comment on the evidence supporting Akurgal's date; according to Kenneth Sams, Akurgal based his date for the K–III pottery "primarily on stylistic comparisons with Greek art and on the ancient historical dates for Midas" (1994: xxix, with references; see also Rose and Darbyshire forthcoming). Thus Young's ceramic comparisons, which should have provided a second independent source to confirm a date based on historical reasoning, instead relied on the same historical framework.

Young concludes that:

> It is tempting to see in the conflagration which destroyed the Phrygian buildings at Gordion the effects of the destructive raid of the Kimmerians which took place at the beginning of the seventh century [696/5 or 676 BC]; and for this the commonly accepted dating of the pottery of Tumulus III and the traditional date of the Kimmerian invasion are in agreement (Young 1956a: 263).

A shift from a "Kimmerian hypothesis" (1956a: 264) to Kimmerians as probable cause for the destruction came in the 1956 season when Young excavated Megaron 1 and most of Megaron 2 (1957: 320). After this time the only real objections to Young's chronology seem to have come from skeptical graduate students who are reported in the oral traditions of Gordion to have sometimes muttered disparagingly about the Kimmerian invaders, with little or no impact on their teacher.

More problematic for Young was the dating of the massive (re)building program in the "second level" which was "tentatively" attributed to the Persians in 1950 (Young 1950: 198). Also referred to as the "archaic city" the deposits assigned to this phase were badly disturbed, but had produced pottery and wall tiles that Young dated to the sixth century, presumably on the basis of Greek parallels (1951: 7; 1953: 25). By 1955, the second level had become "the city of Achaemenian times" (Young 1955: 1; see also Young 1956b: 17–18). The scale of the buildings recovered in the first years of excavation and the amount of labor required for their construction taken together with their stratigraphic position sandwiched between a settlement attributed to the Phrygians and the more securely dated Hellenistic levels must have made this link between archaeological remains and the historical record too obvious for discussion.

But what about the chronological gap in the Citadel Mound sequence between the Phrygian destruction and a rebuilding by the Persians in the mid-sixth century BC, or what Young came to call the lacuna (1956a: 264)? He states that:

> A few sherds and other objects of the intervening 150 years have been found above the clay, but they cannot be associated with any seventh or early sixth century habitation levels and seem to be strays. No sherds or objects of the "lacuna period" have been found below the clay. Yet a number of the tumuli dug contained burials of the mid-seventh century and later, and there was a building on the smaller habitation mound to the southeast [the Küçük Höyük] which was seemingly destroyed in the conflict between Croesus and Cyrus in 547/6 BC. The Gordion of the Lydian period between ca. 690 and 550 has evaded us thus far, though it seems unlikely that the main site was entirely deserted over this long period (Young 1956a: 264).

Not everyone agreed with this interpretation of the finds from the mound. The continuity in plan between the destruction level and rebuilding had been noted as early as 1959 by G. Roger Edwards, Young's colleague at Penn and acting field director for three excavation seasons that were aimed at clearing away "post-Kimmerian" deposits to facilitate excavation of the Destruction Level (Fall 1951, Fall 1955, and 1958). To acknowledge this architectural continuity he suggested the term Middle Phrygian for the rebuilding or Persian city. Edwards also sought to explain the Lacuna material, which he saw as "a theory worth testing, that following the destruction of Gordion by the Kimmerians the king of Phrygia and his government may have established themselves" in the walled area between the Citadel Mound and the Küçük Höyük (Edwards 1959: 264). The theory may have been Edwards's, but it is more likely that he was influenced by (or convinced by) Machteld Mellink, who had been working on the walls of the Küçük Höyük fortification system during the 1958 season. Mellink found three phases of construction in the Lower Town walls. In one area the earliest wall was built of mud bricks with charcoal inclusions, and in a report to Young, she states that she was "tempted" to see these bricks as made from "Cimmerian destruction rubble" (Mellink 1958: 2, partially quoted in Edwards 1959: 264); the second and third wall phases she dated to the seventh and early sixth centuries—i.e the Lacuna Period. In an article summarizing research at Gordion that was also published in 1959 Mellink reconstructs the sequence of events that she thought most likely based on her fieldwork:

> After the violent destruction of the city of Midas, the Cimmerians continued to keep the country in turmoil. On the east [sic] side of the citadel of Gordion the Phrygians rapidly built an interim fortification system. Instead of the stone ramparts that had been the pride of previous generations, they put up a mud brick wall about 20 feet thick and 35 feet high (1959: 105–106).

Left open was the date of reconstruction on the Citadel, referring only to "the rebuilt city, as we know it for the sixth century B.C." (Mellink 1959: 106). In later publications she was more explicit:

> If we can confidently associate the burnt and looted citadel with the final stages of Midas' rule, we are at a loss for the historical identification of the kings who rebuilt the town. The reconstruction, thoroughly Phrygian in type, must have taken place during the rise of Gyges and the Mermnad dynasty of Lydia [i.e. second quarter of 7th century] (Mellink 1991: 630; see also 646, 649).

The idea of Phrygian continuity suggested by his colleagues was ignored by Young until the report on the 1963 season. At this point he acknowledges the similarity in plan between the two Citadels but he then turns to the date of the rebuilding:

> It might be supposed, and it has been suggested, that there was no long hiatus between the settlements, and that the later one should be dated to the seventh century rather than to Persian times. For the date of the laying down of the clay and the building of the city above it there is little direct evidence: a few sherds as late as the mid-sixth century found in the surface of the clay seem to be balanced by occasional seventh-century sherds found immediately above it (Young 1964: 284).

In the end he did slightly modify his position, allowing at least the possibility of Lydian sponsorship:

> The planning and building of the archaic city was a huge undertaking, presumably beyond the capacity of a younger Midas or Gordius with only the resources of a much-reduced kingdom. It required the backing of a Cyrus (or perhaps an Alyattes), supplementing from a wider realm the skill and labor that were locally available (Young 1964: 285).

It was still the "Persian city" in 1968, but in his last published comment on the date of the rebuilding, Young accepted a date in the first half of the sixth century, "probably by Alyattes the Lydian King" (1976: 360; see also Young 1968: 231). This date for what came to be formally named the Middle Phrygian period was accepted by DeVries in his summary of the 1969–1973 seasons:

> The first half of the sixth century makes good historical sense as the time for the rebuilding, for by then the Lydians had driven the Kimmerians out of Anatolia. The rebuilding may well have needed the approval of the Lydians, whose empire now included Phrygia, but in spite of this and in spite of the fact that a Lydian garrison is likely to have been installed in the subsidiary citadel of Küçük Höyük, the project is not likely to have been a Lydian one. The great care in duplication of the long-ruined buildings from the great days of the Phrygian kingdom suggests rather a Phrygian undertaking. If we can take seriously the anecdotal indications in Herodotos that the Phrygian royal house still survived in the sixth century, we might suspect that it was behind the rebuilding of the ancestral seat of power (DeVries 1990: 392).

In 1987 I spent the summer reading all of the preliminary reports and studying photos of Young's excavation at Gordion. DeVries had completed most of his preliminary report on the 1969 and 1973 seasons, and was at Gordion in 1987 to provide information relevant to the proposed new field project. I accepted DeVries's attribution of the rebuilding program to the Phrygians based on the clear continuity in the organization between the elite quarter of the Early and Middle Phrygian citadels: architecture is one of the tools that political leaders use to demonstrate power, and architectural style often carries a strong symbolic meaning. At the same time, it would be hard to understand why Lydian leaders stationed in a military or minor political outpost would set out to replicate the ancient Phrygian palace quarter; as representatives of a powerful state from the west one would expect them to replicate their own architectural and organizational style, if for some reasons they set out to build an elite residential area at Gordion.

DeVries's concept of a cultural revival that spurred a local dynasty to replicate their own past architectural glories seemed reasonable, but a revival and rebuilding after a hiatus of 100 years seemed an unnecessary stretch of the imagination, and it was hard to see why this dynasty, which supposedly had little political and economic power, was able to mobilize the labor that was required for a truly massive building project. A consideration of the labor required to cover the entire area of the Citadel Mound (ca. 8.5 ha) with an earth fill up to five meters deep and a rubble fill up to 9.5 meters deep in the Early Phrygian gateway, had been fundamental to Young's historical argument, as it would be for any field archaeologist who has had to move large quantities of dirt during excavation. Since I was trained as an anthropologist and Near Eastern prehistorian, the archaeological record came first in my thinking, and rather than shift the date, I started to question the supposed weakness of the Phrygians after the Kimmerian destruction that had been inferred from an absence of historical documentation for the seventh century BC. From my point of view, the fact that there were no names mentioned in the few texts that had survived could not overturn the archaeological evidence for considerable labor force and a political power able to mobilize them in the Gordion region in the seventh century. As discussed in the next section, my doubts about the accepted date for the beginning of the Middle Phrygian period grew when we began new fieldwork.

New Fieldwork and How the Old Chronology Fell Apart

As part of our research and planning season in 1987, Robert Dyson and I carried out a study of the large area of Early Phrygian buildings excavated by Young and a photographic survey of the exposed edges/baulks of this "Main Excavation Area." This research gave us a clearer idea of the depositional sequence and raised questions about a construction project that Young had recognized and placed chronologically in the period after the fire (1974: 284). Working from the Gordion notebooks, DeVries had realized that this building activity had in fact taken place before the Destruction Level fire. Specifically he pointed to a terrace and a drain system just inside the Early Phrygian gate (which had been built from a demolished entrance to the gate, the Polychrome House) and a second parallel terrace to the northeast which had signs of burning (The Unfinished Project in DeVries 1990: 387–388, Fig. 22). As newcomers, it looked to Dyson and me as if the group of structures that could be associated with this aborted project included not just terraces, but a large megaron (Building C) and perhaps even the filling of the great gate.

The very existence of the Unfinished Project had historical and chronological implications. First, the Phrygians could not have been under military threat at the time of the fire since they would hardly have started to remodel their fortification system under such conditions. Second, the fills, terraces, and drain built before the fire could be seen not as an aborted project, but as the first stage in the Middle Phrygian rebuilding. This new building project employed thick clay fills, deep rubble foundations and ashlar block walls—the style of masonry typical of the rebuilt citadel—and established a new ground level for the area inside the city gate that was used in Middle Phrygian times. If the Phrygians had begun to raise the level of the citadel before the fire, and simply continued after it, one could argue strongly against any significant hiatus between destruction and rebuilding. I still did not question the date of the Early Phrygian Destruction Level, but had significant doubts about the date of the subsequent rebuilding.

Figure 5: View of the Terrace Buildings excavated by Young at the end of the 1989 season. The two men working are standing in the main room of Terrace Building 2. The room adjacent to the left is the newly excavated anteroom or TB 2A. The stratigraphic section running through TB2A shows the Early Phrygian/YHSS 6A Destruction Level at the bottom, overlain by the thick layer of fill with massive rubble foundations of the Middle Phrygian/YHSS 5 rebuilding.

Figure 6: All of the Terrace Buildings had artifacts preserved in situ on the building's floor. In TB2A the most common finds were pottery vessels and tools associated with textile production including many doughnut shaped clay loom weights as well as spindle whorls.

The 1988–89 stratigraphic sounding provided evidence that the process of reconstruction after the Early Phrygian/YHSS 6A fire began almost immediately, at least in Terrace Building 2 (Figures 5–8). Above the floor deposit of artifacts and charcoal lay reeds and burnt organic material that may have been stored on the roof; together these strata can be identified as primary debris from the fire. This primary deposit, which slopes down from the area adjacent to the walls to the center of the room, is capped by a heterogeneous layer that includes thin sloping layers of clayey soil that could represent erosion from the highest points of the burned debris (including the walls) over a short interval of time. Within weeks or months, however, the Phrygians began to level the walls of Terrace Building 2 to a single course, throwing flat stone slabs and chunks of burnt bricks into the center of the room to provide a level platform. It is at this point in the stratigraphic sequence that we can see a significant lapse of time, long enough to produce a hard erosion surface over the leveled building(s). Eventually a thick fill was laid over the prepared leveled surface, and rubble foundations for a Middle Phrygian/YHSS 5 structure with ashlar walls (Building I:2) were set into the fill (Voigt 2007).

Figure 7: Section through the burned deposit in Terrace Building 2A showing the burned deposit and primary collapse at the bottom, stone blocks and other materials thrown in during cleanup after the fire, and the sharp juncture between cleanup and the layer of relatively clean fill representing the start of rebuilding.

With this sequence, the scenario proposed by DeVries had to be discarded. The Phrygians did not abandon Gordion after the fire, but almost immediately began cleaning up, and it now made a great deal of sense to see the Unfinished Project and the Middle Phrygian reconstruction as part of one sequence. Postulating a very short hiatus between fire and reconstruction also solved two minor questions that were raised by the "long hiatus" theory. First, a practical consideration: I had always wondered how a remnant

Figure 8: After we excavated within Terrace Building 2A, the stone cleared from the walls after the fire was clearly visible as a layer of tumbled rock including rubble from the core of walls as well as nicely cut flat stones from the wall faces. This trench represents the first cut within TB2A; at the top of the photo, the stone deposit has been cleared down to the building floor.

Phrygian (or Lydian) dynasty had managed to duplicate the Early Phrygian/YHSS 6A settlement plan from 100 year old ruins. DeVries (following Young 1964: 284–285) answered this query by saying that the Early Phrygian wall stubs were still visible, providing a general guide to construction—an answer that seemed possible but unlikely to me. Second, if the walls were still visible, I wondered why no one returned to Gordion to poke through the rubble, looking for anything from tools to valuables, not to mention building material. The only Early Phrygian/YHSS 6A building that had been disturbed soon after the fire was Megaron 4, a relatively new construction that was built on a platform of clay.

The YHSS also provided secure evidence for a date in the seventh century for the initial occupation of a semi-subterranean room that was set into clean, gray Middle Phrygian clay and was clearly part of the post-fire construction project (Voigt 1994: 274, Fig. 25.3.2–4, 25.6.3–4, 25.7.1; Voigt and Young 1999: 205,

Fig. 9–11). This room, attached to and built at the same time as Building I:2 (contra DeVries 2005: 43), produced ten bronze fibulae from floors, fill after abandonment, and pits cut into the fill. The stratigraphically earliest examples from this cellar were comparable to fibulae dated to the early seventh century in secure contexts elsewhere. Gordion examples came from the South Cellar, a Middle Phrygian house under Tumulus H and Tumulus S-1 (Caner 1983, Types GI, JII,1 and KI,2; see also DeVries 2007: Fig. 27). An additional Middle Phrygian fibula that came from just above the clay layer in a trench on the western side of the mound could be assigned to Caner Type F2; this dated to contexts placed in the late eighth to seventh century by Caner. There was no material in either the I:2 cellar fill or the many pits dug into it that required a date in the sixth century, and not one piece of Lydian (or "lydianizing") pottery in a very large ceramic sample. There was, however, one small sherd from a Greek import that proved critical in the re-examination of material from the Young excavations (see below).

Data collected between 1987 and 1997 through excavation and an intensive surface survey of the fields surrounding the Citadel Mound (Figure 9) provided a different measure of changing economic and political power. The settlement of Gordion was relatively small in Early Phrygian/YHSS 6A times, consisting of the walled palace quarter which lies under the eastern half of the present mound, and a low settlement to the west (Figure 10; Sams and Voigt 1995: 374–375). We found no evidence for occupation in the area of the Lower Town, but near the western edge of the vast Outer Town was yet another pithouse in Operation 22 (Figure 10) (Voigt and Young 1999: 219, Fig. 23; Sams and Voigt 1995: 376–377, Figs. 10–12). This structure (which may be attached to a "normal" brick building at surface level) produced a good sample of plain gray pottery, some of which Sams considered "transitional" between the Early and Middle Phrygian assemblages from Young's excavations. The evidence is slim, but this modest structure foreshadowed the Middle Phrygian/YHSS 5 expansion of Gordion when the settlement covered an area of about one hectare. It may also point to the area where at least part of the population of the city lived while the Middle Phrygian construction project was underway.

As for the Citadel Mound and the Lower Town, excavation showed that the scale of the construction process at the start of YHSS 5 was enormous (Figure 11). Not only were deep deposits of fill laid over the palace quarter on the eastern half of the Citadel Mound, but on the other side of a street that ran along the western wall of the palace quarter, a second high mound was built of clean clay fill that was more than 5m deep; built on this newly created surface were substantial houses with stone walls and pebble floors (Voigt et al. 1997: 5–6; Voigt and Young 1999: 207–211, Fig. 12–13; see also Voigt 2007). Within the Lower Town, a terrace was built inside the fortification walls. Two kinds of structures were found in the area's samples: monumental architecture with ashlar walls and rubble foundations to the east and domestic buildings of mud brick on foundations of small stones to the west (Voigt 2005: 23–35; Voigt and Young 1999: 211–214, Fig. 14–22).

To summarize, based on the information collected between 1988 and 1997, we could place the construction of the Middle Phrygian city at least as early as the first quarter of the seventh century BC, and could say with confidence that its earliest occupants were Phrygian speakers who left pottery marked with alphabetic graffiti in early YHSS 5 contexts (Henrickson 1974: Fig. 10.7b). But, despite the fact that the seventh century city was not just surviving but truly thriving, there were no historical records for Gordion or its rulers for this time period, an absence that seemed extraordinary, but that we reluctantly accepted as one of the quirks of the textual record.

Back to Basics: Rethinking Old Data and Obtaining New Dates

In excavating Terrace Building 2A, we collected large quantities of wood as well as charred seeds (Figures 12–14). Dendrochronological dates on 256 samples from large beams that were presumably structural resulted in a date of 850 BC for the latest preserved ring. The TB2A timbers were not the first dendrochronological samples collected from the Destruction Level; Young had collected wood from several of the megarons and the service buildings to the west but the TB2A samples were the first very large group from a single room (Kuniholm and Newton in Rose and Darbyshire forthcoming). A ninth century date for the construction of TB2A seemed far too early, so we grumpily ascribed the 160 year discrepancy with the 700 BC Kimmerian Destruction to the reuse of structural timbers.

The real break-through in overturning the old chronology was made by DeVries who was studying Greek imports from the 1988–89 sounding. By 1997, DeVries had been convinced by the stratigraphic evidence that the rebuilding took place immediately after the Early Phrygian Destruction; he also had recognized an Early Greek sherd from the YHSS 5 deposit in the Building I:2 cellar. The sherd was tiny, but it fit a pattern. The presence of eighth century Greek imports at Gordion was to be expected in the Destruction Level, but instead they were found in strati-

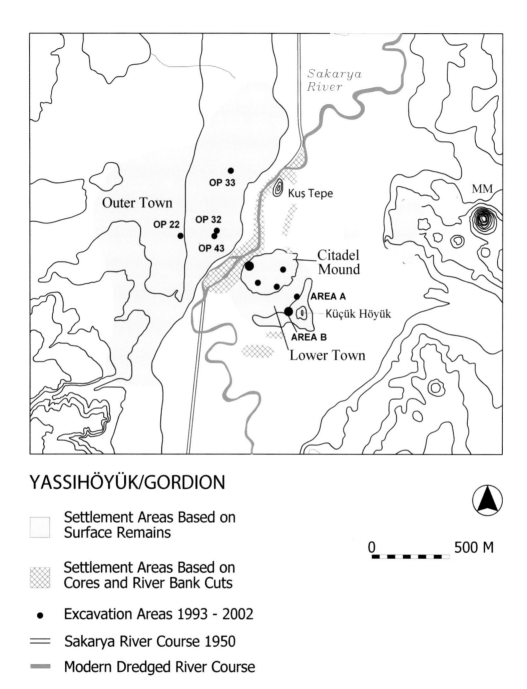

Figure 9: As a site, Gordion has three major topographic zones: the central high mound or Citadel, an adjacent walled Lower Town, and a more extensive Outer Town. During ancient times the Sakarya River flowed to the east of the site, separating it from suburbs and tombs on the slopes further east (Figs 10-11). In addition to intensive surface survey which determined the greatest extent of settlement, soundings were used to establish periods of occupation within the topographic zones.

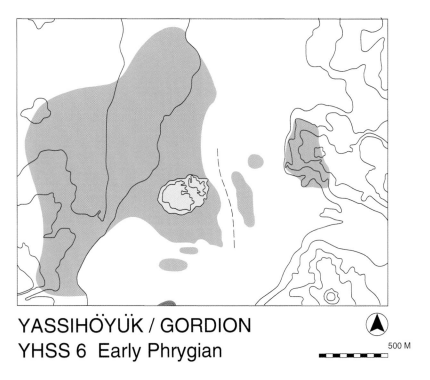

Figure 10: The settlement area during Early Phrygian/YHSS 6 times is shown in yellow, while the maximum area of settlement at Gordion during ancient times is marked in grey. Note the location of the Sakarya River in antiquity, marked by a dashed line.

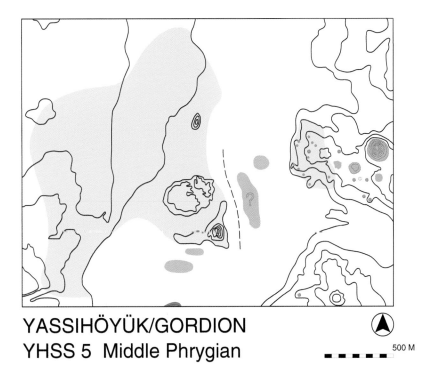

Figure 11: Settlement Area during Middle Phrygian/YHSS 5 times is shown in yellow. This map shows our current understanding of the site, with most of the burial mounds or elite tombs (marked in red) built during 8th to 6th centuries BC.

Figure 12: In addition to many scattered fragments of wood, a part of a wall or the ceiling was found articulated on the floor of TB2A.

Figure 13: Peter Kuniholm and Shana Tarter wrapping one of the large TB2A timbers for shipment to Cornell.

graphic contexts above the clay, a fact that had always worried Ken Sams (1979: 47). DeVries began a re-evaluation of the larger sample of seventh century Greek imports from the Young excavations, looking specifically at the contexts in which this material had been found.

A close reading of excavation notebooks placed material that dated to the late eighth century in good contexts within the Middle Phrygian/YHSS 5 citadel. The most important of these contexts was the "South Cellar," where the Greek imports were associated with pottery that was similar to that recovered from early Middle Phrygian/YHSS 5 contexts found in the new excavations. DeVries found that an Attic red-figure sherd dating to the mid-fifth century that had been used by Young to date the entire South Cellar came from a disturbed and badly excavated area (DeVries 2005: 37; see also Young 1966: 268–269, Pl. 74, Fig. 3–6). Once he decided that the Corinthian Late Geometric kotyle from the cellar actually dated one phase of the cellar's use, DeVries had a new *terminus post quem* of 735–720 BC for this Middle Phrygian context—which had fibulae nearly identical to some from the Building I:2 cellar.

Figure 14: Some of the small pots found in TB2A were full of seeds. When possible, the seeds from a single pot were collected by hand for possible ^{14}C dating.

In 1998 DeVries gave a paper at the annual meetings of the Archaeological Institute of America which accepted a date "not far from 700 BC" for the Middle Phrygian rebuilding and proposed a new historical explanation for both Early Phrygian destruction and reconstruction. Still convinced that the rebuilding could not have taken place when "Kimmerians were creating havoc in central and western Anatolia," he placed the fire before 700, attributing it to an Assyrian incursion into Anatolia in 710 or 709 BC (DeVries 1998: 397). In this scenario Midas was defeated but his political power was not destroyed by Sargon and it was Midas who rebuilt the citadel in the latter part of his reign. This reconstruction had little support from Assyriologists and was soon abandoned by DeVries, leaving the destruction unexplained but dated before 700 BC. The Kimmerian horde was finally defeated.

In 1999 I gave a lecture on Gordion at Cornell and talked with Kuniholm about the possibility that the Destruction Level was earlier than 700. Although we had carefully collected seeds from the Destruction Level in 1989, it had seemed pointless to spend the money to run them if they did indeed date around 700 BC because of the "Hallstatt Plateau" in which almost all ^{14}C dates for some four centuries come out about 2450 BP. Now, with reasonable doubt cast on that date and a strong suspicion that the Destruction Level might be significantly earlier based on ceramics, radiocarbon dating might be highly productive. In January 2000, DeVries, Sams, and I met in Philadelphia to discuss ways of resolving the chronological problem based on artifact comparisons. Both DeVries and I reminded Sams that in his publications on Early Phrygian ceramics, there were some indications of a date earlier than 700 for painted pottery (Sams 1974; 1978). An even firmer date could be obtained for a group of distinctive kantharos kraters found in Megaron 3 and TB1 which had good parallels with a Greek form that dated to the second half of the eighth century (Sams 1994: 78–79). By the end of our conversation Sams suggested a date of around 750 BC for the Destruction Level based on his own meticulous study of the ceramics.

With this new provisional date, Peter Kuniholm offered to obtain funding to run some of the charred seeds collected from Terrace Building 2A and to ask Bernd Kromer if he would be willing to run them in his lab at Heidelberg, the lab that ran the wiggle matching dates for the Aegean Dendrochronology Lab. In late 2000 the samples were sent to Kromer, and in a few weeks Peter got an email, asking why we had been concerned about a possible date within the Hallstatt Plateau: the samples had produced significantly earlier readings.

When the radiocarbon determinations were calibrated, the dates for the Early Phrygian/YHSS 6A Destruction Level were consistent, placing the fire around 800 BC (Figure 15). This was half a century earlier than our guess dates based on pottery comparisons, but compatible with other artifact comparisons (e.g. sculpture) that had been explained away (see Rose and Darbyshire forthcoming). To answer questions raised about the calibration based on statistical grounds (Keenan 2004) additional short-lived samples (seeds from TB2A and samples of reed from the roof of the building) were sent to Heidelberg. The new results from seeds confirmed the ca 800 BC date for the fire, and placed the construction of the building a few

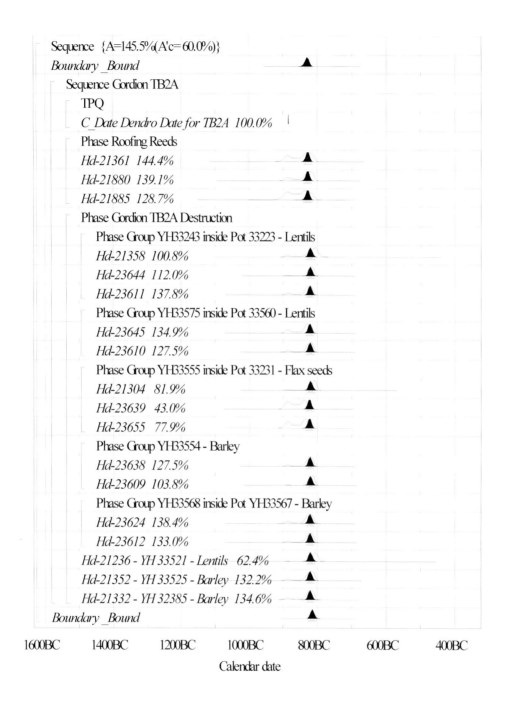

Figure 15: Terrace Building 2A sequence of radiocarbon data from the *terminus post quem* of the last preserved tree ring in the building, to the age of the roof reeds, to the short-lived samples found in the destruction level of the building. These data from the Heidelberg Radiocarbon Laboratory and their analyses are presented and discussed in detail in Manning and Kromer forthcoming. The hollow histograms show the raw (un-modeled) calibrated calendar age ranges for each sample. The black histograms show the modeled calendar calendar age ranges given the archaeological sequence information as shown. The upper and lower lines under each distribution show respectively the 1σ (68.2%) and 2σ (95.4%) confidence ranges for the modeled ages. Calibrated age ranges were determined using the IntCal04 calibration dataset and the OxCal software (see Manning and Kromer forthcoming for details and discussion). The YHSS Phase 6A Destruction Level data on short-lived samples all lie in the few decades before and around 800 BC (and nowhere near 700 BC) (modeled or un-modeled). The agreement index (the % number before each of the distributions shown) compares the final (posterior) distribution (the solid histogram) against the original (hollow) distribution. If the former is unaltered the index value is 100%. The value rises above 100% where the final distribution overlaps only with the very highest part of the prior distribution. In contrast, an agreement index below 60% indicates disagreement with the model (and insufficient overlap of the distributions) at about the 5% level of a chi-squared test. The overall agreement index for the sequence is shown at the top—again a score greater than 60% indicates that the model surpasses an approximate 95% confidence level.

decades earlier, supporting the date obtained through dendrochronology (Figure 15).

A "bonus" came when DeVries went back to check on radiocarbon dates for short-lived samples from Megaron 3 that had been run at the Pennsylvania lab (Kohler and Ralph 1961); when calibrated with IntCal, these samples are consistent with the results from the TB2A seeds (Table 1).

The final step in establishing a new chronology was to re-examine the archaeological evidence from the Early Phrygian Destruction Level. To answer critics who saw the radiocarbon results as inconsistent with artifactual evidence, DeVries began a systematic study and seriation of fibulae and bronzes from the Destruction Level and tumuli that had been dated to the eighth century; his results are published in two papers that were originally presented in Turkey (DeVries 2007, 2008). Continuing the process of reanalysis, Sams has recently presented a paper on the Early Phrygian painted pottery. One striking pattern on the Citadel Mound is that one of two styles of "Early Phrygian Painted Ware" (Brown-on-Buff) can now be shown to occur only in the rebuilt city, a clear indication of continuity that parallels the truncation and resumption of the Unfinished Project to the Middle Phrygian construction project (Sams 2007; see also Sams 1994: 166–173).

Conclusion

The most important argument in favor of the new dating of the Early Phrygian/YHSS 6A destruction and the Middle Phrygian/YHSS 5 rebuilding is that it resolves not only internal problems with the Gordion chronology but also makes sense out of ties to other sites, especially for the late ninth and eighth centuries. Good parallels for Middle Phrygian ceramics and other artifacts can be found at Boğazköy, Kaman Kale Höyük, and Kerkenes to the east, as well as to Midas City and Tumulus D at Bayındır/Elmalı to the southwest. Precise statements about the timing of the expansion of the Phrygian polity and the nature of relationships with specific sites and regions must be based on more complete artifactual analyses and comparisons than we have at present, but the most satisfying aspect of the new dating is that the difference in the size and splendor of the Early and Middle Phrygian settlements now makes sense with what little we know of Gordion from history. Archaeology tell us that Iron Age Gordion provided its first evidence for political and economic power in the form of a relatively small fortified settlement in the early 9th century, ornamented with sculpture that emulated Neo-Hittite themes and style; the city expanded until the 8th century BC when it reached its maximum size and supported its most elaborate public buildings. This Middle Phrygian settlement corresponds in date to the time when texts tell us of Gordius and his son Midas, the latter a skilled ruler who extended his influence as far as the borders of Assyria.

Returning to the broader issues raised at the beginning of this paper, one of the things that the Gordion redating makes clear is the importance of examining assumptions made while interpreting both archaeological and historical evidence. Evidence from the past is rarely consistent, and there are always data that have to be "explained away." It is not all that unusual for the explained-away or discarded evidence to become the foundation for a new interpretation, as has clearly happened at Gordion. Those of us working at the site have been lucky in that Rodney Young was generally quite clear as to what the evidence was, what he chose to emphasize, and what he chose to disregard—a habit he inculcated in his students. Revising a "certainty" in the history of Anatolia and dispensing with Kimmerian hordes has been a difficult but stimulating task that has resulted in an intellectually satisfying conclusion.

Acknowledgments

Excavation and survey at Gordion since 1988 has been supported by grants from the National Endowment for the Humanities (NEH, a US federal agency), the Social Science and Humanities Research Council of Canada, the National Geographic Society, the Royal Ontario Museum, the Kress Foundation, the IBM Foundation, the Tanberg Trust, the Somerville Fund. Gifts were also received from the following generous donors: Mr. Michael R. DeLuca, Mrs. Jean Friendly, Mr. Peter Paanakker, Mrs. Noel B. McLean, and anonymous. All modern archaeological field research at Gordion (1950-2006) has been sponsored and supported by The University of Pennsylvania Museum of Archaeology and Anthropology. The College of William and Mary has been a co-sponsor of the excavation and survey since 1991, and The Royal Ontario Museum co-sponsored work carried out between 1994 and 2002.

I would also like to thank G. Roger Edwards, Crawford Greenewalt, Ellen Kohler, Peter Kuniholm, Machteld Mellink, Oscar White Muscarella, and G. Kenneth Sams who generously shared their time, their records, their memories, and their opinions on the Rodney Young excavations. Without the advice and support provided by Keith DeVries and Robert H. Dyson, I would never have come to Gordion to experience the unique joys and frustrations of working on a site that never fails to provide a physical and intellectual challenge, and would certainly not have taken on the problem of the Gordion chronology.

References

Caner, E. 1983. Fibeln in Anatolien I. *Prähistorische Bronzefunde XIV*: 8. Munich: C.H. Beck'sche Verlagsbuchhandlung.

Çilingiroğlu, A., and Sagona, A. 2007. *Anatolian Iron Ages 6. Ancient Near Eastern Studies Supplement 20*. Leuven: Peeters Press.

DeVries, K. 1980. Rodney Stuart Young 1907-1974. In K. DeVries (ed.), *From Athens to Gordion: The Papers of a Memorial Symposium for Rodney S. Young*: xv–xix. *University Museum Papers 1*. Philadelphia: The University Museum.

DeVries, K. 1981. Introduction. In R.S. Young, et al. *Three Great Early Tumuli*: xxxv–xxxvii. *Gordion Excavations, Final Reports 1. University Museum Monograph 43*. Philadelphia: The University Museum.

DeVries, K. 1990. The Gordion Excavation Seasons of 1969-1973 and Subsequent Research. *American Journal of Archaeology* 94: 371–406.

DeVries, K. 1998. The Assyrian Destruction of Gordion? *American Journal of Archaeology* 102: 397 (Abstract).

DeVries, K. 2005. Greek Pottery and Gordion Chronology. In L. Kealhofer (ed.), *The Archaeology of Midas and the Phrygians: recent Work at Gordion*: 36–55. Philadelphia: University of Pennsylvania Museum of Archaeology and Anthropology.

DeVries, K. 2007. The Date of the Destruction Level at Gordion: Imports and the Local Sequence. In A. Çilingiroğlu and A. Sagona (eds.), *Anatolian Iron Ages 6. Ancient Near Eastern Studies Supplement 20*: 79–102. Leuven: Peeters Press.

DeVries, K. 2008. The Age of Midas at Gordion and Beyond. *Ancient Near Eastern Studies* 45: 30–64.

DeVries, K., Kromer, B., Kuniholm, P.I., Sams, G.K., and Voigt, M.M. 2003. New Dates for Iron Age Gordion. *Antiquity* 77 (296). http://antiquity.ac.uk/ProjGall/devries/devries.html.

Edwards, G.R. 1959. Gordion Campaign of 1958: Preliminary Report. *American Journal of Archaeology* 63: 263–268.

Henrickson, R.C. 1994. Continuity and Discontinuity in the Ceramic Tradition of Gordion during the Iron Age. In A. Çilingiroğlu and D.H. French (eds), *Anatolian Iron Ages 3: The Proceedings of the Third Anatolian Iron Ages Colloquium held at Van, 6-12 August 1990*: 95–129. *The British Institute of Archaeology at Ankara Monograph 16*. London: British Institute of Archaeology at Ankara.

Keenan, D.J. 2004. Radiocarbon Dates from Ion Age Gordion are Confounded. *Ancient West and East* 3(1): 100–103.

Kohler, E.L., and Ralph, E.K. 1961. C-14 Dates for Sites in the Mediterranean Area. *American Journal of Archaeology* 65: 360–363.

Körte, A. 1897. Kleinasiatische Studien II. Gordion und der Zug des Manlius gegen die Galater. *Mitteilungen des Deutschen Archäologischen Instituts, Athenische Abteilung* 22: 1–51.

Körte, G., and Körte, A. 1904. Gordion: Ergebnisse der Ausgrabung im Jahre 1900. *Jahrbuch des Deutschen Archäologischen Instituts, Supplement 5*. Berlin: G. Reimer.

Kuniholm, P.I., and Newton, M.W. Forthcoming. Dendrochronology at Gordion. In C.B. Rose and G. Darbyshire (eds.) *The Chronology of Iron Age Gordion*. To be published by the University of Pennsylvania Museum of Anthropology and Archaeology.

Manning, S.W., and Kromer, B. Forthcoming. Radiocarbon dating Iron Age Gordion, and the early Phrygian dstruction in particular. In C.B. Rose and G. Darbyshire (eds.) *The Chronology of Iron Age Gordion*. To be published by the University of Pennsylvania Museum of Anthropology and Archaeology.

Mellink, M.J. 1958. Küçük Hüyük 1958. Report labeled "to RSY." Manuscript in the Gordion Archives, University of Pennsylvania Museum of Archaeology and Anthropology.

Mellink, M.J. 1959. The City of Midas. *Scientific American* July 1959: 100–109.

Mellink, M.J. 1991. The Native Kingdoms of Anatolia. *The Cambridge Ancient History (2nd ed.)* III(2): 619–655.

Muscarella, O.W. 2003. The Date of the Destruction of the Early Phrygian Period at Gordion. *Ancient West and East* 2(2): 225–252.

Rose, C.B., and Darbyshire, G. (eds.). Forthcoming. *The Chronology of Iron Age Gordion*. To be published by the University of Pennsylvania Museum of Anthropology and Archaeology.

Sams, G.K. 1974. Phrygian Painted Animals: Anatolian Orientalizing Art. *Anatolian Studies* 24: 169–196.

Sams, G.K. 1978. Schools of Geometric Painting in Early Iron Age Anatolia. *Proceedings of the Xth International Congress of Classical Archaeology*: 227–236. Ankara: Türk Tarih Kurumu Basımevi.

Sams, G.K. 1979. Trade in First Millennium Gordion. *Archaeological News* 8: 45–53.

Sams, G.K. 1994. *The Early Phrygian Pottery. The Gordion Excavations, 1950-1973: Final Reports IV* (2 vols). *University Museum Monograph 79*. Philadelphia: The University Museum.

Sams, G.K. 2005. Gordion Over a Century. In L. Kealhofer (ed.), *The Archaeology of Midas and the Phrygians: Recent Work at Gordion*: 10–21. Philadelphia: University of Pennsylvania Museum of Archaeology and Anthropology.

Sams, G.K. 2007. The Impact of the New Gordion Chronology on our Understanding of Early Phrygian Pottery. Paper delivered at conference, The Archaeology of Phrygian Gordion. The University of Pennsylvania Museum of Archaeology and Anthropology, April 20-22, 2007.

Sams, G.K., and Voigt, M.M. 1991. Work at Gordion in 1989. *XII Kazı Sonuçları Toplantısı, Vol. I*: 455–470. *Excavation Reports of the XIIth Annual Symposium on Archaeological Research in Turkey, Ankara, 1990*.

Sams, G.K., and Voigt, M.M. 1995. Gordion Archaeological Activities, 1993. *XVI Kazı Sonuçları Toplantısı, Vol I*:369–392. *Excavation Reports of the XVIth Annual Symposium on Archaeological Research in Turkey, Ankara, 1994*.

Stuckenrath, R., Coe, W.R., and Ralph, E.R. 1966. University of Pennsylvania Radiocarbon Dates IX. *Radiocarbon* 8: 348–385.

Voigt, M.M. 1994. Excavations at Gordion 1988–89: The Yassıhöyük Stratigraphic Sequence. In A. Çilingiroğlu and D.H. French (eds.), *Anatolian Iron Ages 3: The Proceedings of the Third Anatolian Iron Ages Colloquium held at Van, 6-12 August 1990*: 265–293. *The British Institute of Archaeology at Ankara Monograph 16*. London: British Institute of Archaeology at Ankara.

Voigt, M.M. 2005. Old Problems and New Solutions: Recent Excavations at Gordion. In L. Kealhofer (ed.), *The Archaeology of Midas and the Phrygians: Recent Work at Gordion*: 22–35. Philadelphia: University of Pennsylvania Museum of Archaeology and Anthropology.

Voigt, M.M. 2007. The Middle Phrygian Occupation at Gordion. In A. Çilingiroğlu and A. Sagona (eds.), *Anatolian Iron Ages 6. Ancient Near Eastern Studies Supplement 20*: 311–334. Leuven: Peeters Press.

Voigt, M.M., DeVries, K., Henrickson, R.C., Lawall, M., Marsh, B. Gürsan, A., and Young, T.C. Jr. 1997. Fieldwork at Gordion: 1993-1995. *Anatolica* 23: 1–59.

Voigt, Mary M. and Young, T. Cuyler, Jr. 1999. From Phrygian Capital to Achaemenid Entrepot: Middle and Late Phrygian Gordion. *Iranica Antiqua* 34: 192–240.

Young, R.S. 1950. Excavations at Yassihüyük-Gordion 1950. *Archaeology* 3: 196–201.

Young, R.S. 1951. Gordion—1950. *University Museum Bulletin* 16(1): 3–20.

Young, R.S. 1953a. Making History at Gordion. *Archaeology* 6: 159–166.

Young, R.S. 1953b. Progress at Gordion, 1951–1952. *University Museum Bulletin* 17(4): 2–39.

Young, R.S. 1953c. Where Alexander the Great Cut the Gordian Knot. *Illustrated London News* Jan. 3: 20–23.

Young, R.S. 1955. Gordion Preliminary Report, 1953. *American Journal of Archaeology* 59: 1–18.

Young, R.S. 1956a. The Campaign of 1955 at Gordion: Preliminary Report. *American Journal of Archaeology* 60: 249–266.

Young, R.S. 1956b. Gordion. *Anatolian Studies* 6: 17–23.

Young, R.S. 1957. Gordion 1956: Preliminary Report. *American Journal of Archaeology* 61: 319-331.

Young, R.S. 1964. The 1963 Campaign at Gordion. *American Journal of Archaeology* 68: 279–292.

Young, R.S. 1966. The Gordion Campaign of 1965. *American Journal of Archaeology* 70: 267–278.

Young, R.S. 1968. The Gordion Campaign of 1967. *American Journal of Archaeology* 72: 231–242.

Young, R.S. 1976. Gordion. In R. Stillwell (ed.), *Princeton Encyclopedia of Classical Sites*: 360. Princeton University Press.

Young, R.S. 1981. *Three Great Early Tumuli. Gordion Excavation Reports 1. University Museum Monograph 43.* Philadelphia: The University Museum.

Article submitted July 2007

The End of Chronology: New Directions in the Archaeology of the Central Anatolian Iron Age

Geoffrey D. Summers

Abstract: *Improvements in the Iron Age chronology of the Central Anatolian Plateau, together with a growing consensus on matters of terminology, have permitted archaeologists to turn to other equally challenging issues. At the same time, developments in remote sensing, combined with the growing potential of GIS analysis, are providing opportunities to develop archaeological investigations in new directions. These include the complex relationships between material culture, states, and empires. One aspect of particular concern is the foundation and demise of capital cities in the Iron Age. This paper will seek to demonstrate that a balanced combination of remote sensing, GIS analysis, targeted excavation, art history, and epigraphy is able to provide a rounded view of cultural and political centers (capital cities) in places and regions which lack coherent historical evidence. The results are unexpected, requiring the development of explanatory theory against a background of cultural complexity.*

It has been Peter Ian Kuniholm's unique achievement that dendrochronology has brought about an "End to Chronology" for the highlands of Central Anatolia, an end which will very soon be extended to the entire Aegean and Ancient Near East. Of course there are minor problems to be sorted out and refinements to be made. We would all like more dates and even greater precision. It frustrates archaeologists that dendrochronolgy requires the right sort of tree, trees that grow slowly and are not affected by local stream flow. For statistical reasons secure cross-matching requires a certain number of rings, preferably no fewer than 100, and if the felling date is to be calculated, with the bark as well. For a firm archaeological date, not merely the date of the death of one branch or tree, but multiple samples from the same secure context are needed. And after all that the dendrochronological dates are for the cutting or felling of the trees, and not for the thick layers of burnt destruction from which samples might be recovered together with, at sites like Kültepe Kanesh, Gordion, or Kerkenes, a wealth of cultural material which is what archaeologists would very often like to date. Art historians, some of whom have been slow to accept the new dates, because of their implications, can plead that timbers must have been reused, or that something went wrong in the laboratory (e.g. Muscarella 2003), but such arguments are hard to sustain as evidence accumulates.

It has been alarming that over the years the floating dendro sequence originally fixed to Tumulus MM at Gordion, then slowly and cautiously calibrated to calendar years (e.g. Kuniholm 1993), has shifted significantly; the last time by 22 years (Manning et al. 2003). But we are there now, or as almost there as we are likely to get. It has been a magnificent achievement! Archaeologists, and their students, can now focus their attention on the real business of archaeology, or argue about what the real business of archaeology should be, without getting stuck in the quagmire of chronology.

I will confine my remarks to the Central Highlands of Anatolia in the Iron Age. The arrival of the Phrygians, and doubtless other nameless groups, into Anatolia can perhaps be seen as representing a significant intrusion of western culture into the western fringes of the Ancient Near East, a process which was to reach its climax with Alexander's creation of the Hellenistic Empire. An eastern expansion of Phrygian culture has been brought to light by our own work at Kerkenes (Figure 1).

When, in 1993, we began our own work at the Iron Age capital on the Kerkenes Dağ in the center of Turkey—very probably ancient Pteria, just 50 kilometres from the earlier Hittite capital of Hattuša—received wisdom still maintained that the Late Bronze Age empire of the Hittites, along with Mycenaean

Figure 1: Map showing the location of Kerkenes.

Figure 2: Kerkenes from the northwest. The high tor left of centre is the Kale. Much of the ruined stone wall can be seen.

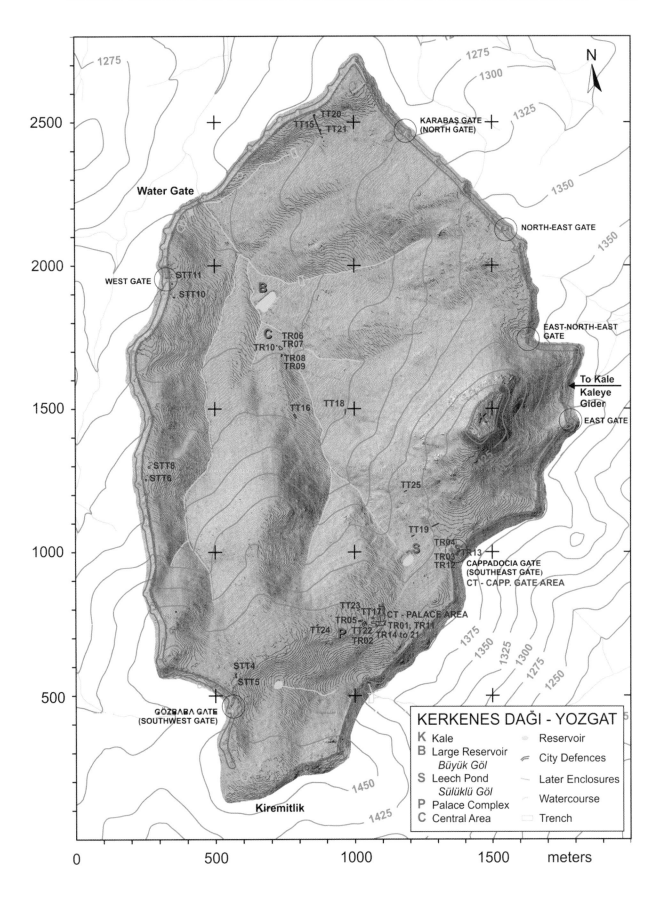

Figure 3: Map of Kerkenes made from 1,400,000 GPS readings.

Figure 4: The northern portion of the site photographed from the hot air balloon in 1993. The north tip is at upper left.

and other literate Mediterranean civilisations, had collapsed around or soon after 1200 BC. This collapse, the cause or causes of which remain hotly debated, was followed by a Dark Age. All the known Hittite centers appeared to have been destroyed and abandoned with no archaeological evidence of settlement in the Anatolian Highlands for some 300 years (Bittel 1970; Ward and Joukowsky 1992; and see now Hawkins 2002 and Fischer et al. 2003). Within the great bend of the Kızılırmak, the classical Halys River, it seems that there was no urban center until the rise of Alishar Höyük, probably on the ruins of the Hittite city of Ankuwa. This modest mound, excavated by Hans Henning von der Osten and Erich Schmidt for the Oriental Institute at Chicago in the late 1920s, hardly, I thought, filled the void between Phrygia (in the west with its capital at Gordion) and Urartu (in the east with its capital at Tushpa, modern Van). There were and are no Iron Age, Neo-Hittite, hieroglyphic inscriptions in the region that had once been the center of Imperial Hittite power (Hawkins 2000). Kerkenes, the largest pre-Hellenistic city in Anatolia, though known was hardly mentioned and rarely visited. Erich Schmidt, in 1928, dug 14 test trenches in five days (Schmidt 1929) and, with chronological astuteness and exemplary brevity, reported by Marconigram to James Henry Breasted that Kerkenes was "Post Hittite Pre Classical" (Summers and Summers 1998). In choosing to work at Kerkenes I had assumed, as had others, that the apogee of this great metropolis would have occurred in the eighth century BC, at the time when King Midas was on the throne of Phrygia and the Kingdom of Urartu was expanding across the Trans-Caucasus. But by the end of the third day of that first season it had become clear that the city at Kerkenes was to be dated to a period not long before the Persian conquest; the entire city, from foundation to abandonment, fell within part of the period known in the jargon as Alishar V; more, precisely, into the late seventh and first half of the sixth century BC. Around the time that we were starting other things were happening in the archaeology of the Plateau; the Dark Age was beginning to stir. At Hattuša Jürgen Seeher was beginning a series of campaigns on the Büyükkaya that were to reveal levels of occupation spanning the entire Dark Age (e.g. Genz 2004 with references) while at Gordion Mary Voigt (this volume) began meticulous excavations into Dark Age (or Early Iron Age) and Early Phrygian layers as well as reinvestigating crucial aspects of the Destruction Level. Light was being shed on the Dark Age at these earlier capitals while Kerkenes was opening a new chapter in the archaeology of what could be termed the Middle Iron Age (the Archaic of the Aegean and related areas, Summers 2008).

The Dark Age has not gone away. Now, thanks in no small part to Peter Kuniholm's work and inspiration, we know much more about it, about its settlements and its material culture, and about its chronology. In the Central Highlands of Anatolia there was a complete breakdown of complex, literate, urban life. What followed was very different. There was, it is true, a Neo-Hittite revival from the southern banks of the Kızılırmak reaching all the way into North Syria, but this seems to have been slow to emerge (Hawkins 2000; Melchert 2003). North and west of the Kızılırmak, in a zone stretching from the western border of Sivas Province to the Highlands of Phrygia, Hittite culture and civilisation seems to have vanished. It is not that the northern plateau was depopulated, although there may have been few "Hittites"—whatever that means—who remained for any time north of the

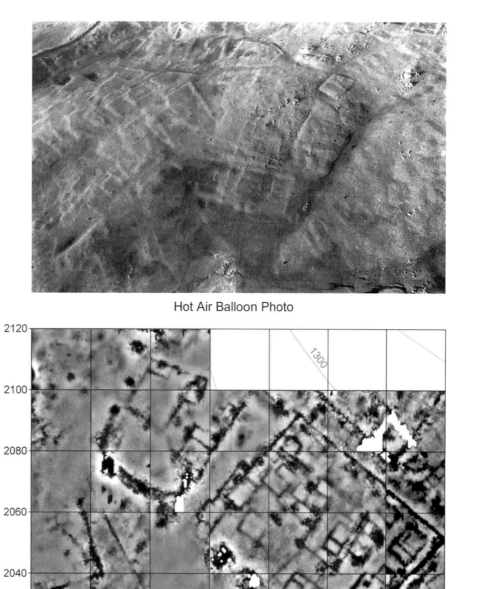

Figure 5: Structures within the prominent urban block in the centre of this hot air balloon photograph, above, taken in 1993 were revealed by resistivity survey, below, with also streets and urban features. The grids measure 20 x 20m.

river which is, after all, only three or four days' walk from Hattuša. What happened, how and why it happened, will be studied and theorised about for many years to come. More evidence will be required before any kind of consensus can be reached.

I would like to stress that what followed the collapse was both dynamic and complex. On the other hand it was neither urban nor literate, surely lacking the sophistication of what had been before as well as of what was to come. It is ironic that as new cursive alphabetic scripts emerged, ultimately it seems from Phoenicia, resulting in a rapid spread in literacy (witnessed, for instance, by graffiti on pottery), the technological advance of writing materials led to the replacement of virtually indestructible clay by perishable materials such as parchment and reusable wax on wooden boards. For Iron Age Anatolia there is no equivalent to the Late Bronze Age tablet archives or the libraries of King Midas's contemporaries in Assyria and beyond. Phrygian inscriptions provide little if any historical information (Brixhe and Lejeune 1984) while the writings of later Greeks, from Herodotus onwards, are notoriously difficult. The correspondence of the Phrygian kings will, I fear, never be found. We are left, then, with the evidence of archaeology, and it is to that which I now turn.

One major reason for choosing to work at Kerkenes was that the site lent itself to a program of Remote Sensing. The 2.5 km^2 city (Figures 2 and 3), enclosed by 7 km of strong stone defences was looted, burnt, and abandoned, with the result that almost the entire city lies immediately below the surface, not obscured by later settlement (e.g. Branting and Summers 2002; and Summers 2006).

Various methods of remote sensing have been employed including, from, as it were, the top down: satellite imagery; photography from a manned hot air balloon (Figures 4 and 5a); photography with a tethered blimp; close contour differential GPS employing a strategy of kinetic or continuous data collection (seen in the background of Figure 3), with three remote receivers and one base station; geophysical survey (Figure 5b), which maps buried features; geomagnetic survey using a fluxgate gradiometer over the entire city; electrical resistivity survey over selected areas and comparative experimentation with electromagnetic induction survey (von der Osten-Woldenburg 2004). The results achieved from resistance survey have greater clarity than those from the gradiometer, but to survey the entire site by this means would take not 10 but 50 years to complete. A new generation of electromagnetic induction or conductivity meters has considerable promise.

All of this has been presented elsewhere and the results are all available on the project web site

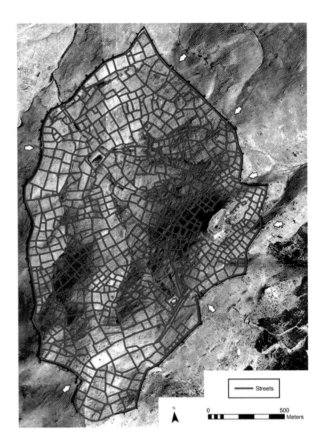

Figure 6: The network of streets at Kerkenes with the city defences and positions of the seven city gates indicated (Scott Branting).

(www.kerkenes.metu.edu.tr) and in preliminary publications (e.g. Summers 2007). My purpose here is to introduce some of the innovative ways in which we are looking at and using this data. Firstly, it gives us a plan of the entire city at the time of its destruction. For the first half of the 6th century this is unique, Sardis and other contemporaneous capitals in Turkey being smothered by later periods. We can see: the pattern of streets (Figure 6); the very uneven distribution of the 7 city gates (Figure 7); the citadel or acropolis, a steep, waterless granite tor that was a place of refuge rather than the site of the chief public buildings (Summers 2001)—more like Sardis or Priene than Hattuša or Nineveh; the urban blocks enclosed by terrace walls to make more level platforms for building, surely indicating a degree of centralized planning and distribution of urban space; and lastly the sophisticated system of water management.

It ought to be possible to analyse all of this data in a GIS, a Geographical Information System. In theory there is no reason why this ancient city cannot be analysed with GIS in some of the same ways that modern city planners and administrators do (Figure 8). And we are hopeful that new, more specifically archaeolog-

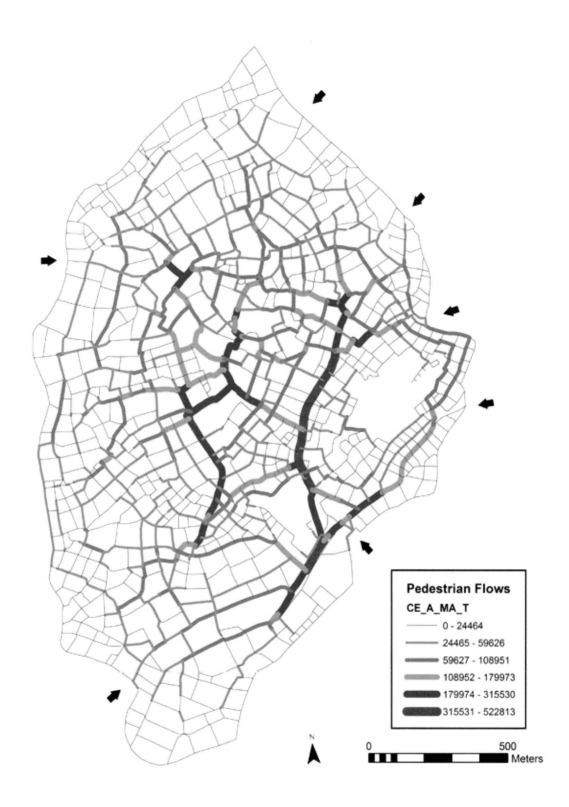

Figure 7: Predictive mode of pedestrian traffic flows at Kerkenes with the seven city gates indicated (Scott Branting).

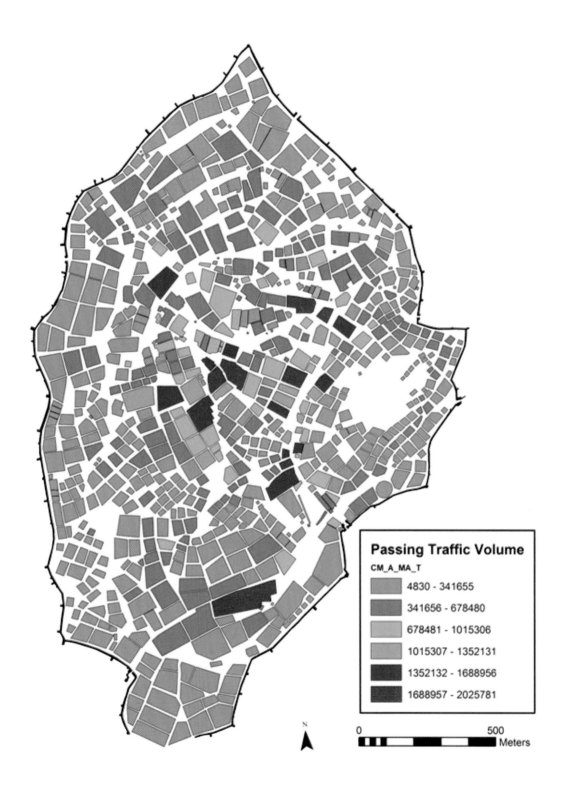

Figure 8: Predicted pedestrian traffic volumes produce a hierarchy of urban units (Scott Branting).

ical, questions will be developed. At Kerkenes, then, the intention is to employ the powerful analytical tools offered by GIS as well as routinely archiving all spatial data and producing good graphics (Figure 9). It will also be possible to make virtual reality reconstructions of parts of the site, and one day perhaps the entire city. I imagine that youngsters will be able to play being Croesus' henchmen, taking simulated rides through the city and putting buildings to the torch. There are two serious sides to this. One is getting the archaeology over to the public, and particularly to the local public through schools, museums, our visitors' centre, and so forth. The second is that in attempting to create such simulations we encounter unforeseen issues, architectural and spatial. Thus our knowledge and understanding are increased while new avenues of research are developed. We still have a very long way to go, but I can show you some of the first tentative steps.

Nahide Aydin completed a master's degree at the University of Mississippi in which she developed methods of computerised pattern recognition of satellite imagery (Aydin 2004). In a nutshell Nahide used plans of the northern end of the site made from the GPS data and from the geophysical survey to teach a computer how to recognise certain characteristics. The computer was then able to apply what it had been taught to other parts of the city. While the archaeological results of this pioneering study were limited, it was shown that this could be done and there is clearly potential for development and refinement. It will be immediately obvious that these techniques will have many applications outside archaeology.

Figure 9: GIS at Kerkenes is used for analysis as well as of for graphic representation. This experimental image makes use of data from balloon photography, intensive GPS topographic simulation, COMMA and geophysical survey.

One of the biggest problems in this and other studies introduced here turns out to be that of creating data bases which can be used in GIS. Spatial analysis requires defined spaces. When we look at some of the geophysical images we can see many buildings (Figure 5b), often in great detail. On these it is not too difficult to draw four sides of each building, or to define the boundaries of the urban blocks. But in other areas there is less clarity with the result that no two students will draw the same plan. In yet other sectors of the site little more than the general patterning can be seen. Initial studies undertaken at METU were over-ambitious in their attempts to draw a complete plan of all the structures in the northern portion of the city. It proved to be frustrating work. Even deciding which spaces were roofed and which were open was often impossible. Sights were lowered. It was seen that a first stage would be to define the urban blocks (Figure 10). This was more successful and the visual results very pleasing but real analysis has not yet been undertaken. At the same time a pilot study involved putting individual walls (rather than spaces) into a GIS. This too was shown to have potential but even a small part of the site took much time and encountered problems of interpretation (Aydin 2001; Aydin et al. 2002; Baturayoğlu 2002).

Scott Branting wrote his doctoral thesis at Buffalo on Transportation GIS at Kerkenes. Scott made a GIS data base of the streets which included such factors as the steepness of slope. Based on a study of pedestrians in the university city of Cambridge, England, Scott then looked at origins and destinations of streets, including city gates, urban catchments, major water sources and so forth. The result is a predictive model of traffic volumes (Branting 2004). Based on a variety of modern studies, factors such as time, energy, gender, and age can be introduced to produce models for different sets of people. The model identifies urban places of importance, as defined by the volumes of traffic going to and from them (Figures 6–8). These predictions can then be tested by reference to the remote sensing data. The predicted hierarchy of streets can also be tested by examining the street surfaces themselves and compaction in the subsurface matrix (Figure 11). In this study we are moving on beyond the static evidence of what became buried at some moment in time to look at the dynamics of people living in the urban environment, looking at time and energy, at preferences with urban space and at transit through the urban boundaries at the city gates. Perhaps I can be forgiven for saying that only the first steps have been taken in this highly innovative approach.

Finally, in a recently completed master's thesis at METU by Nurdan Atalan Çayırezmez, she looked at the relationships between the line of the city defences

Figure 10: A hot air balloon photograph (above) of the north end of the city with an initial attempt at defining urban blocks in the northern sector of the city at Kerkenes (below).

Figure 11: Scott Branting taking a sample from an ancient street for micromorphological examination of street surfaces and compaction in the subsurface matrix.

and gates and the topography. It was possible to demonstrate that the defences follow topographic divides and, in addition, that they enclose water catchments, partially explaining both the location and the exceptional size of the city (Çayırezmez 2006). Viewshed analysis of the city gates goes some way to explaining their uneven distribution to which factors such as access to agricultural land and seasonal variations in optimal routes could be added (Figure 12).

Remote Sensing and GIS analysis are starting to revolutionise some aspects of the archaeology of certain types of ancient city (Figure 13). The methods and approaches that we are using at Kerkenes will not work everywhere, and indeed they are not all appropriate to each site. I hope, however, to have demonstrated some of the innovations and potentials. The technology continues to develop at a great pace. Much of what we have done could now be done faster and more efficiently with newer hardware and ever-improving software. At the start I was, perhaps, over optimistic. I thought that the results would permit excavation to be limited to very small trenches, carefully placed to answer specific questions. There was no reason to embark on the destructive and expensive business of large-scale excavation. However, when we started to excavate we had the most unexpected surprises.

The city defences were not unfinished, but stood preserved to 5m, perhaps less than half their original height (Figure 12a). These defences were built entirely of stone and were deliberately cast down before the abandonment. Public architecture of an unsuspected and unknown order has been uncovered. Buildings were provided with pitched roofs covered with thatch, not the flat mud roofs of the Near East. Sumptuous ivory was recovered (Dusinberre 2002). Inscriptions and grafitti were in Old Phrygian (Brixhe and Summers 2006), rather than Aramaic or Luwian. Statuary and relief sculpture was both unexpected and unique (Draycott et al. 2008). And the iconography, rows of idols like semi-iconic battlements on top of the towers or platforms flanking the entrance to a palatial complex, would have looked truly awesome (see for the moment, Summers and Summers 2006).

I would end on a more personal note. I have observed in several countries in the Near East that digging for written clay tablets usually turns out to be a futile exercise, both disappointing and frustrating. At Kerkenes, where tablets are not to be expected, digging for carbonised wood for dendrochonology has turned out to be equally disappointing and frustrating. When we did find a substantial chunk (of the right sort of wood), with 197 annual growth rings, it fitted "neither consistently nor convincingly" with anything. It turned out that we had but the core of a huge post, probably one of the wooden columns which were seated in recesses almost a metre in diameter. Elsewhere the fire had raged with such intensity that there was nothing but ash. There are now more samples in the laboratory here. Our fingers are crossed, and we will keep looking. I would like to thank Peter for his friendship and support, for sharing in the frustrations, for being part of a team. He is always eager to share knowledge and ideas, to teach and to encourage. The truth is, of course, that I would rather find ivories, sculpture, and inscriptions than lumps of charcoal. So, I suspect, would Peter! But the value of an unshakable chronology, of certain dates, over the uncertainties of stylistic dating—not to mention the ferocity with which views are held and opposed—is inestimable.

Acknowledgments

Françoise Summers prepared the illustrations and suggested many improvements to the text. I would like to acknowledge the help and support of all members of the Kerkenes team over the last 14 years. Firstly my co-directors: my wife Françoise and Dr. Scott Branting, Director of the CAMEL Lab at the OI, and Prof. David Stronach for five seasons of collaborative fieldwork. Of our sponsors, fully listed on the project web site, I would single out the National Geographic Society, the Loeb Classical Library Foundation, The Joukowsky Family Foundation and the Rolex Award for Enterprise all four of which offer support for research regardless of nationality or country of residence of the investigator, and an anonymous donor. It is a pleasure to thank the staff of the General Directorate in Ankara and the Yozgat Museum as well as the Governorate of Yozgat and District Office of Sorgun to-

Figure 12: The Cappadocia Gate (above) has been selected for this example of viewshed analysis showing the territory that can be seen from an urban vantage point (Nurdan Atalan Çayırezmez).

Figure 13: The city wall and streets (Tuna Kalayci and Scott Branting).

gether with Mayors of Yozgat and Sorgun. Of the local sponsors Yibitaş, Yibitaş Lafarge and Mr. Erdoğan Akdağ have been exceptional in their support. Field work is carried out under the auspices of the British Institute of Archaeology at Ankara; the Oriental Institute of Chicago University has recently become the supporting institution.

References

The project web site, www.kerkenes.metu.edu.tr, contains much imagery, annual reports, the *Kerkenes News* and other resources.

Aydin, Z.N. 2001. An archaeometric study of the urban dynamics at Kerkenes Dağ based on the integration of geomagnetic data and GIS. MSc Thesis, Department of Archaeometry, Middle East Technical University, Ankara.

Aydin, Z.N. 2004. The application of multi-sensor remote sensing techniques in archaeology. MA Thesis, University of Mississippi.

Aydin, N., Toprak, V., and Baturayoğlu, N. 2002. The Geophysical Survey of an Iron Age City in Central Anatolia: Kerkenes Dağ. In J. Albertz (ed.), *Proceedings of the XVIII. International Symposium, CIPA 2001, Potsdam (Germany)*: 516–523. Berlin: CIPA.

Baturayoğlu, N. 2002. The Survey and Documentation of the City Walls and Cappadocia Gate of the Iron Age Settlement on Kerkenes Dağ in Central Anatolia. In J. Albertz (ed.), *Proceedings of the XVIII. International Symposium, CIPA 2001, Potsdam (Germany)*: 100–107. Berlin: CIPA.

Bittel, K. 1970. *Hattusha*. New York: Oxford University Press.

Branting, S.A. 2004. Iron Age pedestrians at Kerkenes Dağ: an archaeological GIS-T approach to movement and transport. PhD Dissertation, SUNY Buffalo.

Branting, S., and Summers, G.D. 2002. Modelling terrain: the Global Positioning System (GPS) survey at Kerkenes Dağ, Turkey. *Antiquity* 76: 639–640.

Brixhe, C., and Lejeune, M. 1984. Corpus des inscriptions paléo-phrygiennes. *Éditions Recherche sur les Civilisations* "mémoire" 45, Paris: CNRS.

Brixhe, C., and Summers, G.D. 2006. Les inscriptiones phrygiennes de Kerkenes Dağ (Anatolie Central). *Kadmos* 45: 93–135.

Çayırezmez, N.A. 2006. Relationships between topography and Kerkenes (Turkey), a GIS analysis. MSc Thesis, Middle East Technical University, Graduate School of Social Sciences, Ankara.

Draycott, C.M., Summers, G.D., and Brixhe, C. 2008. *Kerkenes Special Studies 1: Sculpture and Inscriptions from the Monumental Entrance to the Palatial complex at Kerkenes Dağ, Turkey*. Chicago. OIP 135.

Dusinberre, E.R.M. 2002. An Excavated Ivory from Kerkenes Dağ, Turkey: Transcultural Fluidities, Significations of Collective Identity, and the Problem of Median Art. *Ars Orientalis* 32: 17–54.

Fischer, B., Genz, H., Jean, É., and Köroğlu, K. (eds.) 2003. Identifying Changes: The Transition from Bronze to Iron Ages in Anatolia and its Neighbouring Regions, *Proceedings of the International Workshop Istanbul, November 8–9, 2002*. Istanbul: Türk Eskiçağ Bilimleri Enstitüsü.

Genz, H. 2004. Büyükkaya I. Die Keramik der Eisenzeit, *Boğazköy-Hattuša XXI*. Mainz and Rhein: Ph. von Zabern.

Hawkins, J. 2000. *Corpus of Hieroglyphic Luwian Inscriptions. Volume I: Inscriptions of the Iron Age*. Berlin and New York: W. de Gruyter.

Hawkins, J.D. 2002. Anatolia: The end of the Hittite Empire and After. In E.A. Braun-Holzinger and H. Matthäus (eds.), *Die nahöstlichen Kulturen und Griechenland an der Wende vom 2. zum 1. Jahrtausend v. Chr*: 143–151. Möhnesee-Wamel: Bibliopolis.

Kuniholm, P.I. 1993. A Date-List for Bronze and Iron Age Monuments Based on Combined Dendrochronological and Radiocarbon Evidence. In M. Mellink, E. Porada, and T. Özgüç (eds.), *Aspects of Art and Iconography-Anatolia and Its Neighbors: Studies in Honor of Nimet Özgüç*: 371–373. Ankara: Türk Tarih Kurumu.

Manning, S.W., Kromer, B., Kuniholm, P.I., and Newton, M.W. 2003. Confirmation of near-absolute dating of east Mediterranean Bronze-Iron Dendrochronology. *Antiquity* 77 No 295 http://antiquity.ac.uk/ProjGall/Manning/manning.html.

Melchert, H.C. (ed.). 2003. *The Luwians, Handbuch der Orientalistik 68*. Leiden: Brill.

Muscarella, O.W. 2003. The Date of the Destruction of the Early Phrygian Period at Gordion. *Ancient West & East* 2: 225-252.

Osten-Woldenburg, H. von der. 2004. An Algorithm for the Numeric Combination of Geophysical Mappings of Archaeological Sites. Poster: 10th European Meeting of Environmental and Engineering Geophysics, Utrecht. *Near Surface 2004* (online publication of EAGE subscribers).

Schmidt, E.F. 1929. Test Excavations in the City on Kerkenes Dagh. *American Journal of Semitic Languages and Literatures* 45: 221–274.

Summers, F., Atalan, N., Aydin, N., Başağaç, Ö., and Uçar, G. 2003. Documentation of Archeological Ruins and Standing Monuments Using Photo-Rectification and 3D Modeling. In M. O. Altan (ed.), *Proceedings of the XIXth International Symposium CIPA 2003, New Perspectives to Save Cultural Heritage, Antalya, Turkey, 30 September–4 October, 2003*: 660–668. CIPA, Turkey.

Summers, G.D. 2001. Keykavus Kale and Associated Remains on the Kerkenes Dağ in Cappadocia, Central Turkey. *Anatolia Antiqua* 9: 39–60.

Summers, G.D. 2006. Aspects of material culture at the Iron Age Capital on the Kerkenes Dağ in Central Anatolia. *Ancient Near Eastern Studies* 43: 164–202.

Summers, G.D. 2007. Public Spaces and Large Halls at Kerkenes. In A. Çilingiroğlu. and A. Sagona, (eds.), *Anatolian Iron Ages 6: The Proceedings of the Sixth Anatolian Iron Ages Colloquium Held at Eskişehir, 16–20 August 2000*: 241–259. Leuven: Peeters.

Summers, G.D. 2008. Periodisation and Terminology in the Central Anatolian Iron Age: Archaeology, History and Audiences. *Ancient Near Eastern Studies* 45: 202–217.

Summers, G., and Summers, F. 1998. The Kerkenes Dağ Project. In R. Matthews (ed.) *Ancient Anatolia*: 177–194 and plates 29–30. London: British Institute of Archaeology at Ankara.

Summers, G., and Summers, F. 2006. Kerkenes 2006, *Anatolian Archaeology* 12: 32–33 and covers.

Ward, W.A., and Joukowsky, M.S. (eds.). 1992. *The Crisis Years: The 12th Century B.C. from Beyond the Danube to the Tigris*. Dubuque, Iowa: Kendall Hunt.

Article submitted March 2007

The Rise and Fall of the Hittite Empire in the Light of Dendroarchaeological Research

Andreas Müller-Karpe

Aegean and Near Eastern Archaeology has changed significantly as a result of the research of Peter Ian Kuniholm and his team. His life work has contributed most notably to our knowledge of relative and absolute chronology in the eastern Mediterranean area.

In this paper, though, I would like to focus on another aspect of tree-ring analysis: the reconstruction of palaeoclimate as evidenced by variations in tree-rings' thickness.

More than thirty years ago LaMarche drew attention to the temperature sensitivity of bristlecone pines from Campito Mountain in California (LaMarche 1974), and this tree-ring record was later combined with analyses of foxtail pines from other locations in California. Within the second millennium BCE record, these sequences show a tendency to wider tree-rings, indicating rising temperatures between the mid 17th and early 15th centuries. Clearly the climate during this period was more conducive to the growth of these tree species. Although some of the worst years—perhaps those related to volcanic activity (Kuniholm et al. 1996; Manning et al. 2001)—actually fall in the second half of the 17th century, the decisive point is a general tendency toward warmer temperatures that can be identified through at least 150 years (Figure 1).

In the 13th and 12th centuries BCE just the opposite effect can be seen: the growing conditions for the trees deteriorated, which for the region in question suggests a dramatic reduction in precipitation. Similar effects have been noticed in other regions as well.

Mike Baillie then compared the Californian with the Anatolian and Fennoscandian tree-ring records (Baillie 1998), suggesting that rising temperatures and increased rainfall in the late 17th and 16th centuries and the downturn in the 13th and 12th centuries are likely global phenomena.

In fact, the late 17th and 16th centuries witness the rise of the Hittite Empire in Anatolia, while its decline began in the late 13th century and its end

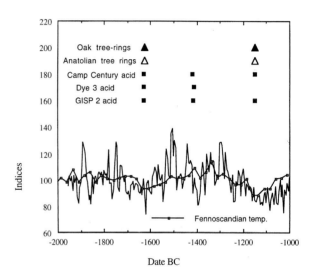

Figure 1: Tree-ring record with a 5-year resolution composed of the dates from California, Fennoscandia, and Anatolia (after Baillie 1998).

can be dated to the beginning of the 12th century. Is the striking consonance with the tree-ring record mere coincidence?

Baillie himself connected the flourishing of the New Kingdom in Egypt and the Shang Dynasty in China with the warm episode (Baillie 1998: 54). But are the growing conditions of trees and of empires identical? Should we return to the old theory of Ellsworth Huntington, who wrote already in 1915 in his famous book *Civilization and Climate*, according to which cultural change is related primarily to climate change? While this view still has many supporters (e.g. Hsü 2000), I do not think that the explanation for cultural development is so simple.

The emergence and decline of the great empires of the old world has always been a central topic of archaeological and historical research. Different models have been discussed, but no consensus has been achieved. Still, it can be considered certain that one factor which must have strongly influenced the devel-

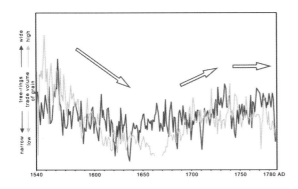

Figure 2: North- and West-German tree-ring record (darker line) in comparison with the volume of grain traded at the Cologne market (lighter line) from AD 1540 to 1790 (after Schmidt and Gruhle 2005).

opment of early agrarian cultures is the climate, since any rural economy is heavily dependent on soil and climate.

In a recently published study Burghart Schmidt and Wolfgang Gruhle demonstrated the connection between the growth of tree-rings and the volume of grain traded in western Germany between 1540 and 1790 CE (Schmidt and Gruhle 2005).

In the first phase of about a century (Figure 2, left side) tree-rings became continuously smaller, during which time a decline in grain traded at the Cologne market can be seen. The trend changes around 1670, when the growing conditions for oaks improved, though as a consequence of the Thirty Years War grain production recovered only some time later. During the 18th century (Figure 2, right side) the correlation between the tree-ring record and the grain trade is seen once again.

This striking connection between the results of dendroarchaeological research and the historical data of agricultural production might indeed aid one in understanding the phenomenon of the rise and fall of empires. An increase in food production means more people can be nourished, and in early societies this is the main condition for the growth of population, for an increase in the birth-rate. More people means more soldiers, more power and better chances of constructing an empire. On the other hand a decrease in food production may cause famine and social instability. A good example is the situation in Late Roman times. The crisis years in the second half of the 3rd and the 4th century CE are well documented in written sources, and unsurprisingly, the West German oak chronology shows on average smaller tree-rings between 250 and 400 CE (Schmidt and Gruhle 2005: 306, 309–312).

Based on the comparison of crops and tree-rings in the 16th to 18th centuries CE mentioned above, the decrease in grain production in Late Roman times can be estimated at about 15–25%. Such a rate might not have been a great problem for the Romans themselves. They had huge silos (horrea) and recourse to importing grain from Africa and Egypt. The peoples of central and northern Europe beyond the limes, however, did not enjoy such luxuries. For them the best chance of gaining access to the grain they needed was to launch raids into the Roman Empire and to loot the horrea. As is well known this marked the beginning of the Migration Period and the beginning of the end of the Roman Empire.

It may be suggested, then, that dendroarchaeology can indeed help us to understand the rise and fall of empires. Of course, every historical situation is unique. The agricultures of Egypt and Southern Mesopotamia are based on irrigation and are thus hardly comparable. But in Anatolia we might also expect a connection between tree-rings and crop production. Indeed changes in climate would have had an even greater effect on vegetation on the Anatolian Plateau than in other areas. The higher the region, the more severe are the effects of climate variations, the most obvious indicator of which is the lowering and the rising of the timberline.

The heartland of the Hittites lies at an altitude of about 800 to 1500m above sea level. It was densely occupied beginning with the Chalcolithic Period. Even the higher, mountainous regions were settled from at least the late 4th millennium (Schoop 2005). The first cities, however, which emerged in the 19th–18th centuries BCE, are found only in the lower regions of the plateau not higher than about 1200m (Figure 3).

The picture changes with the rise of the Hittite Empire. Kuşaklı-Sarissa, a Hittite foundation dated, in large part on the basis of Peter Kuniholm's dendrochronological results, to about 1530 BCE, was erected at an altitude of 1600 to 1650m above sea level (Müller-Karpe 2002a, b). A nearby sanctuary lies at 1900m. As far as I know never before was a city built at such an altitude in Anatolia.

What happened during the 16th century? What circumstances made it possible to sustain a city at such an altitude? The present climate in the region allows only short and dry summers, with about five months of severe winter. Only two months without any frost!

The situation in other parts of the Land of Hatti is only slightly better. At least thirty and as many as a hundred days or more of snow and severe frost are the norm. The summers are hot and dry, and drought is a chronic problem. The annual mapping of regions that have suffered from severe drought in the 20th century

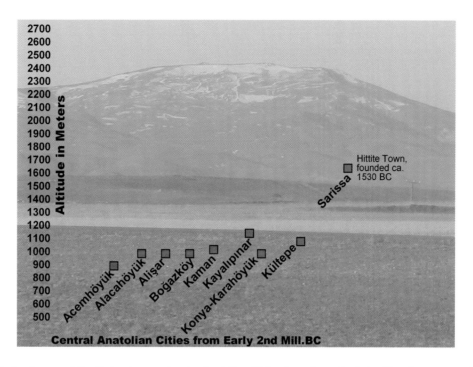

Figure 3: Altitude of some of the first urban centres in Anatolia. Cities from the early 2nd millennium BCE are known up to an altitude of about 1200m above sea level. In the Hittite Kingdom, during a period of more favourable climatic conditions, cities were erected in higher regions.

always includes the highlands of central Anatolia, but normally not the coastal regions such as the Aegean (Figure 4). Clearly, the Hittite Empire developed in an ecologically risky region, near the critical limits. Central Anatolia was much more affected by climate fluctuations; the heartland of the Hittites has significantly lower precipitation.

Before artificial fertilizers came into use, Turkey suffered a bad harvest on average every fourth year and a catastrophic crop failure every seventh to tenth year, as can be gathered from official statistics from the early years of the Turkish Republic (Hütteroth and Höhfeld 2002, 88).

If two years of drought followed one after the other, even greater famine was the consequence: In 1873 and 1874 about a quarter of a million people died in Anatolia from hunger and disease caused by two crop failures. During this drought up to 82% of cattle and 96% of sheep, goats, and chickens died. The Hittite cuneiform texts also mention such catastrophic years of famine (Klengel 1974).

How can we explain the emergence of an empire as important and powerful as the Hittites in an ecologically risky region? I think three main factors are of importance. The first relates to economic and technical development. The Hittites were the first people in Anatolia to build dams and water reservoirs extensively, developing techniques to become more independent of irregular rainfall.

In the vicinity of Kuşaklı-Sarissa we were able to find no fewer than six such dams and associated artificial basins, one of which we excavated completely (Hüser 2007). The results were astonishing indeed. These Hittite dams are not just an amassing of earth, but evince a sophisticated construction. The northwestern dam at Sarissa has a core of clay, impermeable to water (Figures 5 and 6). Both sides of the dam are paved with large blocks of stone. To prevent frost damage, the blocks on the slope were set into a layer of gravel. Even today earthen dams are built in this way. The dam is connected to the northwestern city gate, a typical Hittite construction with two massive towers, one on each side of the gateway, and there is no doubt that they are contemporary.

According to the dendrochronological results produced by the Malcolm and Carolyn Wiener Laboratory for Aegean and Near Eastern Dendrochronology, the timber for the construction was cut in about 1530 BCE. Evidence for the date of its demise comes in the form of sherds from a Late Helladic III A 2 jar found in the destruction layer.

The great temple, Building C on top of the acropolis of Kuşaklı, was erected at the same time as the dam and gate (Kuniholm and Newton 2002). This suggests that the entire city was founded in the early twenties or thirties of the 16th century BCE, during the Old Hittite Period. The gates, water reservoirs, and main buildings were erected according to a mas-

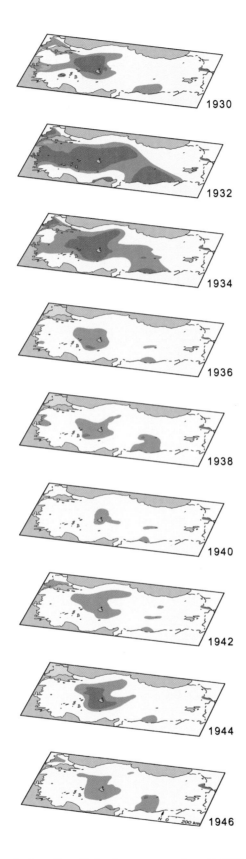

Figure 4: Annual mapping of regions in Anatolia suffering from a severe drought 1930–1947. Mostly Central Anatolia is affected by drought and crop failure, while costal regions are much more stable (after Hütteroth and Höhfeld 2002: 89, fig. 36).

Figure 5: Air photo of the northwestern city-gate of Kuşaklı-Sarissa (construction dendrochronologically dated to ca. 1530 BCE) and the Hittite dam in front of it.

ter plan oriented to the cardinal points of the compass (Figure 7).

Several artificial reservoirs have also been discovered at Hattuša during the last few years (Seeher 2002a), indeed more than would have been necessary for irrigation, suggesting that they functioned, rather as a reserve of water supply during the dry summers for the inhabitants of the city, for livestock, and for nearby gardens.

In addition to the water stores provided by the reservoirs, these Hittite towns were supplied with fresh water. In Sarissa two long pipelines conducted water from the mountains into the town (Figure 7). Two of the dams served simultaneously as aqueducts crossing the small valleys outside the city wall. Without such sophisticated water supply systems it would not have been possible to develop an urban culture in this region.

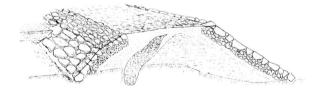

Figure 6: Construction scheme of the Hittite northwestern dam in Kuşaklı-Sarissa.

Another innovation of Hittite times was the large grain silo, normally dug into the ground, then covered with earth. The silos found on the so-called Büyükkaya in the Hittite capital, for instance, had a capacity of up to 155 tons of grain each. Another silo, from the northwestern slope of the Büyükkale, could hold as much as 7000 to 9800 tons, enough to feed 23,000–32,000 persons for a year (Seeher 2006a). The silos were always dug within a town at high, dry locations, e.g. on the Büyükkaya in Boğazköy or at the top of the mounds of Kaman-Kalehöyük and Alacahöyük or on the Acropolis and the southernmost point of Kuşaklı-Sarissa (Mielke 2001).

We know of no comparable silos from the Early or Middle Bronze Ages in Anatolia. The earliest examples can be dated to the 16th century BCE. Together with other innovations, these silos, and the large stocks of grain they made possible, enabled the rise of the empire. The filling of these silos allowed the Hittites to compensate for crop failures and avoid famine, as well as to supply large armies, a principal prerequisite for any military expansion.

The appearance of bronze sickles should also be mentioned in this context. The spread of these new reaping tools revolutionized the harvest in the second millennium BCE (Figure 8). Alongside these economic and technical developments, a second factor, which at least contributed to the rise of the Hittite Empire, was the changing toward more favorable climatic conditions in the second half of the 17th century and continuing into the 16th century BCE. Not only the dendroclimatological record but also recent pollen analyses show a trend toward warmer temperatures.

We had the good fortune to find a small lake in the immediate vicinity of Kuşaklı-Sarissa with excellent conditions for sedimentation and pollen preservation. Using a floating coring platform (Figure 9), Walter Dörfler and his team from Kiel University were able to extract a 6m long sequence of sediment covering the entire Holocene (Dörfler et al. 2000). Time scales were produced on the basis of ten ^{14}C-datings, most of them using the AMS-method on concentrated pollen suspension.

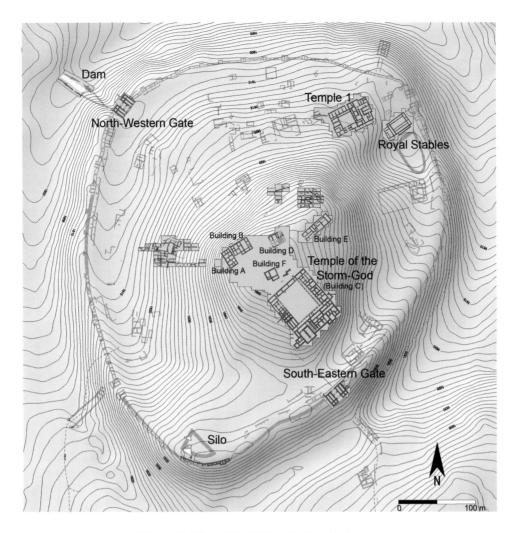

Figure 7: Plan of the Hittite city Kuşaklı-Sarissa.

The first 2 to 2.5m corresponds to the first half of the second millennium BCE. According to Dörfler's investigations, the vegetation in the third and early second millennia was dominated by pine, after which a sharp decline in pine pollen and an increase in the pollen of deciduous trees, primarily oak, can be observed. This indicates greater humidity, higher temperatures, and a rising timberline in this mountainous region.

In the same phase the pollen of cereals and other cultivated plants increased too, reflecting the flourishing period of the Hittite Empire very well. These results provide a basis for an enhanced understanding of why the Hittites were able to found a city like Kuşaklı-Sarissa at such an altitude. The environmental conditions were more favorable, due to the positive climatic development. This situation made a surplus in agricultural production possible, so that even in mountainous regions an urban population could exist and prosper.

The third factor for the rise of the Hittite Empire was of course politics and military. Without charismatic and powerful kings and their armies no empire would have been built. Also a centralized and efficient bureaucracy was important, and evidence for such a centralized bureaucracy is constituted, for instance, by impressions of the same royal seal not only in the capital, but also in Kuşaklı-Sarissa and Tarsus.

The archaeological picture of the beginnings of this empire can be supplemented with reference to several further observations.

Most of the earlier settlements from the so-called Middle Bronze Age or Old Assyrian Trade Colony Period, such as Kültepe, Konya-Karahöyük, Acemhöyük, and Alişar, were abandoned or at least lost their importance in the late 18th or the first half of the 17th century BCE. Only some few of the urban centers show a continuity of settlement, one of them being Kayalıpınar on the Kızılırmak River in eastern Cappadocia. Excavations began at this mound in 2005 (Müller-Karpe 2006), and it quickly became

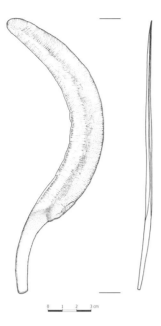

Figure 8: A Hittite bronze sickle from Kuşaklı-Sarissa.

Figure 9: Suppitassu Gölü above Kuşaklı-Sarissa with a floating platform for coring.

clear that the transition from the Middle to the Late Bronze Age is well represented.

During the 19th–18th century BCE a first urban settlement with domestic buildings was established. The walls of this level 4 are rather modest and made of mud-brick (Figure 10). In addition to pottery and seal impressions typical of this period, a fragment of an Old Assyrian tablet was found, part of a contract of sale for a female slave. The town represented by this layer was destroyed by fire.

In the succeeding Old Hittite Kingdom a monumental public building (level 3) with much larger walls was erected in this area, partly destroying the earlier ruins (level 4). The structure might have been a palace, since the portions excavated thus far have yielded several royal seal impressions (Figure 10) as well as some Hittite and Hurrian tablets, mention-

ing "the King" several times, and it thus seems that Kayalıpınar was probably some kind of a royal city. With the rise of the Old Hittite Kingdom the layout of the settlement and architecture changed completely. Figure 11 shows the pre-kingdom domestic structures and the Hittite public buildings A (below) and B (in the upper part) of levels 4 and 3. Both of the latter buildings have several phases which cover almost the entire Hittite period. In one of the late phases we were fortunate enough to discover a large limestone block with a relief of a goddess (Figure 12).

But like other important centers of this period in Anatolia, the Hittite settlement of Kayalıpınar came to an end during the late 13th or early 12th century BCE. The question why all these Hittite centers found their end at about the same time remains unanswered.

The exact nature of the decline of the Hittite Empire, sudden collapse or gradual deterioration towards an inevitable end, has long been debated, and unambiguous archaeologically evidence has not been forthcoming. The former excavators of Boğazköy interpreted the fact that some of the 29 temples located in the Upper City were given up before the final days of the capital as evidence for the rapid decline of the empire (Neve 1992: 33; Seeher 2002: 169). A part of them was given up earlier. Now, however, it is known that the abandonment of these temples should be dated long before the late 13th century (Müller-Karpe 2003; Seeher 2006b), and thus that it can not be taken as a evidence for a sudden destruction. On the other hand, the end of most Hittite settlements is marked by a burned level, indicating violence attributable either to an exterior enemy or an internal struggle.

Climatic deterioration during the 13th and 12th centuries BCE would have at least exacerbated the internal as well as external problems of the Hittite Empire. In a region like the environs of Kuşaklı at about 1600m above sea level even a slight tilt of the climate toward lower temperatures and less rainfall would have had a severe negative effect on living conditions, especially agriculture. Deteriorating weather conditions certainly did not cause the collapse of the Empire directly, but they likely influenced the circumstances which led to its end. As we know from Merneptah's Great Karnak Inscription food shortages were a great problem in the late years of the Hittite Empire. The Pharaoh had to send grain "to keep alive that land of Kheta" (Bryce 1998: 365; Drews 1993: 79).

Years of severe drought may well have caused an agricultural crisis not only for the Hittites themselves but also, and perhaps even more seriously, for their hostile neighbors in Anatolia, the Kaska peoples and

Figure 10: Air photo of the Kayalıpınar excavations in 2006.

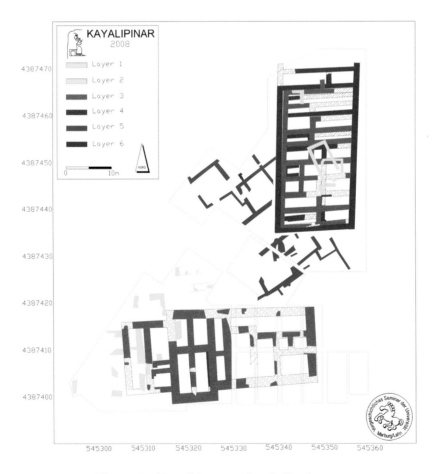

Figure 11: Plan of the excavations in Kayalıpınar.

Figure 12: Hittite goddess from Kayalıpınar.

other tribes who had already threatened the existence of the Hittite Empire several times before.

These semi-nomadic people probably had no large grain silos for bad years like those of the Hittites, and no Egyptian Pharaoh would have sent ships with food "to keep them alive." The collapse of the West Roman Empire can be attributed to internal weakness and external pressure, re-enforced by climate deterioration, as evidenced by the written sources in combination with recent dendroarchaeological research (Schmidt and Gruhle 2005: 305–312); and it seems that one might draw a historical parallel with the comparable phenomena witnessed at the end of the Hittite Empire in Late Bronze Age Anatolia, although more research is needed to more fully understand the complex situation.

References

Baillie, M.G.L. 1998. Evidence for Climatic Deterioration in the 12th and 17th centuries BC. In B. Hänsel (ed.), *Mensch und Umwelt in der Bronzezeit Europas*: 49–56. Kiel: Oetker-Voges.

Bryce, T. 1998. *The Kingdom of the Hittites*. Oxford: Clarendon Press.

Dörfler, W., Neef, R., and Pasternak, R. 2000. Untersuchungen zur Umweltgeschichte und Agrarökonomie im Einzugbereich hethitischer Städte. *Mitteilungen der Deutschen Orientgesellschaft* 132: 367–380.

Drews, R. 1993. *The End of the Bronze Age. Changes in Warfare and the Catastrophe ca. 1200 B.C.* Princeton University Press.

Hoffner, H.A., Jr. 1989. The Last Days of Khattusha. In W.A. Ward and M.S. Joukowsky (eds.), *The Crisis Years: The 12th Century B.C. From Beyond the Danube to the Tigris*: 46–52. Dubuque, IA: Kendall/Hunt.

Huntington, E. 1915. *Civilisation and Climate*. New Haven: Yale University Press.

Hsü, K.J. 2000. *Klima macht Geschichte. Menschheitsgeschichte als Abbild der Klimaentwicklung*. Zürich: Orell Füssli.

Hüser, A. 2007. Hethitische Anlagen zur Wasserver-und-entsorgung. *Kuşaklı-Sarissa 3*. Rahden/ Westf.: Marie Leidorf.

Hütteroth, W.-D., and Höhfeld, V. 2002. *Türkei*. Darmstadt: Wissenschaftliche Buchgesellschaft.

Klengel, H. 1974. "Hungerjahre" in Hatti. *Altorientalische Forschungen 1*: 165–174.

Kuniholm, P.I., Kromer, B., Manning, S.W., Newton, M., Latini, C.E., and Bruce, M.J. 1996. Anatolian tree rings and the absolute chronology of the eastern Mediterranean, 2220–718 B.C. *Nature* 381, 6583: 780–783.

Kuniholm, P.I., and Newton, M. 2002. Dendrochronological Investigations at Kuşaklı/Sarissa. In A. Müller-Karpe et al. (eds.), *Untersuchungen in Kuşaklı 2001. Mitteilungen der Deutschen Orientgesellschaft* 134: 339–342.

LaMarche, V.C., Jr. 1974. Palaeoclimatic Inferences from Long Tree-Ring Records. *Science* 183: 1043–1048.

Manning, S.W., Kromer, B, Kuniholm, P.I., and Newton, M.W. 2001. Anatolian tree-rings and a new chronology for the east Mediterranean Bronze-Iron Ages. *Science* 294: 2532–2535.

Mielke, D.P. 2001. Die Grabungen an der Südspitze. In A. Müller-Karpe et al.(eds), *Untersuchungen in Kuşaklı 2000. Mitteilungen der Deutschen Orientgesellschaft* 133: 237–243.

Müller-Karpe, A. 2002a. Kuşaklı-Sarissa: A Hittite Town in the "Upper Land." In K.A. Yener and A. Hoffner Jr. (eds.), *Recent Developments in Hittite Archaeology and History. Papers in Memory of Hans G. Güterbock*: 145–156. Winona Lake, Indiana: Eisenbrauns.

Müller-Karpe, A. 2006b. Kuşaklı-Sarissa. Kultort im Oberen Land. In Kunst- und Ausstellungshalle der Bundesrepublik Deutschland (ed.), *Die Hethiter und ihr Reich. Das Volk der 1000 Götter*: 176–189. Stuttgart: Konrad Theiss.

Müller-Karpe, A. 2003. Remarks on Central Anatolian Chronology of the Middle Hittite Period. In M. Bietak (ed.), *The Synchronisation of Civilisations in the Eastern Mediterranean in the Second Millennium B.C. II. Proceedings of the SCIEM 2000—EuroConference, Haindorf 2nd–7th of May 2001*: 383–394. Wien: Verlag der Österreichischen Akademie der Wissenschaften.

Müller-Karpe, A. et al. 2006. Untersuchungen in Kayalıpınar 2005. *Mitteilungen der Deutschen Orientgesellschaft* 138: 211–247.

Neve, P. 1992. Hattuša—Stadt der Götter und Tempel. Neue Ausgrabungen in der Hauptstadt der Hethiter. *Antike Welt* 23. Sondernummer.

Neve, P. 1999. Die Oberstadt von Hattuša. Die Bauwerke I. Die Bebauung im zentralen Tempelviertel. *Boğazköy—Hattuša XVI*. Berlin: Gebr. Mann.

Schmidt, B., and Gruhle, W., with contributions from Zimmermann, A. and Fischer, T. 2005. Mögliche Schwankungen von Getreideerträgen—Befunde zur rheinischen Linienbandkeramik und Römischen Kaiserzeit. *Archäologisches Korrespondenzblatt* 301–315.

Schoop, U.-D. 2005. Das anatolische Chalkolithikum. *Urgeschichtliche Studien 1*. Remshalden: Bernhard Albrecht Greiner.

Seeher, J. 2002a. Die Ausgrabungen in Boğazköy-Hattuša 2001. *Archäologischer Anzeiger* 2002: 59–78.

Seeher, J. 2002b. *Hattuscha-Führer*. 2nd ed. Istanbul: Ege Yayınları.

Seeher, J. 2006a. Chronology in Hattuša: New Approaches to an Old Problem. In D.P. Mielke, U.-D. Schoop, and J. Seeher (eds.), *Strukturierung und Datierung in der hethitischen Archäologie. BYZAS 4*: 197–213. Instanbul: Ege Yayınları.

Seeher, J. (ed.) 2006b. Ergebnisse der Grabungen an den Ostteichen und am mittleren Büyükkale—Nordwesthang in den Jahren 1996–2000. *Boğazköy—Berichte 8*. Mainz: Philipp von Zabern.

Article submitted July 2007

Aegean Absolute Chronology: Where did it go wrong?

Christos Doumas

It is true that the excavations at Akrotiri, Thera, have not contributed much to promoting dendrochronology in the Aegean region. This is due to neither neglect nor indifference. On the contrary, Peter Kuniholm was constantly waiting for my phone call to come and collect material for analysis. Unfortunately, our material, mainly pieces of charcoal from household contexts, was quite inadequate for this purpose. Nevertheless, Peter came to Akrotiri once we thought we had found charcoal of sufficient size; but, alas, once again the sample was worthless for dendrochronology. It is an honor for me to be here among so many eminent colleagues, to pay my humble tribute to Peter Kuniholm, a small token in gratitude for what he has contributed to archaeology. And I thank the organizers of this symposium for giving me this opportunity to do so.

As David Carr has remarked, "just as we have no experience of space except by experiencing objects in space, so we experience time as events, things that take or take up time" (Carr 1991: 25). This means that time cannot be perceived independently but in association with specific events. And an event has a beginning and an end; "present and past function together in the perception of time somewhat as do foreground and background or focus and horizon in spatial perception" (Carr 1991: 21). But "narratives, whether historical or fictional, are typically about, and thus purport to represent, not the world as such, reality as a whole, but specifically human reality" (Carr 1991: 19).

In archaeological terms, strata or cultural horizons represent past "events," the decipherment, understanding, and interpretation of which is the task of the archaeologist. And if in narratives "any significance, meaning, or value ascribed to events is projected onto them by our concerns, prejudices, and interests, and in no way attaches to the events themselves" (Carr 1991: 19), this is even more valid in the case of past "events" recorded in material form.

In order to define the beginning, the middle, and the end of an archaeological "event," to place it in time, various methods have been devised and agreed upon, among them stratigraphy and typology. By using these as instruments the Three Ages chronological system was devised and its further subdivisions into periods, phases, etc. With this indeed ingenious system, the sequence of geological strata could be associated with cultural "events," thus facilitating the study of early human history. In Aegean archaeology, this system has helped to establish a universally accepted cultural sequence and relative chronology. Through interpretation of Aegean artifacts found in Egypt and the Near East, or vice versa, synchronisms were established, on the basis of which an absolute chronology for the Aegean sequence has been sought. This conventional chronology was unquestionably accepted and used until, in the 1960s, scientific dating methods were introduced into archaeology.

Among the great archaeological "events" in the Aegean is the Bronze Age eruption of the Thera volcano. Its effects, if properly studied, may help us to understand many other "events" before and after it. Hence the great interest of the scientific world in the excavations at Akrotiri as soon as they were started by Spyridon Marinatos in 1967 (Marinatos 1968). We should not forget that it was Marinatos who first associated the Thera eruption with widespread destructions in Crete (Marinatos 1939). And it was also he who at the First International Scientific Congress on the Volcano of Thera (held in September 1969) pronounced that "the catastrophe (in Crete) was contemporaneous and happened at a moment in which the LM IA ceramic style was still prevailing although a little number of minor vases were already painted in LM IB style. This meant chronologically that the catastrophe happened soon after 1500 BC. However, we cannot decide if we can date this stylistically advanced pottery 10 or 20 years after 1500 BC, and it seems to me almost impossible to descend as low as 1470 BC" (Marinatos 1971: 404). It is obvious that these excavations offered Marinatos a unique opportunity to crosscheck and evaluate his theory. Indeed, evidence from the 1971 and 1972 digging seasons revealed two destruction horizons, for which he proposed the dates 1550 or 1560 BC and 1500 BC respectively. He attributed the latter destruction to an earthquake that immediately preceded the eruption of the volcano (Marinatos 1973: 28–29).

Meanwhile, the first attempts to date the eruption through scientific methods were made, and I had the privilege of being the first to collect and send samples for radiocarbon dating to the Philadelphia Museum Laboratory (Michael 1976; 1978), with strict instructions from Marinatos to avoid providing information of stratigraphical or chronological character. The stormy reaction of archaeologists to the high date for the eruption (17th century BC) triggered scientists to intensify their efforts, as is evident from the overview presented by Peter Kuniholm at the Third Thera Congress in 1989 (Kuniholm 1990). This effort still goes on and new generations are involved just as eagerly in the game, to judge from the impressive and continuous stream of literature on the issue. What is interesting, and for some embarrassing, is the fact that for nearly four decades scientific laboratories persist in giving a high date for the eruption, within the 17th century BC (Kutschera and Stadler 2000).

As soon as the first radiocarbon dates for the eruption appeared, a real tug-of-war began between scientists and archaeologists, their flag being high and low chronology respectively. Stuck in the middle of this game, and feeling as if between clashing rocks, I have been trying to remain impartial, thinking that, since there is no argument about the sequence of the cultural phases, the duration of each phase in absolute calendar years is an issue for which the definite answer is still pending. For, as a non-scientist, I did not feel competent to argue about or dispute the scientific dating methods. And this is still my position. Regardless of any reservation one may have about high chronology, it is difficult to reject outright laboratory results without presenting strong scientific arguments about possible errors or omissions.

As recently as 2000, Hammer, presenting his results from the analysis of the Greenland Ice Core data, repeated, once again, that "the 1645 BC ice event could very well be due to the Thera eruption of 1646 BC" (Hammer 2000: 36). At the same time Kutschera and Stadler confirmed that the impressive database of 30,000 radiocarbon dates as well as the dendrochronological data available suggest a date for the eruption within the 17th century BC (Kutschera and Stadler 2000: 70–71; also Manning et al. 2006). Even more recently, April this year [2006], the dates obtained from the outer ring of an olive tree branch found standing in the pumice layer in the Thera quarries were published (Friedrich et al. 2006). These dates are between 1621 and 1605 BC (68% confidence) or between 1627 and 1600 (95% confidence).

On the other hand, I must admit, I have not always been happy with the arguments presented by my fellow-archaeologists, at the other end of the tug-of-war, arguing against the high chronology for the eruption. For, instead of being well-founded and convincing, these arguments are often either simplistic or based on assumptions and guesses. For example, echoing Marinatos's views, Sir Denys Page wrote in 1970: "The leading authorities on the vulcanology of Santorini...maintain that the interval between the great pumice-layer and the great ash-layer may be quite long—decades, possibly, according to Professor Georgalas; and we are very glad to hear him saying so. But others are confident that the visible phenomena require not more than days or at most weeks; and they give us no pleasure, but much pain. For if we are going to connect the desolation of Crete with the action of the volcano, we shall need at least a couple of decades between the fall of the pumice and the fall of the ash" (Page 1970: 15). From this quotation it becomes apparent that the aim was to connect the disasters in Crete with the eruption of the volcano in Thera, because the opposite would 'give us no pleasure'! I hope that nobody in this audience believes that the degree of pleasure constitutes a serious argument.

The assumption that the volcano was responsible for the destructions in LM IB Crete led to the next step, to the invention of Page's "couple of decades between the fall of the pumice and the fall of the ash" (Page 1970: 15). However, squeezing a period of time between the seismic destruction and the eruption (Money 1973; Warren 1984: 492; Warren 1990/1991: 30) was no easy task without ignoring or even distorting the archaeological evidence (Doumas 1974).

As has already been mentioned, synchronization between Egypt and the Aegean Bronze Age is mainly based on typological and stylistic comparisons between artifacts found in respective archaeological contexts. However, it has to be borne in mind that although such artifacts bear witness to contacts between different areas, they do not constitute an undisputed evidence of synchronization of the respective cultures and certainly cannot provide absolute dates for the duration of cultural phases. Equally, the absence of characteristic foreign elements from the context of a specific site, e.g. Cypriot White Slip I-Ware from early strata of Tell el-Dab'a (Bietak 1998: 321), does not exclude their earlier production at home and, therefore, does not constitute a strong argument of chronological importance.

Similarly, artistic styles originating in one area can be adopted in another area with considerable delay. In this respect the comments of Nicolas Coldstream, the leading scholar on Greek Geometric pottery, made in 1965 about the Theran Geometric style are worth mentioning. As he pointed out, "the vase painting of Thera is reputed to be the most backward in the Cyclades. Her potters were slow to learn a Geometric style in the eighth century BC, and slow to forget it in the seventh" (Coldstream 1965: 34). He understands this 'conservatism' not as a result of isolation but as due to "local pride in a handsome architectonic style which the Therans were slow to learn and slow to abandon." Even when this style was exhausted from within and the potters introduced orientalizing motifs, Coldstream comments that "even then the system of decoration was not radically changed" (Coldstream 1968: 188).

Typology and style can indeed help in the study of the developmental course of an artifact or of an artistic element. But they cannot provide absolute chronology nor even define the length of a period or a phase. Different types or decorative elements may co-exist in the same area without marking historical events or social or economic changes. On the other hand, historical events may have no effect at all on the typology of artifacts. For, as Paolo Matthiae has remarked, in Mesopotamian archaeology, if one takes into consideration "aspects of the material culture, particularly pottery, it does not seem that one can make an objective break between Middle Bronze I and Middle Bronze II" (Matthiae 2000: 138). The same observation has been made on the Aegean island of Ikaria in recent times, where three major socio-economic transformations have taken place in the course of the last four centuries, yet have left few signs in the material culture record, especially in what archaeologists call "moveable finds" (Kapetanios in press).

Coming back to the historical chronologies of Egypt and Mesopotamia, as a non-specialist, I have to rely on what specialists have to say about them. Of the three—high, middle and low—historical chronologies most scholars show preference for the middle one. But, as has been pointed out, the compromise among the scholars to use the middle chronology does not make it correct or accurate (Mellaart 1979: 6). And this is still valid, as modern scholarship shows. Kitchen, stating that Egyptian chronologies can be improved "by use of synchronisms with other ancient Near Eastern states, especially Mesopotamia from c. 1400 BC onwards, and occasionally (only occasionally) by use of a tiny handful of astronomical data," indirectly admits that these chronologies are not satisfactory (Kitchen 2000: 39). Moreover, skepticism is expressed even by specialists in Near Eastern archaeology, since absolute chronology is mainly based "on a wider knowledge of the contrasting evidence provided by different chronological systems, built, in general terms, over the analysis of astronomical data combined with king lists and historical synchronisms" (Matthiae 2000: 136). Yet, they consider all the available king lists "either incomplete or highly corrupt...insufficient as the sole base of building the framework for the reconstruction of history" (Brein 2000: 54). As far as the astronomical dating is concerned, scholars are adamant that "at the moment it seems to be impossible to gain reliable fixed absolute dates of Egyptian history by means of astrochronology alone and it is justified to doubt conventional dates primarily based on the heliacal rising of Sôthis" (Brein 2000: 56). Moreover they add that, although lunar observations of the crescent may be calculated in relation to the chronologies of Sirius (Sothis), "historical correlation with these dates is even more difficult" (Firneis 2000: 59), since the association of the rising of Sôthis with Sesostris III is another compromise (Mellaart 1979: 7).

Taking into consideration the above views of specialists in Egyptian and Near Eastern archaeology, it becomes obvious that, if it is difficult to establish absolute dates in the Egyptian historical chronology, it is even more difficult to correlate such dates with Aegean sequences. Certainly historical chronology cannot be ignored, but it cannot be taken as dogma and should be used with caution. For we must not forget that it was on the grounds of this 'historical chronology' that the Thera eruption was dated to 1200 BC (Pomerance 1970: 16) and that on the grounds of such 'historical evidence' the 30th of April 1483 BC was proposed as the exact date for the Thera eruption (Gödicke 1988: 174; 1992: 61)!

It is true that "scientific belief is even in principle only valid as long as it can resist attempts to show that it is wrong" (Tattersall 2002: 6) and that "in science it is no crime to be wrong, unless you are (inappropriately) laying claim to the truth" (Tattersall 2002: 9). I hope that nobody here is laying claim to the truth; I cannot even imagine that we shall ever meet the truth; what we are all simply trying to do is to get as close to the truth as possible. And this symposium gives us the opportunity to re-examine our position, as far as absolute chronology is concerned, and to seek the points where we may have been wrong.

It seems to me that the problems in the archaeological absolute dating of the Thera eruption are methodological ones. For example, instead of comparing similar things, i.e. scientific dates obtained from both Egypt and the Aegean, we are trying to marry Egyptian historical chronology, with all its ambiguities, with Aegean scientific absolute dating. I am not aware

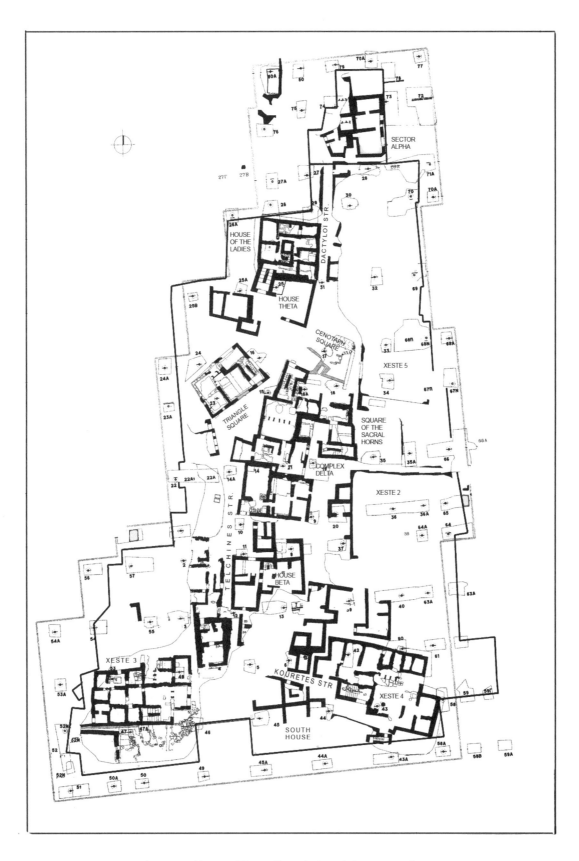

Figure 1: Akrotiri, Thera. Ground plan of the excavated area.

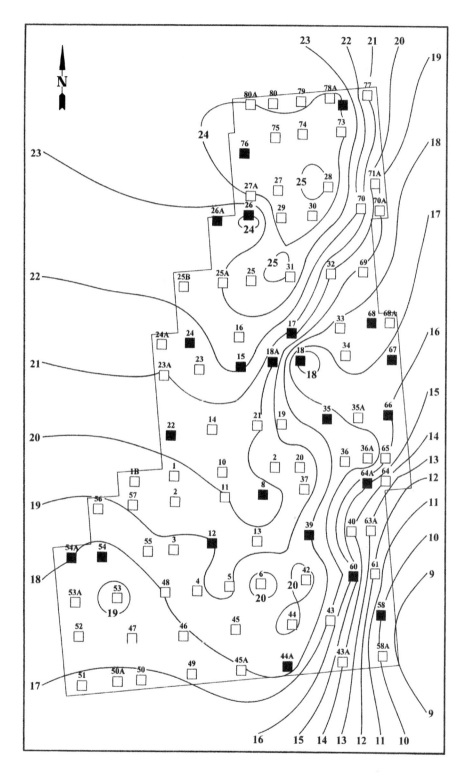

Figure 2: Akrotiri, Thera. The pre-eruption relief: the location of the pillars for the new shelter are indicated by small squares of which the black ones show the presence of rock-cut tombs.

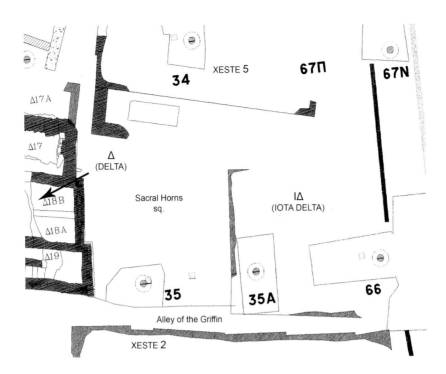

Figure 3: Akrotiri, Thera. Plan of the Square of the Sacral Horns.

if scientific dates have been obtained for all periods of Egyptian history and if these dates have been compared with historical ones. Such a comparison would be necessary before any attempt to match Egyptian chronological sequences with the Aegean ones.

Another methodological problem may lie in the chronological system applied for the distinction of periods in early Aegean history. There is no doubt that the division of European prehistory into three ages—Stone, Bronze, and Iron Age—was an ingenious instrument for studying cultural evolution (Daniel 1981: 102–107). Based on the typological classification of artifacts, this system was also introduced in Aegean archaeology (Daniel 1981: 131) and it was Sir Arthur Evans, the excavator of Knossos in Crete, who dubbing the Cretan Bronze Age as Minoan subdivided it into the Early, Middle, and Late Minoan periods, thus transplanting to the Aegean the division of the Egyptian civilization into the Early, Middle, and New Kingdom. Later on A.J.B. Wace and C. Blegen extended the system to the rest of the Aegean, and introduced the terms Early, Middle, and Late for the Cycladic and Helladic cultures (Cadogan 1978: 209). The weak points of the system were pointed out right from the beginning and scholars appeared cautious in applying it. Their main argument was that the definition of periods through typology and development of pottery styles totally ignored the archaeological evidence of events that left vivid marks in the ground, such as destructions. Moreover, the distinction of some phases was sometimes based on pottery styles of local character, and therefore did not represent time spans (Platon 1968: 1). Another argument was that although the three-period system may be good for the classification of archaeological finds, it is not for the study of cultural evolution, for which the study of man's economy would be more appropriate (Daniel 1981: 107).

Despite the caution and skepticism, throughout the 20th century the three-period system has been applied in Aegean archaeology, and has indeed been a useful instrument. Nevertheless, pottery styles, even if they characterize a whole period, can never provide absolute chronologies. And this is a factor that should always be borne in mind. It would seem more objective if the division of early history into periods were based on undisputed archaeological evidence, such as settlement patterns or technological innovations, but even such a system cannot establish an absolute chronology and the duration of each period will always be a matter of debate.

As early as 1933 Nils Åberg suggested the subdivision of the Cretan Bronze Age into periods following the architectural history of the so-called palaces (Åberg 1933: 148ff.). Thus, the periods Prepalatial (extending from Evans's Early Minoan I to Middle Minoan Ia), the Protopalatial (Middle Minoan Ib-Middle Minoan IIIa), Neopalatial (Middle Minoan IIIb-Late Minoan IIIa), and Postpalatial (Late Minoan IIIb-c) were recognized (Platon 1961: 671). This subdivision was further improved and better documented by Pla-

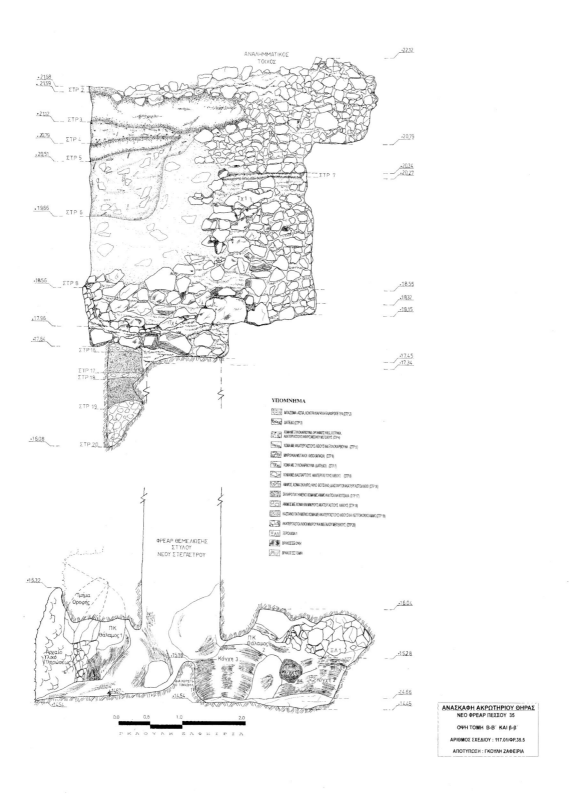

Figure 4: Akrotiri, Thera. The shaft of pillar No 35: stratigraphical section.

Figure 5: Akrotiri, Thera. The Pithos of the Griffins from pillar shaft No 35.

Figure 6: Square of the Sacral Horns: The Alley of the Griffin (the first—Early Cycladic I—terrace).

ton (Platon 1968). More recently, efforts have been made to establish a new terminology for the Early Bronze Age sequence in the Cyclades mainly based on the development of settlement patterns. Thus, a period of "Farmsteads" has been proposed for the Early Cycladic I; of "Hamlets" for the early phase of Early Cycladic II; of "Nucleated Villages" for the advanced stages of Early Cycladic II; and of "Harbour Towns" for the Early Cycladic III/Middle Cycladic I (Doumas 2002: 22–25).

Recent works at Akrotiri have enabled us to follow an uninterrupted sequence from Early Cycladic to Late Cycladic and to make some useful observations. About 150 shafts were excavated for the foundation of the pillars for the new shelter of the site. Through these shafts, dug down to the bedrock, we gained not only a good picture of the ground relief on which the settlement was founded, but also of the entire stratigraphy of the site, enabling us to study the site's history (Figure 1).

Twenty-seven rock-cut chambers, scattered along the gentle slopes of a low promontory (Figure 2), were found filled in with debris up to their ceiling. Use of these chambers seems to have stopped towards the end of the Early Cycladic II period, since no pottery later than the Kastri group was included in the fill. There are strong indications that these rock-cut chambers originally belonged to an Early Cycladic cemetery, which was cancelled after it was decided to expand the city over that area (Doumas in press).

The best picture of the stratigraphic sequence came from the shaft for pillar 35 (Kariotis 2003) in the Square of the Sacral Horns (Figure 3). Two rock-cut chambers were found side by side at different levels, following the slope (+14.75m and +15.75m respectively above sea level, Figure 4). The fill of the east and lower chamber yielded EC II and IIIA pottery (Kariotis 2003: 420 fig. 5), whilst the pottery from the fill of the west chamber was exclusively EC III (Kariotis 2003: 422 fig. 6). This picture suggests the earlier abandonment of the east chamber.

About 2.20m above the floor of the west chamber, a building was erected, the floor of which was found at +17.94m above sea level. From the floor of this building early Middle Cycladic pottery was recovered, including an imported EM III "light-on-dark" sherd (Kariotis 2003: 424 fig. 8). Almost at the same level (+17.88m above sea level) was found the floor of another building, the walls of which were preserved up to 0.85m. Pottery associated with this building included EC and early MC sherds (Kariotis 2003: 425 fig. 10). Obviously both buildings were constructed quite early in the Middle Cycladic period. After the remains of these buildings were covered with a layer of debris, another building was erected, of which only remains of two walls were found defining a floor of beaten earth

Figure 7: Akrotiri, Thera. Square of the Sacral Horns: the entrance door of delta 19.

Figure 8: Akrotiri, Thera. Early Cycladic metallurgical firedogs.

Figure 9: Akrotiri, Thera. Early Cycladic II/ Middle Cycladic I wheel-made pots.

about 1.50m above the floor of the previous buildings (c. +19.50m above sea level). Of the three under layers of this floor, the two lower ones yielded late Middle Cycladic pottery as well as MM II B and III A pottery (Kariotis 2003: 426 fig. 11), while the top one gave sherds also of late Middle Cycladic pottery, some of which could be considered as heralding the Late Cycladic I "dark-on-light" style (Kariotis 2003: 427 fig. 12).

The contents of this building included a number of vases in rather good condition gathered at its SW corner: among them are 11 ribbed vases (Kariotis 2003: 430 fig. 16), an early type of a tall nippled ewer (Kariotis 2003: 431 fig. 18) as well as a monumental cylindrical pithos decorated with two enormous griffins in the bichrome style (Figure 5; see also Kariotis 2003, 428-431 fig. 14). The co-existence of these types of pottery emphasizes not only the rapid changes in pottery styles but also the uninterrupted continuity of the local tradition.

This late Middle Cycladic building was abandoned and buried under a deposit of débris about 1m thick, which produced late Middle Cycladic pottery types decorated with motifs heralding both shapes and designs of the following early Late Cycladic phase (Kariotis 2003: 432 figs. 21–22). This fill seems to have resulted after a change of the urban plan, perhaps following a seismic destruction. At about +20.79m above sea level an open space was created between what we now call Xeste 5 in the north, Xeste 2 to the south, Building Iota Delta (IΔ) to the east and Building Delta (Δ) to the west (Figure 3).

A small retaining wall erected about one meter north of Xeste 2 was founded on a Middle Cycladic floor, leaving sufficient space for the "Alley of the Griffin" (Figure 6). This wall retained a fill ranging in thickness between 0.60 and 1.20m, thus forming a north-south sloping terrace. Large quantities of LC I pottery mixed with little MC and very little EC were included in this fill (Kariotis 2003: 435 figs. 23–24), suggesting the date for the creation of the Alley of the Griffin as well as for the arrangement of the entrance and staircase of Delta 19 (Figure 7).

A second terrace, about 0.70m high, was created further north, leaving free the access to the entrance to Delta 19 (Figure 8). Mature Late Cycladic I and LM IA pottery characterized the fill of this terrace (Kariotis 2003: 438 fig. 26).

Study of the stratigraphy in the shaft for pillar 35 (about seven meters thick) in association with the respective pottery, suggests uninterrupted continuity of occupation from at least the EC II until the final destruction of the city by the volcanic eruption. In almost all levels later pottery styles co-existed with earlier ones, making it difficult to distinguish either the beginning or the end of the so-called Middle Cycladic period, which seems to have been of a much shorter duration than we have imagined so far, and was marked by rapid developments in both building activity and pottery styles.

This fast development does not seem to be independent of the major technological innovations that characterize the end of the Early Bronze Age II (Early Cycladic II) One of these innovations is the replacement of arsenical copper with tin bronze alloys (Gale et al. 1984: 30–31; Muhly 1977: 75; 1985a: 283–285; 1985b: 132; 1999: 18–20; Pernicka et al. 1990: 275; Stech and Piggott 1986: 56–57). To the same period are dated the earliest known pot bellows, from Chrysokamino in the Gulf of Mirabello, Crete (Betancourt 1997; Muhly 2004: 287), as well as the abundant saddle-shaped firedogs from Akrotiri, Thera, obviously designed to facilitate the thrusting of the tuyère under the burning charcoal. Another technological innovation of the time, or perhaps slightly later, was the introduction of the potter's wheel, as the early Middle Cycladic pots indicate (Figure 9).

Metallurgy does not seem to be the only economic activity which benefited from novel ideas toward the end of the 3rd millennium BC. It is perhaps not fortuitous that at about this time viticulture and wine-making technology were diffused throughout the Aegean (Doumas 2006a), and the pomegranate made its first appearance as an exotic fruit in the Aegean world (Asouti 2003: 477; Doumas 2006b).

Akrotiri, strategically located on the maritime route between Cyprus and Crete, was destined to play a decisive role in the copper trade. Hence, perhaps, its rapid development and subsequent acquisition of wealth (Doumas 1997). If this rapid development is not a feature of Akrotiri only but characterizes the entire Aegean region, it is possible that the duration of the Middle Bronze Age was also shorter than is believed. Could this mean that the Late Bronze Age might have started earlier than estimated so far, thus making our conventional chronology more compatible with the one provided by scientific methods?

References

Åberg, N. 1933. *Bronzezeitliche und früheisenzeitliche Chronologie. Band IV: Griechenland. Archaeology monograph 18.* Stockholm: Kungl. Vitterhets Historie och Antikvitets Akademien.

Betancourt, P.P. 1997. The Copper smelting site at Chrysocamino, Crete. *Proceedings of the 1st International Conference on Ancient Greek Technology.* Thessaloniki: 51–54.

Bietak, M. 1998. The Late Cypriot White Slip I-Ware as an obstacle of the High Aegean Chronology. In M.S. Balmuth and R.H. Tycot (eds.), *Sardinian and Aegean Chronology: towards the resolution of relative and absolute dating in the*

Mediterranean. Studies in Sardinian Archaeology V: 321–322. Oxford: Oxbow Books.
Bietak, M. (ed.) 2000. *The Synchronization of Civilizations in the Eastern Mediterranean in the Second Millennium* BC *(SCIEM I)*. Vienna: Verlag der Österreichischen Akademie der Wissenschaften.
Brein, G. 2000. Astrochronology and Ancient Egyptian Chronology. In M. Bietak (ed.) *The Synchronization of Civilizations in the Eastern Mediterranean in the Second Millennium* BC *(SCIEM I)*: 53–56. Vienna: Verlag der Österreichischen Akademie der Wissenschaften.
Cadogan, G. 1978. Dating the Aegean Bronze Age without radiocarbon. *Archaeometry* 20: 209–214.
Carr, D. 1991. *Time, Narrative and History*. Bloomington: Indiana University Press.
Coldstream, N. 1965. A Theran Sunrise. *Bulletin of the Institute of Classical Studies* 12: 34–37.
Coldstream, N. 1968. *Greek Geometric Pottery*. London: Methuen.
Daniel, G. 1981. *A Short History of Archaeology*. London: Thames and Hudson.
Doumas, C. 1974. The Minoan eruption of the Santorini volcano. *Antiquity* 48: 110–115.
Doumas, C. (ed.) 1978. *Thera and the Aegean World 1*. London: Thera and the Aegean World.
Doumas, C. 1997. Τελχίνες. In Γ. Χατζησάββας (ed.), *Η Κύπρος και το Αιγαίο στην αρχαιότητα*. Λευκωσία: 79–84.
Doumas, C. 2002. *Silent Witnesses: Early Cycladic Art of the Third Millennium* BC. New York: Alexander S. Onassis Public Benefit Foundation (USA).
Doumas, C. 2006a. Σταφύλι και κρασί στη Θήρα εδώ και τρισήμισυ χιλιάδες χρόνια. *Ampelos 2003—International Conference on Grapevines, 5–7 June 2003, Santorini*: 1–11. Athens. Heliotopos Conferences.
Doumas, C. 2006b. Ουκ εν τω πολλώ... *ΑΛΣ* 4: 57–60.
Doumas, C. 2008. Chambers of Mystery. In K. Boyle and G. Gavalas (eds.), *Ορίζων: A δλλοχυιυμ ον τηε Πρεηιστορψ οφ τηε °ψςλαδες 25-28 Μαρςη 2004*. Cambridge: McDonald Institute for Archaeological Research.
Firneis, M.G. 2000. Heliacal Sirius-Dates and First Lunar Crescent Dates Depending on Geographical Latitude for the use in Absolute Chronology. In M. Bietak (ed.), *The Synchronization of Civilizations in the Eastern Mediterranean in the Second Millennium* BC *(SCIEM I)*: 58–59. Vienna: Verlag der Österreichischen Akademie der Wissenschaften.
Friedrich, W.L., Kromer, B., Friedrich, M., Heinemeier, J., Pfeiffer, T., and Talamo, S., 2006. Santorini Eruption Radiocarbon Dated to 1627–1600 BC. *Science* 312: 548.
Gale, N.H., Stos-Gale, Z.A., and Gilmore, G.R. 1984. EBA Trojan metal sources in the Cyclades. *Oxford Journal of Archaeology* 3.3: 23–44.
Goedicke, H. 1988. The Northeastern Delta and the Mediterranean. In E.C.M. Van den Brink (ed.), *The Archaeology of the Nile Delta, Problems and Priorities*: 165–175. Amsterdam: Netherlands Foundation for Archaeological Research in Egypt.
Goedicke, H. 1992. The Chronology of the Thera/Santorini Explosion. In M. Bietak (ed.), *Ägypten und Levante III*: 57–62. Vienna: Verlag der Österreichischen Akademie der Wissenschaften:
Hammer, C.U. 2000. What can Greenland Ice Core data say about the Thera Eruption in the 2nd millennium BC? In Bietak (ed.) *The Synchronization of Civilizations in the Eastern Mediterranean in the Second Millennium* BC *(SCIEM I)*: 35–37. Vienna: Verlag der Österreichischen Akademie der Wissenschaften.
Kapetanios, A. (in press). Ο χώρος και ο χρόνος στη δυτική Ικαρία. Communication in the Symposium held in Ikaria: 1–5 September 2006.

Kariotis, S. 2003. Ακρωτήρι Θήρας. Μια πρώτη ανάγνωση της στρωματογραφικής ακολουθίας στην Πλατεία Διπλών Κεράτων. In A. Vlachopoulos and K. Birtacha (eds), *Αργοναύτης: Τιμητικός τόμος για τον Καθηγητή Χρ. Γ. Ντούμα*: 419–444. Αθήνα: Καθημερινή Α.Ε.
Kassianidou, V. 2003. The Trade of Tin and the Island of Copper. In A.R. Giumlia-Mair and F. Lo Schiavo (eds.), *Le problème à l'origine de la métallurgie. BAR, International Series* 1199: 109–119. Oxford: Archaeopress.
Kitchen, K.A. 2000. Regnal and Genealogical Data of Ancient Egypt: The Historical Chronology of Ancient Egypt, A current Assessment. In M. Bietak (ed.) *The Synchronization of Civilizations in the Eastern Mediterranean in the Second Millennium* BC *(SCIEM I)*: 39–52. Vienna: Verlag der Österreichischen Akademie der Wissenschaften.
Kuniholm, P.I. 1990. Overview and assessment of the evidence for the date of the Eruption of Thera. In D.A Hardy and A.C. Renfrew (eds.), *Thera and the Aegean World III. Volume Three: chronology*: 13–18. London: The Thera Foundation
Kutschera, W., and Stadler, P. 2000. 14C Dating for Absolute Chronology of East Mediterranean Cultures in the Second Millennium BC with Acceler. Mass Spectrometry. In M. Bietak (ed.), *The Synchronization of Civilizations in the Eastern Mediterranean in the Second Millennium* BC *(SCIEM I)*: 68–81. Vienna: Verlag der Österreichischen Akademie der Wissenschaften.
Manning, S.W., Bronk Ramsey, C., Kutschera, W., Higham, T., Kromer, B., Steier, P., and Wild, E. 2006. Chronology for the Aegean Late Bronze Age. *Science* 312: 565–569.
Marinatos, S. 1939. The Volcanic Destruction of Minoan Crete. *Antiquity* 13: 425-439.
Marinatos, S. 1968. *Excavations at Thera I*. Athens: The Archaeological Society at Athens.
Marinatos, S. 1973. *Die Ausgrabungen auf Thera und Ihre Probleme*. Vienna: Verlag der Österreichischen Akademie der Wissenschaften.
Matthiae, P. 2000. Studies in the Relative and Absolute Chronology of Syria in the Second Millennium BC. In M. Bietak (ed.), *The Synchronization of Civilizations in the Eastern Mediterranean in the Second Millennium* BC *(SCIEM I)*: 136–139. Vienna: Verlag der Österreichischen Akademie der Wissenschaften.
Mellaart , J. 1979. Egyptian and Near Eastern chronology: a dilemma? *Antiquity* 53: 6–18.
Michael, H.N. 1976. Radiocarbon Dates from Akrotiri on Thera. *Temple University Aegean Symposium* 1: 7–9.
Michael, H.N. 1978. Radiocarbon dates from the site of Akrotiri, Thera, 1967-1977. In C. Doumas (ed.), *Thera and the Aegean World I*: 791–795. London: Thera and the Aegean World.
Money, J. 1973. The destruction of Akrotiri. *Antiquity* 47: 50–53.
Muhly, J.D. 1977. The Copper Ox-hide ingots and the Bronze Age Metals Trade. *IRAQ* 9: 73–82.
Muhly, J.D. 1985a. Sources of Tin and the Beginnings of Metallurgy. *American Journal of Archaeology* 89: 275–291.
Muhly, J.D. 1985b. Beyond Typology: Aegean Metallurgy in its Historical Context. In N.C. Wilkie and W.D.E Coulson (eds.), *Contributions to Aegean Archaeology: Studies in Honor of W.A. McDonald*: 109–141. Minneapolis: Center for Ancient Studies.
Muhly, J.D. 1999. Copper and Bronze in Cyprus and the Eastern Mediterranean. In V.C. Piggott (ed.), *The Archaeometallurgy of the Asian Old World*: 15–25. Philadelphia: University Museum.
Muhly, J.D. 2004. Chrysokamino and the Beginnings of Metal Technology on Crete and in the Aegean. In. L.B. Day, M.S. Mook, and J.D. Muhly (eds.), *Crete Beyond the*

Palaces: Proceedings of the Crete 2000 Conference: 283–289. Philadelphia: INSTAP Academic Press.

Page, D.L. 1970. *The Santorini Volcano and the destruction of Minoan Crete.* London. The Society for the Promotion of Hellenic Studies, Suppl. Paper No. 12.

Pernicka, E., Begemann, F., Schmitt-Strecker, S., and Grimanis, A.P. 1990. On the Composition and Provenance of Metal Artefacts from Poliochni on Lemnos. *Oxford Journal of Archaeology* 9.3: 263–298

Platon, N., 1961. Chronologie de la Créte et des Cyclades à l'âge du bronze. *Bericht über den V. Internationalen Kongress für Vor- und Frühgeschichte, Hamburg vom 24. bis 30. August, 1958*: 671–676. Berlin: Verlag Gebr. Mann.

Platon, N. 1968. Τα προβλήματα χρονολογήσεως των Μινωικών ανακτόρων. *Αρχαιολογική Εφημερίς*: 1–58.

Pomerance, L., 1970. The final collapse of Santorini (Thera): 1400 BC or 1200 BC? *Studies in Mediterranean Archaeology* 26. Göteborg: Paul Åströms Förlag.

Stech, T., and Piggott, V.C. 1986. The Metals Trade in Southwest Asia in the Third Millennium BC. *IRAQ* 48: 39–64.

Tattersall, I. 2002. *The Monkey in the Mirror. Essays on the Science of what makes us human.* San Diego: Harvest.

Warren, P.M. 1984. Absolute dating of the Bronze Age eruption of Thera (Santorini). *Nature* 308: 492–493.

Warren, P.M. 1990/1991. The Minoan civilization of Crete and the volcano of Thera. *Journal of the Ancient Chronology Forum 4*: 29–39.

Article submitted February 2007

The Thera Debate

The final group of papers in this volume offers current assessments of the important topic of the date of the Minoan eruption of the Thera/Santorini volcano in the mid second millennium BC. Did this event occur in the mid to late 17th century BC (the "high" chronology) or the later 16th century BC to around 1500 BC (the "low" chronology)? Much hangs on this question for understanding the history of the Aegean and Mediterranean in the Late Bronze Age. It is perhaps the most controversial topic in the field at present, and has been a source of active scholarly debate since 1939 when S. Marinatos first posited a date of 1500 BC, and especially since the 1960s when Marinatos began the modern excavations at Akrotiri, which have continued for the last quarter century under Christos Doumas. A series of Thera Conferences over the years has not resolved the issue. For readers unfamiliar with the earlier part of the controversy, its first several decades were reviewed in Kuniholm 1990 (see page xvi for reference. See also Doumas, this volume).

Two publications appeared in 2006 in *Science* by Friedrich et al. and Manning et al. which stimulated a new phase of scholarly exchanges on the date of the Thera/Santorini eruption. At the Cornell conference Malcolm Wiener was invited to offer his assessment of the debate. Bernd Kromer, Walter Kutschera, and Sturt Manning included discussion of the recent 2006 papers (of which they had been co-authors), and of current work since, in their presentations at Cornell. In the papers that follow here, Malcolm Wiener provides a detailed survey of the debate and a critique of the *Science* articles in his "Cold Fusion: The Uneasy Alliance of History and Science" starting on page 277. Papers by Friedrich et al. and Manning et al. then follow (on pages 293 and 299 respectively). These articles each assess the debate to the date of writing, and address various of the points raised by Malcolm Wiener in his conference presentation.

There then follows a very current discussion, starting on page 317, with Malcolm Wiener's reply to the Friedrich et al. and Manning et al. papers in this volume followed by two short comments added at proof stage by Walter Friedrich (page 327) and Sturt Manning (page 327). A final response by Malcolm Wiener, beginning on page 329, concludes the volume.

This group of papers thus provides readers with a stimulating exchange of ongoing views and scholarly disagreements surrounding this important but controversial topic as current as of AD 2009.

Cold Fusion: The Uneasy Alliance of History and Science

Malcolm H. Wiener

Abstract: *The goal of establishing a secure second millennium* BC *absolute chronology linking Egypt, the Near East, Cyprus and the Aegean world is as elusive as it is important. Communication between scholars of ancient texts and archaeologists, on the one hand, and physical scientists, on the other, is often marked by lack of understanding of the nature and degree of uncertainty in data from other disciplines. This paper examines the reliability of data from the fields of Egyptian and Near Eastern texts and archaeology, Egyptian astronomical dating, and interconnections between Egypt, Cyprus, and the Aegean during the 16th and early 15th centuries* BC, *in comparison with ice-core, tree-ring, and, in particular, radiocarbon dating.*

Fifty years ago C. P. Snow, the distinguished British scientist, novelist, and senior civil servant, wrote an article and delivered a BBC lecture on "the two cultures" (1959). Snow described Science and the Humanities as living in mutual ignorance and disdain. Scientists, said Snow, generally regarded humanistic research as trivial and akin to postage-stamp collecting, while most humanists, who would be ashamed to be found ignorant of one of the lesser sonnets of Shakespeare, could profess unashamedly their ignorance of the Second Law of Thermodynamics.

Unfortunately, the problem of mutual incomprehension identified by Lord Snow seriously affects chronological studies today. I can count on the fingers of one hand the number of historians and archaeologists who have made a serious attempt to visit the appropriate laboratories and become familiar with the potential contribution and accompanying uncertainties of the relevant sciences—glaciology, dendrochronology, radiocarbon dating, astronomy (Egyptian and Babylonian)—or to grasp the essence and understand the uncertainties of Bayesian statistics as applied to radiocarbon dates. Conversely, I know of no physical scientist who has even attempted to master the essentials of Egyptian-based textual plus astronomical dating and its Near Eastern correlates or the archaeologically established interconnections with the Aegean. I have heard one distinguished physical scientist state at a conference that chronological evidence from the hard sciences is critical because archaeologists have no better means of dating than highly subjective judgments of the duration of pottery styles, and another scientist ask at a conference why historians believe any texts, since people always lie. Conversely, some archaeologists and art historians say that they ignore scientific analyses because scientists' assertions change so frequently. Grist for this mill was provided by erroneous initial geographic sourcings of both metals and ceramics; the inaccurate initial chronological placement of the Anatolian floating tree-ring chronology and the wood from the Uluburun shipwreck; the proposed Theran eruption date of 1628 BC, announced with great confidence, but subsequently disavowed; the claim, since disproved, of conclusive similarities in the chemical composition of glass particles in the Greenland ice core and Theran tephra, and erroneous claims based on ^{14}C. (Of course "science progresses by correcting its mistakes" [Dawkins 1998].) The problem is compounded by boundaries

---

It is a pleasure to offer this small tribute to Peter Kuniholm, friend and scholar. His willingness to dare, his determination, his dedication to students, his physical courage in the face of injury, and his unsurpassed work ethic have created the field of Near Eastern and Aegean dendrochronology.

I am grateful to Sturt Manning and Peter Kuniholm for advice and assistance, given wholeheartedly notwithstanding their knowledge that the views expressed would differ from, and in some cases be critical of, their own. I am grateful as well for the invaluable editorial assistance of Jayne Warner, Erin Hayes, Jason Earle, and Catriona McDonald. I also thank Manfred Bietak, Harriet Blitzer, Paolo Cherubini, Douglas Keenan, Oliver Rackham, Paula Reimer, Steven Soter, and Peter Warren for their contributions of information, corrections, and editorial suggestions.

within the two cultures. Many Aegean prehistorians, for example, know little of Egypt and the Near East, while many physical scientists know little or nothing about other scientific approaches to chronology or about the degree of reliability of various Egyptian astronomical dates. Recent and ongoing controversies regarding the validity of historical (archaeo-textual) approaches on the one hand, and scientific methods on the other, to the critical question of the chronology of the Middle and Late Bronze Age in Egypt, the Near East, and the Aegean illustrate perfectly the conflict of the two cultures.

I. The Textual, Archaeological, and Egyptian Astronomical Evidence

A. Egypt and the Near East

Egypt presents an essentially complete textual record back to the accession of Tuthmosis III between 1479 BC and 1468 BC, buttressed by three astronomical observations (Krauss 2007; Wiener 2003: 365; 2006a: 319; 2006b; 2007). Prior uncertainty concerning individual reigns in the Third Intermediate Period (c. 1100–650 BC) following the end of the New Kingdom have been largely resolved through the work of Kenneth Kitchen and others on textual evidence, and the chronology of the period is buttressed by one Egyptian astronomical observation (Wiener 2006b; Krauss 2007) and particularly the comparison of the description of the campaign of Egyptian pharaoh Shoshenq I in Israel and Judah on the walls of his temple with the biblical account of the invasion of Shishak (925–923 BC) in the fifth year of Rehoboam, whose dates can be closely estimated by counting back from the great battle of Qarqar in 853 BC during the reign of Ahab recorded in the Assyrian annals. There is not much wiggle room back to the beginning of the New Kingdom between 1539 BC and 1525 BC, and only a small amount, in all probability not more than a generation, back to an astronomical determination in the seventh year of the reign of Sesostris III (Wiener 2006b), whose date is disputed to a certain extent, but which must fall between c. 1875 and 1830 BC. A large majority of Egyptologists support the date advocated by Luft (2003: 202) of c. 1866 BC, but Krauss has argued for a date of 1831–1830 BC (Krauss 2006: 448–450). Textual and archaeological evidence, such as the work of Bennett on the prosopography of the 13th Dynasty (Bennett 2006; Wiener 2007), appears to support the Luft date of c. 1866 BC. Wells and Bennett (Wells 2002; Bennett, pers. comm. of 28 April 2007) have suggested that some of the recorded dates may constitute predictions rather than observations; if so, the highest possible date is still unlikely to be earlier than c. 1875 BC. Even the highest date is difficult to reconcile with any proposal to raise Egyptian chronology for the Second Intermediate Period by a century in order to accommodate a small number of tenuous ^{14}C determinations.

The Egyptian textual record includes not merely king lists, records of high officials, inscriptions on buildings, temple records, and records of the life spans of the sacred Apis bulls, but also private documents telling us how long individuals served under different pharaohs. An official named Ineni, for example, tells us how long he served under Amenophis I, Tuthmosis I, Tuthmosis II, and Tuthmosis III, taking us back to before 1500 BC. Recent work on earlier periods, including studies of the prosopography of the priests of El-Kab and on the Hyksos of the Second Intermediate Period, as well as 12th Dynasty texts, clarifies the period back to the 1866–1830 BC astronomical date (Shortland et al. 2005; Bennett 2006).

In the Near East, Assyrian records are continuous back to 911 BC. Before that, we have the information contained in hundreds of thousands of baked clay tablets used for record keeping. Studying these tablets, Brinkman more than 30 years ago was able to construct separate chronologies for Assyria and Babylonia—which he believed accurate, to within at most a dozen years—back to about 1425 BC. (1972). Brinkman subsequently noted that differences in calendars created a further source of uncertainty, amounting to a maximum of three years per century (comment at the 2006 Cornell Conference in honor of P. I. Kuniholm). Brinkman's original reconstruction indicated an eight-year overlap in the 12th century BC in the reigns of Ninurta-apil-Ekur in Assyria and Meli-Shipak in Babylonia (Brinkman 1972: 272–273; 1976: 31–33; 1977). In 2001, a German excavation in Assyria found a record of a letter from one of these rulers addressed to the other, thereby confirming the existence of an overlap in these reigns (Frahm 2002). Finally, the Near Eastern chronologies are linked to the Egyptian through correspondence between Near Eastern and Egyptian rulers and courts, particularly in the 14th century BC (Moran 1992).

A recent presentation of radiocarbon determinations from Tell el-Dabʿa in the Nile Delta produced dates a century too early for the historical dates of the Tuthmoside period (Kutschera and Stadler 2003; Kutschera, pers. comm. of 7 June 2005; Wiener 2006b). Tuthmoside dates, however, are connected to Amarna dates (where according to the Tell el-Dabʿa measurements historical and radiocarbon dates agree) by a secure succession of rulers within the same dynasty. The dates of the period when Egypt was ruled from the short-lived capital of Amarna are historically well anchored by the many tablets found there, recording the correspondence of the Egyptian pharaohs Amenophis III, Akhenaten, and Tutankhamen with many Anatolian and Levantine rulers whose dates are closely known, as well as by Egyptian texts. The moral of the story is not that the historical dates for the Tuthmoside era are wrong, but that some radiocarbon determinations produce accurate chronological dates, and others do not (Wiener 2006b; and see below).

B. The Aegean

Our historical chronology has taken us back securely to the beginning of the New Kingdom in Egypt, which is tied in many ways to the end of the LM IA period in the Aegean and a Theran eruption between c. 1545 and 1495 BC. In part the case rests on specific objects—for example the famous (or infamous) fragmentary Cypriote White Slip I bowl from beneath the tephra level on Thera, which in the view of most scholars could not have been made before c. 1560 BC, and was probably made significantly later. Moreover, time must be allowed for the bowl to have been brought from Cyprus to Thera, broken and repaired in antiquity, and then buried by the eruption (Merrillees 2001). We know about the chronological horizon of White Slip I because of the thousands of sherds of Cypriote pottery—including some White Slip I and its chronological predecessors, Proto White Slip and White Painted III, IV, and V—found in various contexts in Egypt (Bietak and Hein 2001) and the Near East (Bergoffen 2001; Fischer 2003: 265), and in Cyprus in contexts including Minoan LM IA pottery (Eriksson 2001: 62). Can the one Cypriote example from Thera be much earlier in time than all the other White Slip I sherds in datable contexts, which in turn fall into place in the Cypriote sequence behind Proto White Slip and its predecessors?

The argument does not rest on White Slip I pottery alone, but on a long sequence of Cypriote wares which can be found at sites on Cyprus such as Maroni (Cadogan et al. 2001: 75–88), and in very similar stratification at Tell el-Dab'a in Egypt, at Ashkelon in Israel, and at Tell el-'Ajjûl in Gaza (Bietak and Hölfmayer 2007). The appearance in Shaft Grave V at Mycenae, containing burials of the Late Helladic I period prior to or around the time of the eruption of Thera (Graziadio 1991: 434–436, 433 table 4; Dietz 1991: 248–249), of a calcite jar which in Egyptian typological comparanda analysis fits in the early 18th Dynasty (and at the earliest, at the end of the Hyksos period, although the parallels then are less close) further supports the historical chronology, particularly since the jar in all likelihood stopped in Crete on its journey to Mycenae and was modified by a spout in a fashion typical of Minoan adaptation of Egyptian stone vessels (Warren 2006: 307–308; Bietak and Hölfmayer 2007). A jug with a strap handle (Athens NM 592) from Shaft Grave IV of Late Helladic I date further supports this analysis (Warren 2006: 305–308).

Next, we have the sequence of Aegean bronze vessel shapes, which follow one another in Aegean archaeological strata and in their depictions on the walls of Egyptian tombs of known date (Matthäus 1995). Can each of these have been copied by Egyptian artists 50–75 years after they were superseded in the Aegean? Similarly, Egyptian imports, copies, and depictions in tombs of Minoan rhyta, with chronological changes in shape following Minoan examples (Koehl 2000; 2006: 342–345, 358) also lend support to a mature LM IA eruption date around 1525 BC, plus or minus 25 years at most. If LM IA ends before 1600 BC as proposed by Manning et al. (2006a: 569), such rhyta must survive as heirlooms in fixed chronological sequence with a delay in each case of about 75 years. The tombs of viziers in the reign of Tuthmosis III also show the arrival of other grand gifts of LM IB style. Moreover, six Aegean pots from good Egyptian contexts and three Egyptian pots from clear Aegean contexts are all consistent with conventional archaeo-textual dating (Warren 2006). At Knossos the lid of an alabaster jar was found with the cartouche of the fourth Hyksos ruler, Khyan, whose Middle Minoan III context has recently been reexamined after challenge but reaffirmed (Macdonald 2003: 40; Warren, pers. comm. of 4 February 2005). The proposed Aegean Long Chronology based on putative radiocarbon determinations would, on the other hand, require the lid to arrive much later in Cretan terms, in Late Minoan IB. While an attempt has been made to reconcile all the archaeological evidence with dates roughly a century higher, the argument strikes most mainstream Aegean archaeologists specializing in chronology as somewhat bizarre (Bietak 2003; 2004).

II. The Scientific Evidence—Ice Cores, Dendrochronology, and Radiocarbon Dates

What, then, is the putative scientific case for earlier dates, i.e., the source of the unease in the alliance of history and science? A recent article in *Science* magazine asserts that, apart from radiocarbon dating, there is evidence from ice cores and tree rings for a date 75–100 years earlier than archaeological dating for the Theran eruption (Friedrich et al. 2006a). There is in fact no such evidence.

A. Ice-Core Dating

As to ice-core dating, 1) the initial claim of a rare-earth element europium anomaly in both the Greenland ice around 1645 BC and Theran tephra (Hammer et al. 1987; Hammer et al. 2001) was withdrawn by the investigators (Hammer et al. 2003: 93); 2) subsequently it became clear that major differences in the bulk components of the Greenland ice and the Theran tephra made a common source practically impossible (Keenan 2003), and that the trace elements were not closely comparable (Pearce et al. 2004; Keenan 2003; Wiener 2007); and finally, 3) it was shown that the published chemical composition of the ice-core indication was much closer to the composition of an eruption of Aniakchak, a volcano in the Aleutian Chain which on independent evidence is believed to have erupted in the 17th century BC, than to Thera (Pearce et al. 2004). Moreover, wind patterns make it far more likely that the ejecta of an Alaskan volcano would reach Greenland than ejecta from Thera, and, in any event, there is no reason why every Northern Hemisphere eruption should leave an acid signal in every square meter of the Greenland ice (Wiener 2003; Robock 2000 and pers. comm.; Robock and Free 1995). In sum, to date there is no ice-core evidence for the Theran eruption.

B. Dendrochronology

There is at present no direct dendrochronological evidence for dating the Theran eruption either. The key sequence of logs from Porsuk near the Cilician Gates, 800km due east of Thera, shows a growth spurt of indeterminable cause around 1640 BC, an impossibly early date for the Theran eruption on archaeo-textual grounds (and significantly earlier than the date proposed by the recent radiocarbon analysis of a Theran olive branch covered in tephra discussed below). The Porsuk tree-ring sequence largely ends in 1573 BC, and hence is not relevant to the discussion of a later, more historically appropriate, date for the eruption. Apparent correlations of ice-core and tree-ring events in the same year or two at various places occur at several dates, including 1571–1570 BC and 1525–1524 BC (Wiener 2006a: 320–323; see also Salzer and Hughes 2007, which refers to an event so far observed only in trees in Arizona, California, and Nevada in 1544 BC), but the locations of the putative eruptions responsible for the suspected climate-forcing events are presently unknown. Examination of the chemical composition of the Greenland ice-core laminations corresponding to the small acid spikes of those years may be appropriate, if possible. The Cornell Tree-Ring Laboratory has begun efforts to source minute particles in tree rings. Perhaps one day we may have good evidence from ice-core analysis or dendrochronology.

C. Radiocarbon Dating

Arguments based on radiocarbon dating today form the principal challenge to the chronology based on texts, archaeological interconnections, and Egyptian astronomy for the beginning of the Aegean Late Bronze Age and the Theran eruption. Dates for the eruption older by 90 to 120 years than conventional dating have been proposed by some specialists in radiocarbon dating (Hammer et al. 2003: 87; Manning 1999: 335; Friedrich et al. 2006a). The years 1645 BC ±4, 1628 BC, or more generally 1627–1600 BC have been heralded (but in some cases subsequently abandoned) as the date of the eruption on the basis of radiocarbon dating (in some cases combined with presumed ice-core or dendrochronological evidence). All such attempts encounter 1) the inherent difficulty of radiocarbon measurement, 2) the problematic nature of calibration and the resulting uncertainty in calibration curve data, and 3) the obdurate obstacle presented by the oscillation of the calibration curve during the late 17th and 16th centuries BC.

C.1) Inherent problems including pretreatment, regional variation, intra- and inter-year variation, seasonal variation, climate effects and contamination by ^{14}C-deficient carbon

High-precision Accelerator Mass Spectroscopy (AMS) radiocarbon laboratories today can measure the radiocarbon content of a single sample to within a 60-year range BP (Before Present), prior to calibration against a decadal measurement of a tree with rings of a known dendrochronological date (Manning 2006–2007: 60; pers. comm. of 16 September 2008). Repeated measurements may narrow the range. Pretreatment of samples, particularly with regard to the removal of humic acid, may occasionally present difficulties, even for modern high-precision laboratories.

Measurement differences between high-precision labs continue to exist. For example, "[o]verall, comparing the Oxford (OxA) versus Vienna (VERA) data on the same samples (using the pooled ages for each individual laboratory where they re-measured the same sample, thus $n=17$), we find an average offset of -11.4 ^{14}C years. The standard deviation is, however, rather larger than the stated errors on the data would imply at 68.1. This indicates that there is an unknown error component of 54.5 ^{14}C years" (Manning et al. 2006b: 5). Moreover, "the possible likely typical unknown error component of around 14 ^{14}C years found between Oxford and Vienna is about as good as can be expected in such an inter-comparison given the typical level of offsets found in inter-laboratory comparisons even between the high-precision laboratories" (Manning et al. 2006b: 5. For inter-laboratory differences generally, see Scott 2003; Reimer et al. 2004). A mean offset of 27 ±2 ^{14}C years on samples divided between Heidelberg and Seattle was unfortunately never resolved; rather the differing data sets were combined in the calibration curve (Reimer et al. 2004: table 1). Other examples of significant unresolved inter-lab differences on measurements of the same wood exist. For example, earlier measurements of bristlecone pine in Tucson were subject to reexamination by the radiocarbon laboratories in Heidelberg, Groningen, Pretoria, and Seattle and produced a mean difference of 37 ±6 ^{14}C years (Reimer et al. 2004: 1033). Radiocarbon determinations still produce "outliers" with some frequency, with occasional measurements a century apart on samples divided between two or more high-precision labs, as in the Turin Shroud measurements of samples divided between Arizona, Oxford, and Zurich (Taylor 1997: 84–85).

The decadal measurements of the calibration curve necessarily mask to some degree both intra-year as well as inter-year variability. The intra-year difference in radiocarbon ages between the summer high and winter low has been said to vary generally between 8 and 32 radiocarbon years (Housley et al. 1999: 167; Levin et al. 1992: 503–518; Levin and Hesshaimer 2000: 69–80. The analyses cited are based on simplified models utilizing postindustrial measurements, which are affected by industrial emissions [Suess effect]. Keenan [2004] argues that measurements of preindustrial samples show that the difference can be far larger than 32 years).

Intra-year seasonal differences in measurements may be compounded by regional climate differences, which in turn may be magnified greatly during periods of climate change. The difference between the late winter–early spring growing season for seeds in Egypt and the late spring–early summer growing season for European trees would push the Egyptian determinations toward dates older than actual dates after calibration based on determinations from German and Irish oaks, for example. Whereas the German oaks of the calibration curve lie at around 50° latitude and the Gordion logs and Theran seeds and trees at around 40°, Cairo in Egypt lies at 30° latitude, and the growing season in Egypt includes the winter–spring ^{14}C minimum (Keenan 2004). Indeed, even many staunch advocates of the superiority of radiocarbon-based chronologies agree that determinations from the Gordion logs of the Anatolian floating tree-ring chronology give calibrated dates quite different from determinations from German and Irish trees on which the calibration curve is based for what are believed on substantial grounds to be the same decades at the end of the ninth and first half of the eighth centuries BC. They attribute the result to changes in solar radiation and a consequent cold period latening growing seasons in Anatolia in the period (Kromer et al. 2001: 2531; Manning et al. 2001: 2533–2534; Manning et al. 2006c). Whether a cold period occurred at any point in the decades preceding the Theran eruption is of course unknown.

A recent report that for the period AD 1600–1800 Turkish pines and Irish oaks show a seemingly trivial average difference of 1.02 years (Manning et al. 2006c) is potentially misleading with respect to the risks of regional variation, since in many decades the measurements differ by 20–30 years, sometimes in one direction and sometimes in the other, the opposing directions of the differences resulting in the seemingly trivial average difference. Moreover, the average difference should have been stated as 1.02 ±s years (where s = the standard deviation), a somewhat different matter. (I am grateful to D. Keenan for calling this point to my attention.)

Although the calibration curve was initially based on the premise that ^{14}C was distributed evenly in the earth's atmosphere (Reimer 2004), differences in radiocarbon measurements of decadal determinations of trees of the same known dates from the Northern and Southern Hemispheres—amounting to a mean difference of 41 ±14 years over the past 900 years, but with a variation between 8 and 80 years—have led to the recent creation of a separate calibration curve for the Southern Hemisphere. An Intertropical Convergence Zone is believed to act as a curtain between the hemispheres, preventing or delaying the mixing of atmospheric elements. (I am grateful to S. Manning for his reminder in this regard.)

While the principal reason for the lack of convergence of radiocarbon determinations between the hemispheres may be understood, the underlying cause or causes of the differences between Northern and Southern Hemisphere ^{14}C measurements and their relative significance are unclear. More of the Southern than the Northern Hemisphere is covered by water which retains ^{14}C-deficient carbon; accordingly, it has been suggested that this ^{14}C-deficient carbon is supplied to the atmosphere, and from the atmosphere to trees and plants (Lerman, Mook, and Vogel 1970; Knox and McFadgen 2001: 87). Radiocarbon measurements of marine mollusks from the Atlantic Ocean give dates typically 400 years older than their true ages because ^{14}C-deficient deep water supplies some of the carbon dissolved in the upper layers of the ocean (Facorellis, Maniatis, and Kromer 1998). The correction for this "reservoir effect" is typically 400 years, but it varies with location. It has also been proposed that the diffusion throughout the Southern Hemisphere via cold-water currents flowing northward of ^{14}C-deficient carbon from a source in the Weddell Sea in Antarctica may be a factor in the Southern vs. Northern Hemisphere difference (B. Kromer, pers. comm. of 9 November 2002). Living (or at least recently deceased) penguins in Antarctica appear to be 800 years old based on ^{14}C measurements (B. Kromer, pers. comm. of 9 November 2002). Such deep water reservoirs of terrestrial carbon have not been replenished with ^{14}C from the atmosphere and hence have older radiocarbon ages. Upwellings of ^{14}C-deficient carbon during El Niño or ENSO episodes have also been proposed as a causal agent of the hemispheric difference (Stuiver and Braziunas 1993: 296). Attention lately has focused on periodic warming and cooling cycles in the Pacific Ocean at approximately 60-year intervals (S. Manning, pers. comm.; Ministry of Environment, Government of British Columbia 2002). A hypothesis has been proposed to explain the phenomenon: because more of the Southern Hemisphere is covered by water, the cooling cycle may be more intense, and when the warming phase comes, more ^{14}C-deficient carbon may be released into the atmosphere than in the Northern Hemisphere in general.

Is the Southern Hemisphere phenomenon relevant to radiocarbon measurements from Thera? Certainly the ratio of water to land in the Aegean Sea is high, but is there evidence of a ^{14}C-deficient carbon source or sources, and a suggested mechanism or mechanisms for introducing ^{14}C-deficient carbon into the atmosphere? First, Thera and indeed the whole Aegean is notorious for vents containing ^{14}C-deficient carbon. Geothermal areas are known in the northern and central Aegean as well as along the Hellenic Volcanic Arc. A recent occurrence near the island of Melos was described as follows: "Every fumarole on the shore blew out. And the sea boiled as the gas came out with such force. Stunned fish came to the surface" (P. R. Dando, as quoted in Pain 1999: 41). Another major source of old carbon exists 5km NNE of Thera. After a visit to Thera in 1884, the traveler James Theodore Bent reported that the water in the caldera was almost at boiling heat in parts and contained bubbles of vapor, as a result of a recent minor eruption, and that the sulfur content was sufficient to remove the barnacles from ships' hulls (Bent 1966: 118).

On Thera itself, a study by F. Barberi and M. L. Carapezza concludes that "24 points of anomalous soil gas release or concentration have been identified...half in the northern area and half in the southern one" (1994: 340). The most recent detailed study (McCoy and Heiken 2000) reports that "manifestations of volcanism and concomitant hazards remain today with fumaroles, seismic activity, hydrothermal springs, and higher concentrations of helium and CO_2 in soils" (43) and that "high concentrations of helium and CO_2 are present in soils on central Thera" (48). Moreover, in volcanic areas groundwater may be saturated with ^{14}C-deficient CO_2, which may then diffuse through soil and into the air. Tests in Tuscany have shown that where groundwater reaches the surface at natural springs, nearby trees may give elevated ^{14}C ages (Saurer et al. 2003). (I am grateful to S. Soter for calling this article to my attention.)

Agricultural activity can release ^{14}C-deficient gases, and aquifers can contain significant amounts of ^{14}C-deficient carbon (Mörner and Etiope 2002: 193). Fumaroles send ^{14}C-deficient carbon into the atmosphere where it can be absorbed by the leaves of plants and trees. The soil gas in geothermal areas has elevated concentrations of old CO_2. Plants are known to acquire small amounts of carbon directly from the soil through their roots (Stolwijk and Thimann 1957; Skok et al. 1962; Geisler 1963; Splittstoesser 1966; Arteca et al. 1979; Yurgalevitch and Janes 1988).

One study of current short-lived plant material from Thera whose true age was about one year provided radiocarbon ages of 1390 and 1030 years before present (Bruns et al. 1980: 535 table 2). The plants were located near vents of ^{14}C-deficient carbon, which the plants had absorbed. The old carbon effect disappeared beyond a distance of 250m (Bruns et al. 1980: 534 fig. 1). Investigations of fields of ^{14}C-deficient carbon both in southern and northern Italy (Rogie 1996; Chiodini et al. 1999; Rogie et al. 2000; Cardellini et al. 2003) have shown much more widespread geographical effects, however, and there is no way of estimating the amount of such carbon released by the precursory activity of the Theran volcano preceding the major LC I/LM I event. (These examples illustrate the difference between measurement accuracy and date accuracy, which is another matter altogether and one not necessarily captured by stated radiocarbon error bands.)

Some have argued that the presence of ^{14}C-deficient carbon in samples would cause irregular effects and so cannot explain a purported pattern of radiocarbon determinations 80–120 years earlier than historically derived dates for the Theran eruption. No such pattern exists, however. Most determinations on Theran samples fit easily within the oscillating portion of the calibration curve (see below). For example, the most recent set of measurements of seeds from jars buried in the Volcanic Destruction Level on Thera, the best context imaginable, divided between the VERA laboratory in Vienna and the OxCal laboratory in Oxford, produced dates that were compatible with one exception with the historical chronology (Manning et al. 2006a;

2006b). The one exception formed half of a pair of seeds from the same jar, one of which provided an anomalously high ^{14}C age. (Given the more than two-sigma spread between the radiocarbon ages of the two seeds, a statistical case for exclusion of both determinations as incompatible could be made.) In the brief period of perhaps months between the preliminary major earthquake and release of gases and the final eruption, the chance of release of old carbon was in all likelihood heightened. The one other ^{14}C determination (Bronos 1a: Manning et al. 2006b) giving an anomalously high radiocarbon age was collected from an insecure context in the 1970s and the ^{14}C age measured then. The very few samples which provide earlier radiocarbon ages vary substantially; as a result, there exists a risk of biasing results by including in the database a sample with 1% ^{14}C-deficient carbon which adds about 80 years while excluding as an "outlier" a sample with 3% ^{14}C-deficient carbon which adds about 240 years. Moreover the ^{14}C-deficient carbon problem need not explain all of the small number of "early" ^{14}C determinations, given the other sources of uncertainty described above, including regional, seasonal, and climatic variation, all of which tend to produce measurements on Aegean samples higher after calibration than true dates. Other ^{14}C measurements, such as those from Trianda on Rhodes or Miletus in Anatolia, were taken from wood which may not preserve its original outer rings or which may have been in use for an unknown period, in either case making it impossible to date the context from the age of the wood.

C.2) Calibration curve problems in general

The calibration curve, by which radiocarbon measurements are converted to calendar date ranges through comparison with radiocarbon measurements of tree-ring segments of known date, is sometimes viewed as fixed and immutable. In fact the calibration curve is a fragile construct, "not a curve, but a probability band" (Manning 1995: 128), whose application requires both judgment and caution. The uneasy relationship between the 11-year and 25-year sunspot cycles and the decadal calibration curve data is one source of uncertainty. Such cycles are detectable within the curve (Attolini et al. 1993; Buck and Blackwell 2004: 1101). The resulting effect appears to vary over time and location. "Pacific Northwest Δ^{14}C values... contain an 11-yr cycle with an average amplitude of 1.40 ±0.16‰(ca. 11 ±1 ^{14}C yr). This amplitude differs significantly from the 11-yr cycle amplitude of 4.8 ±0.6‰(ca. 39 ±5 ^{14}C yr) found in Russian trees (Kocharov 1992) between AD 1600 and AD 1950" (Stuiver 1993: 68). (I am grateful to D. Keenan for reminding me of these citations.) In retrospect, it might have been preferable to construct a calibration curve based on 11-year segments matching the 11-year sunspot cycle. The IntCal98 curve relies largely on earlier, less accurate measurements, often on samples not subject to modern pretreatment regimes, while the IntCal04 calibration curve combines older measurements with more recent measurements by high-precision laboratories. Even with respect to modern determinations, however, the IntCal04 curve combines measurements from separate laboratories that are significantly different for the same decade. The difference in the case of a particular decade of old vs. new measurements is over 50 radiocarbon years in a number of cases, and recent German oak measurements differ from the IntCal04 determinations by up to 70 radiocarbon years (Manning et al. 2006c). The report of the 19th International Radiocarbon Conference of April 2006 concludes that "each group of researchers who provide data with potential utility for radiocarbon calibration curve estimation do their best to quantify their own internal sources of error and uncertainty and to report these in standard form. What they do not and cannot do is allow for sources of error or uncertainty that they are completely unaware of" (Bronk Ramsey et al. 2006: 792).

Questions include how much emphasis to place on the central area of a distribution given all the uncertainties of ^{14}C determinations discussed above, and the significance of the duration of the intersection between the radiocarbon age span obtained from a sample (e.g., a seed) and the radiocarbon segments of the calibration curve, once the fact of an intersection at two different areas of the calibration curve is observed.

Any single measurement results in a two-sigma (95.4%) probability band that is twice the width of the one-sigma band, e.g., ±20 at one sigma, ±40 at two sigma. Repeated measurements, however, may reduce the two-sigma range via statistical inference. Accordingly one frequently finds ranges stated in the nature of 1621–1605 BC at one sigma and 1627–1600 BC at two sigma, 95.4% probability. The statement of such ranges (particularly given the appearance of exactitude to the first decimal) frequently puzzles, and potentially misleads, humanists not trained in statistics, who assume that "95.4% accuracy" refers to date probability rather than measurement probability. In order for date probability to match measurement probability, it would be necessary for the calibration curve to be exact, which it is not; for the calibration algorithm that converts radiocarbon ages into dates to be exact, which it is not; for pretreatment and other laboratory procedures to be foolproof; and for offsets and variation to be absent. If, however, either the seed or other sample or the relevant decadal segments measured for the calibration curve contain humic acid or are affected significantly by inter-laboratory offsets, intra- or inter-annual variation, regional variation, or seasonal variation sometimes amplified by climate change, repeated measurement will not eliminate the problem, and accordingly measurement-indicated probability will diverge significantly from calendar date probability. Where ^{14}C-deficient carbon is present the divorce is total, for it is the date of the mingled carbon rather than the date of the sample which has been measured and provided with a probability estimate.

The international committee responsible for the preparation of the calibration curve published in 2004 (IntCal04) concluded that because the initial measurements of decadal or duodecadal segments of the calibration curve on wood of known date from long-lived German and Irish oaks were made decades ago with less sophisticated equipment and methods than exist today and depended on a very few measurements for each decade—some of which have since been recognized as erroneous (Wiener 2003: 382; 2007)—the Gaussian bell-curve-derived estimates of measurement accuracies, adopted initially as a default position in the absence of an agreed data-driven standard, should be multiplied at the one-sigma range by 1.3 for the Seattle lab and 1.76 for the Belfast measurements on German oak, for example. The IntCal04 Committee further decided to limit the impact of error in any particular decadal measurement by smoothing the calibration curve through incorporating into each decadal determination the measurements of the nearest 100 data points or observations, whether these observations came from repeated measurements of the same decade from the same piece of wood, the same decade from other pieces of wood, semi-decadal measurements, or measurements from individual annual rings within a decade. Accordingly the time span incorporated into each decadal determination can vary significantly depending on the density of the observations at a given point (Buck and Blackwell 2004: 1100. I am most grateful to Dr. Paula Reimer, the chair of the IntCal04 Committee and Director of the ^{14}CHRONO Centre for Climate, the Environment, and Chronology at the Queen's University Belfast, for clarifying this matter for me). The decision to smooth the calibration curve in this manner proved controversial, and it may be that in the next iteration of the calibration curve, expected in AD 2010–2011, the Random Walk model on which the IntCal04 curve is based will be replaced by a Markov Chain model resulting in less smoothing, particularly since many more decadal determinations from various high-precision laboratories will be available. The resultant modifications generally will not be substantial, but even small changes to critical decades in and surrounding the oscillating portion of the calibration curve in the 17th–16th centuries BC may be relevant to the debate about the dating of the Theran eruption, for example. The IntCal04 Committee also noted that for wiggle-matching purposes the actual data represented by the superseded IntCal98 curve, however flawed, might be preferred to the artificially smoothed date bands, and that "in the case of wiggle-matching of tree-ring sequences, the method is sometimes being pushed to the limits in all respects" (Reimer et al. 2004: 1037. For other critiques of the methodology of wiggle-matching, see Whitelaw 1996; Cavanagh quoted in Wiener 2003: n. 148; Wiener 2003).

The procedures which connect measurements to calibrated dates, such as those contained in the OxCal program, also deserve consideration here. Martin Aitken, the former Deputy Director of the Oxford Research Laboratory for Archaeology and the History of Art, noted that, "conversion to calendar date is confusing because of the irregular form of the calibration curve; the difficulty of translating error limits from one time-scale to the other is particularly acute and here we are inevitably in the hands of the statisticians" (Aitken 1990: 93). Unfortunately, statisticians using different approaches or models on the same data may produce significantly different results. A recent study of the dating of the Iron Age I to Iron Age II transition at sites in Israel (Mazar and Bronk Ramsey 2008) reports that two different Bayesian models (labeled C2 and C3) provided different date ranges, and warns of the sensitivity of Bayesian models to outliers, even where only one measurement at one site is at issue. The inherent, intractable nature of the difficulty presented is discussed in detail in Manning (1995: 127–129). Figures 1A and 1B illustrate the effect of vigorous smoothing on a reported calibrated radiocarbon date in one instance (Buck et al. 2006: 285; Buck was the statistician for the IntCal04 calibration curve project).

The degree of likelihood that the true calendar date of the sample measured falls within the years represented by the peak in Figure 1B remains at issue. In any event, neither statistical method is sufficient to capture the potential impact of factors such as variation exacerbated by climate shift or the presence of ^{14}C-deficient carbon in a sample.

C.3) The oscillating calibration curve of the late 17th and 16th centuries BC

The oscillation of the calibration curve results in similar radiocarbon ages corresponding to calibrated dates at about 1610 and around 1535–1525 BC, whether

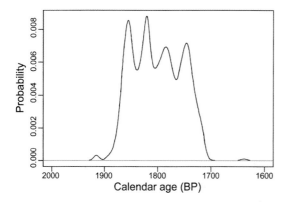 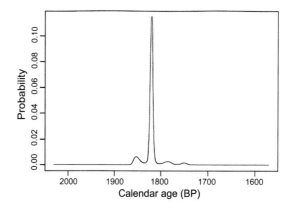

Figure 1: Posterior density for true calendar age θ_0 of a sample with radiocarbon age 1870 ±30. On left, (Figure 1A) Density from the traditional piecewise linear interpolation of the IntCal98 data. Right, (Figure 1B) Density after Buck et al. 2006 Random Walk/Bayesian modeling.

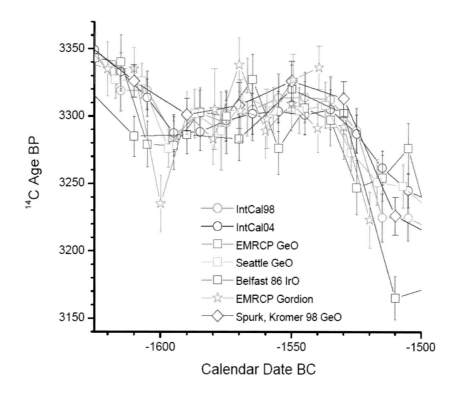

Figure 2: Comparison of the IntCal98 and IntCal04 calibration curves against the two major underlying datasets (the Seattle UWTEN data on German Oak [GeO] for this period, and the Belfast 1986 data on Irish Oak [IrO]) and against the EMRCP data for German Oak and Gordion Juniper, and previous Heidelberg German Oak data revised in 1998.

one uses, for example, 1998 measurements on German oaks or recent measurements on German oaks and the Anatolian logs from Gordion, as shown in Figure 2 (I am most grateful to S. Manning for forwarding this data and depiction). The 1998 German oak measurements were only about 10 radiocarbon years apart for calibrated ages of 1610 and 1530 BC (and this is reflected as well in the IntCal98 measurements for 1615 and 1535 BC); the East Mediterranean Radiocarbon Intercomparison Project (EMRCP) 2006 German oak measurements gave closely similar radiocarbon ages for determinations from the decades centered on 1605, 1555, and 1505 BC and separately for 1585, 1575, and 1535 BC, and the Project measurement of the Anatolian junipers from Gordion gave similar radiocarbon ages at 3335 BP for 1620, 1570, and 1540 BC, and only a minor difference of about 10 radiocarbon years between 1600 and 1520 BC. Both IntCal04 and the Seattle German oak measurements show nearly identical radiocarbon ages at 1595 and 1525 BC.

While the Gordion measurements are more geographically relevant to the dating of samples from Thera than Irish or German oak determinations, nevertheless Thera and Gordion lie in separate meteorological regimes, with Gordion strongly affected by winds from the northeast (Keenan 2002: 237 fig. 1; Reddaway and Bigg 1996: fig. 3). Location-dependent radiocarbon effects may differ accordingly. In connection with the data depicted in Figure 2, it is worth noting that research in the course of establishing the 1998 calibration curve disclosed a mean difference of 24.2 ±6 years between Belfast measurements of Irish oak and Seattle measurements of German oak for the critical years 1700 to 1500 BC (Wiener 2007). Manning et al. in their discussion of a ^{14}C measurement of a three-year tree twig preserved in the Volcanic Destruction Level on Thera which produced a date consistent with the historical chronology, observe that the sample may "reflect short-term higher amplitude and/or frequency variation in atmospheric ^{14}C ages not seen in the IntCal04 record for the period, which is both based on 10-year growth samples and smoothed" (2006b: 12).

C.4) Recent proposals to overcome the oscillation obstacle and their defects

Two articles in the 28 April 2006 issue of *Science* magazine address the problem posed by the oscillation. One (Manning et al. 2006a) focuses on radiocarbon determinations from before and after the period of oscillation to support a high chronology, long favored by Manning. Unfortunately, for the pre-oscillation period the article can only offer ^{14}C measurements which in each case provide at best a *terminus post quem* date.

A piece of wood found at the Minoan site of Trianda on Rhodes produced rather incoherent determinations for its three decadal segments, with 80 years separating the central points of adjacent decades and outer rings which gave earlier dates than the inner rings. There is no way of knowing how many years separated the felling of the tree from the burial of the object—piece of furniture, shelf, beam, or whatever—from which the chunk of wood derived.

The second pre-oscillation data point is a piece of wood covered in Theran tephra from what the excavator, W.-D. Niemeier, believes was a throne in a shrine area at the Minoan site of Miletus on the coast of Anatolia. Whether the outer preserved ring represents what is called a "waney edge" indicating a point close to the felling date is not entirely clear (Bronk Ramsey et al. 2004: 327; Manning et al. 2006b), but in any event the period of time between the felling of the tree and the eventual destruction of the throne (or perhaps chair, chest, shelf, or beam) is unknown.

The Supporting Online Material for the *Science* article also republishes ^{14}C determinations from short-lived samples from the Theran Volcanic Destruction Level. Most fall within the oscillating portion of the calibration curve, as noted above. The few which do not each pose problems of context or measurement as noted above.

At the post-oscillation end, the article republishes dates from two Late Minoan IB destruction levels considerably later than the mature Late Minoan IA eruption on Thera. A prior claim (Manning et al. 2002: 741) that there existed a "unique explanatory solution" for the fact that the two sets of determinations on seeds from Myrtos-Pyrgos and Chania gave quite different dates for what were assumed (without any independent justification) to be destructions within the same or a few years of each other, namely that both destructions occurred within a steeply-sloped portion of the calibration curve, has been abandoned (Bronk Ramsey et al. 2004: 328) after it was noted (Wiener 2003: 391–392) that the vertical segment involved was an illusion based on a single erroneous measurement, pre-high-precision, in one laboratory. In fact, most LM IB destruction deposits give far later dates than those cited in the article, centering on 1460–1440 BC and thus are consistent with the historically well-established chronology (Soles 2004). An absolute date for the end of Late Minoan IB prior to the beginning of the reign of Tuthmosis III in Egypt between 1479 and 1468 BC, as proposed by Bronk Ramsey et al. (2004: 328–329), would indeed appear unlikely to historians conscious of the close links between LM IB Crete and Tuthmoside Egypt (see above). Finally, even a date around 1490 BC for a point in LM IB—even late in LM IB—would not in any event necessarily contradict a date for the Theran eruption between 1540 and 1525 BC.

The seriation analysis, where the time intervals between the various phases cannot be established independently, itself is highly dependent statistically on the correctness of the "boundaries" incorporated in the seriation—such as "Boundary 'End or Final LM IA to Start LM IB'; Event 'Early to Mid/Mature LM IB'; Boundary 'Early/Mid LM IB to Later LM IB'" (Manning et al. 2006b: 10)—many unrecognizable to archaeologists. (Potential hazards in the application of Bayesian analysis to radiocarbon dates are considered in Steier and Rom 2000 and Wiener 2003.) Of course data can be marshaled in various ways, for example by combining or alternatively discarding inconsistent measurements of samples or of segments of the calibration curve. "Bayesian analysis is not a 'cure-all'; it has costs, not least the specification of the prior. This is not easy and even in those situations where we think we are not making any strong assumptions, there may be hidden complications" (Scott 2000: 181). Indeed, the result obtained by Manning et al. is dependent on what data is included and what excluded. In fact, it would seem possible to apply "quasi-Bayesian" or "seriated sequence" analysis to achieve a result consistent with historical chronology by limiting the data bank to short-lived samples (i.e., excluding potentially old wood, as in the measurements from Trianda and Miletus employed in Manning et al. 2006a); eliminating determinations from samples of uncertain stratigraphy and/or displaying inchoate data, as in the Trianda sample; discarding inconsistent measurements of samples of known same date, as in one critical Theran Volcanic Destruction Level measurement; and employing instead as Bayesian boundaries the earliest observed existence of datable objects found in specific Aegean contexts, such as the White Slip I bowl from the Volcanic Destruction Level at Thera, or the Egyptian alabaster lid of a bowl bearing the cartouche of the Hyksos ruler Khyan in what seems clearly to be a Middle Minoan III deposit (P. Warren, pers. comm. 4 February 2005; Macdonald 2003: 40–41) at Knossos (rather than in a Late Minoan IB deposit as the Long Aegean Chronology would require). In short, the analysis presented in the Manning et al. article in *Science* and accompanying online data does not in itself present a persuasive case for drastically modifying the historically determined chronology for the Theran eruption and the Late Minoan IA period.

The second *Science* article presents radiocarbon determinations from four contiguous segments of a Theran olive branch found covered in tephra from the eruption (Friedrich et al. 2006a). Olive wood is difficult to date by standard dendrochronological methods, but an examination of the olive branch by X-ray computer tomography is said to have detected 72 indications, which the investigators assume represent annual rings, with a maximum counting error of ± 16 rings for the entire estimated 72-ring sequence, but only ± 3 rings for the final estimated 12-ring segment. The authors further state that a 50% difference in ring count would only increase the calibrated date limits of the radiocarbon determinations of the final segment (1621–1605 BC at one sigma, 1627–1600 BC at two sigma) by a decade. The Supporting Online Material for the article notes that if the IntCal98 calibration curve which utilizes the actual unsmoothed decadal measurements is employed, then the lower limit of the latest segment may drop down to c. 1575 BC at two sigma. It is accordingly pertinent that recent high-precision measurements of the calibration curve data from the Gordion logs, geographically the closest calibration curve data to Thera, give both 1) inconsistent measurements for the decade centered on 1580 BC and 2) a 1580 BC measurement indistinguishable because of the oscillating curve from the decade centered on 1530 BC (see Figure 2). Indeed, the radiocarbon ages of the two later segments of the olive branch encompass this oscillation whichever calibration curve is employed. The earlier segments of the rings, however, provide ^{14}C ages which on first impression fall before the oscillating portion of the calibration curve and thus offer support to a higher chronology, assuming no significant difference in ring count, the absence of climate/regional distortion, and the absence of ^{14}C-deficient carbon in the earlier segment(s) of the branch, always a risk in the Theran environment.

A preliminary question arises from the uncertainty as to the number of calendar years represented by the rings or partial rings of an olive branch observed through X-ray computer tomography. The article assumes without discussion that the rings were formed annually, but Cherubini has observed that "olive wood, as the wood of other Mediterranean evergreen species (e.g., *Quercus ilex*, *Quercus suber*, *Arbutus unedo*), may have one ring per year (at sites characterized by clear winter seasonality), two rings per year (at sites characterized by extreme dry conditions during summer and cold winter), no rings (at sites with very mild winter)" (pers. comm. of 19 April 2007, for which I am most grateful). Cherubini et al. (2003: 129) note that:

> ...[M]editerranean tree rings have seldom been used for dendroecological, dendroarchaeological or dendroclimatological purposes (Serre-Bachet 1985)....The main reason for this deficiency is the inability in many cases to identify clearly and date tree annual rings. Although the verification of the annual nature of tree rings is necessary

for dendrochronological studies, the seasonal patterns of wood production are not yet well understood in plants lacking annual rings (Gartner 1995).

Of course if the Theran olive branch rings represent seasonal events two consequences follow: 1) the approximately 36 years thus putatively represented no longer present a progression of radiocarbon dates across seven decadal determinations; and 2) the purported wiggle-match to seven decades of the calibration curve disappears. If the earlier two segments of the olive branch are compressed into about a decade and a half as the consequence of rings formed seasonally rather than annually, then the risk of a wayward individual decadal data point in the calibration curve emphasized by the IntCal04 Committee arises.

The question of the number of years represented by the rings is also critical to the determination of the range of the probability bands around the estimated dates, for (as the authors acknowledge) it is only the presumed existence of annual rings that provides the "known time gaps between the samples [which] are the key to the high precision of the obtained calibrated ages" (Friedrich et al. 2006b). If the time gaps are uncertain because of the existence of seasonal rather than annual rings, the absence of rings in cultivated trees and/or the difficulty of interpretation of images of olive tree branches obtained by X-ray tomography, then the stated probability bands would require concomitant expansion. (Whether the OxCal program algorithms are sufficient to cover putative marked irregularities in ring counts is unclear.) As noted above, the central point of the distribution should not be given undue emphasis.

Perhaps the principal question respecting the olive branch stems from the propensity of olive trees to retain dead branches. Oliver Rackham, coauthor of *The Making of the Cretan Landscape* (1996) and *The Nature of Mediterranean Europe* (2003), has kindly provided the following comment in this regard:

> I don't follow the argument that the last growth ring of the wood specimen was contemporary with the eruption. The authors describe it as a 'branch,' but the pictures indicate a shattered radial fragment of a stem or major branch at least 40 cm in diameter. As we all know, many olive trees bear dead branches and fragments of branches, and I would not rule out the possibility that some of these might last 100 years after they died. The tree itself may have been alive when it was buried, but not all its limbs were necessarily alive or even recently dead (pers. comm. of 11 May 2008).

Harriet Blitzer, the leading specialist in the ethnography of preindustrial Cretan agricultural practice and author of "Agriculture and Subsistence" in *The Plain of Phaistos* (2004) concurs, stating that:

> Certain parts of a mature tree may die and other parts of the same tree may continue to grow and bear fruit. The decision to prune the dead branches is based in part on the overall structure of the tree (its stability and balance) and on whether the dead sections prove an obstacle to further growth in other parts of the plant. In many cases, among older trees, there are massive dead branches that have been left untouched for the above reasons. In those instances, the remainder of the tree is alive, growing and producing fruit (pers. comm. of 23 July 2008; see also Blitzer forthcoming).

It is worth noting that the radiocarbon date of 1613 ±13 proposed for the last segment would fit exactly the textual cum archaeological date for the Seismic Destruction Level at the beginning of LC I, an event which could have caused the death of the branch.

Unfortunately, the Theran olive branch covered in tephra from the eruption is for the moment that dreaded scientific phenomenon, a singleton. Both intensive remeasurement of the existing branch (preferably by a different radiocarbon laboratory) to determine whether the initial measurements are replicable and the location and measurement of an additional branch or branches are critical desiderata.

Conclusions

1. Establishing dates for the massive eruption of the volcano on Thera and for the Late Minoan IA period, the acme of Minoan civilization, is of prime importance for understanding both the internal development of Minoan civilization and its relations to the cultures of Egypt and the Near East.

2. A rich interlocking of texts, archaeology, interconnections between societies, and Egyptian astronomical dates limits the eruption, which occurred in a mature or final stage of LM IA, to the range of c. 1545–1495 BC, with the highest likelihood no earlier than c. 1525 BC, a year which also provides evidence of a climate-forcing event in tree rings at various places and in the Greenland ice (Wiener 2006a: 323). Tree-ring and ice-core indications of a volcanic event at 1571–1570 BC (Wiener 2006a: 320) are at the extreme upper limit (and many would say well beyond the upper limit) of an archaeologically conceivable date.

The higher a proposed date above 1525 BC, the harder it becomes to reconcile with the historical evidence; the lower a proposed date below 1525 BC, the harder it becomes to reconcile with the ^{14}C evidence.

3. Neither ice-core nor dendrochronological examination provides direct evidence of the Theran eruption.

4. The sometimes problematic nature of radiocarbon measurement, calibration, and statistical inference concerning their intersection constrains to some degree chronological conclusions based on ^{14}C determinations. The oscillating calibration curve between 1620 and 1520 BC remains a major obstacle to radiocarbon dating for this period. Overly optimistic assumptions about the dependability, accuracy, and precision of measurements of samples and of the boundaries of the calibration probability band against which they are compared are common.

5. Neither the seriated sequence analyses published to date nor the data obtained from the single Theran olive branch covered in tephra yet provide a convincing argument for moving dates for Late Minoan IA and the Theran eruption upward by 70–100 years. "Extraordinary claims require extraordinary evidence," said Carl Sagan (1979: 62), but thus far such evidence in support of an Aegean Long Chronology is lacking.

Accordingly, the separation of "the two cultures" in seeking an absolute chronology for the beginning of the Aegean Late Bronze Age persists.

References

Aitken, M.J. 1990. *Science-based Dating in Archaeology*. London: Longman.
Arteca, R.N., Poovaia, B.W., and Smith, O.E. 1979. Changes in Carbon Fixation, Tuberization, and Growth Induced by CO_2 Applications to the Root Zone of Potato Plants. *Science* 205: 1279–1280.
Attolini, M.R., Cecchini, S., Galli, M., Kocharov, G.E., and Nanni, T. 1993. 400 Year Record of Δ^{14}C in Tree Rings: The Solar-Activity Cycle Before, During and After Maunder Minimum and the Longer Cycles. *Il Nuovo Cimento* 16C(4): 419–436.
Barberi, F., and Carapezza, M.L. 1994. Helium and CO_2 Soil Gas Emissions from Santorini (Greece). *Bulletin of Volcanology* 56: 335–342.
Bennett, C. 2006. Genealogy and the Chronology of the Second Intermediate Period. In M. Bietak (ed.), *Egypt & Time. Proceedings of the SCIEM 2000 Workshop on Precision and Accuracy of the Egyptian Historical Chronology, Vienna, 30 June–2 July 2005. Ägypten und Levante* 16: 231–243. Vienna: Verlag der Österreichischen Akademie der Wissenschaften.
Bent, J.T. 1966. *Aegean Islands: The Cyclades, or Life among the Insular Greeks*. Chicago: Argonaut.
Bergoffen, C. 2001. The Proto White Slip and White Slip I Pottery from Tell el-Ajjul. In V. Karageorghis (ed.), *The White Slip Ware of Late Bronze Age Cyprus. Proceedings of an International Conference Organized by the Anastasios G. Leventis Foundation, Nicosia, in Honor of Malcolm H. Wiener, Nicosia 29th–30th October 1998*: 145–155. Vienna: Verlag der Österreichischen Akademie der Wissenschaften.
Bietak, M. 2003. Science versus Archaeology: Problems and Consequences of Aegean High Chronology. In M. Bietak (ed.), *The Synchronisation of Civilisations in the Eastern Mediterranean in the Second Millennium BC II. Proceedings of the SCIEM 2000–EuroConference, Haindorf, 2–7 May 2001*: 23–33. Vienna: Verlag der Österreichischen Akademie der Wissenschaften.
Bietak, M. 2004. Review of *A Test of Time* by [S.] W. Manning. *BibO* 61: 199–222.
Bietak, M., and Hein, I. 2001. The Context of White Slip Wares in the Stratigraphy of Tell el-Dab'a and some Conclusions on Aegean Chronology. In V. Karageorghis (ed.), *The White Slip Ware of Late Bronze Age Cyprus. Proceedings of an International Conference Organized by the Anastasios G. Leventis Foundation, Nicosia, in Honor of Malcolm H. Wiener, Nicosia 29th–30th October 1998*: 171–194. Vienna: Verlag der Österreichischen Akademie der Wissenschaften.
Bietak, M., and Höflmayer, F. 2007. Introduction: High and Low Chronology. In M. Bietak and E. Czerny (eds.), *The Synchronisation of Civilisations in the Eastern Mediterranean in the Second Millennium BC III. Proceedings of the SCIEM 2000–2nd EuroConference, Vienna, 28th of May–1st of June, 2003*: 13–23. Vienna: Verlag der Österreichischen Akademie der Wissenschaften.
Blitzer, H. 2004. Agriculture and Subsistence in the Late Ottoman and Post-Ottoman Mesara. In L.V. Watrous, D. Hadzi-Vallianou, and H. Blitzer (eds.), *The Plain of Phaistos: Cycles of Social Complexity in the Mesara Region of Crete*: 111–217. Los Angeles: Cotsen Institute of Archaeology, University of California, Los Angeles.
Blitzer, H. Forthcoming. Olive Domestication and Cultivation in the Aegean. *Hesperia*.
Brinkman, J.A. 1972. Foreign Relations of Babylonia from 1600 to 625 BC: The Documentary Evidence. *American Journal of Archaeology* 76: 271–281.
Brinkman, J.A. 1976. A Chronology of the Kassite Dynasty. In *Materials and Studies for Kassite History*, Vol. 1, *A Catalogue of Cuneiform Sources Pertaining to Specific Monarchs of the Kassite Dynasty*: 6–34. Chicago: The Oriental Institute of the University of Chicago.
Brinkman, J.A. 1977. Appendix: Mesopotamian Chronology of the Historical Period. In A.L. Oppenheim (ed.), *Ancient Mesopotamia. Portrait of a Dead Civilization*: 335–348. Chicago: University of Chicago Press.
Bronk Ramsey, C., Manning, S.W., and Galimberti, M. 2004. Dating the Volcanic Eruption at Thera. *Radiocarbon* 46: 325–344.
Bronk Ramsey, C., Buck, C.E., Manning, S.W., Reimer, P., and van der Plicht, H. 2006. Developments in Radiocarbon Calibration for Archaeology. *Antiquity* 80: 783–798.
Bruns, M., Levin, I., Münnich, K.O., Hubberten, H.W., and Fillipakis, S. 1980. Regional Sources of Volcanic Carbon Dioxide and Their Influence on ^{14}C Content of Present-Day Plant Material. *Radiocarbon* 22: 532–536.
Buck, C.E., and Blackwell, P.G. 2004. Formal Statistical Models for Estimating Radiocarbon Calibration Curves. *Radiocarbon* 46: 1093–1102.
Buck, C.E., Portugal Aguilar, D.G., Litton, C.D., and O'Hagan, A. 2006. Bayesian Nonparametric Estimation of the Radio-

carbon Calibration Curve. *Bayesian Analysis* 1(2): 265–288.

Cadogan, G., Herscher, E., Russel, P., and Manning, S. 2001. Maroni-Vournes: A Long White Slip Sequence and its Chronology. In V. Karageorghis (ed.), *The White Slip Ware of Late Bronze Age Cyprus. Proceedings of an International Conference Organized by the Anastasios G. Leventis Foundation, Nicosia, in Honor of Malcolm H. Wiener, Nicosia 29th–30th October 1998*: 75–88. Vienna: Verlag der Österreichischen Akademie der Wissenschaften.

Cardellini, C., Chiodini, G., Frondini, F., Giaquinto, S., Caliro, S., and Parello, F. 2003. Input of Deeply Derived Carbon Dioxide in Southern Apennine Regional Aquifers (Italy). *Geophysical Research Abstracts* 5. [http://www.cosis.net/abstracts/EAE03/09927/EAE03-J-09927.pdf].

Cherubini, P., Gartner, B.L., Tognetti, R., Bräker, O.U., Schoch, W., and Innes, J.L. 2003. Identification, Measurement and Interpretation of Tree Rings in Woody Species from Mediterranean Climates. *Biological Reviews of the Cambridge Philosophical Society* 78: 119–148.

Chiodini, G., Frondini, F., Kerrick, D.M., Rogie, J., Parello, F., Peruzzi, L., and Zanzari, A.R. 1999. Quantification of Deep CO_2 Fluxes from Central Italy. Examples of Carbon Balance for Regional Aquifers and of Soil Diffuse Degassing. *Chemical Geology* 159: 205–222.

Dawkins, R. 1998. *Unweaving the Rainbow: Science, Delusion and the Appetite for Wonder*. Boston: Houghton Mifflin.

Dietz, S. 1991. *The Argolid at the Transition to the Mycenaean Age: Studies in the Chronology and Cultural Development in the Shaft Grave Period*. Copenhagen: The National Museum of Denmark, Department of Near Eastern and Classical Antiquities.

Eriksson, K.O. 2001. Cypriote Proto White Slip and White Slip I: Chronological Beacons on Relations Between Late Cypriote I Cyprus and Contemporary Societies of the Eastern Mediterranean. In V. Karageorghis (ed.), *The White Slip Ware of Late Bronze Age Cyprus. Proceedings of an International Conference Organized by the Anastasios G. Leventis Foundation, Nicosia, in Honor of Malcolm H. Wiener, Nicosia 29th–30th October 1998*: 51–64. Vienna: Verlag der Österreichischen Akademie der Wissenschaften.

Facorellis, Y., Maniatis, Y., and Kromer, B. 1998. Apparent ^{14}C Ages of Marine Mollusk Shells from a Greek Island: Calculation of the Marine Reservoir Effect in the Aegean Sea. *Radiocarbon* 40: 963–973.

Fischer, P.M. 2003. The Preliminary Chronology of Tell el-'Ajjul: Results of the Renewed Excavations in 1999 and 2000. In M. Bietak (ed.), *The Synchronisation of Civilisations in the Eastern Mediterranean in the Second Millennium BC II. Proceedings of the SCIEM 2000–EuroConference, Haindorf, 2–7 May 2001*: 263–294. Vienna: Verlag der Österreichischen Akademie der Wissenschaften.

Frahm, E. 2002. Assur 2001: Die Schriftfunde. *Mitteilungen der Deutschen Orient-Gesellschaft* 134: 47–86.

Friedrich, W.L., Kromer, B., Friedrich, M., Heinemeier, J., Pfeiffer, T., and Talamo, S. 2006a. Santorini Eruption Radiocarbon Dated to 1627–1600 BC. *Science* 312: 548.

Friedrich, W.L., Kromer, B., Friedrich, M., Heinemeier, J., Pfeiffer, T., and Talamo, S. 2006b. Supporting Online Material for Santorini Eruption Radiocarbon Dated to 1627–1600 BC. *Science* 312. [http://www.sciencemag.org/cgi/content/full/312/5773/548/DC1].

Gartner, B.L. 1995. Patterns of Xylem Variation within a Tree and Their Hydraulic and Mechanical Consequences. In B.L. Gartner (ed.), *Plant Stems: Physiology and Functional Morphology*: 125–149. San Diego: Academic Press.

Geisler, G. 1963. Morphogenetic Influence of (CO_2 + HCO_3-) on Roots. *Plant Physiology* 38: 77–80.

Graziadio, G. 1991. The Process of Social Stratification at Mycenae in the Shaft Grave Period: A Comparative Examination of the Evidence. *American Journal of Archaeology* 95: 403–440.

Grove, A.T., and Rackham, O. 2003. *The Nature of Mediterranean Europe: An Ecological History*. New Haven: Yale University Press.

Hammer, C.U., Clausen, H.B., Friedrich, W.L., and Tauber, H. 1987. The Minoan Eruption of Santorini in Greece Dated to 1645 BC? *Nature* 328: 517–519.

Hammer, C.U., Kurat, G., Hoppe, P., and Clausen, H.B. 2001. Recent Ice Core Analysis Strengthen[s] the Argument for a Mid 17th Century BC Eruption of Thera. Extended abstract presented at SCIEM 2000–EuroConference, Haindorf, 2–7 May 2001.

Hammer, C.U., Kurat, G., Hoppe, P., Grum, W., and Clausen, H.B. 2003. Thera Eruption Date 1645 BC Confirmed by New Ice Core Data? In M. Bietak (ed.), *The Synchronisation of Civilisations in the Eastern Mediterranean in the Second Millennium BC II. Proceedings of the SCIEM 2000–EuroConference, Haindorf, 2–7 May 2001*: 263–294. Vienna: Verlag der Österreichischen Akademie der Wissenschaften.

Housley, R.A., Manning, S.W., Cadogan, G., Jones, R.E., and Hedges, R.E.M. 1999. Radiocarbon, Calibration and the Chronology of the Late Minoan IB Phase. *Journal of Archaeological Science* 26: 159–171.

Keenan, D.J. 2002. Why Early-Historical Radiocarbon Dates Downwind from the Mediterranean Are Too Early. *Radiocarbon* 44: 225–237.

Keenan, D.J. 2003. Volcanic Ash Retrieved from the GRIP Ice Core Is Not from Thera. *Geochemistry, Geophysics, Geosystems* 4(11):1097. [doi:10.1029/2003GC000608].

Keenan, D.J. 2004. Radiocarbon Dates from Iron Age Gordion Are Confounded. *Ancient West and East* 3: 100–103.

Knox, F.B., and McFadgen, B.G. 2001. Least-Squares Fitting Smooth Curves to Decadal Radiocarbon Calibration Data from AD 1145 to AD 1945. *Radiocarbon* 43: 87–118.

Kocharov, G.E. 1992. Radiocarbon and Astrophysical-Geophysical Phenomena. In R.E. Taylor, A. Long, and R.S. Kra (eds.), *Radiocarbon After Four Decades: An Interdisciplinary Perspective*: 130–145. New York: Springer.

Koehl, R. 2000. Minoan Rhyta in Egypt. In A. Karetsou (ed.), Κρήτη-Αίγυπτος, Πολιτισμικοί δεσμοί τριών χιλιετιών: 94–100. Athens: Hypourgeio Politismou.

Koehl, R. 2006. *Aegean Bronze Age Rhyta*. Philadelphia: INSTAP Academic Press.

Krauss, R. 2006. Egyptian Sirius/Sothic Dates, and the Question of the Sothis-based Lunar Calendar. In E. Hornung, R. Krauss, and D.A. Warburton (eds.), *Ancient Egyptian Chronology*: 439–457. Leiden: Brill.

Krauss, R. 2007. An Egyptian Chronology for Dynasties XIII to XXV. In M. Bietak and E. Czerny (eds.), *Synchronisation of Civilisations in the Eastern Mediteranean in the Second Millennium BC III. Proceedings of the SCIEM 2000–2nd EuroConference, Vienna, 28 May–1 June 2003*: 173–189. Vienna: Verlag der Österreichischen Akademie der Wissenschaften.

Kromer, B., Manning, S.W., Kuniholm, P.I., Newton, M.W., Spurk, M., and Levin, I. 2001. Regional $^{14}CO_2$ Offsets in the Troposphere: Magnitude, Mechanisms, and Consequences. *Science* 294: 2529–2532.

Kutschera, W., and Stadler, P. 2003. First Results from Sequencing High-Precision ^{14}C Data from Tell el-Dab'a. Paper presented at SCIEM 2000–2nd EuroConference, Vienna, 28 May–1 June 2003.

Lassey, K.R., Manning, M.R., and O'Brien, B.J. 1990. An Overview of Oceanic Radiocarbon. *Reviews in Aquatic Sciences* 3: 117–146.

Lerman, J.C., Mook, W.G., and Vogel, J.C. 1970. ^{14}C in Tree Rings from Different Localities. In I.U. Olsson (ed.), *Radiocarbon Variations and Absolute Chronology. Proceedings of the 12th Nobel Symposium*: 275–301. Stockholm: Almqvist & Wiksell.

Levin, I., Bösinger, R., Bonani, G., Francey, R.J., Kromer, B., Münnich, K.O., Suter, M., Trivett N.B.A., and Wölfli, W. 1992. Radiocarbon in Atmospheric Carbon Dioxide and Methane: Global Distribution and Trends. In R.E. Taylor, A. Long, and R.S. Kra (eds.), *Radiocarbon after Four Decades: An Interdisciplinary Perspective*: 503–518. New York: Springer Verlag.

Levin, I., and Hesshaimer, V. 2000. Radiocarbon—A Unique Tracer of Global Carbon Cycle Dynamics. *Radiocarbon* 42: 69–80.

Luft, U. 2003. Priorities in Absolute Chronology. In M. Bietak (ed.), *The Synchronisation of Civilisations in the Eastern Mediterranean in the Second Millennium BC II. Proceedings of the SCIEM 2000–EuroConference, Haindorf, 2–7 May 2001*: 199–204. Vienna: Verlag der Österreichischen Akademie der Wissenschaften.

Macdonald, C.F. 2003. The Palace of Minos at Knossos. *Athena Review* 3(3): 36–43.

Manning, S.W. 1995. *The Absolute Chronology of the Aegean Early Bronze Age*. Sheffield: Sheffield Academic Press.

Manning, S.W. 1999. *A Test of Time*. Oxford: Oxbow Books.

Manning, S.W. 2006–2007. Why Radiocarbon Dating 1200 BCE Is Difficult: A Sidelight on Dating the End of the Late Bronze Age and the Contrarian Contribution. *Scripta Mediterranea* 27–28: 53–80.

Manning, S.W., Bronk Ramsey, C., Doumas, C., Marketou, T., Cadogan, G., and Pearson, C.L. 2002. New Evidence for an Early Date for the Aegean Late Bronze Age and Thera Eruption. *Antiquity* 76: 733–744.

Manning, S.W., Bronk Ramsey, C., Kutschera, W., Higham, T., Kromer, B., Steier, P., and Wild, E.M. 2006a. Chronology for the Aegean Late Bronze Age 1700–1400 BC. *Science* 312: 565–569.

Manning, S.W., Bronk Ramsey, C., Kutschera, W., Higham, T., Kromer, B., Steier, P., and Wild, E.M. 2006b. Supporting Online Material for Chronology for the Aegean Late Bronze Age 1700–1400 BC *Science* 312. [http://www.sciencemag.org/cgi/content/full/312/5773/565/DC1].

Manning, S.W., Kromer, B., Kuniholm, P.I., and Newton, M.W. 2001. Anatolian Tree Rings and a New Chronology for the East Mediterranean Bronze-Iron Ages. *Science* 294: 2532–2535.

Manning, S.W., Kromer, B., Talamo, S., Baillie, M., Barbetti, M., Friedrich, M., Grudd, H., Kuniholm, P.I., Newton, M., and Pearson, C. 2006c. Investigations of Inter-regional Radiocarbon Comparisons During a Couple of Solar Minima Periods and During One Production Low, and Some Implications for ^{14}C Calibration and Wiggle-Matching. Paper presented at the 19th International ^{14}C Conference, Oxford, 3–7 April 2006.

Matthäus, H. 1995. Representations of Keftiu in Egyptian Tombs and the Absolute Chronology of the Aegean Late Bronze Age. *Bulletin of the Institute of Classical Studies of the University of London* 40: 177–186.

Mazar, A., and Bronk Ramsey, C. 2008. ^{14}C Dates and the Iron Age Chronology of Israel: A Response. *Radiocarbon* 50: 159–180.

McCoy, F.W., and Heiken, G. 2000. The Late-Bronze Age Explosive Eruption of Thera (Santorini), Greece: Regional and Local Effects. In F.W. McCoy and G. Heiken (eds.), *Volcanic Hazards and Disasters in Human Antiquity. Geological Society of America Special Paper* 345: 43–70. Boulder: Geological Society of America.

Merrillees, R.S. 2001. Some Cypriote White Slip Pottery from the Aegean. In V. Karageorghis (ed.), *The White Slip Ware of Late Bronze Age Cyprus. Proceedings of an International Conference Organized by the Anastasios G. Leventis Foundation, Nicosia, in Honor of Malcolm H. Wiener, Nicosia 29th–30th October 1998*: 89–100. Vienna: Verlag der Österreichischen Akademie der Wissenschaften.

Ministry of Environment, Government of British Columbia. 2002. Appendix: Past Trends/Future Projections. *Indicators of Climate Change for British Columbia 2002*. [http://www.env.gov.bc.ca/air/climate/indicat/appendix.html.]

Moran, W.L. 1992. *The Amarna Letters*. Baltimore: The Johns Hopkins University Press.

Mörner, N.-A., and Etiope, G. 2002. Carbon Degassing from the Lithosphere. *Global and Planetary Change* 33: 185–203.

Pain, S. 1999. Vents de Milos. *New Scientist* 103(2197): 38–41.

Pearce, N.J.G., Westgate, J.A., Preece, S.J., Eastwood, W.J., and Perkins, W.T. 2004. Identification of Aniakchak (Alaska) Tephra in Greenland Ice Core Challenges the 1645 BC Date for Minoan Eruption of Santorini. *Geochemistry, Geophysics, Geosystems* 5(3): Q03005, doi:10.1029/2003GC000672.

Rackham, O., and Moody, J. 1996. *The Making of the Cretan Landscape*. Manchester: Manchester University Press.

Reddaway, J.M., and Bigg, G.R. 1996. Climatic Change over the Mediterranean and Links to the More General Atmospheric Circulation. *International Journal of Climatology* 16: 651–661.

Reimer, P.J. 2004. From the Guest Editor. *Radiocarbon* 46: 5–6.

Reimer, P.J., Baillie, M.G.L., Bard, E., Bayliss, A., Beck, J.W., Bertrand, C.J.H., Blackwell, P.G., Buck, C.E., Burr, G.S., Cutler, K.B., Damon, P.E., Edwards, R.L., Fairbanks, R.G., Friedrich, M., Guilderson, T.P., Hogg, A.G., Hughen, K.A., Kromer, B., McCormac, G., Manning, S., Bronk Ramsey, C., Reimer, R.W., Remmele, S., Southon, J.R., Stuiver, M., Talamo, S., Taylor, F.W., van der Plicht, J., and Weyhenmeyer, C.E. 2004. IntCal04 Terrestrial Radiocarbon Age Calibration, 0–26 cal kyr BP. *Radiocarbon* 46: 1029–1058.

Robock, A. 2000. Volcanic Eruptions and Climate. *Review of Geophysics* 38: 191–219.

Robock, A., and Free, M.P. 1995. Ice Cores as an Index of Global Volcanism from 1850 to the Present. *Journal of Geophysical Research* 100: 11,549–11,567.

Rogie, J.D. 1996. Lethal Italian Carbon Dioxide Springs Key to Atmospheric CO_2 Levels. Penn State Earth and Environmental Systems Institute. News and Events: News Archives. [http://www.eesi.psu.edu/news_events/archives/Lethal.shtml].

Rogie, J.D., Kerrick, D.M., Chiodini, G., and Frondini, F. 2000. Flux Measurements of Nonvolcanic CO_2 Emission from Some Vents in Central Italy. *Journal of Geophysical Research* 105(B4): 8435–8445.

Sagan, C. 1979. *Broca's Brain: Reflections on the Romance of Science*. New York: Random House.

Salzer, M.W., and Hughes, M.K. 2007. Bristlecone Pine Tree Rings and Volcanic Eruptions over the Last 5000 Yr. *Quaternary Research* 67: 57–68.

Saurer, M., Cherubini, P., Bonani, G., and Siegwolf, R. 2003. Tracing Carbon Uptake from a Natural CO_2 Spring into Tree Rings: An Isotope Approach. *Tree Physiology* 23: 997–1004.

Scott, E.M. 2000. Bayesian Methods: What Can We Gain and at What Cost? *Radiocarbon* 42: 181.

Scott, E.M. (ed.). 2003. The Third International Radiocarbon Intercomparison (TIRI) and the Fourth International

Radiocarbon Intercomparison (FIRI), 1990–2002: Results, Analyses, and Conclusions. *Radiocarbon* 45: 135–328.

Serre-Bachet, F. 1985. La dendrochronologie dans le bassin méditerranéen. *Dendrochronologia* 3: 77–92.

Shortland, A.J., Ramsey, C., and Higham, T. 2005. Absolute Dates and Egyptian Historical Chronology, a Critical Appraisal. Paper presented at Egypt & Time, a Workshop on Precision and Accuracy of the Egyptian Historical Chronology. Vienna, 30 June–2 July 2005.

Skok, J., Chorney, W., and Broecker, W.S. 1962. Uptake of CO_2 by Roots of Xanthium Plants. *Botanical Gazette* 124: 118–120.

Snow, C.P. 1959. *The Two Cultures and the Scientific Revolution*. New York: Cambridge University Press.

Soles, J.S. 2004. Appendix A: Radiocarbon Dates. In J.S. Soles and C. Davaras (eds.), *Mochlos IC: Period III. Neopalatial Settlement on the Coast: The Artisan's Quarter and the Farmhouse at Chalinomouri. The Small Finds*: 145–149. Philadelphia: INSTAP Academic Press.

Splittstoesser, W.E. 1966. Dark CO_2 Fixation and Its Role in the Growth of Plant Tissue. *Plant Physiology* 41: 755–759.

Steier, P., and Rom, W. 2000. The Use of Bayesian Statistics for ^{14}C Dates of Chronologically Ordered Samples: A Critical Analysis. *Radiocarbon* 42: 183–198.

Stolwijk, J.A.J., and Thimann, K.V. 1957. On the Uptake of Carbon Dioxide and Bicarbonate by Roots and Its Influence on Growth. *Plant Physiology* 32: 513–520.

Stuiver, M. 1993. A Note on Single-Year Calibration of the Radiocarbon Time Scale, AD 1510–1954. *Radiocarbon* 35: 67–72.

Stuiver, M., and Braziunas, T.F. 1993. Sun, Ocean, Climate and Atmospheric $^{14}CO_2$: An Evaluation of Causal and Spectral Relationships. *The Holocene* 3(4): 289–305.

Taylor, R.E. 1997. Radiocarbon Dating. In R.E. Taylor and M.J. Aitken (eds.), *Chronometric Dating in Archaeology. Advances in Archaeological and Museum Science,* Vol. 2: 65–96. New York: Plenum Press.

Warren, P.M. 2006. The Date of the Thera Eruption in Relation to Aegean-Egyptian Interconnections and the Egyptian Historical Chronology. In E. Czerny, I. Hein, H. Hunger, D. Melman, and A. Schwab (eds.), *Timelines: Studies in Honor of Manfred Bietak.* Vol. 2: 305–321. Leuven: Peeters.

Wells, R.A. 2002. The Role of Astronomical Techniques in Ancient Egyptian Chronology: The Use of Lunar Month Lengths in Absolute Dating. In J.M. Steele and A. Imhausen (eds.), *Under One Sky: Astronomy and Mathematics in the Ancient Near East*: 459–472. Muenster: Ugarit-Verlag.

Whitelaw, T. 1996. Review of *The Absolute Chronology of the Aegean Early Bronze Age: Archaeology, Radiocarbon and History* by S.W. Manning. *Antiquity* 70: 232–234.

Wiener, M.H. 2003. Time Out: The Current Impasse in Bronze Age Archaeological Dating. In K.P. Foster and R. Laffineur (eds.), *METRON: Measuring the Bronze Age. Proceedings of the 9th International Aegean Conference/9e Rencontre égéenne internationale, New Haven, Yale University, 18–21 April 2002. Aegaeum* 24: 363–399. Liège: Université de Liège.

Wiener, M.H. 2006a. Chronology Going Forward (With a Query About 1525/4 BC). In E. Czerny, I. Hein, H. Hunger, D. Melman, and A. Schwab (eds.), *Timelines: Studies in Honor of Manfred Bietak.* Vol. 3: 317–328. Leuven: Peeters.

Wiener, M.H. 2006b. Egypt & Time. In M. Bietak (ed.), *Egypt and Time. Proceedings of the SCIEM 2000 Workshop on Precision and Accuracy of the Egyptian Historical Chronology, Vienna, 30 June–2 July 2005. Ägypten und Levante* 16: 325–339. Vienna: Verlag der Österreichischen Akademie der Wissenschaften.

Wiener, M.H. 2007. Times Change: The Current State of the Debate in Old World Chronology. In M. Bietak and E. Czerny (eds.), *The Synchronisation of Civilisations in the Eastern Mediterranean in the Second Millennium BC III. Proceedings of the SCIEM 2000–2nd EuroConference, Vienna 28th of May–1st of June 2003*: 25–47. Vienna: Verlag der Österreichischen Akademie der Wissenschaften.

Yurgalevitch, C.M., and Janes, W.H. 1988. Carbon Dioxide Enrichment of the Root Zone of Tomato Seedlings. *Journal of Horticultural Science* 63: 265–270.

Article submitted October 2008

Santorini Eruption Radiocarbon Dated to 1627–1600 BC: Further Discussion

Walter L. Friedrich, Bernd Kromer, Michael Friedrich, Jan Heinemeier, Tom Pfeiffer, and Sahra Talamo

Abstract: *"I will not be so bold as to express an opinion, in view of my pledge, but I will leave you with the question: I wonder if it may be that the date [of the Minoan eruption] will be within 20 years of 1620 BC?"—Colin Renfrew, "Summary of Progress in Chronology," Thera and the Aegean World (Proceedings of the third congress, Hardy & Renfrew, eds. 1990), Vol. III, p. 242.*

Introduction

The radiocarbon date of the olive branch from Santorini (Friedrich et al. 2006) resulted in discussions among scholars. The following main issues were brought forward:

- Was the tree alive when it was buried by the ashes of the Minoan Eruption?
- Why is the tree preserved while trees in the Akrotiri excavation have decayed?
- Are all the tree rings of the branch preserved and has the irregularity of olive ring growth any effect on ring counting and the calibrated age?
- Is it likely that old CO_2 might have influenced the result of the radiocarbon dating?
- Would the final result be different if older versions of the calibration curve were used?
- Is the ^{14}C calibration based on the OxCal program reliable?

All of these questions are relevant, and therefore we will discuss them in this contribution.

Was the tree alive when it was buried by the Minoan eruption?

The dated branch was part of an olive tree, which had clearly been buried *in situ* in life position by the pumice of the eruption. Thus, the tree was not uprooted and the branches were not found on the ground, but still in life position, imbedded in the pumice matrix, 1–4m above the original soil level. Today, we see

Figure 1: Caldera wall north of Cape Athinios on Thera with the pumice deposits of the Minoan eruption that have a thickness of about 40 meters. About nine hollow structures (molds) contain branches of an olive tree that was buried *in situ*. In the upper left corner a man-made wall from the Bronze Age is visible. In the middle of the photo to the right a mold is seen, where in July 2007 a second olive tree was excavated.

the pumice of the steep caldera wall revealing a lateral cross section of the remaining crown and branches of the tree, while part of the pumice deposits and tree trunk have been eroded away, leaving the cross sections of the proximal ends of approximately nine branches exposed (Figure 1). In the brownish soil section (strongly weathered volcanic tuff) below the pumice, cross sections of molds containing traces of the decayed remains of the roots are seen. We must assume that the tree was alive, as imprints of olive leaves were observed in the precursor layer, which forms a

Figure 2: A. An olive tree growing close to a man-made Bronze Age wall. **B.** In the initial phase of the eruption the southwestern part of the ring island (Akrotiri Peninsula) was hit by a hot volcanic blast. As a result, the leaves of the trees dried up immediately. They fell down and were embedded in the white pumice dust covering the surface. This precursor layer is only 4 cm thick. **C.** In the first phase of the Minoan eruption, falling pieces of pumice almost covered the olive trees with a 4 m thick pumice layer. The still warm pumice charred the trees.

few centimeter thick border line between the soil and the pumice. This eruption precursor formed an up to 7cm thick layer of fine volcanic dust (Figure 2) on the southern part of Thera. The olive leaf imprints are found right inside the dust layer—not under it—which means that a hot cloud of volcanic dust enveloped the olive tree and made its leaves fall to the ground. We also find the fine dust layer on the upper surface of the cylindrical tunnels (molds) that have enveloped the branches. This dust had clearly fallen on the upper side of the branches of the standing tree and now shows the imprints of the bark, typical of olive trees. Even after extensive search in the neighborhood, we found the leaves only in the immediate neighborhood of the tree, indicating that they are unlikely to have originated from hypothetical nearby trees. In addition, about 4 meters from the tree a remnant of a Bronze Age stone wall was observed, indicating that the tree grew near a settlement (Figure 2). If the tree had been killed, for instance by lightning, the inhabitants would most likely have used it for fuel, unless of course the dead tree had deliberately been left standing, perhaps for religious reasons. If so, the bark would probably have been weathered away in not too many years. The precursor layer is also observed in the Akrotiri excavation according to Doumas (1974), followed by the same characteristic pumice layer of the Minoan eruption.

Why is the tree preserved while trees in the Akrotiri excavation have decayed?

The first findings of organic material on Therasia and the Akrotiri peninsula by Fouqué (1879) showed that wood was not preserved, but turned into powder when touched. However, findings closer to the eruption point as for instance in the Fira quarry (Galanoupoulos 1958) and the Karageorghis quarry revealed that wood was charred. Also the olive tree (Friedrich et al. 2006) was lightly charred, indicating that the distance from the eruption point played an important role. The Akrotiri excavation is 7km away from the eruption point. Furthermore, on the site where the olive tree was found, we could observe that only the parts of the tree that were in direct contact with the hot pumice, were lightly charred, but not the roots as they were protected by the surrounding soil. Thus, the roots are only found as fine dust in hollow structures (molds), as is the case with the wood in the Akrotiri excavation.

Has the irregularity of olive growth rings any effect on the calibrated date?

When "events" are dated using dendrodated or radiocarbon dated charred trees the youngest tree-ring is a *"terminus post quem"* date for the event. To get a date close to the event the presence of the last-formed tree-ring, called "waney edge" or "terminal ring" is essential. Its presence implies that the sample is complete to the last year of growth. Even if the presence of the waney edge in subfossil wood is often missing there are ways to estimate the date of the trees' death with some degree of certainty from remnants of sapwood rings, from wood anatomical aspects, or when the dimension of the tree can be estimated, i.e. from stake spur or, regarding the Santorini wood, from the dimension of the hole in the pumice.

Here we present arguments for the presence of the original outermost ring (waney edge) on the olive branch from Santorini:

(1) On the olive section the outermost ring is identical on different radii and is unbroken all around the section, even if the section of the branch is not concen-

Figure 3: **A.** The second phase of the Minoan eruption produced ring-shaped clouds of gas and pumice in an expanding suspension that moved radially away from the crater with high velocity. The uppermost branches of the trees that were not covered by the pumice of the first eruption phase were cut off by the horizontally directed blasts (base surges). **B.** After the third phase of the Minoan eruption the caldera wall was unstable, and as a result huge blocks of the wall became detached and slid down. Thus most of the charred olive trees disappeared, leaving behind only a few molds with branches. **C.** Present situation on the caldera wall showing the molds with branches. The 4cm thick white layer of pumice dust (Bo_0) contains olive leaves. Molds and vestiges of the tree roots are found in the soil. They are not charred because they were protected from the heat by the soil. **Note added in proof**: In July 2007 another olive tree was found only 9m away from the first one. Also that tree is *in situ* with olive leaves lying on the ground beneath the tree, proving that the trees were alive when buried by the pumice of the Minoan eruption.

tric. The non-concentric shape is typical for branch wood because of reaction wood, which stabilizes horizontal branches. If one postulates that due to degradation wood from the outside would have rotted away, it would be highly unlikely that the branch rotted exactly along the same tree ring, especially when the rings are not concentric.

(2) We observed and documented the impression of bark in the pumice inside of the holes where we found the branch. Except for the weathered end of the branch and the mold close to cliff surface, the branch (and neighboring smaller ones) fit fairly tightly in the mold, indicating preservation of the branch surface at the section that we dated, with only a slight shrinkage due to charring. The amount of shrinkage of 7–13% longitudinally and 12–25% radially/tangentially is well established (Schweingruber 1990) and could be confirmed by our carbonization experiments on wood from living olive trees from Santorini.

(3) At the transition from the stem to a branch or from a first order branch to a second order branch the outermost ring follows the surface along to that branch. The outermost ring of the section of the first order branch is therefore identical to the outermost ring of the second order branch. This is what we observed on the olive branch from Santorini, where the outermost ring of a branch (of second order) corresponded to the outermost ring of the first order branch. If we assume that external rings would have decayed, it is extremely unlikely that this degradation would happen exactly along the same tree ring for the two branching levels. The erosion of wood within the same tree ring over larger distances and along different structures (i.e. stem, branches) can be observed only along natural boundaries like the "waney edge," or at strong density boundaries, i.e. the distinct heartwood-sapwood boundary in most oak species, which is in the same ring all along the section (Friedrich and Hennig 1996). In contrast to oak wood, the sapwood boundary in olive wood is irregular. In addition carbonization enhances the resistance to biological degradation and minimizes the differences of wood resistance between hardwood and sapwood. It is therefore very unlikely that the observed uninterrupted outermost tree ring on the olive branch section is a result of a degradation of external wood, but on the contrary we are confident that it really represents the waney edge (terminal growth ring).

Even with the outermost ring preserved, a source of error in the wiggle-match calibration could arise from a too high ring count due to false identification of growth rings or a low count due to missing or unidentified rings. A review of Mediterranean tree growth patterns by Cherubini et al. (2003) has shown that the Mediterranean tree species *Arbutus unedo* L. (strawberry tree) is particularly sensitive to climate changes. It exhibits occasional intra-annual "false" rings, which may in some cases be difficult to identify and could therefore lead to falsely high ring counts, although by no means with two annual rings as a typical figure, let alone three. The same behavior could be suspected for olive trees growing in Santorini, but preliminary studies by one of us (M.F.) on recent olive wood from Sicily has not shown this effect. Ring counts on a polished section were in good agreement with results based on X-ray tomography and with the known age of the specimen. The known problems of indistinct growth ring boundaries in olive wood were overcome using the high resolution 3D-X-ray tomography technique. This technique helps to identify marginal parenchymatic bands, which forms the ring boundary, much better than the traditional optical microscopic analysis.

We are aware that there still may be some missing or false rings in the olive branch section of Santorini, but we are confident that our error estimate of 25% encompasses every conceivable error margin (Friedrich et al. 2006: suppl. material).

Finally, if allowance is made for an exaggerated ring count in the wiggle-match dated Santorini olive branch, the calibrated date would tend to be even earlier than 1627-1600 BC. If, on the other hand, we assume that the ring count has been underestimated by ±25% the calibrated date of the eruption would be only slightly later, i.e. 1635-1591 BC (2 sigma or 95.4% probability) (Friedrich et al. 2006: suppl. material). This is still far from reconcilable with a traditional archaeological date of 1550 BC or later.

Was the tree contaminated by old CO_2?

Since the olive tree grew on a volcanic island, it is also relevant to consider the question of whether the radiocarbon dates might have been influenced by old volcanic CO_2. The Akrotiri excavation, where part of the dated material described by Friedrich et al. (1990) and Manning et al. (2006) was found, shows direct evidence of a long period of volcanic inactivity. Here the Minoan pumice was deposited directly on top of the Cape Riva ignimbrite, which has been dated to 19,000 BC. The distance between the Akrotiri excavation and the crater of the Minoan eruption is about 7km, far away from earlier and present emission points of volcanic gases derived from the active volcanic zone (Figure 4). Studies in Germany (Laacher See) and on Santorini (Palea Kameni) by Bruns et al. (1980) and on the Azores by Pasquier-Cardin (1999) have shown that plants growing close to an emanation point of old CO_2, give falsely old radiocarbon ages. However, their studies also show that the effect is locally restricted to the order of 100 meters or less around strong sources of volcanic CO_2, in agreement with theoretical calculations of atmospheric mixing, and even within a caldera, contributions from volcanic CO_2 were below detection limits for plants growing outside this range (Pasquier-Cardin 1999). Likewise, Shore et al. (1995) tested plants from a site 10km from the rim of the Katla volcano complex in Iceland and saw no detectable effects. Taking into account that (a) the distance between the growth-site of the olive tree and the nearest point on the active volcanic zone is about 3.5km, (b) the distance to the eruption is about 5km, and (c) the tree was found on top of the pre-eruption caldera rim with good air circulation, ensuring both horizontal and vertical atmospheric mixing, it is unlikely that contamination with old CO_2 could have affected the olive tree. Last, but not least, neither faults nor old fumarolic fields or sites with iron oxide deposits were observed in the neighborhood of the tree. Thus, the tree rings must be considered to represent a reliable archive of atmospheric CO_2 in its seven decades of lifetime prior to the eruption.

Also, we exclude a significant local offset of the ^{14}C ages of the tree-ring samples by volcanic CO_2 because in that case it would most likely be impossible to match the measured ^{14}C sequence anywhere to the shape of the calibration curve. We observe a downward slope in our dating sequence (Friedrich et al. 2006), whereas one would expect an upward slope if the eruption took place around 1550 BC and had been contaminated with volcanic CO_2. The aging effect should, if anything, increase due to increased emission in the period up to the eruption.

A final consideration could be possible influence of old CO_2 from groundwater taken up through the roots. However, Tauber (1983) has shown by ^{14}C measurements that even in extreme calcareous soil conditions the uptake of CO_2 from soil carbonate is indiscernible (0.12 ±0.3%). From tree-physiology it is extremely unlikely that there could be any quantitative effect by CO_2 uptake through the roots. The uptake could only occur as CO_2 in water which is transported via xylem (pipes) through the stem to the leaves where wood-cellulose is produced. Because of out-gassing of CO_2 in the xylem on the transport through stem and branches the contribution of CO_2 from root uptake compared to the CO_2 derived from the air through the stomata in the leaves is negligible. (This argument may not apply to herbal, non-woody plants.)

Would the final result be different if older versions of the calibration curve were used?

The recommended ^{14}C calibration curve, IntCal04 shows less high-frequency variation than the previous one, IntCal98, due to different procedures in averaging the contributions to each decadal bin in the ^{14}C data set. Hence, the wiggle-match of the four ^{14}C dates of the olive sections as shown in Friedrich et al., 2006, could be sensitive to the choice of the calibration curve. However, as documented in the supplementary material of Friedrich et al., 2006, after calibration using IntCal98 the high limit of the calibrated age range is within a few years of the result based on Intcal04, and the low limit may extend down to 1575 cal BC.

Is the OxCal program reliable?

It has been argued that the result of the wiggle-match of the ^{14}C dates of the four olive sections may depend on the calibration program in a subjective way, possi-

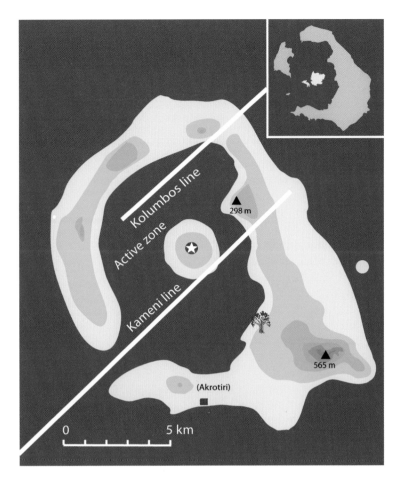

Figure 4: Reconstruction of Santorini prior to the Minoan eruption showing the growth-place of the olive tree and the Akrotiri excavation, away from the active volcanic zone and the eruption point with their possible sources of CO_2 emission. The inserted map shows the present shape of Santorini.

bly suppressing ranges of the probability distribution. For our specific use of the modeling capabilities of Ox-Cal it is important to note that the sequencing model employed here does not introduce any filtering; it only requires the sections to maintain the sequential order within the tree segment as an additional constraint during the calibration. It should also be noted that the calibration programs commonly in use now were rigorously tested for the statistical algorithms that are used to transform the ^{14}C age distribution into the calibrated scale.

Conclusion

In conclusion, we emphasize that the olive branch is not just another sample in line with a multitude of previous (mainly single-year) samples for which radiocarbon dates have been reported. It is unique because it contains ordered information from approximately 72 tree rings with a well established association with the eruption event, unlike more circumstantial evidence such as long-distance correlation with ice cores or climatically affected tree rings.

In particular, we conclude that:

- The tree was alive and standing when it was buried on Santorini by the ashes of the Minoan eruption

- It is unlikely that old CO_2 could have influenced the result of the radiocarbon dating

- The tree is preserved due to light charring due to its proximity to the eruption point, while trees in the more distant Akrotiri site have decayed. Roots and leaves have not been sufficiently preserved to allow radiocarbon dating.

- All the tree rings of the analyzed section of the branch were preserved, including the original outermost ring (waney edge). The well known irregularity of olive ring growth probably has a limited effect on ring counting, and even allowance for substantial errors in ring count would not reconcile the calibrated age with the traditional archaeological eruption date after about 1550 BC.

- Our wiggle-match dating result is roughly independent of the choice of (recent) calibration curves (IntCal04/IntCal98), which will not change the above conclusion.

Because the olive branch is such an ideal sample providing a very direct and precise date based on simple and transparent assumptions and analyzed with well established methodology (calibration), we presently consider it the most solid scientific determination of the timing of the Minoan eruption.

It would clearly be worth while to re-evaluate in detail the current interpretations of the archaeological evidence.

Acknowledgments

We would like to thank Georgios Nomikos from Megalochori for donating us a modern olive tree for comparison with the sub-fossil ones.

References

Allard, P., et al. 1991. Eruptive and diffuse emissions of carbon dioxide from Etna volcano. *Nature* 351: 387–391.

Barberi, F., and Carapezza, M.L. 1994. Helium and CO_2 soil gas emission from Santorini (Greece). *Bull. Volcanol.* 56: 335–342

Bruns, M., Levin, I., Münnich, K.O., Hubberten, H.W., and Fillipakis, S. 1980. Regional Sources of Volcanic Carbon Dioxide and their influence on C^{14} content of present-day plant material. *Radiocarbon* 22(2): 532–536.

Cherubini, P., Gartner, B.L., Tognetti, R., Braker, O.U., Schoch, W., Innes, J.L. 2003. Identification, measurement and interpretation of tree rings in woody species from Mediterranean climates. *Biological Reviews* 78(1): 119–148.

Doumas, C.G. 1974. The Minoan eruption of the Santorini volcano. *Antiquity* XLVIII: 110–115.

Druitt, T.H., Davies, M., Edwards, L., and Sparks, R.S.J. 1999. *Santorini Volcano.* Geological Society Special Memoir. London: Geological Society Publishing.

Fouqué, F. 1879. *Santorin et ses eruptions.* Paris: G. Masson.

Friedrich, M., and Hennig, H. 1996. A dendrodate for the Wehringen Iron Age wagon grave in relation to other recently obtained absolute dates for the Hallstatt period in Southern Germany. *Journal of European Archaeology* 4: 281–303.

Friedrich, W.L., Eriksen, U., Tauber, H., Heinemeier, J., Rud, N., Thomsen, M.S., and Buchardt, B. 1988. Existence of a water-filled caldera prior to the Minoan eruption of Santorini, Greece. *Naturwissenschaften* 75: 567–569.

Friedrich, W.L., Wagner, P., and Tauber, H. 1990. Radiocarbon dated plant remains from the Akrotiri excavation on Santorini, Greece. In D.A. Hardy (ed.), *Thera and the Aegean World III. Vol. 3: Chronology*: 188–196. London: The Thera Foundation.

Friedrich, W.L. 2000. *Fire in the Sea.* Cambridge: Cambridge University Press.

Friedrich, W.L., Kromer, B., Friedrich, M., Heinemeier, J. Pfeiffer, T., and Talamo, S. 2006. Santorini Eruption Radiocarbon Dated to 1627–1600 BC. *Science* 312: 548.

Galanopoulos, A.G. 1958. Zur Bestimmung des Alters der Santorin-Kaldera. *Annales Géologiques des pays Helléniques* 9: 184–185.

Manning, S.W., Bronk Ramsey, C., Kutschera, W., Higham, T., Kromer, B., Steier, P., and Wild, E.M. 2006. Chronology for the Aegean Late Bronze Age 1700-1400 BC. *Science* 312: 565–569.

Pfeiffer, T. 2003. Two Catastrophic Volcanic Eruptions in the Mediterranean-Santorini 1645 BC and Vesuvius 79 AD. PhD Thesis, Aarhus University, Aarhus, Denmark.

Shore, J.S., Cook, G.T., and Dugmore, A.J. 1995. The ^{14}C content of modern vegetation samples from the flanks of the Katla Volcano, southern Iceland. *Radiocarbon* 37(2): 525–529.

Schweingruber, F.H. 1990. *Mikroskopische Holzanatomie 3.* Birmensdorf, Switzerland: Eidgenössische Forschungsanstalt für Wald Schnee und Landschaft.

Tauber, H. 1983. Possible depletion in ^{14}C in trees growing in calcareous soils. *Radiocarbon* 25(2): 417–420.

Article submitted June 2007

Dating the Santorini/Thera Eruption by Radiocarbon: Further Discussion (AD 2006–2007)

*Sturt W. Manning, Christopher Bronk Ramsey, Walter Kutschera,
Thomas Higham, Bernd Kromer, Peter Steier, and Eva M. Wild*

Abstract: *"... There is a lot at stake in the [Santorini/Thera date] debate. Until it is resolved, Warren says, at least for the Late Bronze Age, 'we would have to forget about serious study of the past and relationships between peoples' " (Balter 2006: 509).*

Introduction

The recent publications and findings of Manning et al. (2006a) and Friedrich et al. (2006), and previously those of Manning et al. (2002a), Manning and Bronk Ramsey (2003) and Bronk Ramsey et al. (2004), on the dates of the Santorini/Thera eruption and the Late Bronze Age 1–2 cultural periods in the Aegean, have led to a variety of discussions both in print and informally. Whereas some archaeologists appear willing to engage or consider the new evidence and chronology, others reject them outright and regard radiocarbon as clearly wrong (see e.g. the selection of opinions reported in Balter 2006). Several scholars in the latter camp have in turn sought to suggest reasons why perhaps there is (or must be) something at error with the radiocarbon (^{14}C) data employed in these studies, or the analyses, which might then somehow undermine the (otherwise) clear support these studies and their data currently provide for a "high" Aegean Late Bronze Age chronology. This discussion continues a now long-running thread in the literature from the mid AD 1970s to present, where radiocarbon data have (fairly consistently, but, in recent work, with much greater precision) indicated an earlier date for the eruption and the beginning of the Late Bronze Age than previously thought (and still currently argued) from a conventional or "low" interpretation of the cultural linkages between the Aegean–East Mediterranean–Egyptian cultures (for a few examples, see: Michael 1976; Betancourt and Weinstein 1976; Betancourt 1987; Aitken 1988; Manning 1988; 1999; Hardy and Renfrew 1990; Marketou et al. 2001).

At the Cornell Conference Malcolm Wiener in a typically excellent and wide-ranging lecture raised a number of these concerns from a conventional chronology perspective (and see his paper in this volume for references to his other 2003 and 2006 to present publications, several of which raise some similar issues). It is thus appropriate to address such concerns and to offer some exegesis on the soundness and clarity of the radiocarbon evidence (see also specifically the Supporting Online Materials published with the Manning et al. 2006a paper; Manning and Bronk Ramsey 2003: 124–129; Manning 2007; 2005; and the Friedrich et al. paper in this volume).

The main focus here is the c.100-year conflict between the later to late 17th century BC calendar age ranges calculated for the date of the Akrotiri volcanic destruction level on Santorini/Thera (Manning et al. 2006a), or for the outermost ring of an olive tree killed by the eruption (Friedrich et al. 2006), versus the conventional chronology based dates for the eruption of c.1525 BC or 1520 BC or 1500 BC (e.g. Warren 1984;

We offer this paper in honour of Peter Ian Kuniholm and his three decades of work creating and leading the Aegean Dendrochronology Project and the Malcolm and Carolyn Wiener Laboratory for Aegean and Near Eastern Dendrochronology at Cornell University. The topic is appropriate as Manning first met Peter and Eleanor on Santorini/Thera in 1989, where the Thera date was already (and also) topical and the scholarly field was (seriously) beginning to debate change. Today one is perhaps challenged by the well-known Buddhist saying: "Everything changes, nothing remains without change."

1999; 2006; Hankey and Warren 1989; Wiener 2006; this volume).

1. Quality of the Radiocarbon Data?

See Manning et al. (2006a) and its Supporting Online Material. Several high-quality but different laboratories (with varying procedures and equipment) and/or the same laboratory but with varying procedures and a different accelerator, have repeatedly achieved compatible and consistent radiocarbon evidence for the Santorini eruption date, and for the start of the Late Bronze Age topic. The quality of the results from these laboratories are supported by known age tests (see summary and references in Manning et al. 2006a Supporting Online Material). The Friedrich et al. (2006) paper allows us to add another such set of consonant data. The quality and compatibility of the recent data for the Santorini volcanic destruction level run from AD 2000 onwards as reported by Manning et al. (2006a), from the Oxford and VERA laboratories respectively, are about as good as could be reasonably hoped for given the analytical possibilities available at this time. (NB. If other data are to be brought into such discussions and analyses, then they ideally need to be demonstrated to come from such a high-quality process and from a laboratory with known age test data.)

Thus, if there is a problem with the relevant radiocarbon data, then it must be a general problem that affects all the samples and data, and it must be a problem which is independent of any single laboratory's procedures and equipment. The challenge for those who would dismiss these data is to demonstrate such a general problem/flaw.

2. Volcanic CO_2?

A range of literature has noted (significant) old-age offsets to ^{14}C ages as a result of volcanic emissions, i.e. much-depleted or "old" CO_2 released from a volcano and incorporated into local plants might give falsely old ^{14}C ages. Might such offsets be relevant to the samples analyzed by Manning et al. (2006a) and Friedrich et al. (2006)? There are several reasons to think not (see also Friedrich et al., this volume):

(i) The ^{14}C data available from an olive branch growing on Santorini for around 72 years (give or take whatever ring-counting error may be applied) until killed by the eruption (Friedrich et al. 2006; this volume) demonstrates the exact opposite of the supposed "volcanic effect": local ^{14}C ages (years BP) were falling across the period (Figure 1) with no indication of any volcanic effect (which would manifest itself in much older dates, causing an upwards spike or trajectory in local ^{14}C ages, something which is not seen). Thus no evidence of any substantive volcanic effect is evident. Indeed, the Santorini record very closely matches the mid-latitude northern hemisphere average record (Figure 1). This would be impossible if there was a substantial local volcanic $^{14}CO_2$ effect in operation.

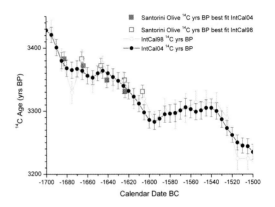

Figure 1: Comparison of the IntCal04 (Reimer et al. 2004) and IntCal98 (Stuiver et al. 1998) ^{14}C calibration curves for the period 1700–1500 BC. Although IntCal98 is a little more noisy, there is general close similarity. The best (peak probability) fit of the ^{14}C data from the Santorini olive branch (Friedrich et al. 2006) is shown against IntCal04 (solid squares) and IntCal98 (hollow squares) (employing here the approximate X-Ray tomography-based ring count calendar spacing of Friedrich et al. 2006 assumed as within 1 year—for further discussion of this point, see Friedrich et al. this volume). There is a 17 calendar year difference; however, both curves place the data in effectively the same calendar region (calculated outermost ring = eruption date range, 6 years after the mid-point of the most recent sample dated, at 2σ, with IntCal04 is 1627–1600 BC, versus 1626–1594 BC from IntCal98).

(ii) As we have noted previously (Manning et al. 2006a Supporting Online Material), the large set of short-lived samples from the final volcanic destruction level on Santorini, despite being of different species and from different findspots, and most likely from different farmers and covering quite a range of different growth locations (Sarpaki 1990), all provide a highly similar set of ^{14}C ages and display no examples with the very significantly older or excess ^{14}C ages typical of volcanic $^{14}CO_2$ contamination in plant samples when present as reported in the literature (e.g. Sulerzhitzky 1971; Bruns et al. 1980; Saupé et al. 1980; Pasquier-Cardin et al. 1999; Olsson and Vilmundardóttir 2000: See Figure 2). And the normal terrestrial plant/tree types involved for the Santorini samples exclude effects observed for e.g. plants growing in geothermal waters (Sveinbjörnsdóttir et al. 1992) or other such unusual cases. (As Friedrich et al., this volume, also note, suggestions of uptake of old volcanic CO_2 via roots are not a plausible possibility for significant contamination.) In detail, all the (n=28) Santorini volcanic destruction level data employed in Manning et al. (2006a) passed a chi-squared test for being compatible with the hypothesis of representing the same ^{14}C age at the 95% confidence level. If some of the samples had been affected by volcanic CO_2, then it is unlikely they could all be successfully combined.

The only counter-argument (although there is no positive evidence) would be to allege that perhaps ALL the Santorini volcanic destruction level data (from a variety of plant types and growing loci) have been equally affected by a minor volcanic aging effect. There are several reasons why this situation would be all but impossible (see iii below especially), but, critically, we can dismiss it because we may note that the ^{14}C ages from the Santorini volcanic destruction level are very similar to those for archaeologically contemporary samples elsewhere in the Aegean where no plausible volcanic effect applies. In particular, we may observe the data from Trianda on Rhodes and Miletos in western Turkey (see Manning et al. 2006a Supporting Online Material Table S1). For example, from Manning et al. 2006a using IntCal04 (Reimer et al. 2004) and OxCal (Bronk Ramsey 1995 and later versions), we find that the weighted average ^{14}C age for a short-lived late LMIA oak twig from Trianda on Rhodes (which had bark: OxA-10643 and OxA-11884) is 3353±25 BP, calibrated calendar age range 1685–1617 BC at 1σ versus the weighted average age for the mature/late Late Minoan IA volcanic destruction level at Akrotiri on Thera from short-lived samples of 3345±8 BP, calibrated calendar age range 1664–1616 BC at 1σ). This is why Manning et al. (2006a) could show that running their dating model excluding all Santorini data (and so any possible/putative volcanic effect) still found that the remaining data required a largely similar chronology (to the one including the Santorini data) (see Table 1). This situation thus vitiates any volcanic CO_2 effect as relevant. The additional olive tree branch samples studied by Friedrich et al. (2006), which came from another site some 3km from Akrotiri, further support the case since they too provide consonant evidence, and do not betray any indication of substantive volcanic CO_2 aging.

(iii) ^{14}C aging effects caused by volcanic CO_2 (see refs. in (ii) above) usually are temporally varying and either spatially specific, or varying. They are not constant through time (and especially not for years and most especially not for several-plus decades). Nor are they constant over wide spatial areas; instead, they are localized, and/or varying across a geographic space where there is degassing or fumarole activity. There is usually rapid decline to undetectable levels over quite small distances from any source (e.g. the volcanic effect was only found at less than 200m distance from a source in the Bruns et al. 1980 study)—though in some cases diffuse gas loss is observed on a wider scale also—but even in such cases this is temporally varying nonetheless (NB. see Pasquier-Cardin et al. 1999). Critically, no study has indicated the possibility of a small, consistent effect which remains stable over periods of even weeks—let alone the months, to years, to decades—that would be necessary to explain the remarkably consistent set of data from the final volcanic destruction level at Santorini from several different species of short-lived plant material from different fields or areas, and from tree-ring material of several years, to some decades, of growth. Instead, measurements on samples affected by volcanic CO_2 invariably do not exhibit consistent values nor just a minor offset; they rather include within the data some substantial variations and some clear *much* older ages. This is not the case for the recent high-quality data from Santorini employed in Manning et al. (2006a) and Friedrich et al. (2006).

A good example can be found in the study of Pasquier-Cardin et al. (1999) based on work in the Azores (see Figure 3) where a significant volcanic aging effect is immediately clear reviewing the data. The Akrotiri volcanic destruction level dataset (and especially the recent AD 2000s dataset) are notable for not exhibiting any such features (see Figure 2). For the Pasquier-Cardin et al. study all but two of the sampled sites were within an active fumarolic area or in areas with diffuse degassing and only one site (PG) was outside the younger caldera area and at least 1km away from marked fumaroles.

This last site yielded in four measurements a total of no measurable increase in age. It is important then to note that the Akrotiri site is over 2km from the reconstructed Minoan caldera edge and some 7km from the believed active vent area of the Minoan eruption (some crops might have been growing closer, of course, some perhaps even further away, but it is highly improbable that all came from within even 0–2km of the active volcanic zone of the Minoan eruption). An average 0 effect scenario would be a reasonable expectation. The other sites in the Pasquier-Cardin et al. study show highly variable and marked offsets—only Site HT shows a much less dramatic pattern—but, even so, the "aging" observed in five samples is 22, 72, 369, 0, and 0 years. No instances of the significant level (369 years) of such variability can be observed in the Santorini data (whereas to extrapolate we might expect at least 2 such samples in the 13 AD 2000s ^{14}C data or 5 such data in the 28 Santorini Volcanic Destruction Level ^{14}C data employed in Manning et al. 2006a), and especially in the high-quality AD 2000s data presented in our study (see Figure 2). We also note that Santorini is a notably

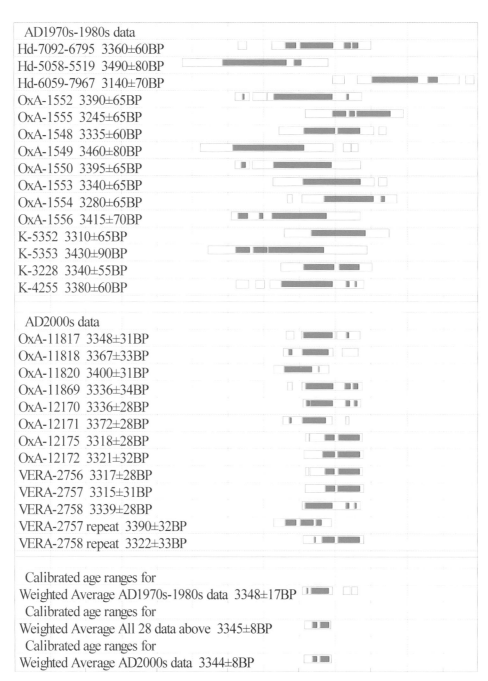

Figure 2: Calibrated (Calendar Years) age ranges for the 28 ^{14}C data from the Santorini volcanic destruction level (VDL) at Akrotiri employed by Manning et al. (2006a) using IntCal04 (Reimer et al. 2004) and OxCal as in the Manning et al. (2006a) paper (see methods described there and in the Supporting Online Material). The green boxes show the 1σ ranges and the hollow boxes the 2σ ranges. All these ^{14}C data can be combined together at the 95% confidence level as in Manning et al. (2006a) demonstrating that the ^{14}C measurements are quite consistent; the recent AD 2000s data presented in Manning et al. (2006a) offer a very tight distribution of ages.

	Transition to Mature LMIA	Felling Date Miletos oak	Akrotiri Volcanic Destruction Level (VDL)	Transition end LMIA to LMIB	Myrtos-Pyrgos Close of LMIB Destruction	Knossos LMII Destruction
Model 1 IntCal04	1737-1673 *1722-1695**	1671-1644 *1664-1652**	1660-1612 *1656-1651 (11.3%)** *1639-1616 (56.9%)**	1659-1572 *1647-1644 (3.1%)** *1642-1603 (65.1%)**	1522-1456 *1517-1491 (58.1%)** *1475-1467 (10.1%)**	1457-1399 *1439-1414**
Ditto with extra 10 ^{14}C yrs uncertainty to short-lived samples	1733-1657 (92.1%) 1655-1647 (3.3%)	1671-1644	1660-1612	1660-1572	1526-1452	1489-1474 (4.1%) 1463-1394 (91.3%)
Model 1 IntCal98	1733-1665	1669-1646	1661-1605	1660-1567	1522-1487 (65.3%) 1482-1451 (30.1%)	1489-1480 (3.6%) 1452-1394 (91.8%)
Ditto with extra 10 ^{14}C yrs uncertainty to short-lived samples	1732-1652	1668-1645	1661-1606	1660-1568	1523-1447	1492-1476 (6.0%) 1459-1390 (88.3%) 1329-1323 (1.1%)
Model 1 *No Santorini data* IntCal04	1728-1643	1672-1645	1668-1585	1661-1553	1522-1456	1487-1481 (1.3%) 1457-1400 (94.1%)
Ditto with extra 10 ^{14}C yrs uncertainty to short-lived samples	1725-1637	1672-1645	1666-1579	1662-1548	1526-1453	1490-1475 (4.4%) 1464-1395 (91.0%)
Model 1 *No Santorini* IntCal98	1727-1641	1668-1645	1665-1582	1661-1547	1523-1487 (66.9%) 1482-1451 (28.5%)	1490-1479 (4.2%) 1452-1394 (91.2%)
Ditto with extra 10 ^{14}C yrs uncertainty to short-lived samples	1726-1636	1668-1645	1663-1577	1658-1546	1523-1448	1492-1476 (7.0%) 1460-1390 (88.4%)
Model 1 adding Friedrich *et al.* (2006) data IntCal04	1737-1673	1671-1644	1654-1649 (3%) 1645-1611 (92.4%) *1σ: 1633-1617**	1626-1562	1522-1457	1487-1480 (1.5%) 1458-1400 (93.9%)
Conventional Archaeological Chronology	1600/1580		1525/1500	1520/1500/1480	1440/1430/1425	1400/1390

Table 1: An extended presentation of the information provided in Manning et al. (2006a) Table 1 (see that publication for details and methods), showing the calendar age ranges determined for several key archaeological events or transitions before, around, and after the Santorini/Thera eruption, but now adding in typical model run outputs for: (i) The Model 1 'No Santorini' data run against the IntCal98 ^{14}C calibration data (Stuiver et al. 1998) set in red. There is very little to no difference when compared to the same using IntCal04 (Reimer et al. 2004); (ii) The models reported in Manning et al. (2006a), and here, also considering an additional 10 ^{14}C yrs BP hypothetical error allowance to all the samples of <10 yrs growth to cover any possible issues of regional, or inter-annual, or local (e.g. very minor volcanic CO_2 effect) factors (shown in blue). Again it can be seen that the possible range of differences is very small and minor and that the data remain entirely consistent with the ranges and conclusions stated in Manning et al. (2006a). Calculated calibrated calendar age ranges BC are shown at 2σ, except for those data marked with an * which are also shown at 1σ. Note each run of the analysis yields very slightly varying outcomes, typically with variations of around 0–2 years. Typical outcomes shown below.

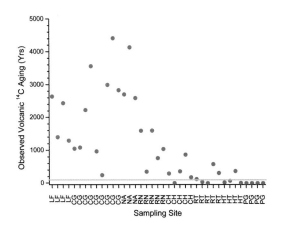

Figure 3: The observed apparent ^{14}C 'aging' caused by volcanic CO_2 reported in Pasquier-Cardin et al. (1999) for measurements made at their sites LF, CG, NA, RN, CH, RT, HT and PG around Furnas Volcano, Azores. All values on or above the cyan line have ≥ 100 ^{14}C years of volcanic aging. It is important to note that all sites except PG were inside the active caldera. LF and CG were in the most active fumarolic areas. NA, RN, CH and RT in Furnas Village (areas of which are stated to have sometimes dangerously high CO_2 levels due to diffuse degassing across the caldera floor on which the village lies) were from areas with clear soil degassing. HT was inside the active caldera, and less than 500m from mapped fumaroles, but with no visible soil degassing. PG was from outside the active caldera and a little over 1km from the closest mapped fumarole.

3. Calibration Curve?

How much radiocarbon calibration information do we have for the period 1700–1500 BC? Sometimes one reads comments which rather imply we have almost no data and it is all rather a best guess given just a few tantalising indicators. For the current IntCal04 calibration curve for the period 3651 Cal BP to 3445 Cal BP (or 1701 BC to 1495 BC) there are 65 data (each of these data comprise a Gaussian probability distribution, thus probability strongly centres on the plotted "point," with 68.2% of the range included within the error bars shown, 95.4% at twice these error bars, and 99.7% at three times these error bars) from 4 different laboratories to cover these 206 calendar years. See Figure 4. While never perfect, this is quite a lot of evidence. Only a few of the data appear to be outliers (e.g. the UB data point at 1500 BC), and just a few have rather too large measurement errors really to be included as "high-precision" (e.g. especially the QL data points at 1571 BC and 1561 BC where the standard measurement errors (SD) are much higher than for all the other data at: 51.6, 53.3, 53.3, and 52).

windy island, invalidating the further counter-factual hypothesis of some stable very minor offset somehow applying over a large area and for a long period. An additional factor to note is that the data in Pasquier-Cardin et al. (1999) are from evergreen plants. In contrast, the Santorini data are from annual plants (and different species) as well as from trees. The annual growth crops should be more susceptible to temporal variability in received CO_2 (versus being near annual averages). But the Santorini annual data are all very consistent. And these data are consistent with the data on the decadal-scale tree-ring samples from Santorini (both those reported by Friedrich et al. 2006, but also e.g. the tamarix twig, K-4255, previously reported by Friedrich et al. 1990). Hence one cannot argue for some "unusual" minor effect that just happened to affect the final harvest since it would also need consistently to affect these multi-year samples (and somehow also leave them all compatible with other contemporary late Late Minoan IA samples from Trianda and Miletos). Hence any substantive volcanic CO_2 effect appears unlikely.

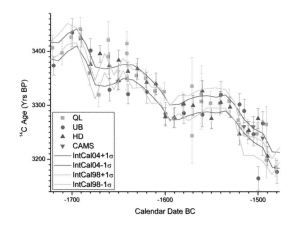

Figure 4: All the constituent ^{14}C data (shown with 1σ error bars) employed to develop the IntCal04 radiocarbon calibration curve for the period 1700–1500 BC shown against the IntCal04 (Reimer et al. 2004) and IntCal98 (Stuiver et al. 1998) 1σ envelopes. Data from: http://www.radiocarbon.org/IntCal04% 20files/IntCal04_rawdata.csv. QL = University of Washington, UB = Queen's University Belfast, HD = Heidelberg, CAMS = Center for Accelerator Mass Spectrometry, LLNL (CAMS) tree-ring. QL and HD samples on German Oak. UB and CAMS samples on Irish Oak.

IntCal04 (Reimer et al. 2004) is then a sophisticated model that best describes the underlying pattern given all the data available for and around a specific calendar interval/point (Buck and Blackwell 2004). The IntCal model weighs the probabilities exhibited around a particular interval and favours the main trend evident and downplays any odd outlying minor values. Where there is a very high intensity of underlying information, IntCal04 better describes the underlying trend than a simple decadal binning approach (as was employed for IntCal98: Stuiver et al. 1998)—see for example Bronk Ramsey et al. (2006: Figure 2)—however, where data are less intense it may be that IntCal04 slightly smoothes out some real underlying variations. Hence, for some applications, such as tree-ring sequence wiggle-matching, IntCal98 may offer slightly better results in some conditions for periods before AD 1510. Both the IntCal04 and IntCal98 curves are shown against the raw calibration data for 1700–1500 BC in Figure 4. IntCal04 can be seen as a best fitting to the data available. IntCal04 clearly better describes the 16th century BC raw data than IntCal98, but, in the 17th century BC there is the question of whether there is real significant underlying variation 1690–1640 BC which IntCal04 is perhaps overly smoothing away, especially when a near-absolutely placed set of decadal data on Juniper from Gordion in central Turkey (Kromer et al. 2001; Manning et al. 2001) seems to exhibit some such extra variation 1685–1655 BC: see Figure 5.

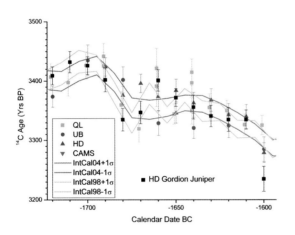

Figure 5: Detail of Figure 4 for the period 1700–1600 BC, adding in the near-absolutely placed Gordion Juniper dataset from the Heidelberg laboratory (Kromer et al. 2001; Manning et al. 2001).

4. Inter-annual and regional ^{14}C variations?

Some have wondered whether the issues of inter-annual and regional variations in ^{14}C levels might somehow be relevant and part of a problem/explanation with regard to Santorini. However, regional effects in the Aegean area (versus the standard northern hemisphere calibration curve and its sources) are demonstrated at even the (ultra-high-precision) 1.5‰ measurement level (whereas most of the data under discussion are at the 3–4‰ measurement level) to only even be detectable at times of exceptionally high ^{14}C production (major solar minima) (Kromer et al. 2001; Manning et al. 2002b; 2005). This situation does not apply in the 17th or 16th centuries BC.

Inter-annual variance is another variable sometimes identified as a possible problem. Annual data do scatter around the longer-term average trend reflected by the IntCal curves. Typically, given any larger set of such data, their average will in fact end up close to the IntCal curve and the variance largely cancels itself out. But, it is fair to suggest this issue could create a little "noise." Stuiver and Braziunas (1998) thus suggested an increase of the sample standard deviation (by 1‰ or 8 ^{14}C years) for single-year samples before calibration: $\sigma_x = \sqrt{(\sigma_x^2 + 8^2)}$ ^{14}C years. In practice, this has no substantive effect regarding the Manning et al. (2006a) study, since it typically means a 0.5 to 1.5 ^{14}C years' increase in the standard deviation for the relevant samples, and this has almost no observable impact on the analyses and outcomes we reported. To demonstrate the robust nature of the findings we reported, we show in blue in Table 1 the outcomes if we add 10 ^{14}C years to each measurement error in our study for samples of <10 years growth (this very large addition—x10 beyond the inter-annual factor suggested by Stuiver and Braziunas (1998)— is to cover all and any of the possible extra and/or not allowed for error factors that anyone has, or might, suggest). In all cases the changes that result are very small and insignificant and merely slightly enlarge the date ranges we reported before. This issue is thus irrelevant.

5. Upwelling?

Keenan (2002) alleged that Mediterranean ^{14}C ages in the BC era are some 100–300 years too old, because, he claimed, of upwelling. This assertion is supported by no positive evidence, and the available evidence in fact contradicts it (Manning et al. 2002b). Where they occur, problems with archaeological ^{14}C samples which appear too old (or for that matter too recent) are instead more likely due to other causes (old or reused/cycled wood, various possible contamination or diagenesis issues that can/could apply, poor associations between dated object and archaeological/historical context, and so on. See also Manning 2006 regarding Egypt).

6. Trend versus Outliers?

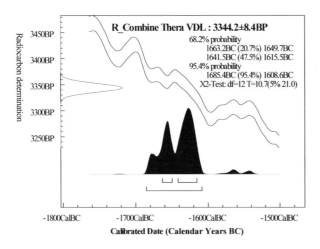

Figure 6: Calibrated calendar probability and age ranges for the weighted average from the 13 measurements on short-lived samples from the final volcanic destruction level at Akrotiri (seeds stored in jars) from the Oxford and VERA laboratories (data from Manning et al. 2006a) using IntCal04 and OxCal. No other data or modelling has been applied.

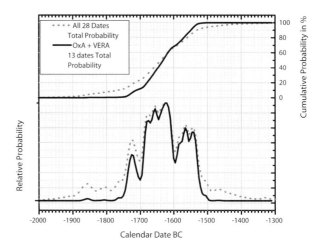

Figure 7: *Lower*—The black line shows the total probability distribution (summed) for the 13 Oxford and VERA dates in Figure 6 (contrast with the narrower weighted average range shown in Figure 6); the grey dotted line shows the total probability (summed) for all the 28 dates on short-lived samples from the final volcanic desruction level at Akrotiri in Figure 2. *Upper*—The black line and the grey dotted line show the cumulative probability in percentage terms for each of the total probability plots shown in the lower section. Thus, for example, over 65% of all probability (both plots) lies *before* 1600 BC, and over 80% of all probability lies before 1570 BC (13 date set) or 1560 BC (28 date set).

The set of short-lived samples from the volcanic destruction level at Akrotiri (Figure 2) represents a set of estimates for the ages of what should be contemporary or very nearly contemporary samples of short-lived growth. This is thus a classic situation where one is considering a population of data all forming estimates of the same underlying value/range, and where it is therefore reasonable and most appropriate to see if it is valid to combine or pool all the data in order to best describe the population observed, and, if so, to do so. The overall trend of the data is the key datum for our interest, and not the variability (as long as this is random variation around a best estimate) nor the extremes of variability in a couple of observations (with the measurements made in the 1970s–1980s having more noise and larger measurement errors due to then available technology). To quantify the original raw range but also the existence of clear trends (which we want to concentrate on), we can consider the total probability displayed by the 28 data (see Figure 7—dotted grey plots upper and lower). Some (tiny at outer ends) probability ranges from just before 2000 BC to the later 13th century BC (overall 3σ, or 99.7%, range is 2026–1262 BC); however, the main probability "peak" area with 41% of all probability is 1693–1600 BC, with the next most likely area with 21.5% of all probability lying 1593–1531 BC. If we consider just the 13 data from Oxford and VERA run in the AD 2000s, then the main area with 49.6% of total probability is 1687–1605 BC and the next most likely area is 1575–1535 BC with 18.6% of total probability: IntCal04 and OxCal with curve resolution set at 5 (see Figure 7—black line plots upper and lower). And, altogether, >68.03% of the total (100%) calendar dating probability lies *before* 1600 BC. And >89.3% of the total calendar dating probability lies before 1550 BC (and >97.9% lies before 1530 BC). Thus claims that the data could also (seemingly it is implied with reasonable probability) offer calendar dates compatible with the low chronology (so c. 1525 BC, 1520 BC, 1500 BC for the Thera eruption) are misleading. The great majority (and most likely) dating probabilities lie *earlier*.[1]

Critically, whereas each individual measurement must have a reasonable measurement error attached to it, the weight of observations in a large set allows us to narrow the effective measurement error and to make our description of the trend in the data more precise. Moreover, as in any set of measurement data, one or two more outlying data are to be expected in any set. Statistical procedures exist to test a hypothesis of whether all the data are consistent (or not) with describing the same age (Ward and Wilson 1978). The Akrotiri volcanic destruction level data in Figure

---

[1] Indeed, *even if* one arbitrarily—since there is no evidence to support doing so—deducted 20 ^{14}C years from each of the 13 Oxford and VERA dates run in the AD 2000s, then still over 80% of the calibrated calendar dating probability lies *before* 1550 BC. Thus, to make a date in the late 16th century BC likely, a very substantial systematic change must be made to all the data, but there is no positive evidence to support such a change.

2 pass such a test at the 95% level (and thus there are no significant outliers in the set). The weighted (that is weighed according the precision of the measurement error specified) average for the recent AD 2000s data on short-lived samples from the Akrotiri final volcanic destruction level, and its calibrated age (with no modeling or other considerations) from the data in Figure 2, is shown in Figure 6. We note that the addition of the previous literature data from Oxford, Copenhagen, and Heidelberg (Figure 2), as in Manning et al. (2006a), makes very little difference: weighted average 3344.9±7.5 BP, 1σ, 1663–1650 BC (21.1% of probability) and 1641–1616 BC (47.1% of probability), and 2σ 1683–1611 BC. (There is no 16th century BC range.)

7. Constraints, and how low could the Santorini age go?

Stratigraphic and archaeological context information from archaeology means that several ^{14}C constraints exist on the dating of the Santorini volcanic destruction level as employed and discussed in Manning et al. (2006a). Data below in this section are from IntCal04 and OxCal (curve resolution set at 5).

(i) *Termini Post Quos*

Several data, and in particular four dendro-^{14}C-dated charcoal samples, set *termini post quos* for the Akrotiri volcanic destruction level. It is worth just highlighting these (again). The oldest is a 30-year sample from pith to bark (NB) from Trianda on Rhodes (sample AE1024) from what was regarded by the excavator as an early Late Minoan IA context (but which others have suggested might be late Middle Bronze Age to Late Minoan IA). All these possible contexts are older to quite a bit older than the mature Late Minoan IA volcanic destruction level. Thus the bark of the sample (cutting/use date) sets a *terminus post quem* for this point (late MBA to early LMIA), but probably somewhat irrelevantly early with respect to the Akrotiri volcanic destruction level. The sample was divided into three 10-year samples and these were each measured twice at Oxford and once at VERA. The data offer an entirely satisfactory analysis (for each decadal sample, and for the weighted average age for each decade against the calibration curve). At the very latest they set a *terminus post quem* (2σ range) for the Akrotiri volcanic destruction at an unknown point after c.1736 BC. The most likely wiggle-match is shown in Figure 8 (showing the good agreement with the calibration curve), although two other older placements are also possible (marked by the green boxes B and C in Figure 8) (in such a case even a longer *terminus post quem* would apply).

The second and third data comprise two branch samples from Akrotiri from Late Minoan IA contexts but before the final volcanic destruction (samples 65/N001/I2 and M4N003). These branches could relate to original use from late Middle Cycladic through to mature Late Minoan IA. However, the samples come from the area of the Late Cycladic I kitchen (Birtacha 2008: 352) constructed early in Late Cycladic I, and one or both likely relate to this time period—though one or both may also relate to later repairs or other modifications/use up to the time of the final volcanic destruction. There is no reason to assume they are contemporaneous; one may be significantly older than the other, or they could be only years or decades apart. One branch (sample 65/N001/I2) with 3 rings (pith to bark) with each ring measured at both Oxford and VERA offers an average ^{14}C age for the 3-year sample combined of 3325±11 BP; the other where the outer (to bark) 6 rings were analysed and again at both Oxford and VERA (sample M4N003) offers an average age 3412±10 BP. Calibrated, sample M4N003 thus sets a *terminus post quem* of 1746–1686 BC at 2σ. (The sample might be assumed to offer a date for an original late Middle Cycladic to early Late Minoan IA use/context.) Sample 65/N001/I2 is later and has a 2σ dating range of 1664–1530 BC in total, with its most likely ranges being 1632–1605 BC (37.4% of probability) or 1575–1536 BC (30.8% of probability). There are two possibilities: (a) the sample dates very shortly before the volcanic destruction level and is for all intensive purposes of the same age as the volcanic destruction level samples, or (b) the sample dates somewhat earlier and reflects a wiggle in the radiocarbon record (to a lower ^{14}C age) as hinted at by some data (see Figures 4 and 5) but smoothed away in IntCal04 (e.g. especially c.1680–1670 BC). In the Manning et al. (2006a) model the only constraint applied was that sample 65/N001/I2 was older (no amount of time was quantified) than the Akrotiri VDL. (A possible change could be made to the Manning et al. 2006 model to allow the 65/N001/I2 sample to be either before or contemporary with the Akrotiri VDL set. The advantage of this is the sample can then enjoy a satisfactory agreement index of e.g. 68.9%, but the impact on the Akrotiri volcanic destruction level date range is very minor, lowering ages by a couple of years only: thus 1σ range with this change of c.1636–1613 BC and 2σ of c.1658–1610 BC.)

The fourth datum is an oak sample from Miletos ending in what was thought likely (but not definitively) to be waney edge (the ring below bark). The sample and context were found covered in Thera/Santorini tephra, so the sample's last ring sets a clear *terminus post quem* for the date of the eruption, by an unknown amount. The sample was

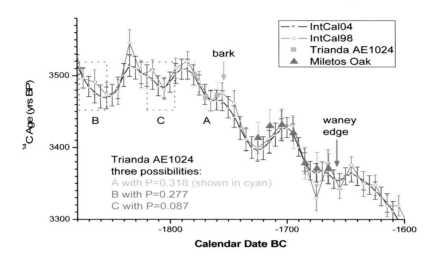

Figure 8: ^{14}C-dendro wiggle-match placements against IntCal04 of (i) the 30-year oak sample (AE1024) from a early Late Minoan IA, or later Middle Bronze Age to early Late Minoan IA, context at Trianda, Rhodes (in cyan); and (ii) the 72-year oak (chair) sample from Miletos found covered by Santorini ash (in magenta). The IntCal98 calibration curve is also shown for comparison. Sample data from Manning et al. (2006a) Supporting Online Material.

divided into seven 10-year samples each measured twice at Oxford (we employ the weighed average for each decade). The date of the final extant ring (and we think likely the waney edge) from the ^{14}C wiggle-match sets a *terminus post quem* for the eruption of c.1673–1643 BC at 2σ and c.1665–1651 BC at 1σ. So an eruption date after this point is required. The wiggle-match of the Miletos sample against the calibration curve is shown in Figure 8, again showing very good agreement.

Overall, whereas the raw weighted average date for the Akrotiri volcanic destruction level by itself at 2σ is 1685–1609 BC (just AD 2000s data) or 1683–1611 BC (all 28 data in Figure 2) (Figure 6), the effect of these *terminus post quos* data incorporated into the Bayesian analytical model employed in Manning et al. (2006a) is to remove part (around 25 or 23 years) of this older possible range, and instead to yield a 2σ range for the Akrotiri volcanic destruction level starting around 1660 BC (Table 1).

(ii) *Termini Ante Quos*

The weighted average calibration for the Akrotiri short-lived sample volcanic destruction set in isolation ends at 2σ confidence at 1611–1609 BC (Figure 6). At present this is the main determining factor for the latest date possible for the set.

There are however two sets of constraining or slightly modifying data. The first is the *terminus post quem* data just reviewed. These have the effect of pushing the Akrotiri volcanic destruction level data as late as possible (since several of these data wish to date down to the mid-later 17th century BC also). Thus if one ignored the post-Late Minoan IA ^{14}C evidence for a moment, and just considered the Late Minoan IA sets, then one finds that the Akrotiri VDL could (maximum case) be pushed down to: 1σ 1629–1610 BC (57.6% of probability), 1567–1562 BC (5.9% of probability) and 1544–1540 BC (4.7% of probability); 2σ 1636–1607 BC (63.6% of probability) and 1576–1534 BC (31.8% of probability). With the data in the Manning et al. (2006a) paper, this represents the maximum "low as can go" scenario, and is only achieved, to note again, by ignoring entirely the Late Minoan IB evidence (which otherwise tries to occupy the mid-16th century BC and so cuts out much of this possible likelihood for the Akrotiri set). The main probability remains in the 1629–1610 BC region (which remains also very compatible with the findings of the Friedrich et al. 2006 paper), but, ignoring the Late Minoan IB evidence, there is 31.8% probability in the mid-16th century BC (1576–1534 BC) which could be seen a territory for a compromise high chronology. But, even at this extreme, the data cannot reach to 1525 BC or 1520 BC or 1500 BC as the conventional chronology posits. Thus the conventional dates remain as incompatible with the radiocarbon evidence.

In reality, we have a second body of evidence also to consider, which provides some limit on the downward movement of the Late Minoan IA evidence. This is the evidence of two sets of ^{14}C data on short-lived samples from subsequent Late Minoan IB contexts. Data on short-lived samples from Late Minoan IB contexts must clearly post-date the Late Minoan IA cultural period, and so they also post-date the eruption of Thera/Santorini (which occurred before the end of the Late Minoan IA period). Short-lived data from the close of (at those sites at least) Late Minoan IB destruction horizons moreover date the mature to later part of the Late Minoan IB period, and not its start or earlier part. Hence the Akrotiri volcanic destruction level samples must be at least substantially older than those samples from the close of Late Minoan IB destruction horizons. Manning et al. (2006a) considered samples from two Late Minoan IB destruction horizons (Chania and Myrtos-Pyrgos) (see previously also Housley et al. 1999). The data from the Myrtos-Pyrgos destruction horizon combine into a weighted average (3229 ± 13 BP) and give a 1σ calibrated age range of 1514–1492 BC (42.5% of probability) and 1476–1460 (25.7% of probability) and a 2σ range of 1522–1451 BC. The Chania data are a little more complicated. The set of eight data from four samples offers a rather wider spread and does not satisfactorily combine into a single weighted average value, indicating either that more than one chronological horizon is included or one or more outlier measurements. Applying a minimum exclusion policy to obtain a satisfactory weighted average, then we find that if just OxA-10320 is excluded then a seven date weighted average is possible (3293 ± 14 BP). The 1σ calibrated range is 1607–1570 BC (41.5%), or 1560–1546 BC (14.9%), or 1540–1529 (11.8%), and the 2σ range is 1615–1520 BC. Overall, the data set is rather older than the Myrtos-Pyrgos one in ^{14}C and calibrated calendar terms, with five of the samples offering radiocarbon ages overlapping with those from the Akrotiri volcanic destruction level. Treating all eight data as a Phase in OxCal (as was done in Manning et al. 2006a), then an event describing the Chania set spans 1590–1483 BC at 1σ and 1628–1447 BC at 2σ.

The entire of the Late Minoan IB cultural phase at each site lies before these destructions, at a date we can call X. (And most recent archaeological work indicates that the Late Minoan IB cultural period was relatively substantial/long.) The *terminus ante quem* for the Akrotiri volcanic destruction level in turn lies before this date X.

The Myrtos-Pyrgos data offer a LMIB close earlier (by 11 to 97 years, using the cited 2σ range above) than the conventional chronology date (see Table 1). But they do not by themselves require a start to the Late Minoan IB period and an end to the Late Minoan IA period necessarily before a date in the 16th century BC (for example, if we assume an overall length for the Late Minoan IB period of 50 to 75 years—short by many current estimates, but perfectly possible—then we could have a start to Late Minoan IB from 1501 to 1597 BC). Thus these data by themselves could work with a mid-16th century BC Santorini eruption date. Although, one has to note that even then a date for the eruption of Santorini around 1525 BC or 1520 BC (late 16th century BC) as proposed by several conventional chronology studies in recent years requires selection only of the later (and less likely) part of the Myrtos-Pyrgos dating range and the assumption of a pretty short Late Minoan IB period (rather in opposition to most current thinking by archaeologists working on the period) (and then selection of the unlikely part of the Akrotiri volcanic destruction date range, etc., the probability of several "unlikely" choices all being correct is of course very, very low).

The inclusion of the Chania data, however, changes the picture substantially. These data wish to occupy much of the 16th century BC. Yet they date samples/contexts after the Santorini eruption. Thus these data imply that whereas, by themselves, the Akrotiri volcanic destruction level data (and the 65/N001/I2 branch) could date either in the later 17th century BC (most likely) or the mid-16th century BC (less likely), however, when then allowing for the Chania data, the reality must be the later 17th century BC option for the Akrotiri samples. The ambiguity, namely that later Late Minoan IA and Late Minoan IB short-lived samples can seem to have fairly similar ^{14}C ages, is because of the history of past natural radiocarbon levels which sees radiocarbon levels (ages) that are quite close when comparing the later 17th century BC versus the mid-16th century BC (see Figure 4).

The model of Manning et al. (2006a) required the Akrotiri volcanic destruction level data set merely to date before the Chania and Myrtos-Pyrgos close of Late Minoan IB data sets. This is an archaeological/stratigraphic requirement. No interval was specified.[2]

8. Could one "fiddle" the Bayesian sequence analysis to get a result consistent with the conventional chronology?

(i) Old wood? The Manning et al. (2006a) analysis was centred on the short-lived samples, but it did include some samples on wood charcoal and (solely) employed them to set *terminus post quem* ranges for their find context. This is entirely valid. Even if the wood in question was re-used, it still sets a *terminus post quem* (just a very early and thus irrelevant

---

[2]For detailed consideration of the Late Minoan IB radiocarbon evidence, including the data from Mochlos in Soles 2004, and for discussion of the archaeological and art-historical evidence and how these can relate in the 15th century BC, see Manning n.d.

one). No bias or possible error is created. And, in reverse, important constraints are added. Thus the wood charcoal samples from clear pre-eruption contexts reviewed in 7(i) above, for example, importantly and usefully set *terminus post quem* ranges which constrain the date of the Akrotiri volcanic destruction level.

(ii) Boundaries? The role and use of boundaries in the OxCal software are described in articles (Bronk Ramsey 1994;1995; 1998; 2001) and discussed and described in the online manual (http://c14.arch.ox.ac.uk/oxcal/help/hlp_contents.html). They are used to define temporal horizons between or surrounding an archaeologically dated event/horizon (like a phase at a site). The boundary values are calculated from the ^{14}C data in the archaeological sequence—they are not independent nor arbitrary extras—and they quantify the relationships between the archaeologically defined sets of information. The labels (and that is all they are) applied to name some of the boundaries seem to cause concern to some scholars. We have merely sought to employ a label which roughly describes the "in-between" cultural time the boundary describes. So, if one has Late Minoan IA ^{14}C data, and then sets of close of Late Minoan IB ^{14}C data, the "in-between" period covers what would be earlier through mature Late Minoan IB before the close of Late Minoan IB destruction horizon. Hence we named this boundary as "End or Final LMIA to LMIB Destructions" in Manning et al. (2006a), and so on.

(iii) Use of "conventional chronology dates" to sort things out? What about if we created some constraints that included the conventional chronology interpretation and included these in the model: for example we say that White Slip I supposedly cannot be found on Santorini/Thera before c.1530 BC because that represents its earliest known secure find at Tell el-Dab'a (and hence the Akrotiri volcanic destruction level must post-date this date, as Wiener, this volume argues—but for why this viewpoint is not valid and indeed demonstrably incorrect, see Manning et al. 2002c; 2006a; 2006b; Manning 2007). Would this then make everything work "properly" and find the conventional chronology? No—all that then happens is that the analysis (if one adds a 1530 BC *terminus post quem* before the Akrotiri volcanic destruction level) finds *zero* possible agreement scores and *zero* analytical outcomes for the ^{14}C data sets in the Manning et al. (2006a) analysis (such is the non-compatibility). The Bayesian sequence analysis thus informs us clearly that this model cannot work. Therefore the incompatibility of the available ^{14}C data versus the conventional chronology for the period c.1700-1400 BC is merely highlighted (again) in a different way.

9. Gaps in our current evidence?

The key missing evidence at present for this debate are one or more sets of ^{14}C data from the early/earlier Late Minoan IB period. If some of the archaeological sites with a sequence of Late Minoan IB horizons could yield a sequence of data across this cultural period then this would be important additional information. These would better clarify what is happening in the 16th century BC. We encourage excavators to look for such evidence.

10. Robustness of the analysis and especially with regard to the Akrotiri volcanic destruction level data?

We have previously discussed and demonstrated that the analyses of the Aegean Late Bronze Age 1-2 sequence are fairly robust (Manning et al. 2006a, especially Supporting Online Material; Manning and Bronk Ramsey 2003: 124–129).

The key observation is that running the Manning et al. (2006a) dating model with NO Santorini data at all (and thus no plausible or possible "special" contaminant factors or other problems which might be suggested or hypothesised—the issue of a lack of any positive evidence notwithstanding), finds an overall chronology, and an inferred date range for the Akrotiri volcanic destruction level, which is similar to (just a little less well defined) than the analysis with the Santorini data included (Table 1). And, critically, this "NO Santorini model" finds a chronology still entirely incompatible with the conventional chronology.

We can consider one further test. What happens if we run the Manning et al. (2006a) Model 1 analysis but with the data for the Akrotiri volcanic destruction level (hypothetically) adjusted by say minus 10 ^{14}C years, and then minus 20 ^{14}C years and so on to minus 100 ^{14}C years? That is to suppose *hypothetically* that there is in fact some "special" factor affecting these data (even though there is no positive evidence yet adduced to demonstrate such an effect). How much adjustment could occur before these data are then incompatible with the rest of the ^{14}C data from the Aegean in the analysis? And how much impact does the adjustment make up to this point of incompatibility on the date ranges found for the Akrotiri volcanic destruction level?

Table 2 shows the outcomes of running Manning et al. (2006a) Model 1 but with −10, −20...−100 ^{14}C years adjustments to the weighted average value for the Akrotiri volcanic destruction level. The analysis shows that a change of −80 ^{14}C years or more means that the Akrotiri data would then not be compatible with the rest of the dating evidence, and thus this sce-

	Model 1 (see Manning et al. 2006a) 1σ range BC	Model 1 (see Manning et al. 2006a) 2σ range BC	Agreement (needs to be ≥60% to reach an approximate 95% confidence level)
Current Akrotiri VDL value 3345±8 BP	1656-1649 (P=0.115) 1640-1616 (P=0.567)	1661-1611	109.9%
-10 ^{14}C years	1634-1610	1662-1605	119.4%
-20 ^{14}C years	1630-1606	1660-1598 (P=0.886) 1590-1560 (P=0.068)	123.8%
-30 ^{14}C years	1627-1600	1632-1556	100%
-40 ^{14}C years	1620-1580	1625-1552	88.2%
-50 ^{14}C years	1613-1583	1620-1553	95.6%
-60 ^{14}C years	1610-1584	1615-1554	108.2%
-70 ^{14}C years	1606-1587	1614-1566 (P=0.938) 1561-1554 (P=0.016)	101.1%
-80 ^{14}C years	1605-1589	1611-1570	52%
-90 ^{14}C years	1604-1590	1610-1568	14.3%
-100 ^{14}C years	1604-1590	1610-1571 (P=0.936) 1528-1518 (P=0.018)	5.3%

Table 2: Robustness/sensitivity of analysis to changes in the Akrotiri volcanic destruction level data? We run Model 1 (see Table 1 above) from Manning et al. (2006a) with the Akrotiri volcanic destruction level at minus 10, minus 20... minus 100 ^{14}C years from the value (3345±8BP) found when employing the current data (as in Manning et al. (2006a), and see Figures 2 and 6 above). We see that the analyses do not offer an acceptable agreement for the Akrotiri data versus the rest of the data for the minus 80 to minus 100 scenarios (shaded grey). Thus an adjustment of this level is entirely incompatible with the rest of the data. However, an adjustment of minus 10 to minus 70 ^{14}C years could be acceptable to the rest of the data in the model (not that there is any reason to make this adjustment). Even at the limits of the 2σ range, this finds a date for the Akrotiri volcanic destruction level older than 1550 BC. The 1σ ranges suggest a date no more recent than c.1580 BC at the extreme. The conventional chronology date range of c.1525 BC to 1500 BC remains incompatible. The notable observation is the robustness of the Akrotiri volcanic destruction level date range. Even a very significant change of minus 30 ^{14}C years still sees the 1σ range stay entirely in the late 17th century BC. And no change that offers a satisfactory analysis in terms of agreeing with the other Late Bronze Age 1-2 data finds outcomes compatible with the conventional chronology (and instead all find data compatible with the so-called 'high' chronology). Data employing IntCal04 and OxCal (curve resolution set at 5). Note each run of the analysis yields very slightly varying outcomes, typically with variations of around 0-2 years. Typical outcomes shown above. A note on agreement indices in OxCal. The agreement index measures how well the measured ^{14}C ages fit to the stratigraphical model, and should be around 100% for a perfect fit. Numbers above 100% indicate that the data fit even better than expected from the measurement uncertainties quoted. This can indicate an overestimation of the measurement uncertainties, but will also happen by chance. An agreement index above 60% corresponds roughly to a confidence level of 95%.

nario is impossible unless one starts to argue that all the other ^{14}C data are also wrong for some other reasons. In particular, the constraining data are now the Late Minoan IB data from Crete (from respectively west and east Crete and from the north and south coasts), where no volcanic or other "special" factor contaminant is plausible. In contrast, a hypothetical adjustment of minus 10 to minus 70 ^{14}C years could be acceptable to the rest of the data in the model if required. A change of -10 to -30 ^{14}C years makes almost no substantive impact. Overall, a -10 to -70 ^{14}C year change, even at the limits of the 2σ range, still requires a date for the Akrotiri volcanic destruction level before 1550 BC. And the main probability area, as indicated by the 1σ ranges, remains supporting a late 17th century BC date through to one in the first two (only) decades of the 16th century BC. (all the 1σ ranges require a date no later than c.1580 BC). The conventional chronology date range of c.1525 BC to 1500 BC remains incompatible even with such very significant *hypothetical* adjustments (for which, *note*, there is at present *no* positive evidence).

Discussion

Radiocarbon provides a substantial body of evidence for the question of the date of the Santorini/Thera eruption, both directly from the volcanic destruction horizon on Santorini/Thera, and in an archaeologically derived sequence of Aegean data from the start of the Late Minoan IA period through to the close of the Late Minoan II period. The radiocarbon outcomes are relatively robust. Despite unsupported suggestions of a number of possible contaminating or complicating factors, no positive evidence to support such concerns exists at present. Counter arguments offered to date rest on suggesting one unlikely minor issue and then adding another possible suggestion of another unlikely factor, and so on in a "shower with purported uncertainty attack," to try to seem to undermine the value of all the radiocarbon evidence. But unlikely x unlikely... x unlikely merely equals very, very unlikely. We have further shown that even if unknown (or undemonstrated) problems might exist with the Akrotiri volcanic destruction level data, or even with the overall body of Santorini ^{14}C data, the remaining radiocarbon evidence from around the Aegean nonetheless yields an overall Late Bronze Age 1–2 chronology, and interpolated date for the Santorini eruption, consonant with the date provided by the samples from Santorini. Thus there can be no significant problem/change, unless one is going to try to argue that all radiocarbon data from the Aegean do not work for some entirely unknown and undemonstrated reason.

Therefore, at this time we regard the likely date for the Santorini eruption as around or very shortly after the temporal horizon defined by the Akrotiri volcanic destruction level short-lived ^{14}C data (representing the last human use of the site before its abandonment and then the subsequent volcanic eruption): i.e. (i) 1σ 1656–1651 BC (P=0.113) and 1639–1616 BC (P=0.569) and 2σ 1660–1612 BC (Table 1); or (ii) 1σ 1636–1613 BC and 2σ 1658–1610 BC (7 (i) above). Within the overall range, somewhere c.1639–36 to 1616–13 BC would appear most likely. The date (c.1627–1600 BC at 2σ) for the outermost ring of the olive tree killed by the eruption and analysed by Friedrich et al. (2006) offers compatible data and mutual support. This date for the outermost ring (and the eruption date inferred from it) should either be contemporary with, or a few years later than, the Akrotiri volcanic destruction level seeds. There is very possibly some added uncertainty surrounding this sample as the exact ring count = annual years equation is at least a little debatable, both because of the difficulty in recognising the "rings" in olive, and because—as Wiener (this volume) comments—it is possible either that more than one apparent ring formed per calendar year in some circumstances, and/or there are missing years as represented in the observed ring record. Friedrich et al. (2006) made a reasonable effort to allow for such possible errors (see their Supporting Online Material Table 2) and Friedrich et al. (this volume) reaffirm their confidence in the approximate dendrochronological assessment of the sample. We nonetheless include some additional (extreme) scenarios of ring count = calendar year errors to assess robustness and sensitivity of this olive sample's analysis in our Table 3 below. The notable observation is that the overall date range placement is relatively stable despite such large error allowances (see Table 3).

Overall, although there are clearly some possible flexibilities in the various data and their analyses, and any date ranges calculated now will change slightly as new data are added and according to the specifics of possible analytical models employed, and so on, nonetheless, there appear good grounds for arguing that such issues in no case appear to make it possible to consider the available (and quite substantial) Aegean Late Bronze Age 1-2 radiocarbon data set in Manning et al. (2006a) and Friedrich et al. (2006) as being compatible with the conventional or low Aegean chronology. Instead, the radiocarbon data and its archaeological sequence analysis are clearly in conflict with the conventional or low Aegean chronology.

A body of literature over the last several years has tried to question the validity of the Aegean-east Mediterranean radiocarbon dates and derived

Ring Count Error Allowance	2σ ranges BC for last ring from IntCal04	2σ ranges BC for last ring from IntCal98
1a. *<1 year error*	1627-1600	1626-1612 (P=0.252) **1607-1594 (P=0.702)**
1b. *25% ring count increase and 25% gap error allowance*	1637-1590	1652-1645 (P=0.031) **1634-1590 (P=0.868)** 1586-1575 (P=0.055)
1c. *50% gap error allowance*	1653-1642 (P=0.047) **1641-1598 (P=0.907)**	1653-1643 (P=0.052) **1636-1591 (P=0.878)** 1584-1576 (P=0.023)
2a. *100% gap error allowance*	1662-1596	1674-1660 (P=0.051) 1658-1642 (P=0.11) **1636-1591 (P=0.755)** 1584-1570 (P=0.038)
2b. *assuming 2 observed rings = 1 calendar year*	1658-1645 (P=0.203) **1640-1607 (P=0.751)**	1658-1653 (P=0.028) 1651-1648 (P=0.018) **1637-1603 (P=0.908)**
2c. *100% missing rings and 50% additional gap error allowance*	1653-1647 (P=0.001) **1637-1594 (P=0.756)** 1586-1550 (P=0.188)	1651-1645 (P=0.015) **1633-1591 (P=0.634)** 1585-1533 (P=0.306)

Table 3: (1a-c) Date for the outermost ring of the Friedrich et al. (2006) olive sample given counting errors on the ring count for each section of (1a) <1 year, (1b) 25% of ring count and gap error, and (1c) 50% gap error allowance employing IntCal04 (Reimer et al. 2004) and IntCal98 (Stuiver et al. 1998) and OxCal. Data taken from Friedrich et al. (2006) Supporting Online Material Table 2 but run again (hence some very small differences compared to their reported results – typical results shown). (2a-c) Date for the outermost ring of the Friedrich et al. (2006) olive sample given counting/calendar errors on the ring count for each section of: (2a) as 1c. but 100% gap error allowance, (2b) 2 observed rings per actual calendar year (i.e. 50% of observed rings actually cases of two rings per calendar year), (2c) 100% missing rings per section (i.e. doubling calendar-year length represented) and 50% gap error allowance—extreme case to try to cover large numbers of supposed missing rings. Data from IntCal04 (Reimer et al. 2004) and IntCal98 (Stuiver et al. 1998) and OxCal. Most probable sub-ranges indicated in bold.

chronology (see Wiener 2003; 2006; this volume, and further references therein; Bietak 2003; Bietak and Höflmayer 2007). The inherent assumption is that the conventional or low archaeological interpretation and chronology is correct (and so there must be something wrong if radiocarbon finds a different result). This critical, sceptical, and forensic commentary forms a very valuable contribution to the field for which the present authors (and others) are grateful, as it ensures care and attention to detail in the radiocarbon work and generally has promoted much higher-quality work. As a result, considerable effort has gone into testing and establishing data quality and analytical robustness in recent radiocarbon work and analyses relevant to Santorini/Thera and the Aegean–east Mediterranean in the mid-second millennium BC (e.g. Bronk Ramsey et al. 2004; Manning et al. 2006a; 2005; 2002b; 2001; Friedrich et al. 2006; Kromer et al. 2001; Manning and Bronk Ramsey 2003; and, in this volume, see the present paper and the paper of Friedrich et al.). At some point, however, the reverse question has to be asked: can the conventional or low archaeological chronology interpretation withstand similarly intense critical and sceptical analyses? Are there some key assumptions or factoids which need re-evaluation? Are there at least plausible alternative archaeological syntheses which are (potentially) compatible with the radiocarbon evidence? (Readers might consider e.g. Merrillees 1971; 1972; 1977; 1992; 2002; 2003; Betancourt 1987; 1990; 1998; Manning 1988; 1999; 2007; Manning et al. 2002c; Manning et al. 2006b; Niemeier and Niemeier 1998; 2000; Niemeier 1990; etc.). As Colin Renfrew commented at the Cornell Conference: something's gotta give.

A final intriguing question which has hung over the last three decades of scholarship is whether the Santorini eruption can potentially equate with the major Greenland ice-core volcanic signal most recently placed at 1642±5 BC (Vinther et al. 2006; 2008) or the major northern hemisphere tree-ring growth anomalies placed c.1650 BC +4/-7 BC (Manning et al. 2001) and 1628/1627 BC (LaMarche and Hirschboeck 1984; Baillie and Munro 1988; Baillie 1990; Grudd et al. 2000)? Clearly the answer is yes, possibly, just in dating terms. The 1628/1627 BC event lies very much in the main dating probability region for the Santorini volcanic destruction level (above). But today there are also several other candidate years in this general time-frame based on recent analysis of the Bristlecone pine data (Salzer and Hughes 2007: esp. pp.64-66), in particular: a package of events 1653-1648 BC including in the Yamal and Finnish tree-ring data series (which could link with the Aegean anomaly c.1650 BC), another 1619-1617 BC, and another anomaly at 1597 BC (two later anomalies at 1544 BC and 1524 BC are outside the likely radiocarbon range for Santorini/Thera). All these events are within the 2σ radiocarbon ranges (above), and some are in the 1σ ranges (above). The 1642±5 BC date is close to the 1σ range for Santorini from the Akrotiri volcanic destruction level, and close to or within some of the possibilities for the Friedrich et al. (2006) olive sample given various possible error allowances. However, at present, there is no clear positive evidence to support any such link. There is also no positive evidence which links the Thera/Santorini volcanic eruption specifically to any of the tree-ring signals (though future potential may exist applying dendrochemical approaches: Pearson et al. 2005; Pearson and Manning, this volume; and see esp. Pearson et al. 2009). A Thera/Santorini association for the 1642±5 BC ice-core signal has also been questioned (Pearce et al. 2004; Denton and Pearce 2008). But Vinther et al. (2008) strongly argue why, in contrast, Thera nonetheless remains a suitable and possible candidate to explain this 1642 ±5 BC volcanic signal. So there may be variously some or no associations, but, strictly chronologically, one or other of these suggested associations might be possible in radiocarbon terms if future evidence does manage to establish any positive connection.

References

Aitken, M.J. 1988. The Minoan eruption of Thera, Santorini: a re-assessment of the radiocarbon dates. In R.E. Jones and H.W. Catling (eds.), *New aspects of archaeological science in Greece: proceedings of a meeting held at the British School of Athens, January 1987*: 19–24. Occasional Paper 3 of the Fitch Laboratory. London: British School at Athens.

Baillie, M.G.L. 1990. Irish tree rings and an event in 1628 BC. In D.A. Hardy and A.C. Renfrew (eds.), *Thera and the Aegean world III. Volume three: chronology*: 160–166. London: The Thera Foundation.

Baillie, M.G.L., and Munro, M.A.R. 1988. Irish tree rings, Santorini and volcanic dust veils. *Nature* 332: 344–346.

Balter, M. 2006. New Carbon Dates Support Revised History of Ancient Mediterranean. *Science* 312: 508–509.

Betancourt, P.P. 1987. Dating the Aegean Late Bronze Age with radiocarbon. *Archaeometry* 29: 45–49.

Betancourt, P.P. 1990. High chronology or low chronology: the archaeological evidence. In D.A. Hardy and A.C. Renfrew (eds.), *Thera and the Aegean world III. Volume three: chronology*: 19–23. London: The Thera Foundation.

Betancourt, P.P. 1998. The chronology of the Aegean Late Bronze Age: unanswered questions. In M.S. Balmuth and R.H. Tykot (eds.), *Sardinian and Aegean chronology: towards the resolution of relative and absolute dating in the Mediterranean*: 291–296. Studies in Sardinian Archaeology V. Oxford: Oxbow Books.

Betancourt, P.P., and Weinstein, G.A. 1976. Carbon-14 and the beginning of the Late Bronze Age in the Aegean. *American Journal of Archaeology* 80: 329–348.

Bietak, M. 2003. Science versus archaeology: problems and consequences of high Aegean chronology. In M. Bietak (ed.), *The Synchronisation of Civilisations in the Eastern Mediterranean in the Second Millennium B.C. II. Proceedings of the SCIEM 2000—EuroConference Haindorf, 2nd of May–7th of May 2001*: 23–33. Contributions to the Chronology of the Eastern Mediterranean II. Vienna: Verlag der Österreichischen Akademie der Wissenschaften.

Bietak, M., and Höflmayer, F. 2007. Introduction: High and Low chronology. In M. Bietak and E. Czerny (eds.), *The Synchronisation of Civilisations in the Eastern Mediterranean in the Second Millennium B.C. III. Proceedings of the SCIEM 2000—2nd EuroConference, Vienna 28th of May–1st of June 2003*: 13–23. Vienna: Verlag der Österreichischen Akademie der Wissenschaften.

Birtacha, K. 2008. 'Cooking' installations in LCIA Akrotiri on Thera: a preliminary study of the 'Kitchen' in Pillar Shaft G5. In N. Brodie et al. (eds.), *Horizon: A colloquium on the prehistory of the Cyclades*: 349–376. Cambridge: McDonald Institute for Archaeological Research.

Bronk Ramsey, C. 1994. Analysis of Chronological Information and Radiocarbon Calibration: The Program OxCal. *Archaeological Computing Newsletter* 41: 11–16.

Bronk Ramsey, C. 1995. Radiocarbon calibration and analysis of stratigraphy: the OxCal program. *Radiocarbon* 37: 425–430.

Bronk Ramsey, C. 1998. Probability and dating. *Radiocarbon* 40: 461–474.

Bronk Ramsey, C. 2000. Comment on "The Use of Bayesian Statistics for ^{14}C dates of chronologically ordered samples: a critical analysis." *Radiocarbon* 42: 199–202.

Bronk Ramsey, C. 2001. Development of the radiocarbon calibration program OxCal. *Radiocarbon* 43: 355–363.

Bronk Ramsey, C., Buck, C.E., Manning, S.W., Reimer, P., and van der Plicht, H. 2006. Developments in radiocarbon calibration for archaeology. *Antiquity* 80: 783–798.

Bronk Ramsey, C., Manning, S.W., and Galimberti, M. 2004. Dating the volcanic eruption at Thera. *Radiocarbon* 46: 325–344.

Bruns, M., Levin, I., Münnich, K.O., Hubberten, H.W., and Fillipakis, S. 1980. Regional sources of volcanic carbon dioxide and their influence on ^{14}C content of present-day plant material. *Radiocarbon* 22: 532–536.

Buck, C.E., and Blackwell, P.G. 2004. Formal statistical models for estimating radiocarbon calibration curves. *Radiocarbon* 46: 1093–1102.

Denton, J.S., and Pearce, N.J.G. 2008. Comment on "A synchronized dating of three Greenland ice cores throughout the Holocene" by Bo M. Vinther et al.: No Minoan tephra in the 1642 B.C. layer of the GRIP ice-core. *Journal of Geophysical Research* 113: D04303, doi: 10.1029/2007JD008970.

Friedrich, W.L., Kromer, B., Friedrich, M., Heinemeier, J., Pfeiffer, T., and Talamo, S. 2006. Santorini Eruption Radiocarbon Dated to 1627–1600 B.C. *Science* 312: 548.

Friedrich, W.L., Wagner, P., and Tauber, H. 1990. Radiocarbon dated plant remains from the Akrotiri excavation on Santorini, Greece. In D.A. Hardy and A.C. Renfrew (eds.), *Thera and the Aegean world III. Volume three: chronology*: 188–196. London: The Thera Foundation.

Grudd, H., Briffa, K.R., Gunnarson, B.E., and Linderholm, H.W. 2000. Swedish tree rings provide new evidence in support of a major, widespread environmental disruption in 1628 B.C. *Geophysical Research Letters* 27: 2957–2960.

Hardy, D.A., and Renfrew, A.C. (eds.) 1990. *Thera and the Aegean world III. Volume three: chronology*. London: The Thera Foundation.

Housley, R.A., Manning, S.W., Cadogan, G., Jones, R.E., and Hedges, R.E.M. 1999. Radiocarbon, calibration, and the chronology of the Late Minoan IB phase. *Journal of Archaeological Science* 26: 159–171.

Keenan, D.J. 2002. Why early-historical radiocarbon dates downwind from the Mediterranean are too early. *Radiocarbon* 44: 225–237.

Kromer, B., Manning, S.W., Kuniholm, P.I., Newton, M.W., Spurk, M., and Levin, I. 2001. Regional $^{14}CO_2$ offsets in the troposphere: magnitude, mechanisms, and consequences. *Science* 294: 2529–2532.

LaMarche, V.C., Jr., and Hirschboeck, K.K. 1984. Frost rings in trees as records of major volcanic eruptions. *Nature* 307: 121–126.

Manning, S.W. 1988. The Bronze Age eruption of Thera: absolute dating, Aegean chronology and Mediterranean cultural interrelations. *Journal of Mediterranean Archaeology* 1: 17–82.

Manning, S.W. 1999. *A Test of Time: the volcano of Thera and the chronology and history of the Aegean and east Mediterranean in the mid-second millennium BC*. Oxford: Oxbow Books.

Manning, S. 2005. Simulation and the Thera eruption: outlining what we do and do not know from Radiocarbon. In A. Dakouri-Hild and S. Sherratt (eds.), *Autochthon: Papers presented to O.T.P.K. Dickinson on the occasion of his retirement*: 97–114. *BAR International Series* 1432. Oxford: Archaeopress.

Manning, S.W. 2006. Radiocarbon dating and Egyptian chronology. In E. Hornung, R. Krauss, and D.A. Warburton (eds.), *Ancient Egyptian Chronology*: 327–355. Leiden: Brill.

Manning, S.W. 2007. Clarifying the "high" v. "low" Aegean/Cypriot chronology for the mid second millennium BC: assessing the evidence, interpretive frameworks, and current state of the debate. In M. Bietak and E. Czerny (eds.), *The Synchronisation of Civilisations in the Eastern Mediterranean in the Second Millennium B.C. III. Proceedings of the SCIEM 2000–2nd EuroConference, Vienna 28th of May–1st of June 2003*: 101–137. Vienna: Verlag der Österreichischen Akademie der Wissenschaften.

Manning, S.W. In press. Beyond the Santorini eruption: some notes on dating the Late Minoan IB period on Crete, and implications for Cretan–Egyptian relations in the 15th century BC (and especially LMII). In D. Warburton (ed.), *Time's Up! Dating the Minoan Eruption of Santorini. Monographs of the Danish Institute at Athens, Vol.10*.

Manning, S.W., Barbetti, M., Kromer, B., Kuniholm, P.I., Levin, I, Newton, M.W., and Reimer, P.J. 2002b. No systematic early bias to Mediterranean ^{14}C ages: radiocarbon measurements from tree-ring and air samples provide tight limits to age offsets. *Radiocarbon* 44: 739–754.

Manning, S.W., and Bronk Ramsey, C. 2003. A Late Minoan I-II absolute chronology for the Aegean—combining archaeology with radiocarbon. In M. Bietak (ed.), *The synchronisation of civilisations in the eastern Mediterranean in the second millennium BC (II). Proceedings of the SCIEM2000 EuroConference Haindorf, May 2001*: 111–133. Vienna: Verlag der Österreichischen Akademie der Wissenschaften.

Manning, S.W., Bronk Ramsey, C., Doumas, C., Marketou, T., Cadogan, G., and Pearson, C.L. 2002a. New evidence for an early date for the Aegean Late Bronze Age and Thera eruption. *Antiquity* 76: 733–744.

Manning, S.W., Bronk Ramsey, C., Kutschera, W., Higham, T., Kromer, B., Steier, P., and Wild, E. 2006a. Chronology for the Aegean Late Bronze Age. *Science* 312: 565–569.

Manning, S.W., Crewe, L., and Sewell, D.A. 2006b. Further light on early LCI connections at Maroni. In E. Czerny, I. Hein, H. Hunger, D. Melman, and A. Schwab (eds.), *Timelines: Studies in honour of Manfred Bietak, Vol.2*: 471–488. OLA 149. Leuven: Peeters.

Manning S.W., Kromer B., Kuniholm P.I., and Newton M.W. 2003. Confirmation of near-absolute dating of east Mediterranean Bronze-Iron Dendrochronology. *Antiquity* 77: 295. http://antiquity.ac.uk/ProjGall/Manning/manning.html.

Manning, S.W., Kromer, B, Kuniholm, P.I., and Newton, M.W. 2001. Anatolian tree-rings and a new chronology for the east Mediterranean Bronze-Iron Ages. *Science* 294: 2532–2535.

Manning, S.W., Kromer, B., Talamo, S., Friedrich, M., Kuniholm, P.I., and Newton, M.W. 2005. Radiocarbon Calibration in the East Mediterranean Region. In T.E. Levy and T. Higham (eds.), *The Bible and Radiocarbon Dating Archaeology, Text and Science*: 95–113. London: Equinox.

Manning, S.W., Sewell, D.A., and Herscher, E. 2002c. Late Cypriot IA maritime trade in action: underwater survey at Maroni-Tsaroukkas and the contemporary east Mediterranean trading system. *Annual of the British School at Athens* 97: 97–162.

Marketou, T., Facorellis, Y., and Maniatis, Y. 2001. New Late Bronze Age chronology from the Ialysos region, Rhodes. *Mediterranean Archaeology and Archaeometry* 1(1): 19–29.

Merrillees, R.S. 1971. The early history of Late Cypriot I. *Levant* 3: 56–79.

Merrillees, R.S. 1972. Aegean Bronze Age relations with Egypt. *American Journal of Archaeology* 76: 281–294.

Merrillees, R.S. 1977. The absolute chronology of the Bronze Age in Cyprus. *Report of the Department of Antiquities, Cyprus*: 33–50.

Merrillees, R.S. 1992. The absolute chronology of the Bronze Age in Cyprus: a revision. *Bulletin of the American Schools of Oriental Research* 288: 47–52.

Merrillees, R.S. 2002. The relative and absolute chronology of the Cypriote White Painted Pendent Line Style. *Bulletin of the American Schools of Oriental Research* 326: 1–9.

Merrillees, R.S. 2003. The first appearances of Kamares ware in the Levant. *Ägypten und Levant* 13: 127–142.

Michael, H.N. 1976. Radiocarbon dates from Akrotiri on Thera. *Temple University Aegean Symposium* 1: 7–9.

Niemeier, W.-D. 1990. New archaeological evidence for a 17th century date of the "Minoan eruption" from Israel (tel Kabri, western Galilee). In D.A. Hardy and A.C. Renfrew (eds.), *Thera and the Aegean world III. Volume three: chronology*: 120–126. London: The Thera Foundation.

Niemeier, W.-D., and Niemeier, B. 1998. Minoan frescoes in the eastern Mediterranean. *Aegaeum* 18: 69–98.

Niemeier, B., and Niemeier, W.-D. 2000. Aegean Frescoes in Syria-Palestine: Alalakh and Tel Kabri. In S. Sherratt (ed.), *The Wall Paintings of Thera. Proceedings of the First International Symposium, Petros M. Nomikos Conference Centre, Thera, Hellas, 30 August–4 September 1997. Vol. II*: 763–802. Piraeus: Petros M. Nomikos and The Thera Foundation.

Olsson, I.U., and Vilmundardóttir, E.G. 2000. Landnám Íslands og C-14 aldursgreiningar. *Skírnir* 174: 119–149.

Pasquier-Cardin, A., Allard, P., Ferreira, T., Hatte, C., Coutinho, R., Fontugne, M., and Jaudon, M. 1999. Magma-derived CO_2 emissions recorded in ^{14}C and ^{13}C content of plants growing in Furnas caldera, Azores. *Journal of Volcanology and Geothermal Research* 92: 195–208.

Pearce, N., Westgate, J., Preece, S., Eastwood, W., and Perkins, W. 2004. Identification of Aniakchak (Alaska) tephra in Greenland ice core challenges the 1645 BC date for Minoan eruption of Santorini. *Geochemistry, Geophysics, Geosystems* 5(3): Q03005, doi:10.1029/2003GC000672.

Pearson, C.L., Manning, S.W., Coleman, M., and Jarvis, K. 2005. Can tree-ring chemistry reveal absolute dates for past volcanic eruptions? *Journal of Archaeological Science* 32: 1265–1274.

Pearson, C.L., Dale, D.S., Brewer, P.W., Kuniholm, P.I., Lipton, J., and Manning, S.W. 2009. Dendrochemical analysis of a tree-ring growth anomaly associated with the Late Bronze Age eruption of Thera. *Journal of Archaeological Science* 36: 1206–1214. doi: 10.1016/J.Jas.2009.01.009.

Reimer, P.J., Baillie, M.G.L., Bard, E., Bayliss, A., Beck, J.W., Bertrand, C.J.H., Blackwell, P.G., Buck, C.E., Burr, G.S., Cutler, K.B., Damon, P.E., Edwards, R.L., Fairbanks, R.G., Friedrich, M., Guilderson, T.P., Hogg, A.G., Hughen, K.A., Kromer, B., McCormac, G., Manning, S., Bronk Ramsey, C., Reimer, R.W., Remmele, S., Southon, J.R., Stuiver, M., Talamo, S., Taylor, F.W., van der Plicht, J., Weyhenmeyer, C.E. 2004. IntCal04 Terrestrial Radiocarbon Age Calibration, 0-26 Cal Kyr BP. *Radiocarbon* 46: 1029–1058.

Salzer, M.W., and Hughes, M.K. 2007. Bristlecone pine tree rings and volcanic eruptions over the last 5000 yr. *Quaternary Research* 67: 57–68.

Sarpaki, A. 1990. "Small fields or big fields?" That is the question. In D.A. Hardy, J. Keller, V.P. Galanopoulos, N.C. Flemming, and T.H. Druitt (eds.), *Thera and the Aegean world III. Volume two: earth sciences*: 422–431. London: The Thera Foundation.

Saupe, F., Strappa, O., Coppens, R., Guillet, B., and Jaegy, R. 1980. A possible source of error in ^{14}C dates: volcanic emanations (Examples from the Monte Amiata district, provinces of Grosseto and Sienna, Italy). *Radiocarbon* 22: 525–531.

Soles, J.S. 2004. Appendix A: Radiocarbon Dates. In J.S. Soles and C. Davaras (eds.), *Mochlos IC: Period III. Neopalatial Settlement on the Coast: The Artisan's Quarter and the Farmhouse at Chalinomouri. The Small Finds*: 145–149. Philadelphia: INSTAP Acaademic Press.

Stuiver, M., Reimer, P.J., Bard, E., Beck, J.W., Burr, G.S., Hughen, K.A., Kromer, B., Mccormac, G., vaan der Plicht, J., and Spurk, M. 1998. INTCAL98 radiocarbon age calibration, 24,000–0 cal BP. *Radiocarbon* 40: 1041–1083.

Stuiver, M., and Braziunas, T.F. 1998. Anthropogenic and solar components of hemispheric ^{14}C. *Geophysical Research Letters* 25: 329–332.

Stuiver, M., and Polach, H.A. 1977. Reporting of ^{14}C data. *Radiocarbon* 19: 355–363.

Sulerzhitzky, L.D. 1971. Radiocarbon dating of volcanoes. *Bulletin of Volcanology* 35: 85–94.

Sveinbjörnsdóttir, A.E., Heinemeier, J., Rud, N., and Johnsen, S.J. 1992. Radiocarbon Anomalies Observed For Plants Growing In Icelandic Geothermal Waters. *Radiocarbon* 34: 696–703.

Vinther, B.M., Clausen, H.B., Johnsen, S.J., Rasmussen, S.O., Andersen, K.K., Buchardt, S.L., Dahl-Jensen, D., Seierstad, I.K., Siggaard-Andersen, M.-L., Steffensen, J.P., Svensson, A., Olsen, J., and Heinemeier, J. 2006. A synchronized dating of three Greenland ice cores throughout the Holocene. *Journal of Geophysical Research* 111: D13102, doi:10.1029/2005JD006921.

Vinther, B.M., Clausen, H.B., Johnsen, S.J., Rasmussen, S.O., Andersen, K.K., Buchardt, S.L., Dahl-Jensen, D., Seierstad, I.K., Siggaard-Andersen, M.-L., Steffensen, J.P., Svensson, A., Olsen, J., and Heinemeier, J. 2008. Reply to comment by J.S. Denton and N.J.G. Pearce on "A synchronized dating of three Greenland ice cores throughout the Holocene." *Journal of Geophysical Research* 113: D12306, doi: 10.1029/2007JD009083.

Ward, G.K., and Wilson, S.R. 1978. Procedures for Comparing and Combining Radiocarbon Age-Determinations—Critique. *Archaeometry* 20: 19–31.

Warren, P. 1984. Absolute dating of the Bronze Age eruption of Thera (Santorini). *Nature* 308: 492–493.

Warren, P.M. 1999. LMIA: Knossos, Thera, Gournia. In P.P. Betancourt, V. Karageorghis, R. Laffineur and W.-D. Niemeier (eds), *Meletemata: Studies in Aegean archaeology presented to Malcolm H. Wiener as he enters his 65th year*: 893-903. Aegaeum 20. Liège and Austin.

Warren, P.M. 2006. The Date of the Thera Eruption in Relation to Aegean-Egyptian Interconnections and the Egyptian Historical Chronology. In E. Czerny, I. Hein, H. Hunger, D. Melman, and A. Schwab (eds.), *Timelines: studies in honour of Manfred Bietak, vol. II*: 305–321. Orientalia Lovaniensia Analecta 149. Leuven: Peeters.

Warren, P., and Hankey, V. 1989. *Aegean Bronze Age chronology*. Bristol: Bristol Classical Press.

Wiener, M.H. 2006. Chronology Going Forward (With a Query About 1525/4 B.C.). In E. Czerny, I. Hein, H. Hunger, D. Melman, and A. Schwab (eds.), *Timelines: studies in honour of Manfred Bietak, vol. II*: 317–328. Orientalia Lovaniensia Analecta 149. Leuven: Peeters.

Wiener, M.H. 2003. Time out: the current impasse in Bronze Age archaeological dating. In K.P. Foster and R. Laffineur (eds.), *Metron: measuring the Aegean Bronze Age*: 363–399. Aegaeum 24. Liège and Austin.

Article submitted July 2007; revised February 2009

Thera Discussion

M. H. Wiener's Reply to the Papers by Manning et al. and Friedrich et al.

The papers by Manning et al. and Friedrich et al. in this volume incorporate a number of optimistic assumptions concerning the accuracy and precision of radiocarbon measurements and of the calibration curve, and provide incomplete citations to the relevant scientific literature regarding reservoir effects, root intake of CO_2, and other matters.

Manning et al. state that "[a] change of -10 to -30 ^{14}C years makes almost no substantive impact. Overall, a -10 to -70 ^{14}C year change, even at the limits of the 2σ range, still requires a date for the Akrotiri volcanic destruction level before 1550 BC" (page 312). This assertion applies to the proffered Bayesian Sequence Analysis as a whole. In order to test this conclusion, it is appropriate to consider each major constituent of the sequence individually, namely 1) the Akrotiri Volcanic Destruction Level (VDL) material; 2) the accuracy and precision of the calibration curve measuring rod; 3) the nature of potential data-distorting factors such as regional/seasonal effects and, in particular, reservoir effects from the uptake of ^{14}C-deficient carbon from the earth, streams/springs, or the sea; 4) the Thera olive branch measurements; and 5) the radiocarbon dates from other Aegean sites. It is important to recall, moreover, that the flatness of the radiocarbon calibration curve between c. 1600 and 1525 BC makes the interpretation of measurements very sensitive to small changes in ^{14}C measurement, with the result that a 20-year difference in measurement may allow an 80-year difference in date.

I. Theran-Seed Sample Measurements

Consider first the radiocarbon measurements from short-lived samples from the VDL on Thera itself. The data is presented in Manning et al. Figure 6. The bottom of the Gaussian bell curve distribution of the measurements of the 13 samples (there depicted at the left as a spearpoint in red) intersects the top of the smoothed IntCal04 probability band c. 1530 BC, with a greater overlap, although still fairly small, if the unsmoothed IntCal98 curve is substituted. IntCal98 "preserve[s] more high-frequency variation compared to IntCal04" (Friedrich et al. 2006b: 2–3). A downward movement of only two to three per mil in measurement, equivalent to 17 to 25 years in the weighted average calibrated date, would point the center of the determinations close to the center of the probability band at c. 1530 BC. Douglas Keenan, a mathematician specializing in the verification of scientific and medical research including studies involving climatology, has kindly provided the following Figure 1 illustrating the effect of such a change of only two per mil (17 years), calibrated by the unsmoothed (and hence more appropriate in this context) 1998 curve. It is worth recalling in this regard that the IntCal04 committee recommended expanding the error bands of the 1998 calibration curve to reflect new estimates of the degree of uncertainty in the Belfast and Seattle laboratories' calibration curve measurements.

1525 BC is clearly within the range of reasonable possibility on these hypotheses, even without considering the potential impact of measurements of the same sample separated by more than one sigma, of uncertainties in the calibration curve, of regional/seasonal variation, or of ^{14}C-deficient carbon, discussed below. Figure 1 on page 566 of the article by Manning et al. in the 28 April 2006 issue of *Science*, reproduced here (Figure 2), presents perhaps a clearer picture, with at least part of each of the Akrotiri VDL dataset sample measurement ranges extending at one sigma below even 3300 ^{14}C age BP, and with central points near 3320 BP, entirely consistent with a calibrated eruption date around 1525 BC.

One should note especially the reference to one-sigma, 68%, probability ranges. Where all the potential sources of uncertainty discussed herein are added to the picture, the reader is perhaps justified in wondering whether the error bars depicted represent much more than a 50/50 bet. It is only a few of the pre-high-precision determinations from the 1970s and 1980s, rejected by Manning in earlier studies, that fail to intersect the 1530 BC portion of the calibration curve at the two-sigma range (Manning et al., page 302).

My thanks are due in particular to Douglas Keenan, Steven Soter, and Peter Warren for helpful comments and rapid responses to my queries.

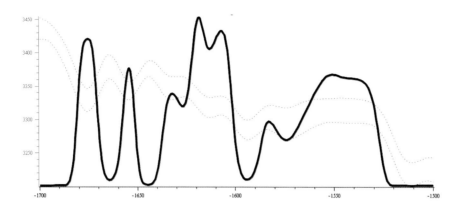

Figure 1: Calibrated radiocarbon determinations of the weighted average of 13 samples from the Akrotiri VDL, offset two per mil = 16.5 ^{14}C years, calibrated against unsmoothed, non-expanded IntCal98 calibration curve (Keenan, pers. comm. of 4 December 2008).

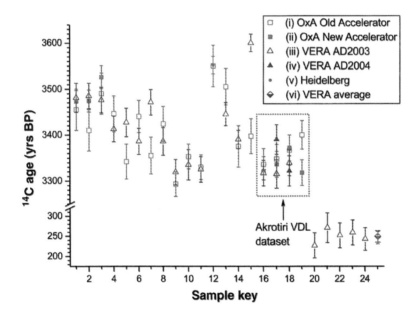

Figure 2: Comparison of ^{14}C age estimates for fractions of identical Aegean samples between (i) Oxford Old Accelerator (samples measured from AD 2000–2002), (ii) Oxford New Accelerator (measured in AD 2003), (iii) VERA (measured in AD 2003), (iv) VERA (measured in 2004), and (v) Heidelberg. The weighted average of the five VERA measurements on a sample of known-age wood are shown (vi) as compared to the Heidelberg measurement of the same sample. Sample key includes the following samples: 1, Trianda AE1024 rings 21 to 30; 2, Trianda AE1024 rings 11 to 20; 3, Trianda AE1024 rings 1 to 10; 4, Akrotiri M4N003 rings 6 to 8; 5, Akrotiri M4N003 rings 3 to 5; 6, Akrotiri M4N003 rings 7 and 8; 7, Akrotiri M4N003 rings 5 and 6; 8, Akrotiri M4N003 rings 3 and 4; 9, Akrotiri 65/N001/I2 ring 3; 10, Akrotiri 65/N001/I2 ring 2; 11, Akrotiri 65/N001/I2 ring 1; 12, Akrotiri M54/2/VII/60/SE>247; 13, Kommos K85A/62D/9:92; 14, Kommos K85A/66B/4:22+23; 15, Kommos K85A/62D/8:83; 16, Akrotiri M31/43 N047; 17, Akrotiri M2/76 N003; 18, Akrotiri M7/68A N004; 19, Akrotiri M10/23A N012; 20 to 24, Çatacık tree rings AD 1640 to 1649; and 25, weighted average VERA Laboratory data (samples 20 to 24) versus Heidelberg measurement of same sample. Samples 20 to 25 also offer a known-age test. All five VERA ^{14}C measurements included the correct calendar age range within their 1σ calibrated ranges (17, 22), as does the VERA weighted average and the high-precision Heidelberg measurement. Error bars indicate 1σ ranges (Manning et al. 2006a: Figure 1).

Here it is appropriate to recall the necessarily irregular nature of the underlying dataset, where decisions or protocols regarding inclusion/exclusion may sometimes produce significantly different results. Of the 13 Theran VDL determinations here at issue, two Oxford measurements of the same sample yielded determinations which did not overlap at the one-sigma range, and the same is true of a seed cluster divided between the Oxford and VERA labs. In each case, the higher (older) reading produced a measurement higher than any of the others in the series. Had a protocol of excluding measurements of the same sample whose one-sigma ranges did not overlap been adopted, the Theran VDL determinations again would appear visually to be consistent with a Theran eruption date of c. 1525 BC. The table of sample measurements provided in Manning et al. (2006b: 39–40) shows that two of the barley samples gave central ^{14}C ages 97 radiocarbon years apart (i.e., OxA-12175 and OxA-1556) and two of the grass pea samples 215 years apart (i.e., OxA-1549 and OxA-1555). Moreover, the one-sigma ranges of the grass pea measurements failed to overlap by 70 ^{14}C years (one sigma for these measurements). Three of these four sample measurements came from the 1989 Oxford study, when preparation protocols and measurement accuracy and precision may have been different. With regard to the post-AD 2000 determinations, two Oxford measurements of one barley sample, M10/23A N012, provided significantly different radiocarbon ages. The removal of humic acid apparently resulted in some shrinkage of an already small sample, to the point where the sample could not be shared with the VERA laboratory in Vienna as originally planned; the small size of the sample may have affected the measurement results. The decision was taken to include both measurements because they fit within the two-sigma range, if only barely (S. Manning, pers. comm. of 29 February 2004, for which I am most grateful). The exclusion of both measurements would lower the average age. The claim in Bronk Ramsey, Manning, and Galimberti (2004: 335) that the seeds in the jars from the Theran VDL "all give a perfectly consistent set of results" defies ready comprehension and requires explanation. To average such data to produce a "coherent weighted average" (Manning et al. 2006b: 6) and then calibrate only the average is statistically adventurous, to say the least. The Ward and Wilson "Case I" statistical method employed to compute the weighted average of these seeds by Manning et al. is explicitly stated in Ward and Wilson (1978: 20–21, 30) as applicable only to measurements known to be of the same date. The application of Case I weighted averaging to this dataset seems questionable. Further sensitivity testing is in order. It would be interesting to know the weighted and unweighted averages of all determinations of ^{14}C ages from seeds from the Theran VDL after all measurements which fail to overlap at one-sigma are removed from the dataset. Of course no amount of measurement and analysis can resolve problems resulting from the potential presence of ^{14}C-deficient carbon in a sample, discussed below.

Decisions as to what constitutes an outlier (sometimes made by laboratory personnel) may result, e.g., in the exclusion of measurements perceived as 250 years too early and the retention of determinations in fact 100 years too early, thereby giving a false sense of uniformity of result. Moreover, while inter-lab comparability of measurements has seen much improvement in recent years, some discrepancies naturally remain. Manning et al. (2006b: 5) report that:

> ...[o]verall, comparing the Oxford (OxA) versus Vienna (VERA) data on the same samples (using the pooled ages for each individual laboratory where they re-measured the same sample—thus n=17), we find an average offset of -11.4 ^{14}C years. The standard deviation is, however, rather larger than the stated errors on the data would imply at 68.1. This indicates that there is an unknown error component of 54.5 ^{14}C years.

Subsequent work has reduced inter-lab discrepancies in general, but many of the calibration curve measurements were made prior to the advent of high-precision AMS laboratories.

Conversely, a calibrated date later than 1520 BC could not be encompassed by even an eight per mil change in reported radiocarbon measurements of samples or calibration curve segments (see below) or some combination of the two, given the slope of the calibration curve after that date (Keenan, pers. comm. of 26 November 2008, for which I am most grateful). Whether the calibration curve is in fact so sharply sloped after 1520 BC is open to question, however, as noted below. An eruption date somewhat after c. 1520 BC, arguably preferable on archaeological grounds, would require some factor other than measurement uncertainty affecting Aegean radiocarbon determinations in this period, such as major distortion in the calibration curve at this point, ongoing upwelling of seawater, or the presence of ^{14}C-deficient carbon in samples (see below).

II. The Accuracy and Precision of the Calibration Curve

The imperfect nature of the decadal calibration curve measurements from trees of known dendrochronological date is apparent, with determinations of adjacent

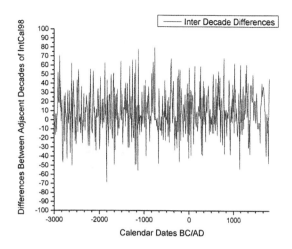

Figure 3: Inter-decade sample ^{14}C age differences (for adjacent decades) in the IntCal98 dataset.

decades regularly producing radiocarbon ages 30 to 70 years apart, as shown in Figure 3. (I am particularly grateful to Sturt Manning for making this unpublished figure available to me, in full knowledge of the use for which it was intended.)

When radiocarbon measurements of ten-year tree segment samples exhibit this amount of noise, notwithstanding the fact that they are far larger in sample size than seeds or seed clusters and are little affected by intra-annual variation in atmospheric radiocarbon (as well as some inter-annual variation, which may average out over a decade), caution concerning the precision of radiocarbon determinations in general is clearly indicated. When the 1998 calibration curve was issued, it was noted that a mean difference of 24.2 ±6 years existed between the Belfast measurements of Irish oaks and the Seattle measurements of German oaks for the critical period between 1700 and 1500 BC (Wiener 2007). Many of the calibration curve measurements still in use were made before the advent of high-precision measurements, and when less was known about best practice in such areas as the pretreatment of samples. Manning et al. Figure 4, reproduced here in Figure 4, is of special interest with respect to the critical decades centered on 1525 BC and 1515 BC. The Belfast laboratory measurements represented by the red circles are significantly above the upper boundary of the one-sigma smoothed 2004 band, and even further outside the unsmoothed IntCal98 band. With regard to the accuracy of the calibration curve in general, we may note as well the very large disparity at 1570 BC of the two measurements from the Seattle laboratory, and also the wayward measurement by the Belfast laboratory shown in Figure 4 at 1500 BC

(but more likely the bi-decadal measurement centered on 1510 BC, subsequently identified as an erroneous measurement by the laboratory [Wiener 2003b: 391–392; Reimer, pers. comm. of 7 February 2003]). As an example of how one erroneous measurement may dramatically affect analysis of the date of the LM IB destructions in Crete and hence the date of the close of LM IA including the Theran eruption date, see Manning et al. 2002. In sum, the calibration curve is by no means a perfect measuring rod.

Finally, as my paper in this volume notes, conversion from radiocarbon measurements to calibrated dates provides additional problems. Manning et al. in their discussion of ^{14}C measurements of a three-year tamarisk tree twig preserved in the VDL on Thera which produced a date consistent with the historical chronology observe that the sample may "reflect short-term higher amplitude and/or frequency variation in atmospheric ^{14}C ages not seen in the IntCal04 record for the period, which is both based on 10-year growth samples and smoothed" (2006b: 12). A seed measurement may represent a growing season of three weeks within a plant whose growing season is between six months and one year, whereas a measurement of a 10-year oak tree segment may, for example, over-represent one or two years with wide rings (reflecting rapid growth from good weather) which do not include the year of the life of the seed. In addition, Theran seeds of course grow in a different climate from European oaks.

With respect to potential regional/seasonal effects on radiocarbon measurements, the authors of the comments are of course correct in noting that there are no marked ongoing differences between the calibration curve decadal measurements from trees from Western Europe and those from Gordion in the 17th and earlier 16th centuries BC. Thera in the Aegean, however, does not lie in the same meteorological zone as Gordion in the central plateau of Anatolia which is affected by strong winds from the northeast (Keenan 2002: 237, Figure 1; Reddaway and Bigg 1996: Figure 3). The effect of the intra-year difference between summer high and winter low radiocarbon measurements would tend to raise slightly the ranges of ^{14}C determinations from Aegean seeds in relation to European trees, but this factor alone is unlikely to affect radiocarbon measurements by more than a decade. As we have seen in the context of the question posed, however, even a shift of a decade or less may matter.

It is surely for these reasons, as well as those discussed below, that Manning wrote:

> [i]t is apparent from the parameters and data for the Thera "problem" reviewed in this paper, that a solution may well be unlikely from the volcanic destruction level radiocar-

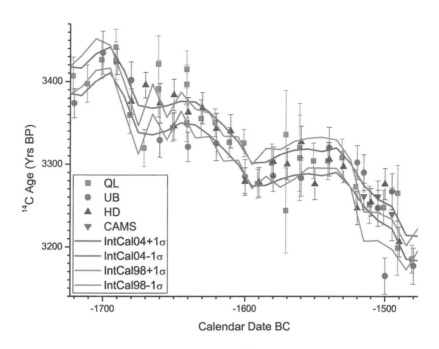

Figure 4: All ^{14}C data employed to develop the IntCal04 radiocarbon calibration curve for the period 1700–1500 BC shown against the IntCal04 and IntCal98 one-sigma envelopes (Manning et al. Figure 4).

bon data alone. The data at hand either indicate strongly, or, in most cases, tend toward, a 17th century solution. However, it is undeniable that not all do, and that the radiocarbon "gap" between 17th century certainty, and 17th/16th century ambiguity, is all of about 20–30 radiocarbon years. This span is about the same as the best measurement precision available today for Accelerator Mass Spectrometry determinations—the source technology for nearly all the modern Thera radiocarbon ages. Hence one is operating on the limits of precision. And even small laboratory offsets, or variations caused in sample pre-treatment regimes, could become relevant in pushing data into, or out of, the ambiguity threshold. Hence we hit an impasse. And a skeptic is justified to be so (Manning 2005: 111–112).

I believe that the foregoing statement of 2005 remains true with respect to the VDL short-lived samples and that the addition of radiocarbon dates from Miletus, Trianda, and Chania in Crete, discussed below, adds little to the solution. The recent olive tree measurements discussed in Friedrich et al. are considered below.

III. Reservoir Effects of ^{14}C-Deficient Carbon

Finally with respect to the ^{14}C determinations from the VDL at Akrotiri, we turn to the potential for reservoir effects from the presence of ^{14}C-deficient carbon (henceforth 'old carbon') in the earth or water. Each one percent of such carbon in a sample would push mid-second millennium ^{14}C measurements back in time by c. 80 radiocarbon years. The authors of the two comments dismiss the problem by asserting that 1) such effects occur only in close proximity to a particular volcano or source of degassing; 2) such effects would necessarily vary by gross amounts; 3) trees and plants do not acquire/retain CO_2 by their roots; and 4) secure radiocarbon dates from the non-Theran sites of Miletus, Trianda, and Chania, not subject to any Theran volcanic field effects, strongly support a 17th-century BC date for the Theran eruption. None of these propositions finds general support in the radiocarbon evidence or the relevant scientific literature.

As to the first, the papers by Manning et al. and Friedrich et al. again cite the studies by Pasquier-Cardin et al. (1999) from the Azores and that by Shore, Cook, and Dugmore (1995) from Iceland of samples collected from the vicinity of volcanoes in areas of strong winds, plus a study on Thera by Bruns et al. (1980) to argue that volcanic effects are quickly dissipated after short distances. The article by Bruns et al. tells us that contemporary short-lived plant material near the volcanic vent gave ages 1300 and 900 years too early and that the effect dissipated after a distance of 250 meters, but does not state exactly what results were obtained beyond 250 meters. In wide areas of Italy non-volcanic CO_2 emissions have marked effects over many kilometers. Italian soil-gas surveys show "that the shape and spatial distribution of the gas anomalies are related to the structural pattern of the area, whereas the magnitude of gas leakage is controlled by the occurrence of deep gas-bearing traps (hydrocarbon reservoirs) and possibly triggered by seismic activity" (Guerra and Lombardi 2001). Radiocarbon dates now recognized by the appearance in the same contexts of datable Greek pottery as 100–300 years too old have long troubled Italian Bronze and Early Iron Age chronology.

Friedrich et al. also state that "Tauber (1983) has shown by ^{14}C measurements that even in extreme calcareous soil conditions the uptake of CO_2 from soil carbonate is indiscernible (0.12 ±0.3%)" (page 296). It is not, however, calcareous soil itself (which may be relatively insoluble, particularly if very old), but dissolved inorganic carbon (DIC, mainly CO_2 and bicarbonate ions) present in the soil of a degassing area which is relevant. Further, limestone decomposition resulting in the release of old carbon may be stimulated by heat and pressure changes at depth during the decades preceding an earthquake and/or eruption. In this regard, it is worth recalling that there is no way of knowing whether the VDL Theran seeds which form a centerpiece of the debate were collected before the earthquake and precursor phase of the eruption (McCoy and Heiken 2000) which drove away the population or during the brief period when work crews and scavengers returned to attempt repairs and recover belongings before the climactic phases of the eruption. If the latter, then the seeds may have been affected by the release of old carbon from the preliminary stage of the eruption. Moreover, an increase in old carbon emissions may have occurred even prior to the initial stages of the eruption. Finally, with respect to the citation in Friedrich et al. to the paper by Tauber on a group of beech trees in New Zealand, the example of unaffected beech trees growing in one particular edaphic situation does not refute all the research showing that other types of trees or plants growing in other edaphic conditions were affected by the presence of old carbon.

Even more surprising is the following assertion by Friedrich et al. (page 296): "From tree-physiology it is extremely unlikely that there could be any quantitative effect by CO_2 uptake through the roots." This simply ignores the substantial body of literature demonstrating the exact opposite. See in particular Cramer (2002) and Ford et al. (2007) and references therein. A good number of other studies have also shown that plants and trees receive some carbon through their roots as well as through their leaves (Stolwijk and Thimann 1957; Skok, Chorney, and Broecker 1962; Splittstoesser 1966; Arteca, Poovaia, and Smith 1979; Yurgalevitch and Janes 1988; Enoch and Olesen 1993; Cramer and Richards 1999. See also Saleska et al. 2007). I know of no study reporting to the contrary. Ford et al. (2007: 375) put the case succinctly with regard to pine trees: "plants can acquire carbon from sources other than atmospheric carbon dioxide (CO_2), including soil-dissolved inorganic carbon (DIC). Although the net flux of CO_2 is out of the root, soil DIC can be taken up by the root, transported within the plant, and fixed...." Similar behavior has been proposed for willow and sycamore trees (Teskey and McGuire 2007; Vuorinen and Kaiser 1997). Oliver Rackham has noted that olive trees in particular spread massive roots in a search for water in dry climates (Rackham 1965–1966). Such groundwater, whether from streams or springs, is a potential source of ^{14}C-deficient carbon. N.-A. Mörner and G. Etiope note that in the "Tethyan belt [which includes the Mediterranean region], high CO_2 fluxes are related to important crustal formations of... carbonate rocks [causing a] high level of CO_2 concentration in ground and groundwater" (Mörner and Etiope 2002: 193. See also the work of Saurer et al. 2003 regarding groundwater effects discussed above). They further report that "the Precambrian bedrock includes stromatolites, marble and other carbonate bearing rocks [which]... may give rise to the escape of CO_2" (Mörner and Etiope 2002: 197). V. R. Switsur stated the general problem succinctly 25 years ago: "For reliable radiocarbon dating it is important to recognize when contamination with radioactive carbon from other sources is possible or probable—not only long rootlets from vegetation but also the percolation of carbon-rich ground water, or bicarbonates from the solution of limestone, even when from sources some distance from the site" (1984: 182). Moreover, it is necessary to consider the possibility that the uptake of soil carbon saturates at a fairly low value to protect the health of the tree or plant (unless the tree or plant is overwhelmed by proximity to a volcanic vent).

Friedrich et al. further state that:

> [t]he uptake could only occur as CO_2 in water which is transported via xylem (pipes) through the stem to the leaves where wood-cellulose is produced. Because of out-gassing of CO_2 in the xylem on the transport through stem and branches the contribution of CO_2 from root uptake compared to the CO_2 derived from the

air through the stomata in the leaves is negligible (page 296).

In fact, as described by Cramer (2002), plant roots themselves convert dissolved inorganic carbon (DIC) from the soil to organic molecules by the enzyme phosphoenolpyruvate carboxylase (PEPc). Cramer and Richards (1999) supplied DIC (labeled with ^{14}C as a tracer) to tomato roots and found that organic carbon in the xylem sap derived from the DIC was sufficient to deliver carbon to the shoot at rates equivalent to 1% and 10% of the photosynthetic rate for plants grown with ambient- and enriched-DIC, respectively. One percent retained in the roots from ambient-DIC lacking ^{14}C would result in an unwarranted increase of 80 radiocarbon years in the measurement of the seed or wood sample in question. Soter (forthcoming) further observes that dense crop canopies sometimes suppress ventilation, allowing CO_2 emitted from the soil to be taken up through leaves by photosynthesis.

In general, the Possible Sources of Error (PSE) in radiocarbon dating appear asymmetrical, with a tendency towards older dates.

IV. The Olive Tree Branch

Let us now turn again to the limb or branch of an olive tree found covered in Theran tephra and analyzed by Friedrich et al. This is indeed important new evidence, which the paper by Friedrich et al. regards as decisive.

Whether the branch was dead or alive at the time of the eruption remains an open question; long-dead branches are frequently observable on Aegean olive trees today (Wiener, this volume, page 288). An olive tree is a significant investment over generations. Olive growers today are conscious that removal of a major branch may damage the tree. In the early Late Bronze Age metal saws may have been less readily available to farmsteads (as distinguished from palaces such as Zakros in Crete where two were found) than they are today, which may have contributed to a reluctance to undertake the effort. Peter Warren, visiting Laconia in April 2008 after the devastating fire of 2007, observed new growth coming from parts of olive trees spared from the burning that had caused the death of other limbs and branches in the same trees (pers. comm. of 3 December 2008, for which I am most grateful). He further notes that the phenomenon of dead branches on living trees is not limited to olive trees, but rather is observable as well on trees such as stag's head oaks in the U.K. (Warren, pers. comm. of 30 November 2008).

The Friedrich et al. article also publishes a hypothetical reconstruction of the presumed location of the major volcanic fault lines prior to the great eruption. Of course no one can say with confidence what the topography of Thera was like prior to one of the greatest eruptions in human history; where streams or springs, under or above ground, may have been situated; where terrestrial sources of old carbon may have been located; or where underwater sources of old carbon capable of causing sea water to boil (Bent 1966: 118) may have existed.

The article by Friedrich et al. lays major stress on the asserted robust nature of the radiocarbon data in response to potential significant alteration in the number of years represented by the ring count, which in olive trees is difficult; moreover, the rings counted may not represent annual markers. The proposed ring count does triple-duty: 1) as an indication of the number of years represented; 2) as a basis for the probabilistic analysis permitting the assertion of narrowly circumscribed error bars of ±13 years; and 3) as the underpinning for the proposed curve-fitting of the successive measurements from four parts of the tree to the IntCal98 calibration curve. Recent work on Thera by Dr. Paolo Cherubini (a senior scientist with the Swiss Federal Institute for Forest, Snow, and Landscape) involved taking specimens of olive trees and examining them by computer tomography. His preliminary conclusion is that even with respect to non-fire- or tephra-blackened specimens it is often difficult to obtain agreement in blind tests as to what constitutes a ring, let alone whether the rings are seasonal or whether some years produce no rings. (I am most grateful to Paolo Cherubini for sharing the preliminary results of his research with me in advance of publication.) Of course if the number of years represented by the Theran VDL olive branch is in major doubt, problems arise, not least with respect to the proposed wiggle-match. It has also been proposed (Manning et al., page 300; Friedrich et al., page 296; 2006a; 2006b) that reservoir effects of old carbon could not be a problem with respect to the olive branch inasmuch as the ^{14}C measurements from the branch produce a downward slope, i.e., they descend in order as they move toward the outermost rings. However, if the intake of old carbon saturates at around 1%, then the 99% of the radiocarbon content absorbed from the atmosphere would still cause the progression to descend in order. For the moment, the Theran olive branch remains that dreaded scientific datum, a singleton, and one providing measurements that are not reproducible. Radiocarbon determinations from more such olive branches could provide a major contribution to the dating controversy. All in all, the radiocarbon determinations obtained by Friedrich et al. from the first VDL olive tree branch constitute significant but certainly not conclusive evidence for the date of the Theran eruption.

V. Radiocarbon Dates from Other Aegean Sites Not Affected by Conditions on Thera

One further major claim is made in connection with the radiocarbon evidence, namely, that whatever special circumstances may exist on Thera, dates unaffected by such factors from Miletus on the Turkish coast, Trianda on Rhodes, and Chania in Crete strongly support an early date for the eruption. Is this claim valid? First, it is good to see in the paper by Manning et al. the straightforward acknowledgment that the radiocarbon dates from a piece of wood with 30 rings from Trianda (Manning et al. page 307) and the wood from a shrine area at Miletus constitute merely "irrelevantly" high *termini post quos*, particularly in light of statements in prior articles suggesting the contrary (Manning et al. 2006a: 566). Two determinations from these sites remain. One from Miletus, consistent only with the traditional historical chronology, is dismissed on the grounds that the excavator, informed of the matter, reconsidered the area where the sample was collected and concluded that there was a significant possibility of later intrusion, and that accordingly the sample could not be firmly associated with the LM IA destruction horizon (Bronk Ramsey, Manning, and Galimberti 2004: 328). Certainly such things can happen, but so can what in archaeological parlance are sometimes called "kick-ups," when material from earlier horizons is carried higher—by leveling during rebuilding, or by digging of trenches for walls, or the creation of storage pits—and mistakenly used to date the higher stratum. The second sample consists of a twig found in a LM IA level at Trianda on Rhodes.

Two measurements of the twig were made at Oxford, yielding ^{14}C ages of 3367 ±39 BP and 3344 ±32 BP at one sigma. The bottom of the average of the one-sigma ranges is consistent with the historical chronology, and of course the two-sigma range is clearly so. Moreover, there is no evidence as to how the twig was used or whether it came to its final resting place as part of a tree harvested earlier. A branch from the Uluburun shipwreck is now thought to be significantly earlier in date than the shipwreck, and perhaps to have been used for many years as packing to cushion the ingots aboard the ship (Wiener 2003a).

Finally we come to the radiocarbon dates from Chania in Crete. The data bank here consists in its entirety of four seed samples, each measured twice. We are told that:

> [t]he set of eight data from four samples offers a rather wider spread and does not satisfactorily combine into single weighted average value, indicating either that more than one chronological horizon is included or one or more outlier measurements. Applying a minimum exclusion policy to obtain a satisfactory weighted average, then we find that if just OxA-10320 is excluded then a seven date weighted average is possible (3293 ±14 BP) (Manning et al., page 309).

Both of the highly experienced excavators of Chania, Dr. Maria Vlazaki and Prof. Erik Hallager, state that all the seeds came from the final LM IB destruction horizon at Chania (pers. comms. provided on short notice, for which I am most grateful), which suggests a problem in the radiocarbon measurements. Manning et al. (page 309) conclude:

> [t]reating all eight data as a Phase in OxCal (as was done in Manning et al. 2006a), then an event describing the Chania set spans 1590–1483 BC at 1σ and 1628–1447 BC at 2σ.

Even the one-sigma range utilizing all eight measurements of the four seed clusters does not necessarily contradict an end of LM IA Theran eruption at 1525 BC in accordance with the historical chronology, and the two-sigma range still less so. The same conclusion applies as well to the somewhat later ^{14}C measurements from Myrtos-Pyrgos on the south coast of Central Crete cited by Manning et al.

Omitted from the Manning et al. analysis of Cretan dates are the radiocarbon determinations from the LM IB destruction at Mochlos on the north coast in East Central Crete (Soles 2004: 147) where the four relevant measurements calibrate to 1500–1435 BC at one sigma and 1510–1425 BC at two sigma (subject to all the caveats stated above, plus the additional factor that the calibration curve is steeply sloped at this point, with the result that a small change in measurement may result in a large change in date). The published dates are consistent with the historical chronology, although potentially acceptable to a Long Chronology as well. Radiocarbon dates from the LM II destruction of the Unexplored Mansion at Knossos and from Kommos (Soles 2004: 148; Manning 1999: 220–223) fall into two sets which do not match, reflecting either problems in measurement or separate LM II destructions; indeed, the Unexplored Mansion is believed to have suffered two destructions in LM II. The earlier set permits a fairly wide span of dates because of an oscillation of the calibration curve here, but includes dates in the 1440–1430 BC range historically appropriate to an early LM II destruction. The later set of LM II destruction dates (3090 ±80 BP [Kommos] and 3070 ±70 BP [Knossos] [Soles 2004, 148]) are consistent with the historical chronology, and perhaps a little less comfortable with a final LM IB destruction

c. 1500 BC, but by the end of LM II, both chronologies essentially join, and the difference is inconsequential. Radiocarbon dates obtained from what the excavator believed were LH I contexts at Tsoungiza near Nemea in the Peloponnese are consistent only with the historical chronology. Manning et al. (2006b) say the context of the samples was "reasonably secure" and that the problem is "unexplained." Surely any attempt to alter dramatically Aegean Late Bronze Age chronology and history on the basis of radiocarbon dates should encompass all the relevant radiocarbon data and include a sensitivity analysis describing what effect small changes at the limit of measurement accuracy would have had on the radiocarbon dates proposed.

In sum, there is no probative radiocarbon evidence from the Aegean ex-Thera for a Theran eruption date earlier than the historical evidence would appear to allow.

VI. The Archaeological Evidence

The concluding paragraph of the Friedrich et al. comment states "[i]t would clearly be worth while to reevaluate in detail the current interpretations of the archaeological evidence" (page 298). Such discussion is constant and ongoing. Twenty years ago I noted that much of the then-known Minoica in Egypt seemed to suffer from some Pharaonic curse, for when the chronological context of an object was clear, the potsherd or other object was somewhat enigmatic, whereas when the nature of the Minoan or Mycenaean piece was clear, the context was uncertain. Of course today we have many more archaeological interconnections, particularly with regard to Cypriot pottery. Every single such object has now been considered at length by numbers of scholars. For example, the critical and, if correct, practically dispositive identification of stone vessels found in the Shaft Graves of Mycenae as New Kingdom Egyptian by Warren and by Bietak has been questioned by Christine Lilyquist at the Metropolitan Museum of Art in New York, inquiring whether the calcite may be Near Eastern or even Aegean, and the inspiration for the shapes found in forms which predate the New Kingdom (Warren 2006: 305–308; Bietak and Höflmayer 2007: 17; Lilyquist 1996: 134–149; 1997: 225–227; pers. comm. of 4 November 2008). Warren notes, however, that large numbers of apparently similar alabaster calcite vessels are found in Egypt, that the Egyptian Nile Valley calcite quarries are well known, that texts show thousands of such alabaster calcite vessels were exported from Egypt, and that the stone of the vessels found at Mycenae seems visually undistinguishable from that used in the Egyptian examples (Warren 2006 and pers. comm.).

The problem is not that *some* of the proposed interconnections would require earlier dates to fit the radiocarbon-based interpretation of the Theran olive branch and perhaps other data, but that *all* the proposed interconnections would need to move. This would include everything from the Khyan lid at Knossos to the vessels in the Shaft Graves, the Aegean metal vessels and their depiction in Egyptian tombs, the Cypriot pottery sequence in the Near East and Egypt, and the similarities in the sequences, based on scarabs as well as pottery, between Tell el-Dab'a and the sites of Tell el-'Ajjul and Ashkelon in the Near East. It is worth noting in this regard that most recent work on Near Eastern chronology has favored the Low Chronology (Pfälzner 2004) or Ultra-Low Chronology (Gasche et al. 1998) for the entire area.

From the early Hyksos Period beginning around 1650–1640 BC onward until c. 1530 BC, there is a constant stream of MC III imports into Egypt, but no LC IA:2 material (Bietak 2000a; 2000b; pers. comm. of 30 November 2008; Eriksson 2003). The great majority of the pumice from the Tuthmoside New Kingdom strata at Tell el-Dab'a is of Theran origin, whereas all the pumice from the well-explored prior Hyksos levels analyzed to date comes from the earlier eruptions of Kos, Gyali or Nisyros in the Dodecanese (Bichler et al. 2003; Bietak 2004: 214–215). The same relationship holds at 21 other sites around the Eastern Mediterranean (M. Bietak, pers. comm. of 6 July 2009; Bietak and Höflmayer 2007: 17). In all, over 400 specimens of pumice and tephra were examined. What is the likelihood that all of the apparent archaeological interconnections or contexts are erroneous, in comparison to the degree of uncertainty inherent in radiocarbon dating as described above?

Manning has attempted to contain the impact of the archaeological data by proposing a division of Cyprus, whereby significant amounts of pottery from the west of Cyprus did not reach Enkomi, the major seaport site in the south-east of Cyprus, for most of a century, and hence did not move on to Egypt (Manning 1999: 119–129; 2001: 80–84; Manning, Sewell, and Herscher 2002: 100–106; Manning and Bronk Ramsey 2003: 112). Marked regionalism accompanied by the appearance of fortifications in a number of places is indeed a hallmark of this period, as Manning notes. (With regard to Enkomi in particular, see Crewe 2007: 153.) Regionalism notwithstanding, objects including Egyptian material did reach the northwest of Cyprus (Bietak 2004, including a trenchant critique of the Manning hypothesis; see also Wiener 2001). Of course it is possible that voyagers sailing to the Near East and Egypt did not appreciate initially the attractiveness of White Slip (WS) I open shapes as trade goods. The chronological argument rests, however, not just in WS I, but in the Egyptian and Near Eastern contexts of a series of Cypriot pottery styles, including Proto WS, consistent with the traditional chronology.

I have considered myself whether an eruption date of 1570 BC (where there is perhaps some evidence of an event in the tree-ring and ice-core record) could be defended archaeologically, but found even 1570 BC, let alone any earlier date, very difficult to square with the archaeological evidence. Moreover, given the oscillating nature of the calibration curve in these decades, an eruption date of 1570 BC is not significantly more compatible with the asserted radiocarbon evidence than a date of 1525 BC.

The article by Manning et al. cites works by Philip Betancourt and Robert Merrillees (page 313) as supporting an early date for the eruption. The articles cited are a decade or more out of date, however. In response to my query regarding his current view, Prof. Betancourt kindly provided the following reply for quotation:

> What I said 25 years ago was that I thought the evidence as we knew it then favored the early chronology. I no longer feel that way. I feel today that the strongest evidence favors the traditional chronology, though I remain open-minded to the possibility that new evidence we do not have yet may shift the balance (pers. comm. of 14 November 2008; I would add that I concur in all respects).

I have also corresponded with Robert Merrillees in this regard. Dr. Merrillees began by observing that in proposing initially a date of "c. 1650 BC" for the beginning of LC I, he meant to indicate that the date in his view should be earlier than the date of c. 1600 BC preferred by most Cypriot specialists then and now, and in particular earlier than dates in the 16th century BC preferred by some, but that c. 1630 BC would do just as well as c. 1650 BC from his standpoint. I replied that 1615–1610 BC would do just as well as the conventional "c. 1600 BC" for the beginning of LM I, particularly inasmuch as LM I must have begun before the transmission of LM I motifs to Mycenaean Greece at the beginning of LH I, and that in any event there is no reason why LC I could not have begun earlier than LM I, with which Dr. Merrillees concurred. Merrillees also does not believe that the White Slip I bowl from the Theran destruction is an early example of White Slip I. Thus there is no inconsistency between the Merrillees Cypriot chronology and the Aegean historical chronology.

Manning et al. further argue that:

> White Slip I supposedly cannot be found on Santorini/Thera before c. 1530 BC because that represents its earliest known secure find at Tell el-Dabʻa (and hence the Akrotiri volcanic destruction level must postdate this date, as Wiener, this volume argues—but for why this viewpoint is not valid and indeed demonstrably incorrect, see Manning et al. 2002c; 2006a; 2006b; Manning 2007). Would this then make everything work "properly" and find the conventional chronology? No—all that then happens is that the analysis (if one adds a 1530 BC *terminus post quem* before the Akrotiri volcanic destruction level) finds *zero* possible agreement scores and *zero* analytical outcomes for the ^{14}C data sets in the Manning et al. (2006a) analysis (such is the non-compatibility) (Manning et al., page 310; and bibliographical references cited therein).

This observation holds true only if one assumes that a two per mil, 16.5-year difference in measurements of a thin data bank is not possible despite the various sources of uncertainty noted. Even without such an adjustment, the bottom of the two-sigma range of measurements when applied to the unsmoothed IntCal98 curve already overlaps the calibration curve at 1525 BC. (The olive branch radiocarbon measurements are a separate matter, considered above.)

It should be noted in this regard that Bietak would now place the earliest secure appearance of Cypriot WS I pottery in Egypt, and probably the Near East as well, in the Thutmosis III period beginning in 1479 BC, rather than in 1530 BC as in the Manning quotation. Bietak adds that he would be prepared to accept a one-generation, 30-year, delay between the creation of the WS I style in Cyprus and its first appearance abroad, or indeed a 50-year delay, but that a delay of 100–150 years (as required by an eruption date of 1613 BC, for example), when set within the context of a chronological series of Cypriot pottery styles present both in Cyprus and abroad, appears outside the bounds of reason (Bietak 2004: 206; pers. comm. of 30 November 2008, for which I am most grateful).

In any event, sherds from a WS I bowl *were* found below the volcanic eruption tephra on Thera. Both an excellent lithograph made from photos and drawings of the sherds exist independently, and these show that the bowl was used, broken and repaired in antiquity before it met its end in the eruption (Merrillees 2001; Manning 1999). In assessing the evidence for the date of the Theran eruption, both the radiocarbon determinations and the Cypriot White Slip I bowl are clearly relevant.

VII. Conclusion

Certainly if there existed only radiocarbon measurements, a substantial but not conclusive preference for an earlier eruption date would follow, particularly in light of the olive tree measurements presented by Friedrich et al. Similarly, if there existed only the historical evidence, a very strong preference for a later date (indeed, perhaps even later than 1525 BC) would follow. Rather than reject either body of evidence, I have sought to inquire whether, in light of the oscillating nature of the calibration curve in the critical century and the inherently problematic nature of radiocarbon dating, there is a point at which both categories of evidence can possibly meet, such as 1525 BC, and if not, which body of evidence is more likely to incorporate a systemic source of error. Surely this is the only path out of the chronological maze. The historical stakes are high, for the chronological solution will determine 1) whether Crete at the height of its pre-eruption florescence at the close of Late Minoan IA, Thera at the close of Late Cycladic I, and Mycenaean Greece at the close of Late Helladic I were in contact with Egypt at the beginning of the New Kingdom and 2) whether Crete in Late Minoan IB was in close contact with the assertive, expansive, and internationalist Egypt of Thutmosis III.

Submitted January 2009

Friedrich et al. Response to M. H. Wiener

In July 2007 a second olive tree was excavated that was, like the first found tree, buried alive *in situ* by the pumice of the eruption. The stem/branch has a length of 183 cm and a diameter of 12–15 cm. Samples of this olive tree are currently being investigated and tested for radiocarbon dating. Together with the man-made Bronze Age wall (Figures 1 and 2, pages 293 and 294) one gets the impression that the trees were part of an olive grove situated close to a settlement on a terrace of the caldera rim of that time. When comparing the new found olive stem/branch with modern olive trees growing less than one kilometer away from the locality we get the impression that the stem/branch could have 40–50 growth rings. Furthermore the finding of a piece of colored pottery of Late Cycladic IA style by the archeologist Nikos Sigalas connects the olive tree site directly to the destruction level of the Akrotiri excavation. Also the second olive tree gives us the impression that the trees were still alive when they were buried by the pumice of the eruption. We can clearly rule out the possibility that a dead branch was used for the radiocarbon dating. As one can learn from the ruins of the Akrotiri city, which is only a few kilometers away, strong earthquakes and precursory blast(s) hit the city, and also the site where the olive trees grew, prior to the main phases of the eruption. As a result all dead branches would have fallen off and be lying on the ground. Also the second olive tree was still standing upright in live position.

Comment on the paper by Malcolm Wiener: The finding of the olive trees is something quite unique and exceptional. As many others, we consider the radiocarbon date of 1627–1600 BC at present to be the most direct and precise for the Minoan eruption of Santorini. Concerning your criticism: We regard criticism as helpful since it sharpens our argumentation. However, we feel that the criticism should be more balanced and also show the weak points of the archaeological chronology.

Concerning the trees: Both olive trees are at present exhibited in a museum on Santorini and the locality where the olive trees grew is still accessible. However, this might change in near future, since houses are being built above the site. We therefore propose a meeting/workshop on Santorini as soon as possible to discuss all remaining issues there. It would also give other laboratories the opportunity to investigate the material.

Submitted November 2008

Manning et al. Response to M. H. Wiener

Malcolm Wiener's "Reply" suggests that "It would be interesting to know the weighted and unweighted averages of ^{14}C ages from seeds from the Theran VDL after all measurements which fail to overlap at one-sigma are removed from the dataset" (page 319).

The Supplemental Figure (Figure 5) here shows the 13 radiocarbon dates (in radiocarbon years BP) on seeds from the final volcanic destruction level run at Oxford and VERA in the AD 2000s period with one-sigma error bars (this is a detailed version of the samples in the dotted box in Wiener "Reply" Figure 2 after Manning et al. 2006a: Figure 1). The samples within each box come from the same sample group. Sample group 19 (Akrotiri M10/23A N012 – OxA-11820, OxA-12175) has two measurements which do not overlap at one-sigma (they do at two-sigma). So let us discount these two data as Wiener proposes. Sample group 17 (Akrotiri M2/76 N003 – OxA-11817, OxA-12170, VERA-2757, VERA-2757r) shows the two Oxford samples overlapping but the two VERA samples do not overlap with each other, although they both overlap with the Oxford samples. We assume Wiener would also like to discount the two

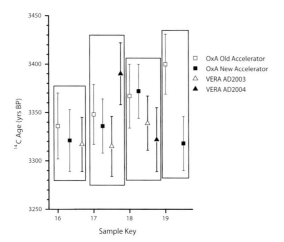

Figure 5: The Oxford (OxA) and Vienna (VERA) radiocarbon dates (in ^{14}C years BP) on short-lived samples from the final volcanic destruction level at Akrotiri (from Manning et al. 2006a: Figure 1). The figure shows in detail the samples indicated by the dotted box in Wiener "Reply" Figure 2. One-sigma error bars are shown. The samples within each box come from the same sample group: as listed in Wiener "Reply" Figure 2 caption, samples 16 to 19. Sample 16 = Akrotiri M31/43 N047; Sample 17 = Akrotiri M2/76 N003; Sample 18 = Akrotiri M7/68A N004; Sample 19 = Akrotiri M10/23A N012.

VERA samples. Let us do so. The other dates each overlap the other dates from their sample group at one-sigma.

We are left then with 9 dates on short-lived samples which are highly consistent and all from the same chronological horizon (and so applying a weighted average is entirely appropriate). The weighted average is 3340 ± 10 BP. The calibrated calendar ranges (from OxCal v.4.0.5, curve resolution = 5) are in Table 1.

IntCal04:	
One-sigma:	1664–1651 BC (16.6%)
	1642–1612 BC (51.6%)
Two-sigma:	1686–1607 BC (88.7%)
	1574–1559 BC (4.0%)
	1551–1539 BC (2.7%)
IntCal98:	
One-sigma:	1683–1668 BC (15.8%)
	1661–1649 BC (13.5%)
	1641–1605 BC (38.9%)
Two-sigma:	1685–1601 BC (75%)
	1576–1531 BC (20.4%)

Table 1: Calibrated calendar ranges for 3340 ± 10 BP.

Again there is no calendar range later than 1600 BC at one-sigma (68.2%) confidence. With IntCal04, there is only 6.7% probability for a date in the 16th century BC within the two-sigma (95.4%) range. With the previous IntCal98 calibration dataset, there is a slightly increased mid-16th century BC range (20.4%

of the total 95.4% two-sigma range). In no case is a date c. 1525 BC or later within even the margins of the two-sigma range.

Wiener asks in addition after the non-weighted average. This is also 3340 BP. Clearly if one then chooses to ignore the standard and appropriate practice of reducing the error on an average from a consistent set of estimates on the same event (as above), and instead merely averages the errors of each of the constituent data, then one will get a wider date range—in this case 3340 ± 31 BP. Even so, the case for a 17th c. BC date remains more likely as shown in Table 2.

IntCal04:	
One-sigma:	1685–1607 BC (59%)
	1571–1561 BC (5.7%)
	1547–1541 BC (3.4%)
Two-sigma:	1730–1719 BC (2%)
	1692–1528 BC (93.4%)
IntCal98:	
One-sigma:	1683–1602 BC (53%)
	1561–1534 BC (15.2%)
Two-sigma:	1690–1523 BC (95.4%)

Table 2: As Table 1 but for 3340 ± 31 BP (note: 0.1% rounding discrepancy in one-sigma IntCal04 range).

In each case the most likely range is in the 17th century BC (the 59% or 53% parts of the one-sigma ranges above). A less likely alternative is in the mid-16th century BC. But by using such a larger error one can claim that dates into the 1520s BC are just possible at the extreme of the two-sigma range of c. 1692/90–1528/23 BC—however, this is special pleading.

In conclusion:

(i) The radiocarbon evidence from the Akrotiri volcanic destruction level is very self consistent (see Manning et al. Figure 2; the chi-squared test is one obvious pointer to this). Given these data come from several laboratories and from different forms of analytical equipment and from different pretreatment regimes, this is a robust finding (and one cannot reasonably allege a measurement issue which somehow affected all the laboratories and their different equipment/methods).

(ii) Any explanation seeking to find a way to nonetheless allow for a later chronology would only work with some kind of a very uniform slight offset over the whole region. Even so, this would not make sense with the good correspondence (graphical fit) of the wiggle-match on the Theran olive branch to the standard northern hemisphere atmospheric radiocarbon record (whether as is, stretched or condensed); nor is there any other positive evidence. But this is at least a testable hypothesis, and one could look to e.g. marine sediment records as a test.

Submitted February 2009

M. H. Wiener's Response to the Friedrich et al. and Manning et al. Responses

Manning et al.'s Response (page 327) to the M. H. Wiener Reply (page 317) clarifies the nature of the data on which their conclusion is based. The heavy emphasis given to the average age of the nine radiocarbon determinations from the three remaining seed or seed-cluster samples serves to obscure the disparities in the underlying data set, both within the radiocarbon measurements of a single sample and between samples. Moreover, the statistical method employed assumes that radiocarbon-dating error is distributed randomly around the center of the dates obtained, whereas there is reason to believe that the error is asymmetrical with a bias toward higher determinations. The response again refers to "the standard and appropriate practice of reducing the error on an average from a consistent set of estimates on the same event..." (page 328). This practice is appropriate under the Ward and Wilson averaging criteria employed if it could be shown that the seeds and/or seed clusters from separate jars 1) were collected at the same time; 2) had lived under the same radiocarbon circumstances with respect to carbon reservoirs (volcanic vents, earth-gas emissions, CO_2 concentrations in groundwater or limestone); and 3) had the same exposure to precursor events, such as are known to have occurred in the period preceding the final eruption. In this connection, Floyd McCoy, the volcanologist engaged in a long-term study of the Theran eruption, notes that ^{14}C-deficient CO_2 gas in the soil commonly leaks upward from a magma chamber prior to an eruption, to the point that such leakage is one of the major signals of an impending eruption used today (pers. comm. of 16 April 2009). The uncertainty in such matters is the cause of the disquiet with respect to the statistical method used in both Manning et al. Further Discussion (AD 2006–2007) (page 299) and the Response to Wiener, supra.

McCoy further comments that in general he finds it "surprising that the potential influence of magmatic CO_2 on ^{14}C dating is not more appreciated... especially on an active volcano such as Santorini" (pers. comm. of 16 April 2009). On this major issue, Manning et al. in their Further Discussion again refer to the Bruns et al. 1980 publication in *Radiocarbon* to support the proposition that no such effect could have been responsible for some of the small dating anomalies between archaeological/historical and radiocarbon dates. That study reported radiocarbon measurements from three contemporary plant samples. Two of the three, located 5 and 10m from an obvious source of volcanic CO_2 lacking ^{14}C, gave ages of 1390 and 1030 years, respectively. The third, located 100m away, provided an anomalous Δ ^{14}C measurement incompatible with the standard correction for the effects of CO_2 release resulting from industrialization (Suess effect) minus the ^{14}C addition resulting from nuclear testing. In any event, the age of the third plant was left blank in the column under "age" in the Bruns et al. study (1980: 535, Table 2). On this single aberrant measurement rests the argument that the radiocarbon dates of samples from Thera will only be affected by volcanic carbon if the sample tested is less than 100m from a major recognizable source. Abundant evidence from various places, such as Italy where large areas of gas emissions throughout the country frequently result in radiocarbon dates 1–300 years too old, indicate just the opposite. Much depends on whether the CO_2 source is a point, a line (fault), or a distributed source. Of course no one can know the topography of Thera prior to the great eruption. Whether because of reservoir effects or for some other reason, the radiocarbon measurements of Theran seeds gave a distance between the central values of two samples of barley found in jars in the same room in the volcanic destruction deposit of 97 ^{14}C years and a distance between two pea samples of 215 ^{14}C years. The claim that this evidence "is very self consistent" (Manning et al. page 328) will puzzle prehistorians unfamiliar with the special statistical vocabulary here employed. The very limited data bank—nine measurements from only three seeds or seed clusters in the latest iteration—may also be noted.

The questions raised concerning the adequacy and accuracy of the measurements are significant in view of the proximity of the Manning et al. average radiocarbon age of 3340 BP ±15 to the top of the one-sigma range of the calibration curve at the archaeologically appropriate date of 1530 BC (particularly given the statement of the IntCal04 Committee that the one-sigma error band required widening and smoothing, with data borrowed from surrounding decades for each decadal determination proposed, because of the imprecision and small number of the radiocarbon determinations available for each decadal segment). Moreover, all radiocarbon probability estimates rest on the implicit but insecure assumptions that the dates obtained are largely unaffected by 1) any reservoir effect, i.e., the effect of ^{14}C-depleted carbon whether from volcanic vents, terrestrial degassing, carbon retained by limestone or groundwater or the upwelling of seawater; and 2) regional and/or seasonal variation, or a combination of the two, notwithstanding the absence of any information regarding the possibility of regional variation in radiocarbon measurements between Aegean plants or trees and inland trees, such as the German oaks of the calibration curve or the Anatolian junipers of Gordion, either in general or at certain periods. A putative solar minima-induced cold period, causing German oaks which grow later in the season than Turkish pine or juniper to absorb less ^{14}C, has been offered as a possible explanation for the large discrepancies in the eighth century BC in radiocarbon dates of tree segments thought to be of the same date.

The foregoing comments apply as well to a new Figure 7 (see page 306) and accompanying paragraph that have been inserted into the Manning et al. Further Discussion, subsequent to my Reply (supra) to that response. The caption states that "over 80% of all probability lies before 1570 BC (13 date set) or 1560 BC (28 date set)." This statement is based on the implicit assumption that the ^{14}C concentration of seeds that grew during springtime on a volcanic island in the Aegean is directly comparable to the ^{14}C concentration of tree rings that grew partly during the summer in a forest in Germany. Again the term "probability" is used within the context of a particular statistical paradigm, whereas the concept of "probability" in general discourse implies that all relevant information, areas of uncertainty, and knowledge insufficiency have been considered.

The chemistry and biology of sky, land, and water is not easy to capture in ^{14}C measurements and statistical probability models. The gaps in our knowledge, the sparseness of our observations in relation to the knowledge we seek, and the insufficiency of our explanations for the anomalies we observe in our measurements should induce caution in our conclusions.

The Friedrich et al. Response to the M. H. Wiener Reply declares that "We can clearly rule out the possibility that a dead branch was used for radiocarbon dating" (page 327). The statement is based on the assertion that all dead branches on all olive trees in the vicinity in question would have been torn off by the precursor earthquake and blast(s) of hot air which struck Akrotiri 7km away, prior to the final stage of the eruption.

There is no basis for the assumption that every dead branch on every olive tree in the vicinity would necessarily have been torn loose prior to the final stage of the eruption. Dead olive tree branches are not easy to remove. Volcanologist Floyd McCoy states, "I doubt that earthquakes would have stripped leaves from trees or removed dead branches—I cannot recall any examples from historic seismic activity. It is unlikely that hot blasts accompanied the precursor eruption; there is no evidence at Akrotiri or in Theran field deposits for such an occurrence" (pers. comm. of 16 April 2009). In fact, Friedrich et al. state that the branch was "in life position" on page 293 of their Further Discussion. (In this case also, material has been added to the initial article subsequent to my Reply. The new material encompasses the whole of the explanatory texts accompanying Figures 1, 2, and 3, which depict the authors' hypothetical reconstruction of events.) With regard to the radiocarbon analyses of the branch, it should be noted that Figure 1 in Manning et al. Further Discussion is based on the stated assumption that the ring-count calendar spacing of the olive tree branch is accurate to within one year. There is no sound basis for this assumption, for as Cherubini and others have noted, olive trees generally produce irregular rings, sometimes seasonal in nature (Wiener supra).

A second branch or limb of an olive tree was found about two years ago. A photograph distributed at a 2007 conference at the University of Aarhus showed a piece of wood so large that it took four people to carry it, a size seemingly sufficient to permit samples to be sent to several laboratories not involved in the examination and publication of the first branch, in order to obtain independent dendrochronological analysis of olive-wood ring counts as well as independent radiocarbon measurements.

The call by Friedrich et al. for an examination of weak points in the archaeological chronology is certainly in order, the more so since few in the physical sciences have sufficient knowledge of texts and inscriptions from Egypt and the Near East, archaeological interconnections, and Egyptian astronomy to form any judgment as to the degree of confidence warranted. A major, decade-long research project under the aegis of the Austrian Academy, The Synchronisation of Civilisations in the Eastern Mediterranean in the Second Millennium BC (SCIEM 2000), has brought together specialists from many countries for this precise purpose, with thousands of pages of analysis published. I personally have two articles in press questioning aspects of the textual/archaeological chronology. Indeed, all should remain open to new discoveries and scientific analyses, refrain from announcing definitive conclusions based on only a part of the data, and follow the evidence wherever it leads.

At the moment, the textual/archaeological chronology seems somewhat more solidly based than the radiometrically based chronology, given the uncertainties noted with regard to radiocarbon dates. Of course new evidence may shift the balance. Time will tell.

Submitted April 2009

References

Arteca, R.N., Poovaia, B.W., and Smith, O.E. 1979. Changes in Carbon Fixation, Tuberization, and Growth Induced by CO_2 Applications to the Root Zone of Potato Plants. *Science* 205: 1279–1280.

Bent, J.T. 1966. *Aegean Islands: The Cyclades, or Life among the Insular Greeks*. Chicago: Argonaut.

Bichler, M., Exler, M., Peltz, C., and Saminger, S. 2003. Thera Ashes. In M. Bietak (ed.), *The Synchronisation of Civilisations in the Eastern Mediterranean in the Second Millennium BC II. Proceedings of the SCIEM 2000–EuroConference, Haindorf, 2–7 May 2001*: 11–21. Vienna: Verlag der Österreichischen Akademie der Wissenschaften.

Bietak, M. 2000a. Datumlines According to the First Appearance of Frequent Artefacts with Wide Distribution (Relative Chronology III). In M. Bietak (ed.), *The Synchronisation of Civilisations in the Eastern Mediterranean in the Second Millennium B.C. Proceedings of an International Symposium at Schloss Haindorf, 15th–17th of November 1996 and at the Austrian Academy, Vienna, 11th–12th of May 1998*: 27–29. Vienna: Verlag der Österreichischen Akademie der Wissenschaften.

Bietak, M. 2000b. Regional Projects: Egypt. In M. Bietak (ed.), *The Synchronisation of Civilisations in the Eastern Mediterranean in the Second Millennium B.C. Proceedings of an International Symposium at Schloss Haindorf, 15th–17th of November 1996 and at the Austrian Academy, Vienna, 11th–12th of May 1998*: 83–95. Vienna: Verlag der Österreichischen Akademie der Wissenschaften.

Bietak, M. 2004. Review of A Test of Time by [S.] W. Manning. *Bibliotheca Orientalis* 61: 199–222.

Bietak, M., and Höflmayer, F. 2007. Introduction: High and Low chronology. In M. Bietak and E. Czerny (eds.), *The Synchronisation of Civilisations in the Eastern Mediterranean in the Second Millennium B.C. III. Proceedings of the SCIEM 2000—2nd EuroConference, Vienna 28th of May–1st of June 2003*: 13–23. Vienna: Verlag der Österreichischen Akademie der Wissenschaften.

Bronk Ramsey, C., Manning, S.W., and Galimberti, M. 2004. Dating the Volcanic Eruption at Thera. *Radiocarbon* 46: 325–344.

Bruns, M., Levin, I., Münnich, K.O., Hubberten, H.W., and Fillipakis, S. 1980. Regional Sources of Volcanic Carbon Dioxide and Their Influence on ^{14}C Content of Present-Day Plant Material. *Radiocarbon* 22: 532–536.

Cramer, M.D. 2002. Inorganic Carbon Utilization by Root Systems. In Y. Waisel, A. Eshel, and U. Kafkafi (eds.), *Plant Roots: The Hidden Half*: 699–714. New York: Marcel Dekker.

Cramer, M.D., and Richards, M.B. 1999. The Effect of Rhizospere Dissolved Inorganic Carbon on Gas Exchange Characteristics and Growth Rates of Tomato Seedlings. *Journal of Experimental Botany* 50: 79–87.

Crewe, L. 2007. *Early Enkomi: Regionalism, Trade and Society at the Beginning of the Late Bronze Age on Cyprus* (BAR International Series 1706). Oxford: Hadrian.

Enoch, H.Z., and Olesen, J.M. 1993. Tansley Review No. 54. Plant Response to Irrigation with Water Enriched with Carbon Dioxide. *New Phytologist* 125: 249–258.

Eriksson, K.O. 2003. A Preliminary Synthesis of Recent Chronological Obervations on the Relations between Cyprus and Other Eastern Mediterranean Societies during the Late Middle Bronze–Late Bronze II Periods. In M. Bietak (ed.), *The Synchronisation of Civilisations in the Eastern Mediterranean in the Second Millennium* BC *II. Proceedings of the SCIEM 2000–EuroConference, Haindorf, 2–7 May 2001*: 411–429. Vienna: Verlag der Österreichischen Akademie der Wissenschaften.

Ford, C.R., Wurzburger, N., Hendrick, R.L., and Teskey, R.O. 2007. Soil DIC Uptake and Fixation in *Pinus taeda* Seedlings and Its C Contribution to Plant Tissues and Ectomycorrhizal Fungi. *Tree Physiology* 27: 375–383.

Friedrich, W.L., Kromer, B., Friedrich, M., Heinemeier, J., Pfeiffer, T., and Talamo, S. 2006a. Santorini Eruption Radiocarbon Dated to 1627–1600 BC. *Science* 312: 548.

Friedrich, W.L., Kromer, B., Friedrich, M., Heinemeier, J., Pfeiffer, T., and Talamo, S. 2006b. Supporting Online Material for Santorini Eruption Radiocarbon Dated to 1627–1600 BC. *Science* 312. [http://www.sciencemag.org/cgi/content/full/312/5773/548/DC1].

Gasche, H., Armstrong, J.A., Cole, S.W., and Gurzadyan, V.G. 1998. *Dating the Fall of Babylon: A Reappraisal of Second-Millennium Chronology*. Ghent: University of Ghent.

Guerra, M., and Lombardi, S. 2001. Soil-Gas Method for Tracing Neotectonic Faults in Clay Basins: The Pisticci Field (Southern Italy). *Tectonophysics* 339: 511–522.

Keenan, D.J. 2002. Why Early-Historical Radiocarbon Dates Downwind from the Mediterranean Are Too Early. *Radiocarbon* 44: 225–237.

Lilyquist, C. 1996. Stone vessels at Kāmid el-Lōz, Lebanon: Egyptian, Egyptianizing, or Non-Egyptian? A Question at Sites from the Sudan to Iraq to the Greek Mainland. In R. Hachmann and W. Adler (eds.), *Kāmid el-Lōz 16, 'Schatzhaus'-Studien (Saarbrücker Beiträge* 59): 133–173. Bonn: Habelt.

Lilyquist, C. 1997. Egyptian Stone Vases? Comments on Peter Warren's Paper. In R. Laffineur and P.P. Betancourt (eds.), *TEXNH: Craftsmen, Craftswomen and Craftsmanship in the Aegean Bronze Age. Proceedings of the 6th International Aegean Conference, Philadelphia, Temple University, 18–21 April 1996 (Aegaeum* 16): 225–228. Liège: Université de Liège.

Manning, S.W. 1999. *A Test of Time: The Volcano of Thera and the Chronology and History of the Aegean and East Mediterranean in the Mid Second Millennium BC*. Oxford: Oxbow Books.

Manning, S.W. 2001. The Chronology and Foreign Connections of the Late Cypriot I Period: Times They Are a-Changin. In P. Åström (ed.), *The Chronology of Base-Ring Ware and Bichrome Wheel-Made Ware. Proceedings of a Colloquium Held in the Royal Academy of Letters, History and Antiquities, Stockholm, May 18–19 2000*: 69–94. Stockholm: The Royal Academy of Letters, History and Antiquities.

Manning, S.W. 2005. Simulation and the Thera Eruption: Outlining What We Do and Do Not Know from Radiocarbon. In A. Dakouri-Hild and S. Sherratt (eds.), *Autochthon: Papers Presented to O.T.P.K. Dickinson on the Occasion of His Retirement* (BAR International Series 1432): 97–114. Oxford: Archaeopress.

Manning, S.W., and Bronk Ramsey, C. 2003. A Late Minoan I–II Absolute Chronology for the Aegean—Combining Archaeology with Radiocarbon. In M. Bietak (ed.), *The Synchronisation of Civilisations in the Eastern Mediterranean in the Second Millennium* BC *II. Proceedings of the SCIEM 2000–EuroConference, Haindorf, 2–7 May 2001*: 111–133. Vienna: Verlag der Österreichischen Akademie der Wissenschaften.

Manning, S.W., Bronk Ramsey, C., Doumas, C., Marketou, T., Cadogan, G., and Pearson, C.L. 2002. New Evidence for an Early Date for the Aegean Late Bronze Age and Thera Eruption. *Antiquity* 76: 733–744.

Manning, S.W., Bronk Ramsey, C., Kutschera, W., Higham, T., Kromer, B., Steier, P., and Wild, E.M. 2006a. Chronology for the Aegean Late Bronze Age 1700–1400 BC. *Science* 312: 565–569.

Manning, S.W., Bronk Ramsey, C., Kutschera, W., Higham, T., Kromer, B., Steier, P., and Wild, E.M. 2006b. Supporting

Online Material for Chronology for the Aegean Late Bronze Age 1700–1400 BC. *Science* 312. [http://www.sciencemag.org/cgi/content/full/312/5773/565/DC1].

Manning, S.W., Sewell, D.A., and Herscher, E. 2002. Late Cypriot IA Maritime Trade in Action: Underwater Survey at Maroni-*Tsaroukkas* and the Contemporary East Mediterranean Trading System. *Annual of the British School at Athens* 97: 97–162.

McCoy, F.W., and Heiken, G. 2000. The Late-Bronze Age Explosive Eruption of Thera (Santorini), Greece: Regional and Local Effects. In F.W. McCoy and G. Heiken (eds.), *Volcanic Hazards and Disasters in Human Antiquity (Geological Society of America Special Paper* 345): 43–70. Boulder: Geological Society of America.

Merrillees, R.S. 2001. Some Cypriote White Slip Pottery from the Aegean. In V. Karageorghis (ed.), *The White Slip Ware of Late Bronze Age Cyprus. Proceedings of an International Conference Organized by the Anastasios G. Leventis Foundation, Nicosia in Honour of Malcolm Wiener, Nicosia, 29–30 October 1998:* 89–100. Vienna: Verlag der Österreichischen Akademie der Wissenschaften.

Mörner, N.-A., and Etiope, G. 2002. Carbon Degassing from the Lithosphere. *Global and Planetary Change* 33: 185–203.

Pasquier-Cardin, A., Allard, P., Ferreira, T., Hatte, C., Coutinho, R., Fontugne, M., and Jaudon, M. 1999. Magma-Derived CO_2 Emissions Recorded in ^{14}C and ^{13}C Content of Plants Growing in Furnas Caldera, Azores. *Journal of Volcanology and Geothermal Research* 92: 195–208.

Pfälzner, P. 2004. Minoan Style in Inland Syria: The Evidence from the Palace at Qatna. Paper presented at the New York Aegean Bronze Age Colloquium, 28 May 2004.

Rackham, O. 1965–1966. Transpiration, Assimilation and the Aerial Environment, Ph.D. dissertation, Cambridge University, Cambridge.

Reddaway, J.M., and Bigg, G.R. 1996. Climatic Change over the Mediterranean and Links to the More General Atmospheric Circulation. *International Journal of Climatology* 16: 651–661. Saleska, S.R., Didan, K., Huete, A.R., and da Rocha, H.R. 2007. Amazon Forests Green-Up during 2005 Drought. *Science* 318: 612.

Saurer, M., Cherubini, P., Bonani, G., and Siegwolf, R. 2003. Tracing Carbon Uptake from a Natural CO_2 Spring into Tree Rings: An Isotope Approach. *Tree Physiology* 23: 997–1004.

Shore, J.S., Cook, G.T., and Dugmore, A.J. 1995. The ^{14}C Content of Modern Vegetation Samples from the Flanks of the Katla Volcano, Southern Iceland. *Radiocarbon* 37: 525–529.

Skok, J., Chorney, W., and Broecker, W.S. 1962. Uptake of CO_2 by Roots of Xanthium Plants. *Botanical Gazette* 124: 118–120.

Soles, J.S. 2004. Appendix A: Radiocarbon Dates. In J.S. Soles and C. Davaras (eds.), *Mochlos IC. Period III. Neopalatial Settlement on the Coast: The Artisans' Quarter and the Farmhouse at Chalinomouri. The Small Finds (Prehistory Monographs* 9): 145–149. Philadelphia: INSTAP Academic Press.

Soter, S. Forthcoming. Radiocarbon Anomalies from Old CO_2 Emitted in Forests and Cultivated Fields. *Radiocarbon* 51.

Splittstoesser, W.E. 1966. Dark CO_2 Fixation and Its Role in the Growth of Plant Tissue. *Plant Physiology* 41: 755–759.

Stolwijk, J.A.J., and Thimann, K.V. 1957. On the Uptake of Carbon Dioxide and Bicarbonate by Roots and Its Influence on Growth. *Plant Physiology* 32: 513–520.

Switsur, V.R. 1984. Radiocarbon Date Calibration Using Historically Dated Specimens from Egypt and New Radiocarbon Determinations for El-Amarna. In B.J. Kemp (ed.), *Amarna Reports* Vol. 1: 178–188. London: Egypt Exploration Society.

Tauber, H. 1983. Possible Depletion in ^{14}C Trees Growing in Calcareous Soils. *Radiocarbon* 25: 417–420.

Teskey, R.O., and McGuire, M.A. 2007. Measurement of Stem Respiration of Sycamore (*Platanus occidentalis* L.) Trees Involves Internal and External Fluxes of CO_2 and Possible Transport of CO_2 from Roots. *Plant, Cell and Environment* 30: 570–579.

Vuorinen, A.H., and Kaiser, W.M. 1997. Dark CO_2 Fixation by Roots of Willow and Barley in Media with a High Level of Inorganic Carbon. *Journal of Plant Physiology* 151: 405–408.

Ward, G.K., and Wilson, S.R. 1978. Procedures for Comparing and Combining Radiocarbon Age Determinations: A Critique. *Archaeometry* 20: 19–31.

Warren, P.M. 2006. The Date of the Thera Eruption in Relation to Aegean-Egyptian Interconnections and the Egyptian Historical Chronology. In E. Czerny, I. Hein, H. Hunger, D. Melman, and A. Schwab (eds.), *Timelines: Studies in Honour of Manfred Bietak* Vol. 2 (*Orientalia Lovaniensia Analecta* 149): 305–321. Leuven: Peeters.

Weinstein, G.A., and Betancourt, P.P. 1978. Problems of Interpretation of the Akrotiri Radiocarbon Dates. In C. Doumas (ed.), *Thera and the Aegean World II*. Vol. *1*: 805–814. London: The Thera Foundation.

Wiener, M.W. 2001. The White Slip I of Tell el-Dab'a and Thera: Critical Challenge for the Aegean Long Chronology. In V. Karageorghis (ed.), *The White Slip Ware of Late Bronze Age Cyprus. Proceedings of an International Conference Organized by the Anastasios G. Leventis Foundation, Nicosia in Honour of Malcolm Wiener, Nicosia, 29–30 October 1998:* 195–202. Vienna: Verlag der Österreichischen Akademie der Wissenschaften.

Wiener, M.W. 2003a. The Absolute Chronology of Late Helladic III A2 Revisited. *Annual of the British School at Athens* 98: 239–250.

Wiener, M.H. 2003b. Time Out: The Current Impasse in Bronze Age Archaeological Dating. In K.P Foster and R. Laffineur (eds.), *METRON: Measuring the Bronze Age. Proceedings of the 9th International Aegean Conference/9e Rencontre égéenne internationale, New Haven, Yale University, 18–21 April 2002* (*Aegaeum* 24): 363–399. Liège: Université de Liège.

Wiener, M.H. 2007. Times Change: The Current State of the Debate in Old World Chronology. In M. Bietak and E. Czerny (eds.), *The Synchronisation of Civilisations in the Eastern Mediterranean in the Second Millennium* BC *III. Proceedings of the SCIEM 2000–2nd EuroConference, Vienna, 28th of May–1st of June, 2003:* 25–47. Vienna: Verlag der Österreichischen Akademie der Wissenschaften.

Yurgalevitch, C.M., and Janes, W.H. 1988. Carbon Dioxide Enrichment of the Root Zone of Tomato Seedlings. *Journal of Horticultural Science* 63: 265–270.